UNITY LIBRARY 8 ARCHIVES
Scientists debate Gaia : the ne
QH 331 .S324 2004
D1060983

Scientists Debate Gaia

The Next Century

edited by Stephen H. Schneider, James R. Miller, Eileen Crist, and Penelope J. Boston

The MIT Press
Cambridge, Massachusetts
London, England

This book was set in Times New Roman on 3B2 by Asco Typesetters, Hong Kong. Printed and bound in the United States of America.

Library of Congress Cataloging-in-Publication Data

Scientists debate gaia : the next century / edited by Stephen H. Schneider ... [et al.].
p. cm.
Includes bibliographical references (p.).
ISBN 0-262-19498-8 (alk. paper)
1. Gaia hypothesis. I. Schneider, Stephen Henry.

QH331.S324 2004
577—dc22 2003061532

10 9 8 7 6 5 4 3 2 1

Contents

III PHILOSOPHY, HISTORY, AND HUMAN DIMENSIONS OF GAIA

IV QUANTIFYING GAIA

V LIFE FORMS AND GAIA: MICROBES TO EXTRATERRESTRIALS

Foreword

The University of Valencia, on the Mediterranean coast of Spain, was founded in 1499 by the Council of the City. After five centuries of development that included periods of intellectual glory and of scholastic obscurity, our university has become a modern European institution. We now celebrate our clear interests in higher education and cultural activism. We enjoy active scholarly, scientific, and other research.

During the period 1999–2002, we organized artistic and other programs to commemorate our five centuries. We published this program, *Cinc Segles* (Five Centuries) in our own language, Valencian (*valenciano*), our way to speak Catalán. This publication, a beautifully produced volume that dealt with history, sports, art, music, and other aspects of the university's heritage, was the main focus of our attention at the festival celebrating our five-hundredth year.

One of the most compelling events that occurred after the book was printed was the Second Chapman Conference on the Gaia Hypothesis, Gaia—2000. We held this international meeting at our world famous botanical garden on the summer solstice of 2000. As a secular scientific institution, our university provided a perfect backdrop to this important scientific congress. The call for papers sponsored by the American Geophysical Union (AGU) led to participation that filled the botanical garden's spanking new auditorium and its state-of-the-art audiovisual facilities with world-class science. The coincidence of our five-hundredth anniversary with this meeting, twelve years after the first Chapman Conference, held in 1988 at San Diego, California, provided excellent opportunity for collaboration between AGU and our university. It led to this book, a forum for discussion of the frontier science of Gaia theory and practice. In retrospect, the Gaia—2000 conference was not only a remarkable local scientific success but also extended far beyond Spain, Europe, and the United States. The lively debate attracted coverage in the Spanish press. The presence in Valencia of scientists such as Dr. James E. Lovelock from the United Kingdom and Dr. Andrei Lapo of the Geological Institute of St. Petersburg, as spokesman for Vladimir Vernadsky's (1863–1945) work, was a privilege for our city and university. This book, as the most tangible result of the meeting, presents outstanding documentation of current international scientific views on Gaia theory.

Finally, I wish to recognize the efforts of institutions and persons, among the many that made Gaia—2000 possible in Valencia. The AGU, the U.S. National Science Foundation, the Caja de Ahorros del Mediterráneo, the Generalitat Valenciana, the Ajuntament de València, and the Comisión Interministerial de Ciencia y Tecnología were the main sponsors. I also acknowledge the contributions of the former Vice Rector of Culture, the biochemist Dr. Juli Peretó; the former Vice Rector of Research, the theoretical chemist Dr. Francisco Tomás; the former Botanical Garden Director, Dr. Manuel Costa; and the indefatigable organizer, the botanist (lichenologist) and the President of the local organizing committee, Professor Eva Barreno. Last but not least, Professor Lynn Margulis's enthusiastic support made it possible for us to host this seminal scientific meeting here in Valencia.

Dr. Pedro Ruiz Torres, Rector
University of Valencia, 1994–2002

Preface

A scientific and intellectual question of fundamental importance to the Earth sciences is the formative role of life in the biosphere's biogeochemical cycles and climate. The Gaia hypothesis, introduced in the early 1970s by English atmospheric chemist Jim Lovelock and American biologist Lynn Margulis, proposes that the interaction between the biota and the physical-chemical environment is large enough in scope to serve as an active feedback mechanism for biogeoclimatologic control. The core of Gaian theory is that life has a substantial, even regulating, impact on the Earth's geochemical cycles and climate—an impact that tends to favor organisms and living processes. This view has met with vocal criticism from both the Earth sciences and the life sciences. Geochemical arguments have been proposed to explain many environmental changes over time without recourse to the role of life. And biologists, especially neo-Darwinians, have argued that the Earth understood as a global ecosystem actively "managing" environmental parameters for the benefit of life as a whole is incompatible with the view of living organisms as competitively and selfishly inclined toward narrowly definable survival and reproductive success.

The possibility of active climatic regulation systems, and the relative importance of feedback processes between organic and inorganic components, clearly needed to be examined in a frank, scientifically rigorous, and richly interdisciplinary setting. To that end the first American Geophysical Union Chapman Conference on the Gaia hypothesis was held in San Diego, California, in 1988. Many of the papers presented at that conference appeared in the book *Scientists on Gaia*, edited by Stephen Schneider and Penelope Boston. Twelve years later a second Chapman Conference on the Gaia hypothesis was held in Valencia, Spain. Most of the chapters in this volume are based on presentations given there.

Why was it important to hold a second AGU Chapman Conference on the Gaia hypothesis, and why is it necessary to publish this second volume? Prior to the 1988 conference, Gaia had been highly controversial within scientific arenas, yet there was little scientific debate. Controversy occurred for at least three reasons: (1) there was outright hostility to the name "Gaia" (a goddess of the Earth metaphor, implying the Earth was alive); (2) there was little or no shared understanding (or even a consistent definition) of the "Gaia hypothesis"; and (3) studying Gaia required strong multidisciplinary training and an interdisciplinary commitment that transcended traditional scientific approaches. The first conference extended the Gaia hypothesis into the mainstream of scientific thinking and debate, and encouraged scientists to work within the field. Major progress in defining and identifying "Gaian" processes was made, inconsistencies were flagged, and new thinking about the coupling of physical and biological systems was offered. Nevertheless, controversies about Gaia still abound today—and many are debated in this volume. Happily, the evolution of interdisciplinary science into the mainstream and the ongoing development of Earth system science have promoted scientific inquiry seeking and elucidating Gaian—and non-Gaian—mechanisms within the Earth system.

Borrowing from philosopher of science Thomas Kuhn's classical analysis, we could say that Gaia is exiting its "revolutionary" phase—of vociferous controversy and ostracism from the scientific establishment—and is entering its phase of "normal," puzzle-solving science. More often, today, the expression "Gaian science" is used in the literature in lieu of the older "Gaia hypothesis." However, along with the evolution to Gaian science has come a sharper focus on what is "controlled"—or not controlled—and by what processes. Moreover, the open exchanges have helped to correct much of the loose language and soft definitions from early days of the debate, and thus Earth science has been enriched by a decade and a half of Gaian science activities.

The three interlinked themes of the Valencia conference were Gaia in time, the role of the biota in regulating biogeochemical cycles and climate, and understanding complexity and feedbacks in the Earth system. The Earth system is dynamic and evolving, yet Gaia can provide stability for long periods of time while undergoing changes in response to external stimuli and/or emerging components within the biota. The biogeochemical cycles of the principal elements of life are intertwined. To understand how they work in the present, have worked in the past, are being modified by human activity, and may change in the future requires several approaches, including the analysis of the geological record, the use of models of varying complexity and realism, and the empirical study of ecosystems, with special emphasis on the formative impact of life on nonliving processes, such as climate, composition of the atmosphere, and cycling of elements.

To address Gaia requires articulating broad themes, in combination with specific questions, that can be scientifically investigated. Three such themes relate to the temporal variability of Gaia, the structure of Gaia, and how to quantify Gaian feedbacks. These can be further divided into clusters of questions that are addressed in the volume:

- How has the global biogeochemical/climate system we call Gaia changed in time? What is its history? Can Gaia maintain stability of the system at one time scale, yet undergo fundamental change at longer time scales? How can we use the geologic record to address these questions?

- What is the structure of Gaia? Are the components of the Earth system determined pragmatically by available disciplinary tools, or are there components that are objectively primary for a valid understanding of Gaia? Put in the language of complex system analysis, are there "emergent properties" of the coupled subsystems (typically studied in isolation by individual disciplines) that give rise to collective behaviors which would not be detected using a purely reductionist approach? If the latter is the case, are the components invariable over time or do they change with the evolution of the Earth system? What are the feedback loops among these different parts of the Gaian system, and are they sufficiently robust to influence the evolution of climate?

- How do models of Gaian processes and phenomena illuminate the real Earth, and how do they help us to understand Gaia? Do results from Gaia's most renowned—and still controversial—model, Daisyworld, transfer to the Earth system? Daisyworld demonstrated how temperature-sensitive "daisies" can regulate climate through an automatic feedback process on planetary albedo (reflectivity). What are the main candidates for "daisies" on real Earth, and how should we be searching for them? How can Gaian mechanisms or emergent properties be investigated using processes or global models of the climate system that include the biota and allow for chemical cycling?

From the outset, a major criticism of the Gaia hypothesis has been that Gaian properties maintaining the Earth system would be unlikely to arise because they imply global scale teleology—or purposefulness—incompatible with a view of natural selection occurring at the level of individual organisms or genes. Gaian scientists respond that biological control is not purposeful, but an emergent property of the complex Earth system. Several chapters in this volume address the relationship between the evolutionary mechanism of natural selection and Gaian processes, arguing for a deep compatibility between the two. Groups of organisms that create physical and chemical conditions advantageous for themselves and their progeny would tend to be favored by natural selection. Such a connection between evolutionary processes and life's "control" over nonliving environments, as well as the scope of such control, need to be demonstrated scientifically (empirically or through modeling) for each case. But just as natural selection is blind and nonpurposeful, so Gaian emergent properties are blind and nonpurposeful—and there is no conflict between the two mechanisms. Life is, moreover, profoundly opportunistic. To the extent that one group of organisms makes use of the waste products of another group, inadvertent "associations" are formed with local or large-scale effects on the nonliving environment; such effects can clearly benefit both groups of organisms without forethought or "altruism" being involved.

Several chapters in the volume address the role of modeling in Gaian science. The Daisyworld model was developed to support the claim that the biosphere can regulate the climate system by modifying the Earth's surface albedo. Critics have argued, however, that there are no real-world counterparts to the black and white daisies of the model. Part IV, "Quantifying Gaia," critically examines and updates the Daisyworld discussion. Daisyworld has inspired Gaian scientists to create models that are more sophisticated, empirically realistic, and/or applicable to previous geological periods. In one chapter, for example, feedback between biota and climate in a simple climate model is demonstrated, and it is shown that as the complexity of the community increases, so does its ability to resist perturbation. In another chapter, a more complex version of Daisyworld is provided, yet many of the results of the prototype still hold. Yet a third chapter demonstrates that the hypothetical introduction of vegetation in a barren region creates climatic effects that are a stabilizing feedback for the vegetation. The Daisyworld model thus has been effective not only in vividly illustrating how the biota can "tune" climate (on the metaphor of a thermostat) but also in spearheading modeling as a productive arena of Gaian science. Ironically, as the "daisies" fade from Daisyworld, insights begin to emerge for the Earth system.

While these are major scientific frameworks for Gaia, nonscientific facets of Gaia continue to be woven into the debates—for better or for worse. The name "Gaia" has haunted the scientific theory and studies that have emerged under its name. In his Introduction, James Lovelock discusses how "Gaia" was resisted—and, to a lesser extent, continues to be resisted—by the scientific community for the offense of naming the Earth's coupled living and nonliving systems after a Greek goddess. But Lovelock notes that only the name remains controversial; the science of Gaia has become increasingly acceptable, even mainstream, under the more respectable label "Earth system science." He boldly outlines a number of predictions of Gaia theory that have generated intriguing new ways of seeing the connections between living and nonliving processes and/or have received strong empirical support: Familiar examples are the key role of life-driven rock weathering in regulating atmospheric carbon dioxide levels and, thus, temperature; the fact that levels of oxygen within a narrow viable range are not a fortuitous phenomenon, but more plausibly an outcome of systemic feedback; and the fact that dimethyl sulfide emissions from the ocean implicate the importance of algae in the formation of cloud cover and, therefore, of climate. However, the direction of that feedback is not clear, and it is at least as likely to be destabilizing as stabilizing at different times and scales. That is precisely the kind of question that Gaian science must address, and why volumes such as this and conferences such as that at Valencia must be repeated.

The interdisciplinary character of Gaia extends beyond weaving the Earth and life sciences in a novel way. The controversial entrance of the Gaia hypothesis into science, in conjunction with its global-level claims, ensured that from the outset history, philosophy, and environmental perspectives have been high-profile aspects of Gaia theory. Continued thinking in these areas is represented in the volume. For example, while the idea of a "living Earth" has been given a novel scientific formulation with Lovelock's Gaia, clearly it is an ancient concept. Its intellectual history in Western thought, as one chapter shows in broad brush, can be traced from the pre-Socratics through early modern science. Another historical chapter considers Charles Darwin's last book as a quasi-Gaian case study demonstrating that earthworms may be "biotic regulators" of geological, physical, chemical, and ecological phenomena. Other chapters use tools from the philosophy of science to explore the connection between Gaia and physical thermodynamics.

Given the global ecological degradation that humans are effecting—the biodiversity crisis and climate change being two key (interrelated) facets of our disruptive impact—Gaian scientists and theorists are often obliged, one way or another, to address the question of the human impact on Gaia. Early on, the Gaian perspective unsettled environmentally concerned people and analysts when two "anti-environmental" messages were seemingly implied: (1) that the Earth system is not "fragile," and Gaia can survive our affronts; and (2) that Gaia will "clean up" our messes, for if we perturb the system, it will respond, via feedback mechanisms, to reinstate homeostasis.

Neither of these ideas, however, accurately reflects how Gaian scientists and analysts view our current environmental predicament. It is true that Gaia, with a history of some 3 billion years, is clearly not a fragile entity. But many individual biotas, along with their ecosystems, that compose the Earth's present-day biodiversity—the organisms which are our evolutionary companions—are clearly imperiled by our activities. Gaia cannot "save" countless species and ecosystems from annihilation following massive habitat fragmentation, exotic species introduction, chemical pollution, or rapid climate change, although it is exceedingly improbable that we could threaten life itself. It is, therefore, not inconsistent to maintain that the Earth system is fragile in some ways, and not in others: it depends on the time scale under consideration, and on what composition one is referring to as the "Earth system."

Moreover, few if any Gaian scientists would wager that the Earth system will respond to "correct" our perturbations. Our global impact on the planet is both substantial and rapid. Two salient examples are an estimated rate of human-driven extinction 100 to 1000 times greater than the background rate; and our 30 percent increase of carbon dioxide levels in the atmosphere in the last 200 years. Such large-scale, swift changes can lead to unpredictable perturbations of the Earth system—perturbations that are not likely to be entirely benevolent toward our species. Moreover, it is also clear that a new Earth system "equilibrium," in the wake of our continued reckless impact, will not provide the quality and wealth of amenities and beauty to which we are accustomed, regardless of the larger per capita Gross National Product that might occur. While those thinking under the auspices of Gaia do not necessarily share an environmental agenda or philosophy, it is safe to say that Gaians wholeheartedly embrace ecologist's Aldo Leopold's famous injunction "to keep all the parts."

We would like to thank all of the contributors to this volume and all of the participants in the Chapman Conference at Valencia for stimulating discussions and insights into the Gaia debate. We thank all of the anonymous reviewers who provided constructive comments that led to substantive improvements to the chapters. We also thank all those organizations and individuals recognized in the last paragraph of Dr. Pedro Ruiz Torres's Foreword to this volume, including the National Science Foundation and the American Geophysical Union. And, finally, we thank Dr. Torres and the University of Valencia for hosting the Chapman Conference during the summer solstice of 2000, and providing us with a beautiful venue including an arboretum and state-of-the art conference facility. We are grateful to our hosts for their gracious hospitality. We hope that this volume helps to provide new ways to frame the debate about Gaia for the scientists of the twenty-first century.

Reflections on Gaia

James Lovelock

In November 1987, the Dahlem Conference in Berlin had the title "The Changing Atmosphere." Just before we started our discussions, a delegate asked me what part Gaia theory had played in the paper by Robert Charlson and his friends on dimethyl sulfide, clouds, and climate. Before I could reply, another delegate interrupted. "Gentlemen," he said forcefully, "we are here to discuss serious science, not fairy stories about a Greek Goddess." I was not amused by his ill-mannered protest, but looking back, I see that there was some sense in it. He was expressing the frustration felt at the time by many scientists who wanted to believe that their science, whether geophysics, chemistry, geology, or biology, said all that there was to say about the Earth. I also felt that we were spending far too much time at meetings arguing about the metaphor and ignoring Gaia science. What we do not seem to have noticed is that the science of Gaia is now part of conventional wisdom and is called Earth system science; only the name Gaia is controversial. The most recent acknowledgment of this was the Amsterdam Declaration, issued by a joint meeting of the International Geosphere Biosphere Programme, the International Human Dimensions Programme on Global Environmental Change, the World Climate Research Programme, and the International Biodiversity Programme on July 13, 2001. The declaration had as its first bullet point: "The Earth System behaves as a single, self-regulating system comprised of physical, chemical, biological and human components." I wonder how many of those who signed the declaration knew that they were putting their names to a statement of Gaia theory.

Let us go back to the 1960s and remind ourselves how separated and reductionist the Earth and life sciences were. In the splendid and authoritative book *Earth*, published in 1974 by Frank Press and Raymond Siever, they shared the general view and stated: "Life depends on the environments in which it evolved and to which it has adapted" (p. 489). In the same period, John Maynard Smith published an equally authoritative book on evolutionary biology, *The Theory of Evolution*. In it he said: "The study of evolution is concerned with how, during the long history of life on this planet, different animals and plants have become adapted to different conditions, and to different ways of life in those conditions" (p. 15). Neither of these eminent representatives of the Earth and life sciences thought it necessary to acknowledge that organisms alter their environment as well as adapt to it. I do not think they realized that the evolution of the organisms and of their environment was a single, coupled process.

It is true that the biogeochemists A. C. Redfield, V. I. Vernadsky, and G. E. Hutchinson had already shown that organisms were more than mere passengers on the planet, and geochemists were aware that organisms in the soil accelerated rock weathering and that weathering was faster in hot climates, but no one saw that these were parts of a global system able to regulate climate and chemistry. Even the clear concise writings of Alfred Lotka, who recognized the Earth as a system in the modern sense as far back as 1925, were ignored.

Since the 1970s, our view of the Earth has changed profoundly, and it is reasonable to ask what caused the change and how much of it is due to Gaia theory. My first thoughts about the Earth as a self-regulating system that sustained a habitable surface came from space research. When NASA set up its planetary exploration program in the 1960s, it forced on all of us a new way of looking at life and the Earth. By far the greatest gift of space research was the view it provided from above—for the first time in human history, we had a God's-eye view of our planet and saw the Earth as it was. To see the Earth this way corrected the myopic distortion inherent in reductionist science. We need reduction in science, but it is not the whole story.

The idea that life, the biosphere, regulated the Earth's surface environment to sustain habitability came to me at the Jet Propulsion Laboratory (JPL) in 1965. It arose from a life detection experiment that sought the presence of life on a planetary scale instead of looking at the details visible on the surface. In particular, NASA's quest to find life on Mars provided me with the opportunity to ask the question Can the existence of life be recognized from knowledge of the chemical composition of a planet's atmosphere? The answer was a resounding yes. This way of thinking predicted in 1965 that Mars and Venus were lifeless long before the Viking landers failed to find life on Mars in the 1970s, but it also drew attention to the extraordinary degree of chemical disequilibrium in the Earth's atmosphere, which led me to think that some means for its regulation was needed. Although they disliked my conclusions about life on Mars, JPL actively supported the early development of Earth system science. In 1968 they invited me to present a paper that included for the first time the notion of the Earth as a self-regulating system at a meeting of the American Astronautical Society. NASA now recognizes the validity of atmospheric analysis as a life detection experiment. Without realizing it, they have taken the science that led to Gaia and made it their new science, astrobiology. By doing this they have brought together under one theoretical view life on Earth and life on other planets.

The next important step was in 1971, when Lynn Margulis and I began our collaboration. Lynn brought her deep understanding of microbiology to what until then had been mainly a system science theory that saw a self-regulating Earth through the eyes of a physical chemist. By stressing the importance of the Earth's bacterial ecosystem and its being the fundamental infrastructure of the planet, Lynn put flesh on the skeleton of Gaia. Selling a new theory is a lonely business, and it was wonderful to have Lynn as a friend who stood by me in the fierce arguments with the neo-Darwinists who were so sure that they were right and we were wrong. And they were right to say that there was no way for organisms to evolve global scale self-regulation by natural selection, because the unit of selection is the individual organism, not the planet. It was not until I made the model Daisyworld that I recognized that what evolved was not the organisms or even the biosphere, but the whole system, organisms and their material environment coupled together. The unit of evolution is the Earth system, and self-regulation is an emergent property of that system.

I was not trying to be perverse when I introduced the metaphor of a living Earth. Self-regulating systems are notoriously difficult to explain, and it was natural to use the metaphor of a living Earth. I saw its apparent capacity to sustain a steady temperature and composition, in spite of solar warming and planetesimal impacts and other catastrophes, like the homeostasis of an animal. This was too much for biologists, and they pounced. Here was the fallacy of Gaia: Lovelock was claiming that the Earth was alive. The idea of the Earth as alive in a biological sense became the strong Gaia hypothesis, and the Earth as a self-regulating system became the weak Gaia hypothesis. By setting up these two straw hypotheses, it was easy for them to demolish the strong, which I had never claimed, and leave me with the weak Gaia hypothesis, doomed to ignominy by the adjective "weak."

There is an intriguing irony here, for it was about this time that the great evolutionary biologist William Hamilton coined the powerful metaphors of the selfish and spiteful genes, which did so much to establish the power of the neo-Darwinism perspective among the public. We all know that the selfish gene is a metaphor as prone to misinterpretation as any in science. But it was introduced in a world where, at least in scientific circles, it faced paltry opposition, and so its proponents were spared endless debates on whether or not a gene could be truly selfish, whether strong selfishness was more coherent than weak selfishness, and so on. Gaia was not so lucky. Call Gaia weak if you will, but let us see which predictions of this weak little theory have been confirmed.

1. That the Earth was, and largely still is, managed by its bacterial ecosystem
2. That the atmosphere of the Archean period was chemically dominated by methane
3. That rock weathering is part of a self-regulating system involving the biota that serves to regulate carbon dioxide in the atmosphere and keep an equable temperature
4. That oxygen levels need regulation within a mixing ratio of 15 to 25 percent
5. That the natural cycles of the elements sulfur and iodine take place via the biological products dimethyl sulfide and methyl iodide
6. That dimethyl sulfide emission from the ocean is linked with algae living on the surface, clouds, and climate regulation
7. That regional climate on the land is coupled with the growth of trees in both the tropical and the boreal regions
8. That biodiversity is a necessary part of planetary self-regulation
9. That mathematical procedures for modeling these systems originated with Daisyworld
10. That life on other planets can be detected by chemical compositional analysis of the planets' atmospheres.

My recollections, looking back on the years since the word *Gaia* was introduced, are not of heated argument and hostility but of friendly disagreement. Even those who most deeply disagreed did so courteously. I am especially grateful to Stephen Schneider and Penelope Boston, who put together the first Chapman Conference in 1988, a time when few saw merit in Gaia theory.

Gaia is a new way of organizing the facts about life on Earth, not just a hypothesis waiting to be tested. To illustrate the use of the theory in this way, let us go back to the origins of life some 3.5 to 4 billion years ago. At that time and before life appeared, the Earth was evolving as terrestrial planets do, toward a state that ultimately would be like that of Mars and Venus—an arid planet with an atmosphere mainly of carbon dioxide. Early in its history the Earth was well watered, and somewhere on it there was an equable climate, so that life, once begun, could flourish. When it did begin, the first organisms must have used the raw materials of the Earth's crust, oceans, and air to make their cells. They also returned to their environment their wastes and dead bodies. As they grew abundant, this action would have changed the composition of the air, oceans, and crust into an oxygen-free world dominated chemically by methane. This means that soon after its origin, life was adapting not to the geological world of its birth but to an environment of its own making. There was no purpose in this, but those organisms which made their environment more comfortable for life left a better world for their progeny, and those which worsened their environment spoiled the survival chances of theirs. Natural selection then tended to favor the improvers. If this view of evolution is correct, it is an extension of Darwin's great vision and makes neo-Darwinism a part of Gaia theory and Earth system science. And for these reasons I do not think that the Earth's environment is comfortable for life merely because by good fortune our planet sits exactly at the right place in the solar system. Euan Nisbet's book *The Young Earth* is a fine introduction to this interesting period of Gaian history.

We have some distance still to travel because a proper understanding of the Earth requires the abolition of disciplinary boundaries. It is no use for scientists with an Earth science background to expect to achieve grace simply by including the biota in their models as another compartment. They have to include a biota that lives and dies and evolves by natural selection and interacts fully with its material environment. It is no use for biologists to continue to model the evolution of organisms on an imaginary planet with an intangible environment. In the real world the evolution of life and the evolution of the rocks, the oceans, and the air are tightly coupled.

Neo-Darwinist biologists were the strongest critics of Gaia theory, and for twenty years they rejected the idea of any regulation beyond the phenotype as impossible. In the past few years they have softened, and now are prepared to accept that there is evidence for self-regulation on a global scale, but still can see no way for it to happen by means of Darwinian natural selection. It took William Hamilton to see this as a challenge, not an objection. With my friend Tim Lenton he produced a paper titled "Spora and Gaia" in which they argue that perhaps the selective advantage for algae in producing dimethyl sulfide clouds and wind is to spread their spores more efficiently. This open-mindedness makes his untimely death so sad. Bill Hamilton changed from being a strong opponent to seeing Gaia theory as a new Copernican revolution. Without his wisdom, I think it will take many years to disentangle the links between organisms, their ecosystems, and large-scale climate and chemistry. I see it as rather like the problem faced by supporters of natural selection in the last century when they were asked, "How could anything so perfect as the eye evolve by a series of random steps?"

References

The Amsterdam Declaration on Global Change. Issued at a joint meeting of the International Geosphere Biosphere Programme, the International Human Dimensions Programme on Global Environmental Change, the World Climate Research Programme, and the International Biodiversity Programme. Amsterdam, 2001.

Betts, R. A. 1999. Self-beneficial effects of vegetation on climate in an ocean–atmosphere general circulation model. *Geophysical Research Letters*, 26, 1457–1460.

Charlson, R. J., J. E. Lovelock, M. O. Andreae, and S. G. Warren. 1987. Oceanic phytoplankton, atmospheric sulfur, cloud albedo and climate. *Nature*, 326(6114), 655–661.

Hamilton, W. D. 1996. *Narrow Roads of Gene Land*. New York: W. H. Freeman.

Hamilton, W. D., and T. M. Lenton. 1998. Spora and Gaia. *Ethology, Ecology and Evolution*, 10, 1–16.

Harding, S. P. 1999. Food web complexity enhances community stability and climate regulation in a geophysiological model. *Tellus*, 51B, 815–829.

Hutchinson, G. E. 1954. Biochemistry of the terrestrial atmosphere. In *The Solar System*, G. P. Kuiper, ed. Chicago: University of Chicago Press.

Lenton, T. 1998. Gaia and natural selection. *Nature*, 394, 439–447.

Lotka, A. 1924 [1956]. *Elements of Mathematical Biology*. Reprinted New York: Dover.

Lovelock, J. E. 1965. A physical basis for life detection experiments. *Nature*, 207(4997), 568–570.

Lovelock, J. E. 1969. Planetary atmospheres: Compositional and other changes associated with the presence of life. In *Advances in the Astronautical Sciences*, O. L. Tiffany and E. Zaitzeff, eds., pp. 179–193. Tarzana, CA: AAS Publications Office.

Lovelock, J. E. 1988. *The Ages of Gaia*. New York: W. W. Norton.

Lovelock, J. E. 1992. A numerical model for biodiversity. *Philosophical Transactions of the Royal Society of London*, 338, 383–391.

Lovelock, J. E., and M. Margulis. 1974. Atmospheric homeostasis by and for the biosphere: The Gaia hypothesis. *Tellus*, 26, 2–10.

Lovelock, J. E., and A. J. Watson. 1982. The regulation of carbon dioxide and climate: Gaia or geochemistry. *Planetary Space Science*, 30(8), 795–802.

Margulis, L., and J. E. Lovelock. 1974. Biological modulation of the Earth's atmosphere. *Icarus*, 21, 471–489.

Maynard Smith, J. 1958. *The Evolution of Life*. London: Penguin Books.

Nisbet, E. G. 1987. *The Young Earth*. Boston: Allen and Unwin.

Pimm, S. L. 1984. The complexity and stability of ecosystems. *Nature*, 307, 321–326.

Press, Frank, and Raymond Siever. 1974. *Earth*. San Francisco: W. H. Freeman.

Redfield, A. C. 1958. The biological control of chemical factors in the environment. *American Scientist*, 46, 205–221.

Vernadsky, V. I. 1945. The biosphere and the noosphere. *American Scientist*, 33, 1–12.

Watson, A. J., and J. E. Lovelock. 1983. Biological homeostasis of the global environment: The parable of Daisyworld. *Tellus*, 35B, 284–289.

Gaia by Any Other Name

Lynn Margulis

We upright, nearly hairless, chatty chimps owe our burgeoning population numbers to our flexible brains and our intense social behavior. All of us can attest to the strength of cultural and linguistic influences between birth and say, twenty years of age; words and symbols are powerfully evocative and may even stimulate violent activity (Morrison, 1999). Examples of symbolic emotion-charged phrases abound. In today's political realm they include "evil Middle East dictator," "HIV-AIDS victim," "neo-Nazi," "genetically manipulated crops," "dirty nigger," "one nation under God," "drug addicts," "white supremacist," "sexual abuser," and many far more subtle others.

Science, ostensibly objective and free of such name-calling, is not immune. Although to most of the contributors to this book, "science" simply refers to an open, successful, international, and cooperative means of acquiring new knowledge by observation, measurement, and analysis, to many outsiders "science" is an emotion-charged term. To some it implies atheism, triviality, lack of patriotism, or willingness to collaborate with huge corporations against their workers. To others a scientist is someone deficient in empathy or lacking in emotional expression or, worse, a supplier of technical know-how complicit in the development of weapons of mass destruction.

Here, following James E. Lovelock's lead in his accompanying piece "Reflections on Gaia," as someone proud to participate in the international scientific effort, I mention the impetus to new investigations. The "Gaia hypothesis" has now become the "Gaia theory" and has given voice to disparately trained researchers over the last few decades.

The very beginning of the Gaia debate, I submit, was marked by a little-known *Nature* paper (Lovelock, 1965). Gaia's middle age, her 40th birthday, ought to be celebrated with appropriate fanfare in or near the year 2005. Such recognition would mark the anniversary of the widespread dissemination of her gorgeous dynamic image. Photographed and made well known by Russell (Rusty) Schweikart and especially as the "blue marble" (the living Earth seen from space) taken by the 1968 circumlunar Apollo 8 team (Frank Borman, Jim Lovell, and Edward Anders), the image generated a gaggle of Gaia enthusiasts. From the beginning Gaia's intimate portrait has been delivered to us by these and lesser fans of outer space, most of whom were interviewed by Frank White (1998). Indeed, close-ups of her green and mottled countenance are newly available in the spectacular full-color, oversized book that reveals Gaia from above (Arthus-Bertrand, 1999).

To me, the Gaia hypothesis, or theory as some would have it, owes its origin to a dual set of sources: the immense success of the international space program that began with the launch of *Sputnik* by the Soviet Union in 1957 and the lively but lonely scientific imagination, inspiration, and persistence of Jim Lovelock. Part of the contentiousness and ambiguity attendant on most current descriptions of the Gaia hypothesis stems from confused definitions, incompatible belief systems of the scientific authors, and inconsistent terminology across the many affected disciplines (for example, atmospheric chemistry, environmental studies, geology, microbiology, planetary astronomy, space science, zoology). Anger, dismissive attitude, and miscomprehension also come from the tendency of the human mind toward dichotomization. In this limited summary whose purpose is to draw attention to several recent, excellent books on Gaia science and correlated research trends, I list the major postulates of the original Gaia statement and point to recent avenues of investigation into the verification and extension of Lovelock's original ideas. I try to minimize emotionally charged rhetoric aptly indulged in and recently reviewed by Kirchner (2002) and to maximize the proximity of the entries on my list to directly observable, rather than computable, natural phenomena. I self-consciously align this contribution to a field ignored by most of today's scientific establishment and their funding agencies, one considered obsolete, anachronistic, dispensable, and atavistic. To me this field in its original form, "natural theology" that became "natural history," should be revived with the same enthusiasm with which it thrived in the 18th and early 19th centuries.

That age of exploration of the seas and lands generated natural history in the same way that satellite technology and the penetration of space brought forth Gaia theory. In fact when Lovelock said, "People untrained ... do not revere ... Geosphere Biosphere System, but they can ... see the word Gaia embracing both the intuitive side of science and the wholly rational understanding that comes from Earth System Science" he makes a modern plan for the return to the respected natural history, the enterprise from which biology, geology, atmospheric science, and meteorology had not yet irreversibly divorced themselves. Is he not explicit when he writes, "We have some distance still to travel because a proper understanding of the Earth requires the abolition of disciplinary boundaries"? For the science itself, although precluded today by administrative and budgetary constraints, the advisable action would be a return to natural history, the status quo ante, before those disciplines were even established. As Lovelock says, and I agree, "We need reduction in science, but it is not the whole story." My point is that yes, I agree, reductive simplification to control one variable at a time is indispensable to scientific inquiry. Yet no reason exists for us not to continue reductionist practices in the context of Gaian natural history. Indeed, the name changes ought not to deceive us about the true identities of our friends. "Astrobiology" is the field of natural history reinvented to be fundable for a wide variety of scientists, whereas "Earth system science" is none other than Gaia herself decked in futuristic garb and made palatable to the "hard rock" scientists, especially geophysicists.

The original Gaia hypothesis primarily involved biotic regulation of three aspects of the surface of the Earth: the temperature, the acidity-alkalinity, and the composition of the reactive atmospheric gases, especially oxygen. Accordingly I tentatively offer an adequate working definition of the Gaia hypothesis that can serve to organize an enormous, unwieldy scientific literature. Gaia, a name that makes our third planet, as Lovelock likes to say, "a personal presence for all of us" refers to the science of the living Earth as seen from space. My definition for the Gaia hypothesis is as follows:

Some 30 million types of extant organisms [strains of bacteria and species of eukaryotes; Sonea and Mathieu, 2000] have descended with modification from common ancestors; that is, all have evolved. All of them—ultimately bacteria or products of symbioses of bacteria (Margulis and Sagan, 2002)—produce reactive gases to and remove them from the atmosphere, the soil, and the fresh and saline waters. All directly or indirectly interact with each other and with the chemical constituents of their environment, including organic compounds, metal ions, salts, gases, and water. Taken together, the flora, fauna, and the microbiota (microbial biomass), confined to the lower troposphere and the upper lithosphere, is called the biota. The metabolism, growth, and multiple interactions of the biota modulate the temperature, acidity-alkalinity, and, with respect to chemically reactive gases, atmospheric composition at the Earth's surface.

A good hypothesis, as Lovelock has noted, whether or not eventually proved right or wrong, generates new experimental and theoretical work. Gaia, defined this way, undoubtedly has been a good hypothesis. Gaian concepts, especially in the 1980s and early 1990s, generated an environmental literature (Lapo, 1987; Lovelock, 1979, 1988; Sagan, 1990; Westbroek, 1991) that extends far beyond the bounds of the traditional relevant subfield of biology: "ecology." Ecology as taught in academic circles has become more Gaian or has faded away.

Of particular interest to me is "new Gaia," newly generated scientific ideas beyond the original statement of the theory. Several are worthy of closer scrutiny by observation, experimentation, and model calculation. New books to which I refer (Lowman, 2002; Morrison, 1999; Smil, 2002; Sonea and Mathieu, 2000; Thomashow, 1996; Volk, 1998) have done us a great service by review and interpretation of jargon-filled incommensurate scientific articles. These authors provide an essential prerequisite for future investigation. In the case of Thomashow (1996), the review is less of the science and more of the history and emotional importance of Gaian concepts in the context of environmental education and ecological understanding.

In this necessarily brief contribution to what Lovelock sarcastically refers to as his "weak little theory," some predictions have been confirmed. Thus, I concur with the ten items on Lovelock's list, but I concentrate on other "new Gaia" aspects of the science. For discussion I especially question the Earth's relation to the phenomenon of continental drift and plate tectonics.

"Surface conditions on Earth," NASA geologist Lowman (2002) writes, "have been for most of geological time regulated by life." Lowman identifies this statement as Lovelock's Gaia hypothesis and claims, "This new link between Geology and Biology originated in the Gaia hypothesis" (p. 272). The Gaia concept leads Lowman to a new perspective on the evolution of the crust of the Earth and to his "unified biogenic theory of the Earth's crustal evolution," which will be defined here.

Lowman's synthesis derives the earliest events in our planet's evolution from those which surely occurred on our lifeless solar system neighbors: the Moon, Mercury, and Venus. The new science of comparative planetology is generated by many studies, especially the use of the superb new tools of space geodesy, satellite measurements of geomagnetism, remote sensing across the electromagnetic spectrum, and analyses of impact craters. This new work leads Lowman to a radically different view of Earth's tectonic history. He posits that the Earth's major concentric layers—the liquid core, the convecting plume-laden mantle, and the cooler, more rigid outer crust—were formed by the same processes that occurred on our neighboring silicate-rich planets. Such planetary and petrologic processes preceded Gaia. The main crustal dichotomy of an Earth divisible into the two regions (generally granitic continental masses and basaltic ocean basins), he argues, was initiated by the great early bombardment scenario of the inner solar system. The Earth, like its neighbors, was so beset by bolides that the crust was punctured and heated time and again. Incessant volcanism was intense on an Earth far hotter and tectonically more active than today. Two-thirds of the primordial global crust may have been removed by the giant impact of a Mars-sized bolide that ejected the debris from which our huge satellite, the Moon, accreted. The so-called lunar birth explosion, he thinks, may have triggered mantle upwelling, basaltic magmatism, and tectonic activities similar to "those of the Moon, Mercury, Mars, and possibly Venus" (p. 279). However, "the broad aspects of the Earth's geology as it is now—continents, ocean basins, the oceans themselves, sea floor spreading and related processes—are the product of fundamentally biogenic processes, acting on a crustal dichotomy formed by several enormous impacts on the primordial Earth."

Lowman goes on to claim, "The fundamental structure of the Earth, not just its exterior and outer layers, thus appears to have been dominated by water-dependent—and thus life-dependent—plate tectonic processes." Life has actively retained water and moderate surface temperatures, not just passively "adapted" to them. In summary of many detailed investigations and their interpretations, Lowman writes:

> The most striking characteristic of the Earth is its abundant water: colloidally suspended in the atmosphere; covering two-thirds of its surface; coating, falling on, and flowing over the remaining one-third; and infiltrating the crust and mantle. It retains this water partly because of the planet's surface temperature but also because the Earth behaves like a living organism that maintains this temperature by a wide variety of feedback mechanisms, many of which are caused by life itself. (p. 280)

Presenting an integrated view of energy flow, oceanography, and climatology with the physics, chemistry and biology of the biosphere we all call home, Vaclav Smil, a distinguished professor at the University of Manitoba in Winnipeg, has written a book that might as well be called *Gaia: The Living Earth from Space*. His immensely learned, highly accessible narrative is that of the true environmentalist. From these new books, coupled with earlier works by Lovelock himself (1979), Morrison (1999), Volk (1998), Westbroek (1991), and Bunyard (1996), enough responsible scientific literature on Gaia exists to fuel college/university-level curricula.

Other new comprehensive and comprehensible contributions to the Gaia debate include the incredibly detailed 400,000-year-old annual ice core record of climatic change and atmospheric CO_2 rise. The story of how international science obtained this fund of Pleistocene data from the central Greenland ice sheet reads like a novel (Mayewski and White, 2002). Another fascinating book, an integration of modern ecological processes and other complex systems determined by the second law of thermodynamics, is in the works for 2004 (Schneider and Sagan, forthcoming). This treatise on energy sees Gaia, even its origin over 3.5 billion years ago, as a part of the tendency of the universe to increase in complexity as energetic gradients are broken down. The sun inexorably loses its heat and light into the cold blackness of space. This temperature and other gradient imperatives generate and sustain organized systems that seem to appear from nothing. These "other-organized" systems, however, enhance thermodynamic, informational, pressure, and other gradient reduction. "Nature," write Schneider and Sagan, "abhors [not just a vacuum] but all gradients." Gaia can be understood as a peculiar, long-lived, expanding, and complexifying "planetary-scale gradient reducer." The history of thermodynamics and this arcane science's ability to describe all manner of energy flow phenomena sheds light on the intimate connection between the physical-chemical sciences and the evolution of life. Furthermore, since the 1970s Gaia theory has continued to draw attention to the mighty microbe, the diverse set of bacterial cells, their communities, and their larger protoctist descendants (Margolis, McKhann, and Olendzenski, 1992). How microbes metabolize and organize into effective, functional communities forms a crucial component of Gaian research.

Gaia theory's original postulates were limited to global temperature, acidity-alkalinity, and the composition of reactive gases of the air. The new Gaia, whatever her name, becomes respectable because postulated explanations for Earth's surface activity require living beings and interrelations between them and the rest of the lithosphere.

Here are just a few scientific queries stimulated by the wily ways of the ancient Earth goddess in elegant modern dress. Without inquisitive prodding, as Jim Lovelock has noted, such questions of the coy Gaian goddess would never have been raised by polite scientific society.

1. Are plate tectonics (i.e., the deep, lateral movements of the lithosphere apparently limited in the solar system to the Earth) a Gaian phenomenon?

2. Is the remarkable abundance of aluminosilicate-rich granite, a crustal rock type unknown elsewhere in the solar system and one that comprises 0.1 percent of the Earth's volume, directly related to the presence of life? Did water flow and oxygen release, so strongly influenced by life over 3 billion years, generate the granitic raised portions of the plates?

3. Is the Earth's distribution of certain metals and other elements, those known to strongly interact with life (e.g., phosphorus, phosphorites, banded iron formation, marine and freshwater iron-manganese nodules), a Gaian phenomenon? Are Archean conglomeratic, organic-rich sedimentary gold deposits related to life?

4. Is the rate of dissolution of vast quantities of salt (sodium chloride) retarded by biological activity (e.g., in the M-layer beneath the Mediterranean sea, the Hormuz basin of Iran, the Texas Permian Basin deposits, and the great German and North Sea Permian zechstein deposits)? In other words, are the worldwide evaporite deposit patterns a Gaian phenomenon?

5. Can long-lasting thermodynamic disequilibria and reactive gaseous chemical anomalies in a planetary atmosphere be taken as a presumptive sign of life?

6. If life is primarily responsible for the enormous differences in the meters (m) of precipitable water on the surfaces of the three silica-rich inner planets (Venus, 0.01 m; Earth, 3000 m; Mars, 0.0001 m), what have been the biological modes of water retention on Earth since the Archean eon?

7. If Earth's surface temperature has been modulated mainly by carbon dioxide, other carbonates, and organic compounds being removed from the atmosphere into limestone, to what extent have chemoautotrophic, anoxygenic phototrophic, and other metabolic pathways of CO_2 reduction supplemented the oxygenic photoautotrophy of cyanobacteria, algae, and plants?

8. Can environmental regulation studies be valid and representational in "mini-Gaia" contained systems that are closed to matter but open to sunlight or other electromagnetic energy fluxes?

No doubt many more such questions might be raised. Indeed, they are raised in several contributions to this book. Let it suffice here for me to claim that the heuristic value of this global concept is unprecedented in modern times. All of us as readers and contributors to *Scientists on Gaia: The Next Century* are profoundly indebted to Jim Lovelock for his intellectual leadership and healthy disdain of "academic apartheid." We cannot be fooled: Gaia's core identity and liveliness will survive her many fancy guises, bold dance steps, cruel deceptions, and name changes. Our Earth by any other name will smell and look and feel as sweet.

References

Arthus-Bertrand, Y. 1999. *Earth from Above.* Harry N. Abrams, New York.

Bunyard, P., ed. 1996. *Gaia in Action: Science of the Living Earth.* Floris Books, Edinburgh.

Kirchner, J. W. 2002. The Gaia Hypothesis: Fact, theory, and wishful thinking. *Climatic Change*, 52, 391–408.

Lapo, A. 1987. *Traces of Bygone Biospheres*. Mir Publishers, Moscow.

Lovelock, J. E. 1965. A physical basis for life detection experiments. *Nature* 207:568–570.

Lovelock, J. E. 1979. *Gaia: A New Look at the Life on Earth*. Oxford University Press, New York.

Lovelock, J. E. 1988. *The Ages of Gaia*. W. W. Norton, New York.

Lovelock, J. E. 2002. Reflections on Gaia. Introduction to the present volume.

Lowman, P. 2002. *Exploring Space, Exploring Earth: New Understanding of the Earth from Space Research*. Cambridge University Press, Cambridge.

Margulis, L., H. I. McKhann, and L. Olendzenski. 1992. *Illustrated Glossary of Protoctista*. Jones and Bartlett, Boston.

Margulis, L., and D. Sagan. 2002. *Acquiring Genomes: A Theory of the Origins of Species*. Basic Books, New York.

Mayewski, P., and F. White. 2002. *The Ice Chronicles; The Quest to Understand Global Climate Change*. University of New England Press, Hanover, NH.

Morrison, R. 1999. *The Spirit in the Gene: Humanity's Proud Illusion and the Laws of Nature*. Cornell University Press, Ithaca, NY.

Sagan, D. 1990. *Biospheres: Metamorphosis of Planet Earth*. McGraw-Hill, New York.

Schneider, E. R., and D. Sagan. Forthcoming. *Energy Flow: Thermodynamics and the Purpose of Life*. University of Chicago Press, Chicago, IL.

Schneider, S. H., and P. Boston, eds. 1991. *Scientists on Gaia*. MIT Press, Cambridge, MA.

Skoyles, J., and D. Sagan. 2002. *Up from Dragons: On the Evolution of Human Intelligence*. McGraw-Hill, New York.

Smil, V. 2002. *The Earth's Biosphere: Evolution, Dynamics and Change*. MIT Press, Cambridge, MA.

Sonea, S., and L. Mathieu. 2000. *Prokaryotology: A Coherent View*. University of Montreal Press, Montreal.

Thomashow, M. 1996. *Ecological Identity: Becoming a Reflective Environmentalist*. MIT Press, Cambridge, MA.

Volk, T. 1998. *Gaia's Body: Toward a Physiology of Earth*. Springer-Verlag, New York.

Westbroek, P. 1991. *Life as a Geological Force*. W. W. Norton, New York.

White, F. 1998. *The Overview Effect: Space Exploration and Human Evolution*, 2nd ed., American Institute of Aeronautics and Astronautics, Reston, VA.

I
PRINCIPLES AND PROCESSES

1
Clarifying Gaia: Regulation with or without Natural Selection

Timothy M. Lenton

Abstract

This chapter is an attempt to resolve current debates about the Gaia theory. Gaia is defined as a type of planetary-scale, open thermodynamic system, with abundant life supported by a flux of free energy from a nearby star. The Earth supports the only known example of a Gaia system. This system's environmental state has been profoundly altered by the presence of life; some of its state variables are remarkably stable; and it responds surprisingly fast to certain perturbations. Such self-regulation comes about through a combination of positive and negative feedbacks, often involving life, and the feedback structure of the system is continually being transformed by evolution. The Gaia theory seeks to explain the development and functioning of Gaia systems, based on the tenets that life affects its environment, organisms grow and multiply (potentially exponentially), the environment constrains growth, and natural selection occurs. Traits are rarely selected for their environmental consequences. Instead, environmental effects are usually "by-products" of more localized selection. Resulting changes in the environment can influence the growth of the responsible organisms (and their compatriots), giving rise to feedback on growth. In cases where environmental changes alter the benefit of possessing the responsible trait, feedback on selection also occurs. Global environmental regulation can readily emerge from both types of feedback. This is demonstrated with a range of examples and models.

Introduction

I began an earlier version of this chapter thus: "The once controversial notion that the Earth is a self-regulating system now appears to be gaining acceptance in the scientific community." That was a red flag to the reviewers, who made it abundantly clear that Gaia continues to generate controversy. No doubt a reading of this volume will convey the same impression. Yet the Amsterdam Declaration on Global Change (Moore et al., 2001), put forward by the heads of four international global change research programs and signed by numerous scientists, states as fact: "The Earth System behaves as a single, self-regulating system comprised of physical, chemical, biological and human components." Although it doesn't say what, how, or why the system regulates, this statement reads like the beginning of an exposition on Gaia. At the very least it indicates that a perception shift toward thinking of the Earth as a system is under way. The controversy lies in the what, how, and why of system functioning, all rather inadequately covered by the term "self-regulating."

It has often struck me that many of the critics of Gaia misunderstand what is being proposed, partly because what is being proposed has been refined over time. Attacks have often been launched on the basis of early versions of the Gaia hypothesis that its originator and supporters had long since abandoned. One of the reasons I got involved in the subject was a sense that the proponents and the opponents were talking at cross-purposes, and that between them lay an unrecognized but fertile common ground. That was a motivation for thinking about "Gaia and natural selection" (Lenton, 1998). In particular, evolutionary biologists' and ecologists' "it cannot work!" dismissal of Gaia missed the fundamental point that regulation does not have to be evolved. Yes, the functioning of the Earth system (which includes life) must be consistent with natural selection occurring within it, but no, this does not mean that planetary-scale regulation has to be the product of natural selection.

My objective here is to clarify the debate about Gaia and natural selection (Lenton, 1998; Dagg, 2002; Kirchner, 2002; Kleidon, 2002; Volk, 2002). I will concentrate on the difference between environmental regulation that is based on by-products of selection and that involving natural selection (Lenton, 1998), which is a distinction that has recently been ignored or misunderstood (Dagg, 2002; Kirchner, 2002; Kleidon, 2002; Volk, 2002). The scope of the Gaia theory is broader than the aspects focused on here (Lovelock, 1988; Lenton, 1998; Lenton and van Oijen, 2002). However, if we are to make progress toward a more principled theory, a clearer understanding of the relationship between Gaia and natural selection is a prerequisite.

Defining Gaia

Before launching into the argument, I will offer some definitions to help clarify the debate. More detailed definitions of Gaia and aspects of self-regulation are given elsewhere (Lenton, 2002; Lenton and van Oijen, 2002).

Gaia is a particular type of open thermodynamic system, planetary in scale with abundant life supported by a flux of free energy from a nearby star. At present we know of only one example of such a system. That is the system at Earth's surface, comprising life (the biota), atmosphere, hydrosphere (ocean, ice, freshwater), dead organic matter, soils, sediments, and that part of the lithosphere (crust) which interacts with surface processes (including sedimentary rocks and rocks subject to weathering). The upper boundary of the system is at the top of the atmosphere, with outer space. The inner boundary is harder to define. A rigid definition that excludes any component of the crust from the system has been suggested (Volk, 1998), but I think this is too exclusive. I suggest that the inner boundary can be taken to depend on the timescale of processes under consideration. For processes that can approach steady state rapidly (in $<10^3$ years), the outer surface of the crust may be considered the boundary of the system. On timescales longer than the recycling of the crust ($\sim 10^8$ years), the system must include the crust, and its boundary extends into the mantle to the depth that rock slabs are subducted. The heat energy source in Earth's interior, like the sun, is not significantly influenced by surface processes and hence is best considered to be "outside" the system (although it is inside the Earth).

I distinguish Gaia from the "Earth system," in that the Earth system includes states before the origin of life and with sparse life, whereas Gaia refers to the system with abundant life. I also distinguish Gaia from the "biosphere" (defined as the region inhabited by living organisms). The boundaries of Gaia are almost certainly wider, because the influence of living organisms extends beyond the region they inhabit. The upper boundary of the biosphere is >50 km above Earth's surface in the mesosphere, where viable microscopic fungi and bacteria have been collected, whereas the exosphere extends above 500 km. The lower boundary of the biosphere is unknown, but organisms have been found to be thriving at depths of a few kilometers in the Earth's crust. It has been suggested that a "deep hot biosphere" could survive independently of surface life. If that is so, it would represent a different system, running off a relatively small amount of free energy.

Stability is a property of some states of many systems. A stable state is one that is returned to for some range of perturbations (alterations in the state variables or changes in forcing) of a system. Stability alone does not distinguish Gaia from other systems, however, because many systems possess stable states. What is of greater interest is the degree of stability of Gaia's state variables, that is, what range of perturbations can they withstand—for example, relative to an Earth system without abundant life (Lenton, 2002).

Regulation describes the return of a variable to a stable state after a perturbation. In this sense it occurs in all systems with stable states. Again, what is of interest for Gaia is the degree of regulation. This can be quantified in terms of the degree to which a given perturbation alters the state of the system (resistance), and the rate at which the system returns toward a stable state (resilience). As above, it is interesting to contrast the regulation of Gaia's state variables with what is predicted for an Earth system without abundant life (Lenton, 2002).

Feedback occurs when a change in a variable of a system triggers a response that affects (feeds back on) the forcing variable. Feedback is said to be negative when it tends to damp the initial change and positive when it tends to amplify it. Feedback may involve a chain of processes; hence the term *feedback loops*. Gaia contains a large number of both negative and positive feedback loops, and the overall behavior of the system is influenced by a combination of positive and negative feedback.

Self-regulation describes a system automatically bringing itself back to a stable state, rather than an external agent imposing regulation or any conscious purpose (teleology) within the system bringing about regulation. It is a popular misconception that self-regulation is synonymous with negative feedback. In complex systems, self-regulation can involve a combination of positive and negative feedback.

Volk (2002) has asked, "How can the Earth ... not be self-regulating?," citing the fact that many systems, including planets without life, return to a stable state after being disturbed. The fact that Gaia possesses stable states may not be remarkable, but the habitable nature of those states (which change over time) and the means by which self-regulation is achieved in such a complex system are fascinating scientific questions. The system not only responds to changes in its environment, but its feedback structure is continually being transformed by evolution occurring within it (Lenton and van Oijen, 2002). The resulting regulation demands an explanation, and the explanation may itself prove to be quite complex.

Gaia Theory

The Gaia theory aims to explain the development and functioning of Gaia systems, and is currently restricted to the one example system described above. Of particular interest are its far-from-equilibrium ordered state, habitability, flourishing of life, self-regulation, and pattern of change over time (Lovelock, 1988; Lenton, 1998). From the outset Gaia has been viewed as a cybernetic (feedback control) system (Lovelock and Margulis, 1974). Important tenets of the theory are the following:

1. Life affects its environment: all organisms alter their environment by taking in free energy and excreting high-entropy waste products, in order to maintain a low internal entropy (Schrödinger, 1944).
2. Growth (including reproduction): Organisms grow and multiply, potentially exponentially.
3. Environment constrains life: for each environmental variable there is a level or range at which growth of a particular organism is maximized and there are conditions under which growth is impossible.
4. Natural selection: once a planet contains different types of life (phenotypes) with faithfully replicated, heritable variation (genotypes) growing in an environment of finite resources, the types of life that leave the most descendants come to dominate their environment.

Growth is intrinsically a positive feedback process (the more life there is, the more life it can beget). The growth of life drives life's effects on its environment to become global in scale. The extreme thermodynamic disequilibrium of the Earth's atmosphere indicates that this has occurred. The fact that life alters its environment and is also constrained by it means that environmental feedback is inevitable. Environmental feedback arises at the local level. With growth driving it, environmental feedback has the potential to become global in scale. In summary:

$(1) + (2) \rightarrow$ global environmental effects

$(1) + (3) \rightarrow$ environmental feedback

$(1) + (2) + (3) \rightarrow$ global environmental feedback.

With global environmental feedback comes the possibility of global environmental regulation. Thus life altering the environment, growth, and environmental constraints taken together are *sufficient conditions* for global environmental regulation to occur. Natural selection is *not* a *necessary condition.* I have termed feedback without natural selection "feedback on growth" or "nonselective feedback," and it can give rise to one type of regulation. We clearly live in a world where natural selection operates. If changes in the environment caused by life affect the natural selection of the responsible traits, then what I have called "feedback on selection" or "selective feedback" also occurs. This can give rise to a different type of regulation.

Misunderstandings of Gaia and Natural Selection

A central misunderstanding of Gaia and natural selection is summed up in the assumption that "life-enhancing effects would be favoured by natural selection" (Kleidon, 2002). This is not a valid generalization, as both Kirchner (2002) and Volk (2002) have pointed out. It is not a generalization I have made (although Lenton, 1998 was cited in Kleidon, 2002). I do think it is true in specific cases, but only when the carriers of the responsible traits benefit more from those "life-enhancing effects" than do noncarriers. Thus we should avoid "making blanket statements that evolution will select life forms that benefit their environments" (Volk, 2002), but also should not fall into the opposite trap of assuming that natural selection never favors traits that enhance their environments. That would be a logical error, and the statement is readily falsified with examples such as nitrogen fixation (see below). Rather, let us think carefully about how and why organisms are altering their environments, how natural selection drove the evolution of the responsible traits, and how the environmental consequences of those traits affect the carriers and noncarriers, and from that try to build an understanding of the resulting feedbacks between organisms and their environments.

According to Kirchner (2002), the Gaia hypothesis proposes that "stabilising, environment-enhancing feedbacks should arise naturally as the result of natural selection acting on individual organisms." This

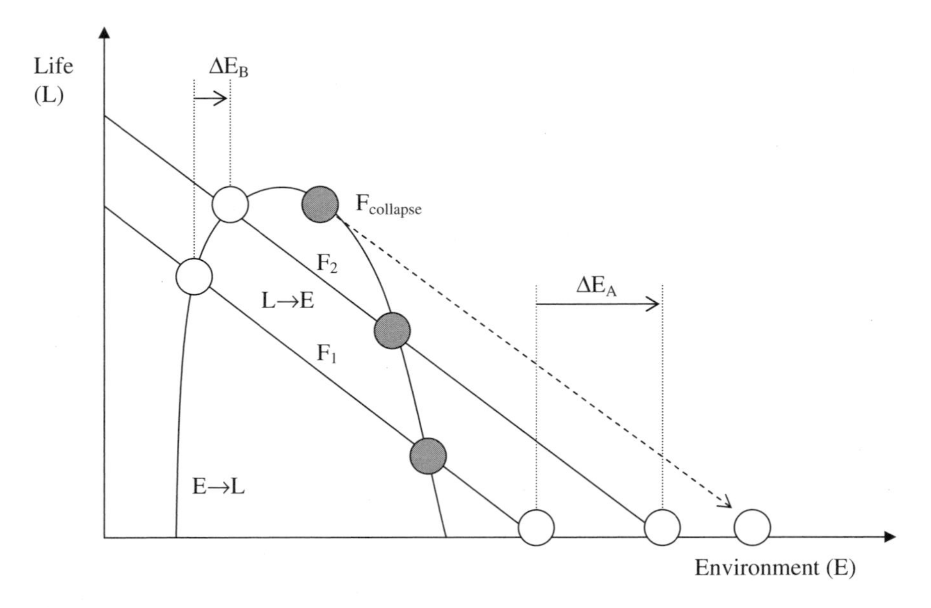

Figure 1.1
A simple coupled system in which "life" and the "environment" are each reduced to one variable (e.g., plant population and temperature). The environment affects life ($E \rightarrow L$), following a bell-shaped curve with upper and lower limits on habitability. Life affects the environment ($L \rightarrow E$), causing a linear decrease in the environment variable with increase in the life variable. The system is subject to external forcing (F) that also affects its state (shifting $L \rightarrow E$). Open circles indicate stable equilibrium points. Shaded circles indicate unstable equilibrium points. The system with life tends to reside in a negative feedback regime on one side of the optimum for growth, and hence it is more resistant to perturbation than it would be without life: An increase in forcing from F_1 to F_2 causes a smaller change in the biotic state of the environment (ΔE_B) than in the abiotic state of the environment (ΔE_A). The system illustrated is bi-stable over a range of forcing. If forcing is increased to a critical point ($F_{collapse}$), the system switches from the biotic to the abiotic state, in a transition dominated by positive feedback. The illustration is based on a Daisyworld with only white daisies (Watson and Lovelock, 1983).

misrepresents the theory on a further count. As Kirchner himself notes, it is misleading to associate "stabilising" and "environment-enhancing." One can say that a particular effect of life is environment-enhancing or environment-degrading (for a particular organism), and one may describe a particular feedback as stabilizing, but it is important to distinguish effects from any resulting feedbacks. This is because a single effect on a single environmental variable can give rise to both positive and negative feedback, when the responsible organisms have a peaked growth response to that environmental variable and the effect is sufficiently strong. If the effect tends to increase the value of the variable (e.g., warming), then it will give rise to positive feedback below the optimum (temperature for growth), and to negative feedback above it. Conversely, if the effect tends to decrease the value of the variable (e.g., cooling), then it will give rise to positive feedback above the optimum (temperature for growth) and to negative feedback below it (figure 1.1).

Kirchner (2002) focuses on just one class of interaction within the Gaia system: that which gives rise to feedback on selection (Lenton, 1998). Abiotic feedbacks and biotic feedbacks on growth that do not involve changes in the forces of selection are also encompassed by the theory. I will now expand upon the distinction between feedbacks that involve natural selection and those that do not.

Feedback Involving Natural Selection

Let us start with the "special case" represented in the original Daisyworld model (Lovelock, 1983) (figure 1.2 a, b). The essence of the model is that a trait (daisy albedo) affects the individual and the environment in the same way. A black daisy captures more solar energy, warming itself and its surroundings. A white daisy reflects more solar energy, cooling itself and its surroundings. Under cool background conditions (i.e., below the optimum temperature for growth) black daisies have a selective advantage. They alter the environment (warm it) in a way that enhances growth but reduces their selective advantage. Under warm background conditions (i.e., above

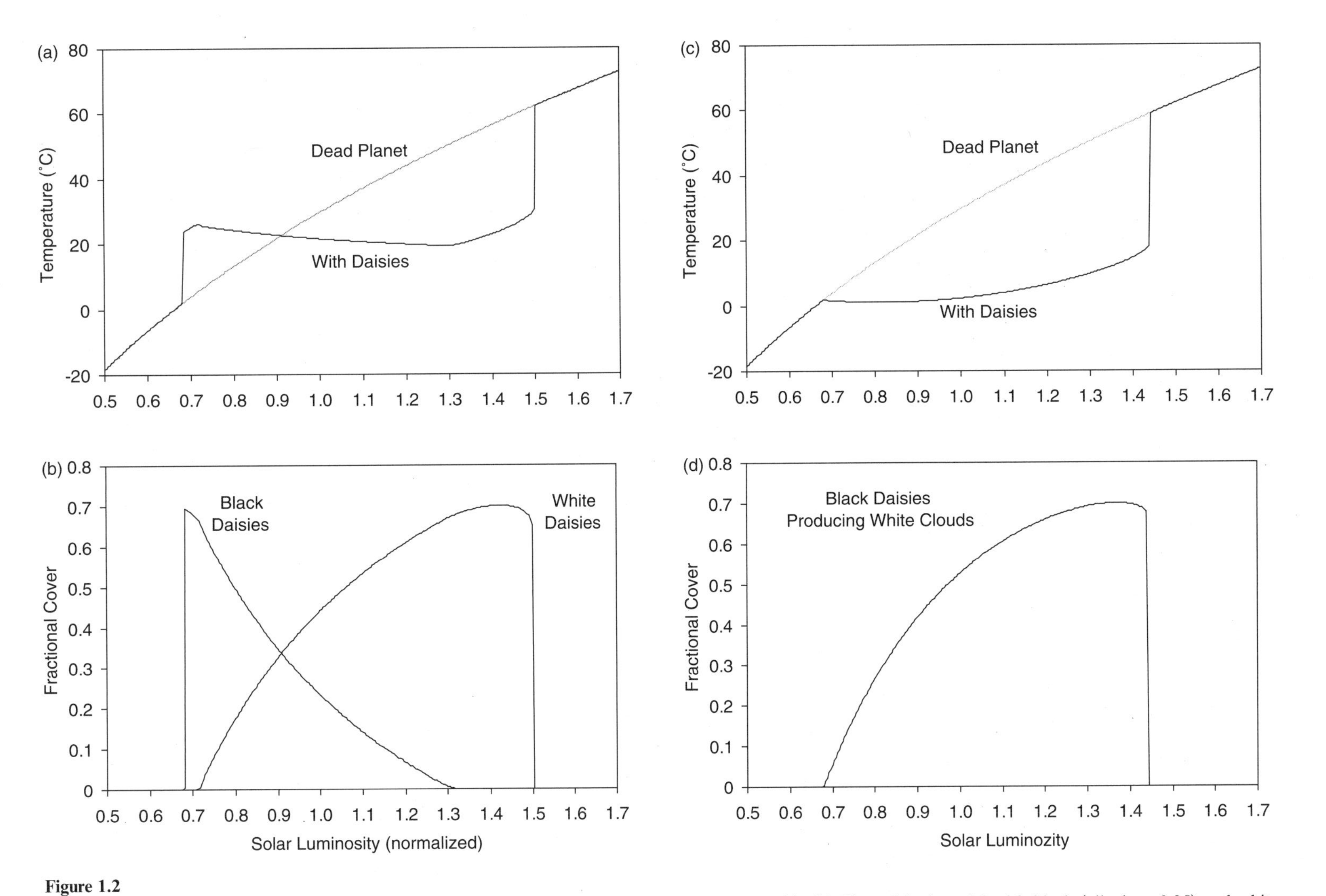

Figure 1.2
Two variants of Daisyworld (Watson and Lovelock, 1983) illustrating two types of regulation: (a), (b) The original model with black (albedo = 0.25) and white (albedo = 0.75) daisies. Negative feedback on selection contributes to regulation toward optimal conditions. (c), (d) A variant with black daisies that produce white clouds (albedo = 0.8). White daisies are still present but they are outcompeted by the cloud-making black daisies, due to positive feedback on selection. Underlying negative feedback on growth generates regulation toward constraints. The feedback relationship between global temperature and growth of the cloud-making black daisies is qualitatively the same as that shown in figure 1.1.

the optimum growth temperature) white daisies have a selective advantage. They also alter the environment (cooling it) in a way that enhances growth but reduces their selective advantage. Hence the description "negative feedback on selection." The result is a self-regulator that maintains global temperature close to the optimum for daisy growth over a wide range of solar forcing (figure 1.2 a, b). When both daisy types are present, as solar luminosity increases, the surface temperature of the planet actually decreases.

Daisyworld demonstrates that it is possible for feedback on natural selection to contribute to generating an optimizing regulator "but only given a very specific assumption embedded in the model" (Kirchner, 2002). The assumption is that a given trait (in this case, daisy albedo) affects the individual and the environment in the same way (Lenton, 1998). As noted above, it is a mistake to generalize that this will always be the case, but it will be true in some specific cases. Critics have speculated that this type of negative selective feedback (where what is good for the individual is good for the global) is rare in the real world. I suggest that we have insufficient knowledge to make such a generalization.

An important real-world example of negative feedback on selection is that arising from the process of nitrogen fixation. Nitrogen fixers increase their own supply of available nitrogen, and ultimately this "leaks out" to increase the amount of available nitrogen in the environment (their surroundings). If nitrogen is deficient in the environment, relative to other essential elements, then organisms that can fix nitrogen gain a selective advantage over nonfixers (as long as the benefits exceed the considerable energy cost of nitrogen fixation). However, as the fixed nitrogen leaks out into the environment (and becomes available to the nonfixers as well), this selective advantage is reduced. Thus systems with nitrogen fixers and nonfixers should tend toward a stable state with deficits in the input of available nitrogen being met by a corresponding fraction of the population fixing nitrogen. If the supply of nitrogen relative to other essential elements is perturbed, the system will respond in a manner to counteract the change. Reduced nitrogen input will lead to more nitrogen fixation; increased nitrogen input, to less nitrogen fixation. This mechanism has been observed in terrestrial ecosystems (e.g., pastures) and some lakes, and is thought to play a key role in the nutrient balance of the global ocean (Tyrrell, 1999; Lenton and Watson, 2000a).

We should also consider the alternative case where a trait affects the individual and the environment in opposite directions. One early variant of the Daisyworld model was based on this premise (Watson and Lovelock, 1983): black daisies were introduced that generate white clouds (figure 1.2 c, d). These black daisies still warm themselves, but they cool the shared environment. The result is that they keep the environment in a cool state in which they have a selective advantage (over white daisies). I have called this positive selective feedback. The important thing to note is that despite inverting the environmental effect, the system still regulates. That is because the population of cloud makers is controlled by negative feedback on growth: their spread is suppressed by their cooling effect. As solar luminosity increases, so does the population of cloud makers, thus stabilizing the planet's surface temperature against changes in solar forcing. This is an example of the second type of regulation, which will be expanded upon in the following section.

In the real world some organisms certainly alter their environment in a way that maintains or promotes their selective advantage. The examples I can think of tend to be relatively localized and biased to terrestrial ecosystems: sphagnum moss acidifying the soil and promoting waterlogging through the formation of an iron pan; plants producing and releasing compounds that are toxic to other species (allelopathy); eucalyptus trees encouraging fires that are fatal to their competitors but not to themselves. An important point here is that positive feedback on selection does not preclude stability: it tends to be constrained positive feedback (rather than runaway positive feedback). This can be because the traits involved are costly in terms of energy, because the range of their environmental effects is limited, and/or because the environmental effects generate negative feedback on growth for both the carriers and noncarriers, as is the case with the cloud-making black daisies in the variant of Daisyworld.

Feedback from By-products of Natural Selection

The same critics who stress that the type of negative feedback on selection present in Daisyworld is rare (or even nonexistent) in the real world sometimes claim that therefore biotic regulation is rare (or nonexistent). This is an error of logic. The implicit but false assumption is that biotic regulation can arise only from traits that have evolved through natural selection based on their environmental consequences. In other words, regulation must be selected for. This fails to recognize that regulation is an emergent property in many systems where there is no active selection for regulation (Lenton and van Oijen, 2002; Wilkinson, this volume). Indeed, many (and perhaps most) globally important biotic feedbacks appear to be based on by-products of selection—for example, those generated by dimethyl sulfide production (Hamilton and Lenton, 1998; Lenton, 1998). This can lead to a kind of regulation different from that in the original Daisyworld model, with the environment maintained in a limiting state, which in turn can be remarkably resistant and resilient to change.

Such regulation occurs because the spread of environment-altering traits is ultimately subject to constraints. The spread of a particular trait is constrained if it alters an environmental variable in a manner that reduces the growth rate of the organisms carrying it. In this case noncarriers are equally affected (i.e., the environmental effect is not selective) and the trait remains selected for, but negative feedback on growth (the spread of the trait) still occurs. Positive feedback on growth can also occur, when the spread of a particular trait alters an environmental variable in a manner that increases the growth rate of the organisms carrying it (together with noncarriers). If the effect of the trait on the environment is sufficiently strong, and there is a peaked growth response to the environmental variable, then the positive feedback regime will be transited and the system will stabilize in the negative feedback regime, with the environmental variable in a limiting state (figure 1.1). This type of feedback is involved in regulating two of the most important components of the atmosphere: oxygen (at an upper limit) and carbon dioxide (at a lower limit).

Oxygen has remained within relatively narrow bounds of approximately 15–25 percent of the atmosphere for the last 350 million years. A number of negative feedback mechanisms have been hypothesized to explain this stability (Lenton and Watson, 2000b), all of which involve processes whose effects on atmospheric oxygen are a by-product of selection to meet other more localized, short-term requirements. One of these processes is the enhancement of the source of phosphorus from biological amplification of rock weathering (Lenton, 2001). This supplies an essential nutrient to the responsible organism. Ultimately increased phosphorus weathering drives increased productivity on land and in the ocean and increased organic carbon burial, which generates a corresponding net source of atmospheric oxygen. However, the spread of traits driving up atmospheric oxygen is subject to constraints. In particular, the process of combustion of organic matter is sensitive to oxygen mixing ratio and partial pressure. If oxygen increases above its present mixing ratio and/or partial pressure, the energy required for ignition decreases, the rate of burning increases, and thus fire frequency increases. This tends to reduce biomass and cause ecological shifts from forests to faster-regenerating ecosystems. This in turn suppresses phosphorus weathering, tending to decrease the oxygen source and generate negative feedback. The thermodynamics of organic matter combustion thus provides a "set point" for oxygen regulation. The set point will vary somewhat with wetness of the climate (moisture content of the fuel), but it is not subject to biological adaptation because evolution cannot circumvent the laws of thermodynamics.

The biological amplification of weathering also enhances the liberation of calcium and magnesium ions from silicate rocks and the subsequent uptake of carbon dioxide when they form marine carbonates. This process has reduced atmospheric CO_2 to a concentration that is limiting the productivity of the majority of (C_3) plants, and the corresponding cooling has also made global average temperature suboptimal for plant productivity (figure 1.3). Thus the system has stabilized in a negative feedback regime, where increases in CO_2 and/or temperature are counteracted by increases in plant-driven silicate weathering rate, and decreases in CO_2 and/or temperature are counteracted by decreases in plant-driven silicate weathering (both on a timescale on the order of 10^5 years). The organisms accelerating silicate weathering are very responsive to changes in atmospheric CO_2 and global temperature, and they have a large effect on these environmental variables; hence the negative feedback on CO_2 and temperature is strong. Although the resulting regulator does not optimize the state of the environment for life, it has other "benefits" of increased resistance (smaller changes in the state of the environment for a given change in forcing) and increased resilience (faster recovery from a given perturbation) (Lenton, 2002).

One aspect of increased resistance is that the feedback system with life can tolerate a wider range of external forcing. In this case, increasing solar luminosity tends to warm the planet and drive a reduction in the CO_2 content of the atmosphere. Previous models have predicted that plant life (and all that depends on it) will perish in ~0.8 Gyr due to CO_2 starvation (Caldeira and Kasting, 1992). Stronger negative feedback delays the loss of CO_2 from the atmosphere, thus extending the life span of the biosphere (up to a maximum of ~1.2 Gyr) (Lenton and von Bloh, 2001).

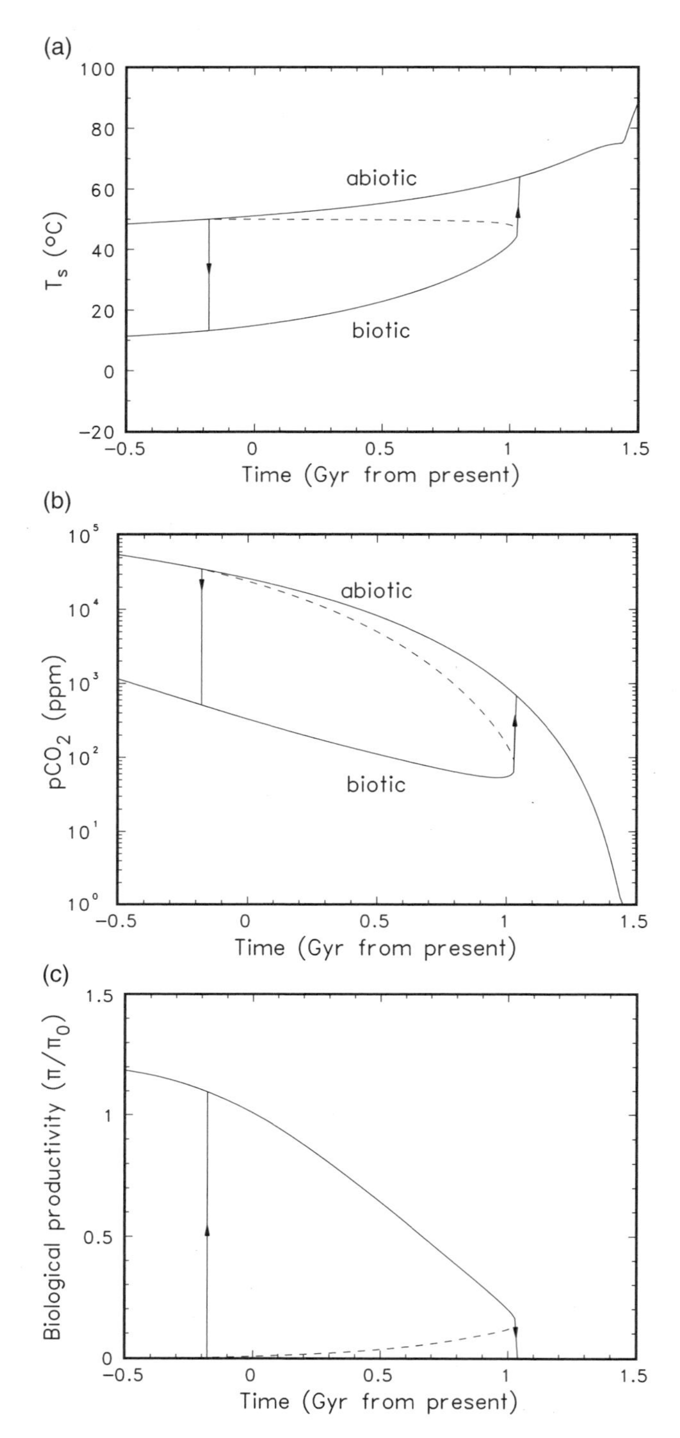

Figure 1.3
State of the Earth with ("biotic") and without ("abiotic") weathering organisms (Lenton and von Bloh, 2001). The model of Caldeira and Kasting (1992) was altered to include a factor of 20 biotic amplification of silicate weathering due to processes other than enhanced soil pCO_2, giving a total amplification factor of ~31. Increasing solar luminosity forces the system. Development of (a) surface temperature (T_s), (b) atmospheric CO_2, (c) biological productivity (normalized to the present value). Solid lines indicate stable equilibrium solutions. The dashed line indicates an unstable equilibrium solution. The arrows show the transitions when the biotic state collapses and when the abiotic state becomes unstable due to the introduction of weathering organisms. The relationship between temperature and growth is qualitatively the same as that shown in figure 1.1. In addition CO_2 limits growth, which is why biological productivity declines into the future.

If the effect of the contemporary land biota on the rate of silicate weathering is sufficiently strong, such as an amplification factor on the order of ~100 (Schwartzman, 1999), then the system becomes bistable. Without biotic amplification of weathering, average surface temperature would be >50°C and weathering organisms perish at ~50°C, making the Earth without them too hot for them to recolonize (figure 1.3).

It has been suggested that evolution can undermine environmental feedback if it leads to adaptation of organisms to prevailing conditions rather than alteration of the prevailing conditions (Robertson and Robinson, 1998). In the case of the long-term CO_2 (and temperature) regulator, the set point can be altered by evolution either of the traits affecting silicate-weathering rate or of the response of the organisms carrying them to CO_2 or temperature. The currently dominant C_3 plants typically cease carbon fixation at a CO_2 concentration below ~100 ppmv. However, it appears that in the last ~10 Myr, declining CO_2 has triggered the proliferation of C_4 plants, which possess a CO_2 concentrating mechanism that allows them to be net carbon fixers down to concentrations of ~10 ppmv, and benefits them in hot, arid conditions by minimizing water loss. The rise of C_4 plants is shifting the set point of the CO_2 regulator to a lower value, and thus extending the range of solar forcing over which the regulator operates. This illustrates the more general points that constraints on adaptation exist for all life forms, but adaptation within constraints can prolong regulation by enabling organisms to tolerate more extreme forcing conditions (Lenton and Lovelock, 2000).

Problems for Regulation Without Selection?

It has been suggested that the existence of disruptive "Genghis Khan" species such as our own may be evidence against any innate regulatory tendencies of the Earth system (Hamilton, 1995). However, "if the connection between environmental good deeds and individual reproductive advantage is only coincidental" (Kirchner, 2002), beneficial effects on the environment may not evolve any more frequently than detrimental effects. Thus we should not be surprised at the emergence of disruptive species. However, the issue is not whether disruptive species *arise* but whether they can *persist*, become globally dominant, and do long-term damage (Longhurst, 1998). This depends, at least in part, on the resulting feedbacks. By their very nature detrimental effects on the environment tend to be self-limiting (they generate negative feedback on growth) and beneficial effects on the environment tend to be self-enhancing (they generate positive feedback on growth). As I have tried to show, negative feedback on growth can be a basis for regulation. Positive feedback on growth, though destabilizing in the sense of amplifying change, is, by definition, enhancing the environment for the responsible organisms; and if the system goes past an optimum for that environmental variable, the feedback becomes negative.

More extremely, it has been speculated that life could in principle drive the Earth into an uninhabitable state (Watson, 1999). I think this is unlikely for the reasons just given (as uninhabitable conditions are approached, negative feedback on the growth of the responsible organisms should stabilize the system within the habitable regime). However, it is possible that an abiotic positive feedback process could take over and drive the system into an uninhabitable or barely habitable state. This may have happened in the past as life evolved new cooling effects, such as stronger amplification of weathering. The danger was that if surface radiative forcing fell below a critical threshold, runaway ice-albedo positive feedback could have totally glaciated the planet (Budyko, 1968). Weaker solar insolation made the Earth more vulnerable to such a catastrophe in the past. It has been suggested that "snowball Earth" conditions occurred repeatedly in the Paleoproterozoic (~2.2 Gyr ago) (Evans et al., 1997) and in the Neoproterozoic (750–580 Myr ago) (Hoffman et al., 1998). If they did occur, they were not fatal to life. Furthermore, in the Neoproterozoic any "snowball" must have been sufficiently "soft" (e.g., with open equatorial oceans) to allow the persistence of metazoans (Hyde et al., 2000). However, the possibility that such events could occur highlights a potential vulnerability of biotic regulation that is operating toward constraints.

Further Possibilities for Selection of Regulation

Thus far I have concentrated on one reason why life-forbidding levels of Gaia's abiotic state variables are difficult to reach: that if an organism is driving an environmental variable toward an uninhabitable state, its growth and spread will be suppressed (by negative feedback) such that the system will stabilize within the habitable regime. A second possibility is that when organisms become so abundant that the side effects of their metabolism become life-threatening on a global scale, different organisms will evolve for which the abundant pollutants and the polluters themselves become resources (Wilkinson, 1999; Lenton and van Oijen, 2002; Wilkinson, this volume). If natural selection has enough substrate (genotypic variation) to operate on, such checks and balances may always be likely to evolve, and make extreme conditions unreachable. This old idea, originating with Adam Smith (Smith, 1776), could explain the likelihood of regulation at livable levels as a side effect of evolution in a sufficiently diverse biota. Something akin to this has been observed in artificial life systems, where closed nutrient recycling and biotic enhancement of gross primary productivity have been found to "evolve" with high probability (Downing and Zvirinsky, 1999).

There are a number of further mechanisms by which regulation could in principle be selected for—for example, if organism-environment assemblages compete with one another and thus are subject to natural selection (Hamilton and Lenton, 1998; Kirchner, 2002). This is the notion of a "superorganism." It has been argued that the temperature-regulating properties of beehives (and ant colonies) have been refined by natural selection operating between colonies (Ehrlich, 1991). (Alternatively, temperature regulation might be explained as an emergent property from individual responses.) The evolutionary notion of a superorganism gets harder to apply when the candidate communities include unrelated individuals (e.g., termite colonies) or multiple species (e.g., peat bogs). That is because this introduces additional scope for selection at lower levels (gene, individual, population) to disrupt the system. It is possible that peat bog communities might function as superorganisms (Hamilton, 1996), with the multiple-species assemblage dispersing together on birds' feet (Hamilton and Lenton, 1998). Successful artificial selection of small soil and aquatic ecosystems has recently been demonstrated in the laboratory, raising the possibility of natural ecosystem selection (Swenson et al., 2000). However, even if one could be convinced that there is a means of evolving regulation through selection at the community scale, this does not get us to the scale of Gaia.

Other cruder forms of selection than heritable-trait-based natural selection can be envisaged that could in principle contribute toward the generation of planetary self-regulation. One suggestion was to make models "of interacting macro systems, for example, some of which are assumed destabilising and some stabilising, and try to show that the former tend to lose out and be replaced by the more stabilising ones before the global system dies" (W. D. Hamilton, personal communication). A further possibility is that of "sequential" selection based on a series of systems originating over time rather than a population of systems coexisting at the same time (R. A. Betts, personal communication). These are topics for future work.

Conclusions

I subscribe to the view (Kirchner, 2002) that "Understanding the Earth system, in all of its fascinating complexity, is the most important scientific adventure of our time. We should get on with it, as free as possible from our preconceptions of the way the world ought to work." I also know from experience that "good holistic thinking is harder than reductionist thinking, so learning to do it properly, even learning what sorts of things we can accomplish with it, is a challenge for the future" (Saunders, 2000). My advice to readers is that if you want to understand Gaia, you have to learn to think in terms of the nonlinear, circular logic of feedback systems. Jim Lovelock has been saying this since the 1960s. Yet too few scientists have grappled with systems thinking (and that includes both critics and supporters of Gaia). Meanwhile, the phrase "Earth system science" is gaining increasing usage. If we are to avoid its being just a fashionable umbrella term with no real substance, we need to promote a new scientific *synthesis*. Unfortunately, the emphasis of the Second Chapman Conference on the Gaia Hypothesis was still on "debating Gaia" (Schneider, 1990). It's time we prioritized building an understanding of Gaia.

Acknowledgments

This chapter has been rewritten a number of times in response to comments on earlier drafts by Jim Lovelock and two prominent critics of Gaia: James Kirchner and Tyler Volk. I thank them for their provocations to make my case clearer. I would also like to thank Dave Wilkinson for his thoughts on how to respond, and an anonymous referee for helping improve the final version. The chapter is dedicated to the memory of W. D. (Bill) Hamilton.

References

Budyko, M. I. 1968. The effect of solar radiation variations on the climate of the Earth. *Tellus*, 21, 611–619.

Caldeira, K., and J. F. Kasting. 1992. The life span of the biosphere revisited. *Nature*, 360, 721–723.

Dagg, J. L. 2002. Unconventional bedmates: Gaia and the selfish gene. *Oikos*, 96(1), 182–186.

Downing, K., and P. Zvirinsky. 1999. The simulated evolution of biochemical guilds: Reconciling Gaia theory and natural selection. *Artificial Life*, 5(4), 291–318.

Ehrlich, P. 1991. Coevolution and its applicability to the Gaia hypothesis. In *Scientists on Gaia*, S. H. Schneider and P. J. Boston, eds., pp. 118–120. Cambridge, MA: MIT Press.

Evans, D. A., N. J. Beukes, and J. L. Kirschvink. 1997. Low-latitude glaciation in the Palaeoproterozoic era. *Nature*, 386, 262–266.

Hamilton, W. D. 1995. Ecology in the large: Gaia and Genghis Khan. *Journal of Applied Ecology*, 32, 451–453.

Hamilton, W. D. 1996. Gaia's benefits. *New Scientist*, 151(2040), 62–63.

Hamilton, W. D., and T. M. Lenton. 1998. Spora and Gaia: How microbes fly with their clouds. *Ethology Ecology and Evolution*, 10, 1–16.

Hoffman, P. F., A. J. Kaufman, G. P. Halverson, and D. P. Schrag. 1998. A Neoproterozoic snowball earth. *Science*, 281, 1342–1346.

Hyde, W. T., T. J. Crowley, S. K. Baum, and W. R. Peltier. 2000. Neoproterozoic "snowball Earth" simulations with a coupled climate/ice-sheet model. *Nature*, 405, 425–429.

Kirchner, J. W. 2002. The Gaia hypothesis: Fact, theory, and wishful thinking. *Climatic Change*, 52, 391–408.

Kleidon, A. 2002. Testing the effect of life on Earth's functioning: How Gaian is the Earth system? *Climatic Change*, 52, 383–389.

Lenton, T. M. 1998. Gaia and natural selection. *Nature*, 394, 439–447.

Lenton, T. M. 2001. The role of land plants, phosphorus weathering and fire in the rise and regulation of atmospheric oxygen. *Global Change Biology*, 7(6), 613–629.

Lenton, T. M. 2002. Testing Gaia: The effect of life on Earth's habitability and regulation. *Climatic Change*, 52, 409–422.

Lenton, T. M., and J. E. Lovelock. 2000. Daisyworld is Darwinian: Constraints on adaptation are important for planetary self-regulation. *Journal of Theoretical Biology*, 206(1), 109–114.

Lenton, T. M., and M. van Oijen. 2002. Gaia as a complex adaptive system. *Philosophical Transactions of the Royal Society of London*, B357(1421), 683–695.

Lenton, T. M., and W. von Bloh. 2001. Biotic feedback extends the life span of the biosphere. *Geophysical Research Letters*, 28(9), 1715–1718.

Lenton, T. M., and A. J. Watson. 2000a. Redfield revisited: 1. Regulation of nitrate, phosphate and oxygen in the ocean. *Global Biogeochemical Cycles*, 14(1), 225–248.

Lenton, T. M., and A. J. Watson. 2000b. Redfield revisited: 2. What regulates the oxygen content of the atmosphere? *Global Biogeochemical Cycles*, 14(1), 249–268.

Longhurst, A. 1998. Too intelligent for our own good. *Nature*, 395, 9.

Lovelock, J. E. 1983. Gaia as seen through the atmosphere. In *Biomineralization and Biological Metal Accumulation*, P. Westbroek and E. W. de Jong, eds., pp. 15–25. Dordrecht: D. Reidel.

Lovelock, J. E. 1988. *The Ages of Gaia—A Biography of Our Living Earth*. New York: W. W. Norton.

Lovelock, J. E., and L. M. Margulis. 1974. Atmospheric homeostasis by and for the biosphere: The Gaia hypothesis. *Tellus*, 26, 2–10.

Moore, B., A. Underdal, P. Lemke, and M. Loreau. 2001. The Amsterdam Declaration on Global Change. http://www.sciconf.igbp.kva.se/Amsterdam_Declaration.html.

Robertson, D., and J. Robinson. 1998. Darwinian Daisyworld. *Journal of Theoretical Biology*, 195, 129–134.

Saunders, P. 2000. The medical model—from geophysiology to physiology and back again. *Gaia Circular*, 3(1–2), 4–9.

Schneider, S. H. 1990. Debating Gaia. *Environment*, 32(4), 5–32.

Schrödinger, E. 1944. *What Is Life?* Cambridge: Cambridge University Press.

Schwartzman, D. 1999. *Life, Temperature and the Earth: the Self-Organizing Biosphere*. New York: Columbia University Press.

Smith, A. 1776. *An Inquiry into the Nature and Causes of the Wealth of Nations*. London: W. Strahan & T. Cadell.

Swenson, W., D. S. Wilson, and R. Elias. 2000. Artificial ecosystem selection. *Proceedings of the National Academy of Sciences of the USA*, 97(16), 9110–9114.

Tyrrell, T. 1999. The relative influences of nitrogen and phosphorus on oceanic primary production. *Nature*, 400(6744), 525–531.

Volk, T. 1998. *Gaia's Body—Toward a Physiology of the Earth*. New York: Springer-Verlag.

Volk, T. 2002. Toward a future for Gaia theory. *Climatic Change*, 52, 423–430.

Watson, A. J. 1999. Coevolution of the Earth's environment and life: Goldilocks, Gaia and the anthropic principle. In *James Hutton—Present and Future*, G. Y. Craig and J. H. Hull, eds., pp. 75–88. Special Publication 150. London: Geological Society.

Watson, A. J., and J. E. Lovelock. 1983. Biological homeostasis of the global environment: The parable of Daisyworld. *Tellus*, 35B, 284–289.

Wilkinson, D. M. 1999. Is Gaia really conventional ecology? *Oikos*, 84, 533–536.

Wilkinson, David. Homeostatic Gaia: An ecologist's perspective on the possibility of regulation. Chapter 6 in this volume.

2
Gaia Is Life in a Wasteworld of By-products

Tyler Volk

Abstract

I offer here an essay of personal history of grappling with Gaia theory, with conclusions. These are as follows: The relatively steady states in the global environment are simply the expected, natural results of a system containing chemical reactions, many of which involve life. There is nothing special about the existence of both positive and negative feedbacks in this system; these are to be expected. Certainly much of the global environment has been, and is being, transformed by life into a state very different from that of a planet without life—but what is this state? I suggest that the global environment is, in essence, a wasteworld: a system of by-products (and their effects). This wasteworld plus life creates a complexly structured dynamical system, because life is not passive. Organisms make metabolic products aimed to ensure their success at living and reproducing, not aimed at transforming or controlling the global environment. But in making these products, organisms also produce by-products, and these often build large-scale environmental side effects. The environmental consequences of the by-products are inadvertent but do create a system with evolutionary and population feedbacks—for instance, the biogeochemical cycles. The dynamics of this system are such that some forms of life alter the environment, and then all other forms within that altered environment must adapt or perish. Life thus shoves the environment around, subject to limits at various extremes. Are there general principles of this wasteworld? At the end, I spell out some directions that I see for the future of Gaia theory.

Introduction

This chapter is a personal essay about Gaia theory. I will reflect a bit on what became for me major lessons about the theory. I hope readers will not be put off by this more informal format. I want to show the evolution of my thinking. Admittedly, I will not attempt a survey of all work in Gaia theory, and thus will necessarily leave out many substantial attempts by other researchers to solve outstanding issues. For this I apologize. I focus on those principles and concepts that have emerged over time as the solid truths I carry around as a foundation for continuing work.

Let me start by thanking James Lovelock and his colleagues, who proposed and promulgated what started as the Gaia hypothesis but has come to be known as Gaia theory. Since the 1980s my thinking has been influenced and inspired by the Gaia theorists.

In the early 1980s I was working on my Ph.D. dissertation. Its published title was *Multi-property Modeling of the Marine Biosphere in Relation to Global Climate and Carbon Cycles* (Volk, 1984, see also Volk and Hoffert, 1985). I now would change one word. Let's say *Biota* rather than *Biosphere*, because I prefer to reserve "biosphere" to encompass all life (biota) and its global environment.

In fact, I take "biosphere" as essentially equivalent to Gaia. I define *Gaia* as the system that includes all Earth's surface life, the soils, oceans (all surface water), and atmosphere (Volk, 1998). I also include in Gaia the active interface of rocks that are contributing on the thousand-year timescale to the chemistry of the rest of the biosphere. Soil, air, and ocean are unified over about a thousand-year ocean mixing cycle, making all life linked into what is essentially a synchronized moment in global biogeochemical evolution.

Tim Lenton (chapter 1, this volume) uses a somewhat more flexible definition of the Gaia system, in which the lower boundary of Gaia deepens as the timescale under consideration expands. I can concur with his reasoning, and I hope the system's boundaries will become clearer the more we are able to elucidate the dynamics of this special system. I also wish to acknowledge another opinion. Axel Kleidon, in his review, suggested that the term *Gaia* is value laden and should be restricted to Lovelock's original concept. In this chapter I will stick with my preferred heuristic of a definition: equating Gaia with the biosphere.

Back to the 1980s. Somehow during this time I became aware of Lovelock's Gaia hypothesis. I read his book *Gaia: A New Look at Life on Earth* (Lovelock, 1972). It was eye-opening. Could he be right? Could salinity, temperature, and atmospheric and ocean chemistry have a single explanation behind their apparent stability over long periods of time? Could this explanation have to do with the global activities of life that maintain those conditions for life's very own benefit? Are there feedback loops in Gaia (Lovelock used the analogy of a thermostat) that hold conditions relatively constant, and possibly even adjust them for thermal and chemical comfort on a global scale? Are we, in fact, living in a gigantic superorganism that in some ways behaves like the manner in which our body creates its remarkable, internal homeostasis?

I also studied two key journal papers (Lovelock and Margulis, 1974; Margulis and Lovelock, 1974). Certainly from the viewpoint of my dissertation on the marine biota, I was open to their persuasive ideas. My work involved modeling marine life at the biosphere scale, aiming to place life in the equations for marine carbon dioxide, phosphate, nitrate, alkalinity, and oxygen. The concentrations of these constituents, varying with water depth, are profoundly affected by life. Life does have an indisputably powerful influence on global ocean chemistry.

As a theoretician I was drawn to any hint of general principles for these global patterns. I recall at the same time also looking into nonequilibrium thermodynamics, as well as chaos theory, nonlinear dynamics, and fractals. What theoretical framework might give insight about the biosphere? Lovelock's Gaia hypothesis seemed particularly relevant because it specifically dealt with life: life as the central organizing factor for Earth's surface system.

Where I subsequently went with the Gaia hypothesis in my work, how my ideas about Gaia developed as a result of early explorations, and how I see the current status of Gaia theory's present and future are the subjects of the body of this chapter.

Positive and Negative Feedbacks from Life's By-products in Steady States

In the mid-1980s, as a newly minted Ph.D. and then professor, I drew inspiration from a paper in *Nature* by Lovelock and Whitfield (1982). They noted that life enhances the chemical weathering of soil minerals. This is because the soil CO_2 level, compared to that of the atmosphere, is elevated by respiration from plant roots, soil animals, and microbes. And weathering takes place in the soil, releasing calcium ions from silicates (to simplify). These ions then flow via rivers to the ocean, where they are precipitated out as calcium carbonate, thus carrying CO_2 into a rock burial, removing it from Gaia (in other words, from the biosphere).

At the same time, I was aware of the important BLAG model of this so-called geochemical carbonate-silicate cycle, named after the authors Berner, Lasaga, and Garrels (Berner et al., 1983). BLAG provided a framework for computing CO_2 levels over multimillion-year time periods. It specified weathering as a function of atmospheric CO_2 (among other factors). But significantly, BLAG did not include life as an active player. As noted, life is a major factor in determining the CO_2 levels where weathering occurs, and thus I was motivated to add life and soil to the BLAG model, to make it more "Gaian" (Volk, 1987).

Constructing a "lively" BLAG model required quantifying a crucial loop: atmospheric CO_2 affects plant growth, which by direct respiration of roots and the supply of photosynthesized organic detritus to soil respirers affects soil CO_2, which in turn affects the weathering rate of soil minerals, which then, via the eventual burial of ocean calcium carbonates, circle back to affect atmospheric CO_2. The loop is a negative feedback, as Lovelock and Whitfield perceptively pointed out. How strong is the feedback, I wondered. Does it create homeostasis—a key term that appears in Lovelock's first book and his early papers with Lynn Margulis?

After formulating the relevant dynamics, I examined how the lively BLAG model responds to an increase in volcanic emissions (a chemical forcing into Gaia from beneath Gaia). How much would atmospheric CO_2 rise with a given extra input flux of volcanic CO_2 in a lifeless BLAG model, compared to the model with life and its lively feedback loop? I found that the biota could roughly perform what I preferred to call "mitigation" rather than "regulation."

By "regulation" I was looking for homeostasis, which meant holding the atmospheric CO_2 and global climate nearly constant in the face of external forcing. "Mitigation," on the other hand, would be less powerful. David Wilkinson has pointed out to me that he regards regulation as any bounded condition in which perturbations amplified by relatively weak positive feedbacks over short times scales are ultimately constrained by negative feedbacks that dominate over longer time periods. But then what are the logical bounds on the use of "regulation"? I don't know. Furthermore, even the lifeless BLAG model contains a negative feedback. Does the existence of any degree of negative feedback warrant the use of "regulation"? Not in my opinion. As I found, the additional negative feedback added by life in the lively BLAG system is relatively weak; thus, at the time I felt (and still feel) more comfortable with "mitigation."

How much is this mitigation? About 30 percent for a reasonably strong forcing. For instance, if volcanoes in the lifeless BLAG model create a new steady-state climate 3°C higher in the forced condition, then with my lively BLAG the temperature increase is only about 2°C. Life mitigates (reduces) the increase by about 1°C. This is significant. But it is definitely not anywhere near perfect regulation.

Why even expect regulation to be perfect? At that time, in ignorance, I had not given enough thought to why we might even theorize about homeostasis as a property of the biosphere. I suppose that in my mind something along the lines of Lovelock's analogy with a thermostat made sense. But what could possibly create the mechanisms of such tight regulation within a biosphere that contained organisms as evolving entities? Eventually I came to realize that negative feedback associated with life, as in the lively BLAG model, is not a property to be expected to follow from evolution any more than positive feedback does. I will explain further.

Trees are not been positively selected by evolution to have roots that pump lots of CO_2 into the soil. Just the opposite: trees whose root cells are more efficient and generate less CO_2 as metabolic waste will be at a reproductive advantage. Along the same lines, grasses are not selected by evolution because, at the end of their growth season, their bodies wither to feed the earthworms and microbes with detritus, so that these soil respirers elevate soil CO_2. The elevation in soil CO_2, which happens to increase chemical weathering, is a result of waste by-products of soil's life-forms. Consider, too, actively released substances such as phosphorus-dissolving enzymes, which are secreted by soil microbes to liberate nutrient ions from soil minerals. The resultant lowering of the steady-state value of atmospheric CO_2 as a consequence of these enzymes is a by-product, a side effect.

Life's influence on atmospheric CO_2 is a collateral effect of chemical weathering in the soil. Again, elevated soil CO_2 is itself a by-product of waste from respiration. There is no evolutionary pressure that drove life to mitigate excursions of atmospheric CO_2, computed in the lively BLAG model. The reason the model's negative feedback is 30 percent just happens to be that way, given the quantitative dynamics of photosynthesis and chemical weathering.

So far I have been discussing the value of a particular negative feedback between life and the global environment. But there is no reason for the sign of feedback between life and its environment to be negative. Schneider (1986) pointed this out in an important paper that questioned some of the logic applied in Gaia theory. At that time it was becoming evident that Earth's Pleistocene Ice Age cycles of about 100,000 years were at least partially due to positive feedbacks between life and the environment. Although the cyclic Ice Ages are still a mystery and the reasons still unresolved why, for example, CO_2 was lower during the cold periods, clearly life was involved somehow in amplifying the pacing signal from the changes in Earth's orbit.

David Wilkinson's point above is apropos here: that positive feedbacks will in general be bounded in extent and time by longer-term, stronger negative feedbacks. I agree, although in the case of Ice Ages the bounds might have been set by the orbital cycles themselves. I do generally agree, however, with his point: Gaia is a system with both types of feedbacks, and the system will tend to settle into stable states bounded by negative feedback.

Thus naturally there will be steady states in which life plays a major role. But the creation is a side effect, an inadvertent consequence of life's by-products. Life creates these states—but life did not evolve to create these states. This logical distinction is crucial. (See also Kirchner, 2002; Volk, 2002). I will deal with the issue of ultimate environmental limits later.

Here is how I have come to see the Gaia system: The biosphere is a vessel containing chemical reactions. There are flows of matter into this vessel (considering input flows from volcanoes and rock weathering). As described in the example of lively BLAG, life can influence the weathering rate, which puts elements from rocks into the circulation of the Gaia system.

In addition, there are exiting flows from the biosphere vessel, primarily by the burial of materials in sediments and by chemical exchanges at ocean vents. These exit fluxes, too, can be influenced by life. But the fluxes are also modified by purely inorganic processes, such as entrapment of solid materials as sediments accumulate, with subsequent passage out of the active biosphere.

What goes on inside the chemical vessel of the biosphere? Some of the internal reactions don't require life—equilibrium-seeking exchanges between bicarbonate and carbonate ions in the ocean, for instance. But as noted and emphasized, life participates in many internal transformations. Specifically, photosynthesis can be written as an equation with reactants and products. Nitrogen fixation can be written as an equation. And so forth. The reaction rates of these processes are not simply driven by the chemical free energy of the reactions alone. As Axel Kleidon has emphasized to me, forms of life actively degrade free energy between input and output, thereby building up free energy inside their bodies as complex chemical compounds. Therein lies the challenge to find simple rules that capture how organisms determine the rates of their metabolic equations. But just the same, as a foundational starting point, life's chemical equations contain reactants that go to products, with the release or addition of energy, just as we find in the abiotic chemical processes of soil, air, or water.

Now, let us consider once more, in conclusion, Wilkinson's idea of a nesting of feedbacks. Assume for comparison two chemical systems or cases. The first case has no life. The second case has life. Each system contains a number of reactions. States are determined by the entering fluxes, the exiting fluxes, and internal reactions. In both systems, the soup of chemicals inside the vessel will likely reach a steady state, given relatively steady input fluxes (these change as forcings, of course, but usually more slowly than the rates of reaching steady states). In dynamical systems of chemicals, other behaviors are possible as well. We might find cycles in the chemical concentrations. We could have chaotic attractors. But whatever the case, the systems generally arrive at distinct, bounded behaviors. This is because there will be regions of phase space for the systems where negative feedbacks dominate and act to bound any fluctuations caused by positive feedbacks. These will be the regions into which such systems will settle.

With life, the behavior is essentially the same as without life: for example, steady states occur in the both the lifeless and lively BLAG models. I emphasize: It's not the situation (at least no one has shown this to be generally true) that without life the system is wild, whereas with life the system is steady. I don't question that life influences the concentrations of the steady state. It's just that we cannot elevate life to a status of creator of the steady state.

Because life participates in causal loops of positive and negative feedbacks, the levels of the steady states will be different when comparing a lifeless "biosphere" to one with life. But it's not the existence of positive and negative feedbacks that makes life special. Causal loops of both positive and negative feedbacks exist as well without life. Life makes the biosphere more complex, certainly. But is that complexity qualitatively different? Are we talking about an entirely different dynamic behavior, such as stable versus unstable, when we compare a planet with life to one without? I don't think so. I haven't seen the evidence.

Change in the Biosphere from the Evolutionary Output of By-products

I now return to personal history. At the 1988 Chapman Conference on the Gaia Hypothesis, I met David Schwartzman of Howard University. Soon after the meeting, having seen my presentation on the lively BLAG model, he called me. We eventually collaborated on a series of papers whose results can be used to look at the issue of life as maintainer of stability or creator of change.

I remember that first phone conversation well. David said that life influences weathering to a much greater degree than in my lively BLAG model. How much, I asked? He said that in addition to my maximum enhancement of perhaps 2 to 3 times via elevated soil CO_2, other biological factors create a weathering enhancement of 100 to perhaps 1000 times over what should be considered the true baseline comparison for Gaia theory—the abiotic rate of weathering. These factors include release of acids by soil organisms, the retention of water by humus (no water, no aqueous dissolution reactions), and the structure of soil as a matrix of fine particles held in place over long enough time periods to be chemically weathered.

By looking at the available data, we estimated that such magnitudes for the overall "biotic enhancement of weathering" were plausible. What did that mean for quantifying life's influence on Earth's atmospheric CO_2 and temperature? We found that life might currently cool the biosphere by about 35°C (Schwartzman and Volk, 1989). This number is interesting because the current global temperature is about 15°C, and our calculated temperature for an abiotic Earth of 50°C (at today's solar flux) is about the upper temperature limit tolerated by any eukaryote lifeforms. So without ancient bacteria that created a biotic enhancement of weathering on a pre-eukaryotic earth (with cryptogamic soils as our reference), then perhaps we would see no elephants, birds, amoebas, or people. It became evident that earlier microbial life created cooler conditions that led to more complex organisms, which presumably needed those cooler conditions to evolve and survive.

We then extended this work to specifically treat the biosphere over time, in which the biotic enhancements to weathering took place in a series of stages (Schwartzman and Volk, 1991; especially see Schwartzman, 1999). In this work we computed a history over 4 billion years of atmospheric CO_2 and global temperature. There were several forcings whose relative impacts we examined. The effects of these forcings were isolated in my book (Volk, 1998); here I discuss the results and what I have gained from them with regard to Gaia theory.

Three major abiotic forcings exist on the geochemical carbonate-silicate carbon cycle. First, the sun has increased in intensity by about 30 percent, which, when isolated in the model, causes a 6°C warming over 4 billion years, from 66°C to 72°C. (The initial temperature is set to be about 66°C, to yield today's state with all the forcings applied, abiotic and biotic.) In addition, there are two major geological forcings: first, the growth of the continents, which especially in Earth's early history added raw area for weathering reactions; second, the decrease in the release of Earth's deep heat, which lessened the CO_2 emitted by volcanism (see discussion in Schwartzman, 1999, for formulations and justifications). Both of these long-term changes act to decrease the steady-state level of atmospheric CO_2 over time, thus gradually cooling the climate. When the forcings from sun, continental growth, and volcanism are combined, as they would be in reality, the two geological forcings "win," with the net result of a 20°C overall cooling, from 66°C to 46°C.

Now add life. As Schwartzman and I have argued, the evolution of life would have caused a progressive series of enhancements in the biological impact on chemical weathering, starting with the earliest prokaryotes on land and ending with modern soils created by as well as supporting vascular plants. With everything combined in the model—sun, geology, and the evolution of life—Earth's temperature progressively moved from 66°C 4 billion years ago to today's 15°C, an overall cooling of 51°C.

Thus, in our scenario, life has not stabilized the planetary surface temperature; it has helped to destabilize it. This isn't the place to review uncertainties and controversies over the weathering formulations, or whether or not methane, rather than CO_2, was important as the early greenhouse gas, or what prelife temperatures were. The point is that we attempted to assemble, in a common geochemical framework, reasonable best estimates for the main effects on CO_2 and temperature.

Does this biologically determined weathering effect on global temperature support Gaia theory or not? That depends on what is meant by Gaia theory.

In the traditional view of Gaia as stabilizer of biosphere conditions (a la Daisyworld), the progressive cooling from weathering enhancement is not Gaia. But to my mind, the cooling of the planet due to the progressive and evolutionarily driven biotic enhancement of weathering is as much "Gaia" as a theory gets because of the intimate coupling of life and environment.

But suppose Gaia theory says that some aspects of the global biota are stabilizing and some are destabilizing. Then I submit that if Gaia exhibits both stabilized and destabilized trends, and that's our conclusion about Gaia theory, then we are not saying much—other than whatever is, is Gaia. If stability is Gaia, and if change is Gaia, then what is Gaia? There is no non-Gaia state, no null hypothesis. Sometimes Gaia is stable, due to some inadvertent negative feedbacks that create steady states (which must exist, as I argued earlier). But life also creates forcings on the biosphere's chemistry by way of evolution. So is that Gaia, too?

Lovelock (2003) has said that "Gaia theory clearly states that the Earth self regulates its climate and chemistry so as to keep itself habitable and it is this that is the sticking point for many, if not most, scientists." (See my comments to Lovelock in Volk, 2003a.) In the Schwartzman and Volk model, the temperature without life varies from 66°C to 46°C. With life the temperature varies from 66°C to 15°C. All right—I admit that at all these temperatures the Earth is "habitable." But "life" did not keep things more or less habitable, because as Schneider (1986) pointed out, it depends on what you are referring to. In this case is it hyperthermic microbes or tundra grasses?

The biosphere is a co-evolved entity consisting of life and what primarily are the by-products of life and the effects of those by-products (such as the waste CO_2 in the soil affecting the weathering rate). It's one big wasteworld. My calculations show that regarding the atmosphere's CO_2, more than 99 percent of the entire reservoir has recently been ejected by a living respirer rather than a volcano. For nitrogen, more than 99 percent has been discharged from living denitrifiers rather than volcanoes. And for methane and many other trace gases, more than 99 percent has been expelled from living prokaryotes rather than volcanoes. The atmosphere is one giant waste dump.

The great influence of life on the environment was one of the key insights of James Lovelock, and I salute him for that. And Tim Lenton has asked me to emphasize that the "production of waste by-products is an inevitable aspect of being alive." But I also will emphasize that I don't see any particular "optimization" or even securing of "habitability" in the numbers above. What I see is that life produces by-products and side effects that can shove the environment around into various chemical states. All organisms, linked in the chemical vessel of the biosphere, must adapt to these states or go extinct.

Why? Why can't organisms direct the global environment for their own benefit? Why can't they put substances into, say, the atmosphere that improve the atmosphere relative to their needs?

The main problem has to do with the dynamics of evolution and the issue of "cheats." Others have recognized this problem for Gaia theory and the resultant need to incorporate what is essentially a concept of by-products into the theory (Lenton, 1998; Wilkinson, 1999; Lenton, this volume). In particular, Ken Caldeira (1989) published a key paper that looked into the problem of cheats and the inability of organisms to be selectively evolved to change the global environment. I incorporated his work into a preliminary view of the biosphere as a wasteworld of by-products (Volk, 1998).

Caldeira attended the 1988 Chapman Conference on the Gaia Hypothesis as a graduate student in my department at New York University. I was later proud to serve on his dissertation committee. At the meeting, Caldeira became interested in the CLAW hypothesis (Charlson et al., 1987), that DMS (dimethyl sulfide) released to the atmosphere by marine plankton creates brighter clouds and cools the Earth, and is somehow connected to a feedback loop that impacts the plankton producers. Did some species of plankton evolve specifically to alter the clouds?

Caldeira analyzed this question. Could certain species of plankton have evolved to create DMS to create clouds above them for their own benefit (for example, perhaps the resulting cooler water stirred up more nutrients for the plankton)? Caldeira wanted to weigh these presumed benefits against the metabolic costs to the plankton of producing the precursor molecule to DMS. He liberally estimated an enhancement of growth from an increased stirring of nutrients. He conservatively tallied the metabolic costs. The finding: no contest. Metabolic costs outweigh climatic benefits by a factor of a billion or more (Caldeira, 1989).

With such a skewed ratio of costs to benefits, mutant cheater plankton would proliferate. They could live mixed in with the DMS producers and derive all the benefits of upwelled nutrients without paying the huge metabolic costs. The lesson: Phytoplankton must synthesize the DMS precursor solely because it benefits their individual growth and reproduction while it is inside their bodies, not because this chemical affects the atmosphere. The precursor has indeed been proven to help cells regulate their ion contents relative to the surrounding salty water. Plankton don't even want to release DMS; it is forced from them in predation by zooplankton or bacteria. The survival-promoting, internal function of the DMS precursor is the reason why the genetic heritage of synthesizing it is passed on by the generations, not because it has a climatic effect when it is altered into a gas spreading across the sky.

What if DMS were in fact not beneficial but detrimental to marine life? More reflective clouds, for example, dampen photosynthetic potential by reducing the light that reaches the surface. In this case, the DMS emitters are actually hurting all the other life in their locale. But the emitters would still keep on emitting because of the huge survival benefits of regulating their internal ions. The numbers are of the same magnitude as before: The climatic detriment would be only a ripple on the ocean of the real evolutionary math going on within the organism. On the Gaian scale, whether DMS as a diffuse gas causes large-scale benefits or detriments (or both) may not matter, because the climatic effects forge intimate links among all organisms living within the DMS-determined climate. A world that is cooler because of DMS would have different climate zones, rainfall patterns, and ocean circulation.

If the world average temperature is now 5°C cooler because of DMS, and 35°C cooler because of the progressive biotic enhancement of weathering, then the whole living world is to some extent adapted to a physical reality vastly influenced by some forms of life. Tens of millions of species are united by DMS and the biotic enhancement of weathering. The situation is awe-inspiring: neither biogenic DMS nor the biotic enhancement of weathering evolved *because* it cooled climate, and yet their existence perpetrated free Gaian effects that profoundly link all life.

To conclude this section, note that the issue of by-products involves levels in space. The result of an organism's metabolism has to be to promote its own reproductive potential. This means it will behoove the creature to keep the effects of its beneficial substances as close to its body as possible. Sending a costly substance out to change the atmosphere is not a good way to do this because the effects become so diffuse and easily shared with others, even halfway around the world. These others, if cheats, are not penalized for the metabolic costs of creating the effects, yet derive the same presumed benefit. We can start citing the different kinds of wastes that organisms produce, and the litany of side effects or by-products will grow substantially. To me, this litany, when eventually understood as a complex system of chemical interactions, is the structure of Gaia; this litany is fundamental to the dynamical shape of the biosphere.

Conclusion: By-products and Cycling Ratios Are Crucial to the Future of Gaia Theory

Although in earlier work I elaborated on the importance of by-products for Gaia theory and explicitly used the language of by-products and side effects (Volk, 1998), as I mentioned above, there were others who noted this phenomenon as well (though using different terminology and sometimes with different logical architecture). Tim Lenton (1998) developed the concept of "growth feedback" versus the stronger Gaia process of "selective feedback." I also wish to acknowledge David Wilkinson (1999) for seeing that the "cheater" problem is solved by considering global environmental effects as side effects.

Specifically, Lenton (this volume) says, "Indeed, many (and perhaps most) globally important biotic feedbacks appear to be based on by-products of natural selection." He goes on to identify one important pattern of Gaian dynamics associated with the effects from by-products: constraints on life that occur as the effects create extreme levels of environmental values which feed back to limit life (and thus those levels). Oxygen serves as an example. I think Lenton would basically agree with me that life's production of oxygen is crucial in creating a steady state of oxygen and that this state has been altered and generally increased over Earth's history into different values over many orders of magnitude. I agree with him that when a very high value of oxygen is reached (approximately not much higher than today's 21 percent), constraints can come into play that limit its further growth.

For example, photosynthesis, which is responsible for oxygen production, is biochemically limited as oxygen rises. Also, if terrestrial photosynthesis becomes limited by increased fires during excursions into higher oxygen, phosphorus weathering (via the biotic enhancement of weathering) decreases, suppressing the supply to marine photosynthesizers and further limiting the production of oxygen (Lenton and Watson, 2000; Lenton, 2001). Thus oxygen might be constrained to about today's value by feedbacks involving life. I agree with Lenton that searching for constraints of this kind will be an important part of future science of Gaia theory. And I emphasize that such feedbacks were not evolved to constrain oxygen. They are by-products.

Lenton (this volume) agrees with me that there is nothing special in the mere existence of steady states in Gaia, that both living and nonliving systems typically reach steady states. Instead, he says, "what is of interest for Gaia is the degree of regulation." Lenton (2002) proposes two main properties to evaluate the degree of regulation: resilience and resistance. Does life make the chemical system of Gaia more resistant to changes? Does life make the system more resilient to changes in that the return to some steady state (not necessarily the original state) is quicker following an external perturbation?

Lenton tentatively answers "yes" to both questions, offering evidence from models and from Earth observations. How can we determine the truth of these proposals, given limited modeling studies? It does seem likely that systems with life could be more resilient (Volk, 2002). This is because shorter turnover times of elements in the reservoirs of ocean, air, and soil are implied by the increased rates of chemical fluxes that life creates in the environment. Shorter turnover times generally mean more rapid returns to steady-state conditions following perturbations. I also allow the possibility that systems with life will be more resistant, but I don't yet see any general reasons why this should be, given a wasteworld containing numerous positive and negative feedbacks that are side effects from by-products. Perhaps the mere fact of an increased number of feedbacks in a system with life ups the probability of regions of phase space in which the system is particularly stable.

I do want to go on record here that I think the word "regulation" should stop being used in Gaia theory. Its definition is too vague and subject to too many interpretations. The word is difficult to use precisely in a way that will be agreed on by everyone. For example, sometimes "regulation" means just the dynamics that create a relatively bounded steady state, a meaning that David Wilkinson used in a communication to me. Sometimes it has been used in discussions of Gaia to refer to a homeostasis of truly beneficial conditions. Sometimes the word is limited to certain types of feedbacks with life compared to feedbacks without life, not in reference to the steady state itself (Lenton and Wilkinson, 2003). Because of this problem of shifting meanings and shades of meanings that change from paper to paper or even within papers, I never use the word.

To move toward my conclusions and to generalize, I submit that it's all by-products. It's all a world of life-forms interacting primarily with each other and their wastes. No one, as far as I am aware, has yet come forth with an example on the Gaia scale of an environmental effect that is selected for (in the evolutionary sense) to exist as a trait that costs the organism something to create. So if Gaia is built from by-products, then where do we stand with regard to Gaia theory?

First, we must throw out any concept that organisms are constructing the environment for their own benefit. Gaia is built from by-products. Lenton (this volume) and Lenton and Wilkinson (2003) agree that this statement about by-products is mostly true, and I see this as a major convergence among some of us involved in Gaia theory. But Lenton and Wilkinson also note that the specific favoring by natural selection of effects that enhance life as a whole might be true in specific cases. While I don't say this is impossible, I predict that the concept of global effects built from by-products will either always be true or so commonly true that the few cases in which the concept is not true will be trivial in our understanding of the dynamics of the biosphere. See Volk (2003b) for a critique of Lenton and Wilkinson (2003).

Second, we should realize that the effects of by-products released for free from organisms will shove the environment around into different states. For example, oxygen levels have varied by probably about six orders of magnitude over Earth's history, and carbon dioxide by three or more orders of magnitude.

Third, following Lenton's lead, we should consider that this shoving around of the environment will be subject to constraints which generally occur toward extremes when life overall is diminished as a result of the change. This diminishment will include not only the life-forms that create the push toward extremes but most other life-forms as well.

Fourth, we should look into Lenton's possibility of increased resistance and resilience as natural consequences of by-products released and merging into a complex system of fairly contained dynamics (the biosphere vessel). Currently I lean more strongly toward the possibility of increased resilience, given the reasons outlined earlier.

Fifth, and finally, we should think about how by-products from certain groups of organisms are used by others. In a point stressed to me by Axel Kleidon, wastes still contain much chemical potential energy—for example, the feces of animals and the sloughed-off parts of plants. The entire human body enters the category of waste at death, but this is far from implying that the corpse is a chemical dead end, as the hordes of bacteria waiting to gobble us well know. Thus wastes to some creatures are actually foods for other organisms able to use the by-products as sources of energy or sources of necessary elements. Wilkinson (2003) and I have been emphasizing this aspect of Gaia, by which creatures become linked in a complex matrix of chemical exchanges with other, different creatures.

My proposal along these lines is to think about the biosphere as composed of organisms grouped in "biochemical guilds" (Volk, 1998). I was influenced by Ron Williams's book *The Molecular Biology of Gaia*, which deserves closer attention from the Gaia community (Williams, 1996). Williams discusses the main molecules of the nitrogen cycle, ranked by the magnitudes of different types of transformations of nitrogen. Glutamine synthetase, for example, ranks foremost in the pantheon of nitrogen enzymes by virtue of the sheer mass of nitrogen that it channels as an essential step in making all the nitrogen-containing compounds essential for life. The second most important molecule is nitrate reductase, key in synthesizing nitrite from nitrate, and the third is nitrogenase, which catalyzes nitrogen fixation.

I was motivated by Williams to think of Gaia's parts as sets of organisms grouped by their key molecular transformations—the "biochemical guilds." These can be conceptualized for different elements by grouping organisms that perform similar chemical transformations. Components of biochemical cycles are then are formed by linked guilds. One biochemical guild produces a by-product waste as output that is the input flow to another biochemical guild. For instance, consider the photosynthesizers and respirers, in which the wastes from each (O_2 and CO_2, respectively) become the gaseous "food" for the other.

Photosynthesizers and respirers together form a binary loop, but to fully map the carbon cycle we would have to add methanogens and methanotrophs, as well as other guilds. The nitrogen cycle is also composed of many biochemical guilds; the major ones are the nitrogen fixers, denitrifiers, nitrifiers, ammonium assimilators, nitrate reducers, and ammoniaficators. One can do the same analysis for sulfur, phosphorus, and all the nutrient elements essential for life. Ultimately, one would have a list of the guilds for all biochemical transformations mediated by life.

In conceptualizing the biosphere as linked guilds, it is clear that this is why global photosynthesis is greatly amplified over what it would be if it were limited to the sum of the fluxes of key elements into the biosphere from below. These supply fluxes are basically from rock weathering, ocean ridge exchanges, and volcanic emissions. For the example of carbon, these sources total about 0.5 billion tons of carbon entering Gaia each year.

Now let us conduct a thought experiment. Assume the products of photosynthesis are unusable. In other words, no molecule of fixed carbon from any plant or algae can be recycled to carbon dioxide by consumers. Thus assume totally indigestible bodies of photosynthesizers, which are all buried upon death and become parts of rocks such as shales. Photosynthesis in this imaginary case could still exist, but its global total would be limited to the flux each year into the biosphere, the 0.5 billion tons of carbon. That is a pittance compared to the actual global value of about 100 billion tons of carbon per year going into the bodies of photosynthesizers. The "extra" 99.5 billion tons comes from respirers.

One can form a ratio between today's value of the flux of a given element into photosynthesizers and the flux into the biosphere—"the cycling ratio" (Volk, 1998). For carbon the cycling ratio is about 200 (100/0.5). For nitrogen it's somewhere on the order of 500 to 1300. For sulfur it's only about 10. Thus the structure of the biosphere can be studied from the viewpoint of the cycling ratio.

The cycling ratio could be a metric for Gaia (Volk, 2002). A metric is a number we might use to compare states of systems involving life, across space (watersheds versus the global ocean) and across time. Perhaps there will be patterns that appear. Does the cycling increase over time as new guilds evolve to use what were once only wastes that exited unutilized from the biosphere system? Steve Schneider (1986) asked how we about might compute the Gaia-ness of Gaia. By biomass? By diversity? I suggest the cycling ratio.

In contrast to the cycling ratio, Kleidon (2002) proposed global productivity as a metric for Gaia. But note that in Kleidon's proposal if the supply from beneath Gaia of some limiting element is suddenly doubled, and photosynthesis doubles as a result, we would call the new state "more Gaia." But I don't think we want to give the label "more Gaia" to such a simple response to what is essentially fertilization from outside the biosphere. "More Gaia" should come about from internal changes in the dynamics of some forms of life coupled to the dynamics of other forms, plus the environment of wasteworld. The cycling ratio as a metric captures that possibility. In addition, the cycling ratio as a metric in a model of Gaia dynamics has been successfully investigated in an evolutionary model by Downing and Zvirinsky (1999).

I don't know if we will find general principles that enable us to make Gaia theory more robust with precise language and hard-core findings. I hope so. In general, I think the quest to further Gaia theory will be helped by clearer attention to the world as it is as a source for data and ideas for models. In particular, we should continue to seek for and test principles via models and by thinking about patterns in Earth history, perhaps investigating Lenton's resistance and resilience, perhaps thinking along lines of an evolutionary drive toward higher cycling ratios by the successive evolution of new biochemical guilds that discover how to use wastes as food, and, finally, perhaps through developing concepts not even yet thought of.

Acknowledgments

I am indebted to Jim Lovelock for his inspiration to think big, for his invitations to the Oxford Gaia conferences, and for his courage to span disciplines. For formally requested comments on the earlier version of this chapter, I thank Tim Lenton, Axel Kleidon, and Lee Kump. I also value the more informal comments I received from Amelia Amon, David Schwartzman, and David Wilkinson.

References

Berner, R. A., A. C. Lasaga, and R. M. Garrels. 1983. The carbonate-silicate geochemical cycle and its effect on atmospheric carbon-dioxide over the past 100 million years. *American Journal of Science*, 283, 641–683.

Caldeira, K. 1989. Evolutionary pressures on planktonic production of atmospheric sulfur. *Nature*, 337, 732–734.

Charlson, R. J., J. E. Lovelock, M. O. Andreae, and S. J. Warren. 1987. Ocean phytoplankton, atmospheric sulfur, cloud albedo and climate. *Nature*, 326, 655–661.

Downing, K., and P. Zvirinsky. 1999. The simulated evolution of biochemical guilds: Reconciling Gaia theory and natural selection. *Artificial Life*, 5, 291–318.

Kirchner, J. W. 2002. The Gaia hypothesis: Fact, theory, and wishful thinking. *Climatic Change*, 52, 391–408.

Kleidon, A. 2002. Testing the effect of life on Earth's functioning: How Gaian is the Earth system? *Climatic Change*, 52, 383–389.

Lenton, T. M. Clarifying Gaia: Regulation with or without natural selection. Chapter 1 in this volume.

Lenton, T. M. 2002. Testing Gaia: The effect of life on Earth's habitability and regulation. *Climatic Change*, 52, 409–422.

Lenton, T. M. 2001. The role of land plants, phosphorus weathering and fire in the rise and regulation of atmospheric oxygen. *Global Change Biology*, 7, 613–629.

Lenton, T. M. 1998. Gaia and natural selection. *Nature*, 394, 439–447.

Lenton, T. M., and A. J. Watson. 2000. Redfield revisited: 1. Regulation of nitrate, phosphate, and oxygen in the ocean. *Global Biogeochemical Cycles*, 14, 225–248.

Lenton, T. M., and D. M. Wilkinson. 2003. Developing Gaia theory: A response to the criticisms of Kirchner and Volk. *Climatic Change*, 58, 1–12.

Lovelock, J. E. 2003. Gaia and emergence: A response to Kirchner and Volk. *Climatic Change*, 57, 1–3.

Lovelock, J. E. 1972. *Gaia: A New Look at Life on Earth*. Oxford: Oxford University Press.

Lovelock, J. E., and L. Margulis. 1974. Atmospheric homeostasis by and for the biosphere: The Gaia hypothesis. *Tellus*, 26, 2–10.

Lovelock, J. E., and M. Whitfield. 1982. Life-span of the biosphere. *Nature*, 296, 561–563.

Margulis, L., and J. E. Lovelock. 1974. Biological modulation of Earth's atmosphere. *Icarus*, 21, 471–489.

Schneider, S. H. 1986. A goddess of the Earth: The debate on the Gaia hypothesis. *Climatic Change*, 8, 1–4.

Schwartzman, D. W. 1999. *Life, Temperature, and the Earth: The Self-organizing Biosphere*. New York: Columbia University Press.

Schwartzman, D. W., and T. Volk. 1991. Biotic enhancement of weathering and surface temperatures on Earth since the origin of life. *Palaeogeography, Palaeoclimatology, Palaeoecology, section of Global and Planetary Change*, 90, 357–371.

Schwartzman, D. W., and T. Volk. 1989. Biotic enhancement of weathering and the habitability of Earth. *Nature*, 340, 457–460.

Volk, T. 2003a. Seeing deeper into Gaia theory. *Climatic Change*, 57, 5–7.

Volk, T. 2003b. Natural Selection, Gaia, and inadvertent by products: A reply to Lenton and Wilkinson's response. *Climatic Change*, 58, 13–19.

Volk, T. 2002. The future of Gaia theory. *Climatic Change*, 52, 423–430.

Volk, T. 1998. *Gaia's Body: Toward a Physiology of Earth*. New York: Springer-Verlag. (2003, paperback. Cambridge, MA: The M.I.T. Press.)

Volk, T. 1987. Feedbacks between weathering and atmospheric CO_2 over the last 100 million years. *American Journal of Science*, 287, 763–779.

Volk, T. 1984. *Multi-property Modeling of the Marine Biosphere in Relation to Global Carbon and Climate Cycles*, Ph.D. dissertation. New York University: University Microfilms #84-21570.

Volk, T., and M. I. Hoffert. 1985. Ocean carbon pumps: Analysis of relative strengths and efficiencies in ocean-driven atmospheric CO_2 changes. In *The Carbon Cycle and Atmospheric* CO_2*: Natural Variations Archean to Present*, edited by E. T. Sundquist and W. S. Broecker, 99–110. Geophysical Monograph 32. Washington D.C.: American Geophysical Union.

Wilkinson, D. M. 2003. The fundamental processes in ecology: A thought experiment on extraterrestrial biospheres. *Biological Reviews*, 78, 171–179.

Wilkinson, D. M. 1999. Is Gaia really conventional ecology? *Oikos*, 84, 533–536.

Williams, G. R. 1996. *The Molecular Biology of Gaia*. New York: Columbia University Press.

3
Models and Geophysiological Hypotheses

Arthur C. Petersen

Abstract

Since the meaning of the Gaia hypothesis is unclear and scientists have problems with the term *Gaia*, the homeostatic Gaia hypothesis is reformulated as a class of "geophysiological hypotheses" that postulate global homeostasis of environmental quantities by the biota. Models are necessary for testing geophysiological hypotheses, but current models are too simple for that purpose. The original 1983 Daisyworld model is criticized with respect to its usefulness as a demonstration tool for homeostatic geophysiological hypotheses. New models of intermediate complexity need to be developed to bridge the gap between the very simple and the very complex models currently available in Earth system science.

Mathematical models play an important role in science. They are used for purposes such as investigating and developing theories or making projections of future changes in a certain domain. Also, models are used to analyze experimental or observational data, for instance, by using them as part of a measurement device. Earth system science, which encompasses all the different scientific disciplines dealing in some way with the Earth as a whole, makes heavy use of many kinds of models. Within Earth system science, a specific class of hypotheses called geophysiological hypotheses, which deal with the role of the biota in the Earth system, can be identified. An example of a geophysiological hypothesis is the Gaia hypothesis, which postulates the global homeostasis of quantities such as temperature by the biota. As I will argue in this chapter, the Gaia hypothesis should not be considered as one single hypothesis, but rather as a class of hypotheses. In the testing of geophysiological hypotheses, models play a crucial role.

One of the reasons models are used in Earth system science is that real, controlled, and reproducible experiments with the Earth system are impossible—and uncontrolled, nonreproducible experiments are undesirable. In the Earth system science literature, model experiments therefore often figure as substitutes for real experiments. A particularly apt rationale for using models in Earth system science, which may be considered typical for this area of science, is the following: "Virtual Earth Systems can be scrutinized safely in order to give deeper insights into the interactions of the various constituents and so avoid irreversible dead-end streets for the evolution of the original planet" (von Bloh et al., 1997: 249). The "irreversible dead-end streets," referring to human influences, imply that both curiosity about the workings of the Earth and concern about the effects of human activities constitute important values for Earth system science.

From a methodological point of view, model experiments are only in some respects similar to real experiments. One of the similarities—which makes people so easily speak of model experiments—is that just as with real experiments, one can intervene in a computer model and subsequently watch what happens. The main difference from real experiments is, of course, that the scientist is interacting with a representation of a material object and not directly with the object itself. In the evaluation of model results, therefore, the reliability of the model (i.e., the scientific quality of the representation of the object of study) is always at issue.

In research related to the Gaia hypothesis, models have played an important role, the archetype being the Daisyworld model (Watson and Lovelock, 1983; henceforth WL83). At the outset of WL83, Andy Watson and James Lovelock stated that they were aware of the reliability problem: "We are not trying to model the Earth, but rather a fictional world which displays clearly a property which we believe is important for the Earth" (Watson and Lovelock, 1983: 284). Still, at the end of their paper the authors argued that one might expect there to be an analogous "temperature stabilization system" for the Earth (i.e., they considered it plausible for negative feedbacks similar to those in Daisyworld to be present in the Earth system). This is what is often considered to be the Gaia hypothesis in its elementary form. This chapter deals with the use of Earth system models to test geophysiological hypotheses. I will first introduce the concept of geophysiological hypothesis and demonstrate that it can be useful for reformulating the Gaia hypothesis. The reformulation builds on the chapter written by James Kirchner for *Scientists on Gaia* (Kirchner, 1991) and on Tyler Volk's book *Gaia's Body* (Volk, 1998). Subsequently, the Daisyworld model, as the archetype of a model to study geophysiological hypotheses, is discussed. I will conclude the chapter with a brief discussion of some methodological issues related to the use of models of different complexities.

Geophysiological Hypotheses

Critical questions have been raised about the possibility of testing the Gaia hypothesis using only models, as well as about the testability of the hypothesis itself (see, e.g., Kirchner, 1991). As a "solution" to this problem, some scientists have offered an alternative way of introducing "Gaia" in science, a way which emphasizes both the metaphorical and the hypothesis-generating character of the term Gaia. According to this view, Gaia researchers use certain paradigm models, such as the Daisyworld model, for elaborating conceptual consequences of general assumptions rather than for testing those assumptions or some precisely stated hypothesis. Of course, for this work to be useful, connections to the real world must be made and specific geophysiological hypotheses that can be tested must be formulated. The fact that this has not happened much does not mean that it is impossible.

An example of what research into Gaia could be taken to mean is given by the biologist Tyler Volk. As he sees it, the term Gaia acts as "a reminder [to science] to think about special properties of the whole" (Volk, 1998: 4). This "whole," named Gaia, includes every part of the Earth that is significantly affected by the presence of life (i.e., soil, atmosphere, and ocean). According to Volk, one should not speak of a Gaia hypothesis. Instead of one hypothesis he prefers to speak of five "directives" for exploring Gaia, summarized by one "prime directive": "Think about how the planet would be different without life or particular forms of life" (Volk, 1998: 27). These directives can act as "generators" for hypotheses that can be tested (Volk, 1998: 27). In line with the term "geophysiology" used by many authors, such as Lovelock and Volk, I propose to name these hypotheses "geophysiological hypotheses." Earth system models are used for exploring these hypotheses and ultimately for testing them. Mathematical models are necessary for this purpose, for to be able to "think about how the planet would be different without life or particular forms of life," an intuitive grasp of all the complex relations between life and its environment will typically not be sufficient.

The term *Earth system science* entails a somewhat impassive reminder to science to think about the integration of all parts which constitute the Earth system. The added value of including a reference to Gaia (a subsystem of the Earth system, the part that is significantly influenced by life) could be that it provides a potentially fruitful unit of analysis at the largest scale (larger than what is classically understood as the "biosphere," which does not include the atmosphere, for example). Furthermore, some scientists like to use the term Gaia because they also consider it to be an expression of gratitude—an emotion typically left unstated in scientific works (see Volk, 1998: 5).

Since many scientists have problems with the term Gaia hypothesis, I will elaborate on the concept of geophysiological hypotheses here. Such hypotheses should be able to capture essential elements entailed by the original Gaia hypothesis and should be testable in principle. One characteristic of a hypothesis is that it stakes a claim that can be falsified. Furthermore, we typically do not refer to largely accepted background knowledge as hypotheses, even though we know that in principle today's background knowledge may be questioned tomorrow. I must admit that I was a bit puzzled by the reference to the Gaia hypothesis in the title of the Gaia conference at Valencia in 2000, since it has long been clear that there is no agreement on its contents. At the 1988 Gaia conference in San Diego, which had a similar title, Kirchner presented a taxonomy of Gaia hypotheses (Kirchner, 1991). The weakest versions of the Gaia hypothesis included in this taxonomy were shown to have been background knowledge for decades (some would even claim for more than a century), and the strongest versions were severely criticized for being seriously flawed as scientific hypotheses. If one takes this situation seriously, the usefulness of referring to one Gaia hypothesis becomes questionable, to say the least. In a charitable spirit, however, I will select a sensible interpretation of the Gaia hypothesis here, and reformulate it in terms of a class of geophysiological hypotheses.

In the research literature on the Gaia hypothesis, Kirchner found the following varieties of the hypothesis: the influential, coevolutionary, homeostatic, teleological, and optimizing. The influential Gaia and coevolutionary Gaia hypotheses (they are weak Gaia hypotheses; the other three are strong Gaia hypotheses) merely describe the well-established fact in Earth system science that the biota and the abiotic world are strongly coupled. The influential Gaia hypothesis asserts that there is an influence from the biota on the environment, and the coevolutionary Gaia hypothesis holds that there are feedbacks (positive and negative) on the biota through (biologically influenced) environmental constraints on Darwinian evolution. The weak Gaia hypotheses belong to the background knowledge of mainstream science: in the present Earth systems science era it does not seem appropriate to refer to these basic assumptions as "hypotheses."

This is not to deny that there was a time at which the number of Earth scientists working within an Earth system science framework was much smaller than today. For instance, this was the case at the beginning of the 1970s, when Lovelock first published the three strong versions of the Gaia hypothesis in a letter to the editors of *Atmospheric Environment* (1972). Since nowadays the weak Gaia hypotheses (entailed by the strong Gaia hypotheses) have become background knowledge, the problems many people still have with Lovelock's early papers mainly have to do with the strong versions of the Gaia hypothesis that he proposed. Two versions proved to be the most controversial: the teleological and the optimizing. The teleological Gaia hypothesis states that the biosphere maintains homeostasis of its environment "by and for itself" (in some sense). The optimizing version specifies, in addition, that the environment is manipulated to be optimal for the biota. As shown by many authors (e.g., Kirchner, 1991), both are flawed as scientific hypotheses and therefore cannot usefully be addressed by Earth system science. The flaws are that the teleological hypothesis is of a metaphysical character and that the optimizing hypothesis does not take into account that optimality—even if it is made clear what is meant—will be different for different organisms. Furthermore, the optimizing Gaia hypothesis seems inherently contradictory, since it violates a basic theorem of systems science, stated as follows by Kirchner (1991: 43): "No homeostatic system can be stable at a point that is optimal for the component supplying the homeostasis."

The only version of the Gaia hypothesis that remains interesting (i.e., adds something extra to the common understanding of Earth system science) and could in principle be scientifically testable—if properly recast as a class of geophysiological hypotheses—is the homeostatic hypothesis, which maintains that the biota influences the environment in a way that causes a homeostasis in the face of a changing external forcing (e.g., keeping a planet's temperature relatively constant under significantly increasing solar forcing, typically over geological timescales of hundreds of millions to billions of years). Testing such a kind of hypothesis is not a straightforward matter, however. First, one needs to distinguish clearly between the concepts of homeostasis and (mathematical) stability, as will be discussed below. Second, one has to specify what is to be considered "homeostasis," what are the "external forcings," and what are the timescales of interest. This is needed, since many quantities can be in homeostatis (e.g., temperature, gas concentrations, etc.), many life processes can be involved, many external forcings can be considered, and many timescales can be chosen. In fact, an infinite number of geophysiological hypotheses can be formulated that all belong to one class which can be associated with the homeostatic Gaia hypothesis. Common to all geophysiological hypotheses is that they (a) pertain to the global scale and (b) consider the difference it makes for the planet that life in general or particular forms of life are present (e.g., for the class of homeostatic geophysiological hypotheses, some quantity is made homeostatic by the biota).

One must be aware of the fact that kinds of geophysiological hypotheses can be generated other than those referring to homeostasis. For instance, one can follow Volk's directives for studying Gaia and formulate geophysiological hypotheses accordingly. Homeostasis is not the only interesting phenomenon involving the biota on the global scale. However, since most publications referring to Gaia until now have dealt with homeostasis, in this chapter I will retain a focus on global homeostasis phenomena.

The Zero-Dimensional Daisyworld Model

The Daisyworld model was introduced by Lovelock in order to mathematically demonstrate that the biota could cause homeostasis without any teleology involved. Just two dynamical equations, describing the area cover of black and white daisies on an imaginary planet, suffice to demonstrate the workings of a geophysiological hypothesis about the homeostasis of temperature with a biosphere consisting of two kinds of daisies. The equations of this zero-dimensional model need not be reproduced here ("zero dimensional" refers to the fact that no spatial dimensions are taken into account—except for a division of the planet into two areas). For the discussion in this

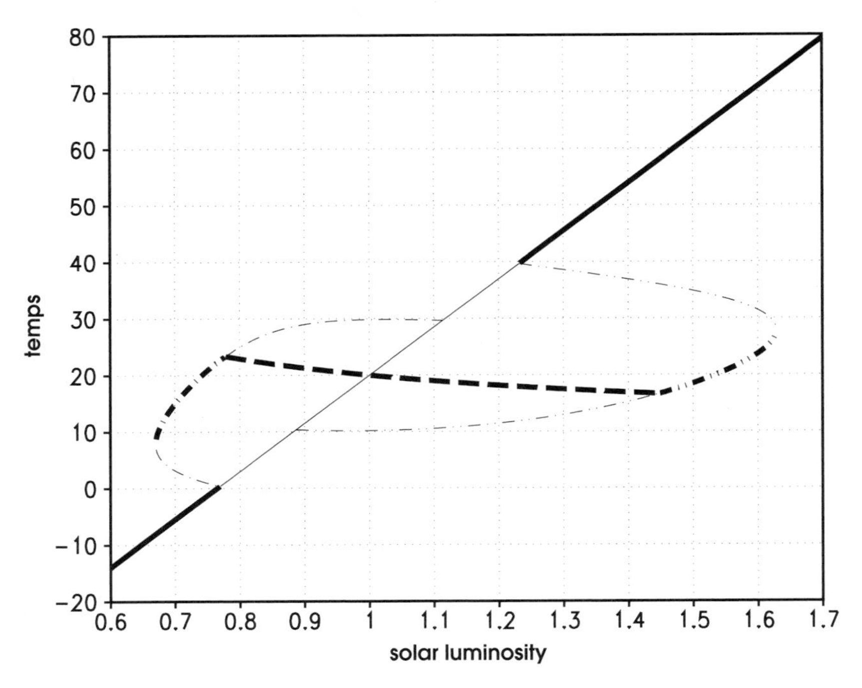

Figure 3.1
The equilibrium temperature (in °C) as a function of the solar luminosity: no daisies (solid line), white and black daisies (dashed line), black daisies only (dot-dashed line) and white daisies only (dot-dot-dashed line). Attracting states are indicated in thick lines, repelling states in thin lines. (From Weber 2001, figure 1, © 2001 Kluwer Academic Publishers.)

chapter it is useful to repeat the basic assumptions of the model:

1. There are two species present on the planet, and they warm differently due to their different colors and hence their different albedo (the fraction of sunlight that is reflected).

2. The growth rate of daisies is a peaked quadratic function of temperature; daisy growth is maximal for one fixed "optimal" temperature.

3. The heat exchange between areas with white and black daisies is implicitly modeled by assuming that the local temperatures are uniform in each of the two areas and that the deviation of an area's temperature from the planet's average temperature in first approximation varies linearly with the area's albedo deviation.

The behavior of Daisyworld under varying solar luminosity is pictorially summarized in figure 3.1 (this figure contains all information about the equilibrium temperatures of Daisyworld, and the figure's elements can be found in numerous other publications on the original Daisyworld model). The equilibrium temperature of Daisyworld (the planet's average temperature in equilibrium) is shown as a function of (slowly) varying solar luminosity (given in dimensionless units, normalized at the luminosity for which the planet's temperature with or without the biota is at the optimal temperature for daisy growth). The parameters used are identical to those of the original WL83 formulation (for graphs of the varying daisy covers the reader is referred to WL83). The main feature of the model's behavior is the large range of stable two-daisy states (from solar luminosities of about 0.8 to about 1.4) for which temperature changes slowly with varying solar luminosity. Actually, the temperature of the planet slowly decreases from 23°C to 18°C over this range, whereas without the presence of daisies it would have risen from 2°C to 57°C.

In WL83 Watson and Lovelock stated that the large "homeostasis" range found in the Daisyworld model was the result of "stabilizing" negative feedbacks related to the peaked shape of the growth-temperature curve for the daisies. Furthermore, they claimed this homeostasis feature was independent of the chosen parameter values. It can be shown analytically, however, that although it is true that the stability (or attractivity) of the system—which refers to the behavior of the system after small perturbations to its state (i.e., to the fractional daisy covers)—

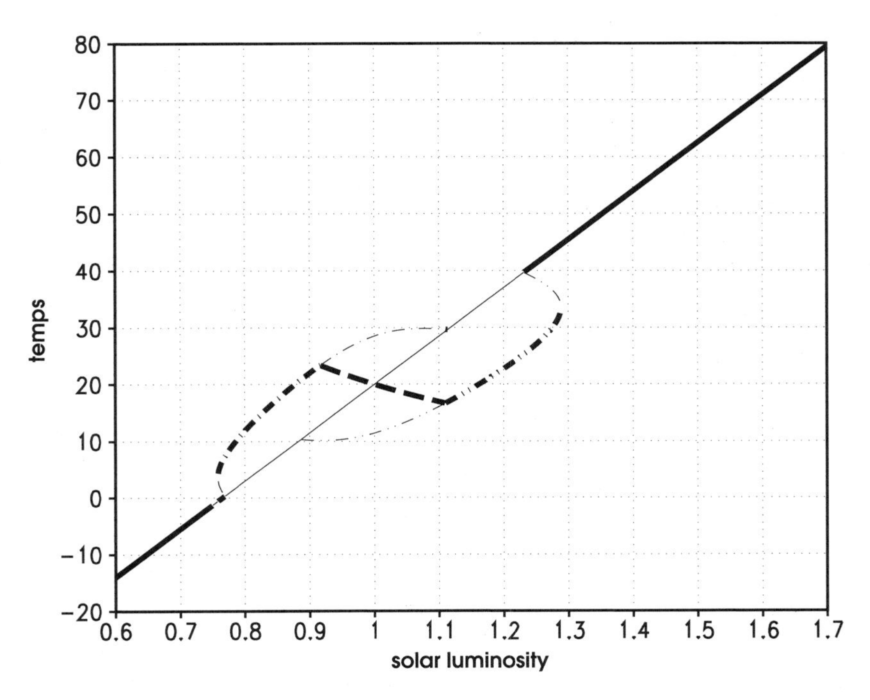

Figure 3.2
As in figure 3.1, but now with a different parameter setting that decreases heat transport. (From Weber 2001, figure 4, © 2001 Kluwer Academic Publishers.)

is due to the peaked shape of the growth-temperature curve (which leads to negative feedbacks), the homeostasis property of the system—which refers to the relative insensitivity to long-timescale external pertubations (in this case the changing solar forcing)—is dependent on the specific values of the parameters in the model (Weber, 2001). Of course, one can refer to this homeostasis as a "stability" with respect to the changing model parameter of solar forcing. The point is that this concerns a different kind of stability. Moreover, one can give examples of Daisyworld models (with similar negative feedbacks but different model parameters than WL83) that are quite sensitive to changes in external forcing (i.e., show only weak homeostasis). One example is shown in figure 3.2. In the version of Daisyworld pictured here, a seesaw behavior of the equilibrium temperature is found for varying solar luminosity. In the relatively small two-daisy range, the equilibrium temperature decreases much faster with changing solar luminosity, compared with figure 3.1. Indeed, the homeostasis property can be said to be significantly weakened in the Daisyworld version shown in figure 3.2.

Why did WL83 not make a clear distinction between homeostasis and mathematical stability, since, as concluded above, the homeostasis of Daisyworld cannot be explained by the stability of equilibrium solutions? My contention is that this has to do with the fact that even for extremely simple models such as the zero-dimensional Daisyworld model, the equilibrium states can more easily be obtained from computer simulation than from performing more tedious analytical calculations. From such analytical calculations one can derive insights—for instance, on the homeostasis property—that are less directly available from computer simulation, as shown, for instance, by Weber (2001). This is not to say that WL83 was critically flawed. It just serves as a reminder that in those rare cases where an analytical approach is feasible, it really adds knowledge to what can be directly concluded from computer simulations. This does not mean that this additional knowledge cannot be obtained from simulations at all; the additional knowledge just follows less directly. For example, when different parameter settings are used in the Daisyworld simulation, the homeostasis property can "empirically" be found to be significantly weakened, whereas in the analytical approach a weakened homeostasis regime can be anticipated from manipulation of the equations. The methodological point here is that determining the sensitivity of simulations to parameter settings (for instance, by doing different runs for different parameter settings) takes a significant amount of work (and often is skipped for this reason), but can sometimes really be necessary in order to prevent drawing sloppy, or even erroneous, conclusions.

Even nowadays, the distinction between homeostasis and stability in the Daisyworld model is still not always clearly made in the literature on Gaia. Take, for instance, Timothy Lenton's review article in *Nature*, "Gaia and Natural Selection" (Lenton, 1998). Lenton writes that the WL83 Daisyworld model "provided a hypothetical example of planetary regulation emerging from competition and natural selection at the level of individuals" (Lenton, 1998: 440). The "competition and natural selection" processes among daisies that Lenton refers to are an interpretation of the simple equations of the Daisyworld model in evolutionary terms. The problem with this presentation of the model is that the homeostasis (or "regulation") property of the model does not simply derive from the fact that competition and natural selection processes are included, but instead depends strongly on the specific parameter setting of the model.

Do Daisyworld-like mechanisms play an important role on the Earth? To provide evidence for the truth of homeostatic geophysiological hypotheses pertaining to the real Earth, it is not enough to show that the Earth system over the period of interest has contained some negative feedbacks involving life. One must also show that these feedbacks were structured in such a manner that they led to a significant dampening of external influences, and thus to homeostasis, and that if the feedbacks involving life had not been present, the environmental conditions would have changed. There remains a difficulty in separating homeostasis from change or, equivalently, in establishing the bandwidth of change which can still be associated with the homeostasis property. It is hard to determine how environmental conditions would have changed if significant dampening had not been present, since this refers to a counterfactual situation—an alternative history of the Earth system, a route not taken. This again demonstrates why models are so important in the field of research in which geophysiological hypotheses are studied. In model worlds one can switch processes on and off and look at what happens under different external forcing scenarios.

Daisyworld is such a virtual world. In order to make a connection between virtual worlds and reality, we must compare model results of homeostasis phenomena to similar phenomena in the real Earth system. To corroborate a geophysiological hypothesis, the model results should be "close enough" to the observations (which usually will be indirect and derived from a combination of evidence and theory if we are interested in geological timescales) and the explanation given using the model should be considered plausible. Furthermore, we should establish the plausibility of the model results for the counterfactual case with an absence of life or particular forms of life.

"Close enough" captures a pragmatically defined notion that allows one to focus only on the processes and timescales of interest, instead of having to give an adequate description of all processes for all timescales. The plausibility of the model assumptions should be evaluated using knowledge from physics, chemistry, biology, and the geosciences.

The plausibility of the assumptions of the zero-dimensional Daisyworld model must be considered low. Especially the second and third assumptions, as summarized above, can be criticized from a biological and a physical perspective, respectively. The second assumption gives the most serious problem: the homeostatic behavior of Daisyworld hinges on the prescription of a fixed optimal temperature. Of course, the evolution of the species (including possible shifts in the optimal temperature) should be considered when focusing on such long timescales. The debate on how mutation should be included in the model is still raging (many ways to do it have been proposed, some leading to increased homeostasis and some leading to a collapse of the homeostasis property), and will probably continue to rage for a long time. The third assumption is also problematic: as was pointed out by Weber (2001), the heat transport assumption of the zero-dimensional Daisyworld model is physically unrealistic compared with how we know we should model the process in the comparably simple models which are used in climate science. For this heat-transport assumption, in contrast to the biological no-mutation assumption, a state-of-the-art hierarchy of models is available which can help in determining the impact of the assumption made in the simple Daisyworld model by comparing the simple model with more complex models, as will be discussed below (the conclusion being that the Daisyworld model is unreliable).

Models of Different Complexity

The Daisyworld model, as such, does not prove anything. For instance, in WL83 no comparisons of model results against real phenomena on Earth were presented. The model is much too simple to adequately describe any real feature of the Earth. It thus is advisable to move toward more complex models to study geophysiological hypotheses (many Gaia researchers, including Lovelock, have been moving in this direction). For comparison, in the climate-change field, both simple and complex models are used for simulating global climate change. Sometimes the importance of the complex modeling approach is overstated at the expense of the simple modeling approach (for examples, see Shackley et al., 1998; Petersen, 2000). However, as is also explicitly recognized in the latest report of the Intergovernmental Panel on Climate Change (Houghton et al., 2001), simple and complex models have complementary roles to play in the science of climate change. Moreover, the gap between simple and complex models is increasingly being closed by models of intermediate complexity (McAvaney et al., 2001; Claussen et al., 2002). For testing geophysiological hypotheses, depending on the timescales of interest, a complementarity of complex and simple models similar to the climate-change field can be assumed to hold. The Daisyworld model, however, must be considered too simple to serve the purpose of testing geophysiological hypotheses, or even the purpose of demonstration (unless one is very clear in making the distinction between stability and homeostasis and acknowledges the sensitivity of the homeostasis property to the parameter setting).

Currently, most modeling work related to geophysiological hypotheses is still done with Daisyworld-like models (i.e., extremely simple models). Therefore, the perceived need to move toward models of higher complexity—as was documented, for instance, in the call for papers for the Gaia 2000 conference—is understandable. However, one must be aware that there are advantages and disadvantages associated with moving toward more complex models. The main disadvantage is that more complex models are more difficult to thoroughly analyze and understand. Also, the burden of rigorously performing sensitivity analyses can become too heavy if the complexity of models is increased. The main advantage of more complex models is that one can try to put more realistic processes into these models, which can thus become more representative for the real Earth system and possibly more reliable.

Are there ways to determine how complex a model should be in order to reliably model a certain phenomenon? One possibility is to systematically compare the results of models of different complexity. I will give an example to illustrate this point. The simple zero-dimensional Daisyworld model has been extended to a two-dimensional version (von Bloh et al., 1997). In this two-dimensional model spatial heat transport is explicitly included through diffusion. The major consequence of adding spatial heat transport is an increase of the homeostasis range. Homeostasis is also improved in the sense that the equilibrium temperature for the two-daisy state varies even less with luminosity (compared with the slight decrease with increasing luminosity for the zero-dimensional model). From a comparison between zero-dimensional and two-dimensional Daisyworlds we can learn something about the sensitivity to the heat transport formulation in the model—but since there is no real Daisyworld to compare the model results against, we cannot say the two-dimensional model is better just because it is two-dimensional; what we *can* say is that the processes in the two-dimensional model are described in a physically more realistic way and that the model may therefore be considered more reliable.

To conclude, my advice to the community of Gaia researchers is to continue the development of more complex models, taking care to avoid the associated pitfalls. I have devoted a considerable amount of space to the Daisyworld model in this chapter because it is a paradigm model that many researchers use as a reference point. But, as we all know, the model is too simple. And its two-dimensional extension, though we can obtain knowledge about some interesting sensitivities to model assumptions, is still far—in terms of its complexity—from the three-dimensional Earth system models used to study short-term climate–vegetation interactions, for example. But the dimensionality of a model is not the only factor determining how realistically processes are modeled. The number of processes included and the level of detail with which processes are represented also determine the realism of a model (see Claussen, 2002). Thus, I do not agree with the claim made by von Bloh and coauthors that their two-dimensional Daisyworld model is "*mid-way* between pure toy models and three-dimensional analogical Earth-System models based on state-of-the-art and geographically explicit simulation modules for the atmosphere, the ocean, the biogeochemical cycles, civilisatory land-use etc." (von Bloh et al., 1997: 261; emphasis added). It is still a very simple world these authors are modeling. We must conclude that the gap between the most complex and the simplest models is still huge in the field of Earth system science. And the most complex models available (general circulation models) cannot be used to study the geophysiological hypotheses that have been proposed under the heading "Gaia hypothesis" (i.e., hypotheses dealing with geological timescales), because these models have been developed in the context of climate-change research in order to deal with timescales of typically less than 1,000 years. Thus, much work remains to be done before the homeostatic geophysiological hypotheses that have been proposed in Gaia research can start to be tested.

Acknowledgments

This chapter was largely written in 2000–2001, while the author worked at the Faculty of Philosophy, Vrije Universiteit Amsterdam. The figures in this chapter have been reprinted with kind permission of Kluwer Academic Publishers.

References

Claussen, M., et al. 2002. Earth system models of intermediate complexity: Closing the gap in the spectrum of climate system models. *Climate Dynamics*, 18, 579–586.

Houghton, J. T., et al., eds. 2001. *Climate Change 2001: The Scientific Basis. Contribution of Working Group I to the Third Assessment Report of the Intergovernmental Panel on Climate Change.* Cambridge: Cambridge University Press.

Kirchner, J. W. 1991. The Gaia hypotheses: Are they testable? Are they useful? In S. H. Schneider and P. J. Boston, eds., *Scientists on Gaia*. Cambridge, MA: MIT Press.

Lenton, T. M. 1998. Gaia and natural selection. *Nature*, 394, 439–447.

Lovelock, J. E. 1972. Gaia as seen through the atmosphere. *Atmospheric Environment*, 6, 579–580.

McAvaney, B. J., et al. 2001. Model evaluation. In J. T. Houghton et al., eds., *Climate Change 2001: The Scientific Basis. Contribution of Working Group I to the Third Assessment Report of the Intergovernmental Panel on Climate Change*. Cambridge: Cambridge University Press.

Petersen, A. C. 2000. Philosophy of climate science. *Bulletin of the American Meteorological Society*, 81, 265–271.

Shackley, S., P. Young, S. Parkinson, and B. Wynne. 1998. Uncertainty, complexity and concepts of good science in climate change modeling: Are GCMs the best tools? *Climatic Change*, 38, 155–201.

Volk, T. 1998. *Gaia's Body: Toward a Physiology of Earth*. New York: Springer-Verlag.

von Bloh, W., A. Block, and H. J. Schellnhuber. 1997. Self-stabilization of the biosphere under global change: A tutorial geophysiological approach. *Tellus*, 49B, 249–262.

Watson, A. J., and J. E. Lovelock. 1983. Biological homeostasis of the global environment: The parable of Daisyworld. *Tellus*, 35B, 284–289.

Weber, S. L. 2001. On homeostasis in Daisyworld. *Climatic Change*, 48, 465–485.

4

Gaia: Toward a Thermodynamics of Life

Eric D. Schneider

Those who envisage a fundamental link between the thermodynamic arrow of energy dissipation and the biological arrow of the greening earth make up a small minority, and stand well outside the main stream of contemporary biological science. But if their vision is true, it reveals that deep continuity between physics and biology, the ultimate wellspring of life.
—Franklin M. Harold, *The Way of the Cell*

Abstract

The chapter you are about to read illuminates an important connecting principle of nature, one which helps explain processes as different as boiling pots of highly organized fluids, hurricanes, pulsating chemical clocks, the origin of life, the development of ecosystems, the direction of evolution, and Gaia itself. In retrospect all these systems are held together by a simple concept drawn from the famous second law of thermodynamics. This simple concept is "Nature abhors a gradient." With this intelligible but contemporary view of thermodynamics, one sees all of life as a thermodynamic dissipative entity residing at some distance from equilibrium, sustained by the ability to degrade energy gradients. The Gaian global ecosystem is involved in tapping the gradient between the hot sun and frigid outer space. Living systems are the result of energy processes which, while building higher complexity locally, are at the same time exporting lower quality energy beyond the confines of the evolving system. Life never violates the second law, and thermodynamic gradient reduction provides living systems with a "final cause," their "go," their "direction." Nature is not just the result of accumulated improbable events, since both abiotic and biotic phenomena obey the same thermodynamic laws. These ideas connect life and nonlife at a fundamental level.

Introduction

Two of the enduring aspects of Lovelock and Margulis's Gaia proposal are that Earth is one large, cybernetic, interconnected ecosystem with an atmosphere whose gas concentrations are not at chemical equilibrium. They proposed that this disequilibrium is caused by living processes. Such a view of Earth as a single, nonequilibrium ecosystem has changed how we look at our planet. In principle, it could help predict future simple ecosystem behavior. But if Earth is a single ecosystem, what are the underlying fundamental laws that govern the actions of whole ecosystems? Where are the Newtonian laws of ecology, the $F = MA$ or the Navier-Stokes equations for ecosystems? What is the answer to such a seemingly simple question as whether such and such an ecosystem would change or stay the same if perturbed?

Ecology is the scientific study of the interactions between abiotic and biotic components in nature that determine the distribution and abundance of organisms. Most of our present ecological knowledge has come from hundreds of years of careful observations of changes in species, populations, and landscapes. However, only since the 1850s years have these observations been synthesized and attempts been made to develop a theoretical framework for the science. Ecology is rich in theory. There is population abundance theory, predator-prey theory, niche and biodiversity theory, and lots of increasingly complex models.

Because of the ubiquity of energy and material flow in all living systems, the study of energy flow and thermodynamics should provide some insights into ecosystem function and development. I plan to show that thermodynamics is a necessary but not a sufficient explanation for life. How important a role thermodynamics plays in living systems will be left to the reader. Ecosystems are the result of boundary conditions like the biotic, physical, and chemical components of nature acting together as open thermodynamic systems with both material and energy flowing through their networks. Analysis of energy flow through ecosystems shows that there are regular transformations in ecosystems, called successions. Until now successions have been predictable only because numerous observations show similar patterns occurring in differing ecosystems. More specifically, succession is the recovery process that occurs when the preceding ecosystem has been wiped out or almost wiped out—for example, by a forest fire, pests, or over fishing. The most widely used example of succession is a cleared field that is left fallow and then goes through a series of plants that follow from grasses to shrubs, to conifer stands that give way to an oak forest after about 150 years. Similar changes take place in all terrestrial ecosystems, be they grasslands; the oak, spruce, and fir forests of the American West; or bacterial and marine plankton communities. All these different ecosystem successions can be characterized as biological variations on an underlying theme of energy transformation. However, each of these successions marches to the beat of a different drummer as it moves at its own pace with its distinct species in an ecosystemic band. Successions are unfolding energetic processes with increasing energy flows, increasing material cycling, increasing efficiencies, and increasing entropy production over time. The phenomenology of ecosystems shows that successions build structures and processes to capture available gradients and to degrade the captured energy as efficiently and completely as possible. Rich in data, ecosystems can provide crucial support for a thermodynamic theory of living systems.

Thermodynamics, from "heat" and "motion" in Greek, is the study of how systems handle and transform energy. Classic thermodynamics studied isolated and more or less closed systems, mechanical systems that became more disordered over time until they either lost their function or became completely random. But living beings are not this kind of mechanical system. We are open systems, organized by the energy and materials flowing through us. Our physical complexity, our abilities to perceive and to reason, are linked to the behaviors of complex energy-driven systems. These systems are not the gods or God of religion imagined in any human sense; nature, rather, in its production of energy-dealing systems, including evolving life, has produced our apelike selves who then imagine nature's productions to be similar to those little devices and machines which we ourselves have put together with our hands. Science has been extremely productive in its imagining of mechanisms. But complexity in nature does not appear by meticulous construction under the guidance of curious and diligent fingers. Complex, cycling, energy-driven systems in nature come spontaneously into being.

Classical Thermodynamics

Initially a study of how to gain mechanical work from heat, thermodynamics has been extended to cover all forms of energy and its transformations in nature. There are four—some say five—laws of thermodynamics. Thermodynamics starts with Sadi Carnot's observation that heat always flows from a hotter system into a cooler one, and that work can be extracted from this gradient. The second law was the first thermodynamic law to be understood as Carnot described the one-wayness, the asymmetry of energetic processes in the universe. Water flows down over waterfalls, never in the other direction. The first thermodynamic law (and the second to be formalized) was presented by Rudolf Clausius and Lord Kelvin as the conservation law. They built on the work of Carnot, who mistakenly believed that heat was an all-penetrating fluid, "calor." Carnot was correct in stating that his "calor" energy was conserved, and that any action in nature changes the distribution of that energy in an irreversible manner. Today we say that total energy, in all its forms, is constant in the universe but its quality is decreasing with time. The third law of thermodynamics evolved out of Antoine Lavoisier's and Joseph Gay-Lussac's work on the development of the pressure, temperature, and volume experiments. They showed that a pressure of any gas contained in a given volume increases or decreases by 1/273 of the initial value for each degree C of its original value. If one starts with a gas at zero degrees C as a reference point, and cools it to −273°C or 0° Kelvin, the gas pressure and molecular motion are predicted to go to zero. Later, Walther Nernst and Max Planck formally developed this idea into the relationship between thermodynamic entropy and 0° Kelvin. At absolute zero the entropy of the system is zero.

The Zeroth law, formalized in 1931, dealt with thermodynamic equilibrium. It plays a pivotal role in the most modern interpretation of all the laws of thermodynamics and provides important underpinning for our general thesis. It was first formally developed by R. H. Fowler and deals with two seminal concepts in thermodynamics: thermodynamic equilibrium and temperature. When a closed thermodynamic system with rigid adiabatic walls reaches a point where independent properties of the system become time-constant, and no further changes can be measured, the system is said to have reached a state of thermal equilibrium.

Moving on to possible future laws, formalization of cycles may be the Rosetta Stone that allows for the integration of nonequilibrium thermodynamics with biology. Indeed, Harold Morowitz, the present-day grandfather of the application of thermodynamics to biology, suggests that cycling be added as a general theorem of thermodynamics. "In the steady state systems, the flow of energy through the system from a source to a sink will lead to at least one cycle in the system" (Morowitz, 1968, p. 33). This is a very general theorem which holds for cycles from biochemical to quantum systems. An example is an autocatalytic, self-reinforcing chemical reaction where compound A produces compound B, and B produces compound C, and C aids in the production of A. If one were to excite this reaction with some outside reactant to make more B, all the components in the cycle would increase. For instance, an increase in B, and thus in C, would in turn reinforce A (an autocatalytic action) and develop a cycle. A more subtle but just as real cycle would be the action in an equilibrium system where a minute fluctuation momentarily lifts a particle to a higher energy level and the particle immediately falls back to its equilibrium state. Morowitz's cycling theorem is the best candidate for a fourth law of thermodynamics because of the ubiquity of cycling in all thermodynamic systems.

Thermodynamic systems can be isolated, closed, or open to the outside world, and are described according to their relationship to the equilibrium state. Classical thermodynamics takes place in idealized, isolated chambers where actions are reversible and carried out very, very slowly. Despite these highly limiting theoretical constraints, thermodynamics grew into a very practical science that brought about the industrial revolution, helped to make more efficient steam engines, explained phenomena like pressure-volume-temperature relationships, and became an integral part of modern cosmology. Because of advances in thermodynamics, our society has been able to liberate itself from dependence on forced animal labor, into a thermodynamic world in which energy is produced in coal and nuclear power plants to manufacture goods, power ships, and run interconnected technological networks.

Thermodynamics is almost synonymous with its second law, all too precipitously derived from the nonrepresentative cases of closed and isolated adiabatic systems. Derived from the observation of steam engines, energy dissipating and matter disorganizing in enclosed settings, thermodynamics remains associated with an irretrievable loss of usable energy and organization over time. This is measured by a quantity, entropy, which increases with time. Entropy, formulated as heat divided by temperature, is associated with the rise of random combinations, disorder, disorganization, and lack of complexity. It rules out perpetual motion machines and shows that there is an energy cost for any and all actions.

Nonequilibrium Thermodynamics

As important as classical thermodynamics is as a science, it takes place in isolated adiabatic containers. There is, however, a more general thermodynamics, one not confined within artificial boundaries. This is the thermodynamics of energy and material flows, of cycling chemical reactions, of complexity, and of life. Beginning with the examination of near-equilibrium, metastable states, thermodynamics cautiously crawled out of the isolated boxes of classical thermodynamics.

The first complete theoretical discussion of open nonequilibrium systems that would reach a metastable or "steady" state was by Yale professor Lars Onsager (Onsager, 1931). His ideas led to two very important insights. The first was that near-equilibrium "reciprocity relationships" exist in which forces and fluxes are coupled. For example, the flow of the water in a pipe (laminar flow) is directly related to pressure, while at the same time the pressure is linearly related to the flow rate. Well-known laws in chemistry and physics agree with the reciprocal relationships of Onsager's near-equilibrium processes. The second significant insight is that these near-equilibrium processes achieve metastability—they don't "die" but "change to stay the same"—at some distance from equilibrium while moving toward a state of minimum specific entropy production.

But the most fascinating thermodynamic phenomena lie beyond the linear near-equilibrium Onsager region—in the unexplored territory that most thermodynamicists call far-from-equilibrium. Like their near-equilibrium cousins, these systems depend upon energy and material flows. But whereas linear processes in the Onsager region vary proportionally with each other, the relationships among processes in far-from-equilibrium systems are nonlinear, cyclical, wild. They were investigated by Ilya Prigogine and his collaborators at the Free University of Brussels; Prigogine, who won the Nobel Prize in 1977 for his accomplishments in nonequilibrium thermodynamics, is recognized mainly for his work on elaborate cyclical chemical reactions and other systems, all of which he calls dissipative systems (Prigogine, 1955).

Prigogine's Brussels research group was among the first to suggest that both linear and nonlinear systems could maintain themselves away from thermodynamic equilibrium. The dissipative systems maintain their stable, low-entropy state by transferring material and energy across their boundaries. The systems of life, connected but separated by membranes, skin, bark, and shell, are examples of such systems. The low-entropy, more organized state within a dissipative structure, living or not, depends on increasing the entropy of the larger global system in which the dissipative structure is embedded. The second law is not violated: overall entropy increases.

Prigogine and his colleagues have shown that dissipative structures self-organize through fluctuations, small instabilities which lead to irreversible bifurcations and new stable system states. Thus the future states of such systems are not predetermined (as are fractals, chaos, artificial life, and other computer simulations). Dissipative structures are stable over a finite range of conditions and are sensitive to fluxes and flows from outside the system. Glansdorff and Prigogine (1971) have shown that these thermodynamic relationships are best represented by coupled nonlinear relationships (i.e., autocatalytic positive feedback cycles), many of which lead to stable macroscopic structures which exist away from the equilibrium state. Convection cells, hurricanes, autocatalytic chemical reactions, and living systems are all examples of far-from-equilibrium dissipative structures which exhibit coherent behavior.

Bénard and Belousov-Zhabotinski Reactions

Some of the most intriguing natural phenomena are highly complex systems that emerge and evolve under simple physical and chemical gradients. Neither generated by computers nor biological in nature, these complex systems are strictly physical and chemical structured processes generated by gradients being degraded according to the second law.

The transition in a heated fluid from conduction to convection is a striking example of emergent coherent organization in response to an external energy input. The bottom of an apparatus filled with a fluid is heated and the upper surface is kept at a cooler temperature—a temperature gradient is induced across the fluid. The initial heat flow through the system is by conduction; energy transfers from molecule to molecule. When the heat flux reaches a critical value—when the temperature gradient is steep enough—the system reorganizes. The molecular action of the fluid becomes coherent. These surface tension-convective structures, hexagonal surface patterns, are called Bénard cells. The coherent kinetic structuring increases the rate of heat transfer and gradient destruction in the system. Form follows function: the complex structuring is more adept at heat transfer and entropy production.

This transition from noncoherent, molecule-to-molecule heat transfer to coherent structure results in excess of 10^{22} molecules acting in concert. From a Boltzmann statistical point of view, this is ridiculously improbable. Yet the occurrence is the direct result of the applied temperature gradient—it is the system's inanimate response to attempts to keep it from equilibrium. Nature resists by creating an unexpectedly complex system.

One of the characteristics of these systems is a kind of self-perpetuation of a cycle—autocatalysis, now known to be at work in physical, chemical, and biological systems. In autocatalysis, the product of a chemical reaction leads to further production of itself, a situation highly reminiscent of biological reproduction. Prigogine proposed a series of equations to simulate autocatalysis; the solutions of these equations gave rise to a periodic cycling of variables. Such periodic cycling is seen in actual chemical systems, the most famous being the Belousov-Zhabotinski (BZ) reaction, named for the Russian chemists who discovered these reactions in the late 1950s and early 1960s (Zhabotinski, 1964, 1974). The Belousov-Zhabotinski reaction is a chemical limit cycle; it rises out of disorder, organizing as it goes. The macroscopic visible structure is the result of billions of molecules acting in concert, in spatial scales of centimeters, many orders of magnitude larger than the molecular size of the compounds in the reaction. Such long-range temporal and spatial correlation of processes is typical of nonequilibrium systems. In the BZ reaction a concentration gradient is set up and patterns of horizontal yellow bands appear as the system feeds off the gradient. As they promote more probable distributions of matter in the area around them, they themselves become more surprising, more improbable.

Classical thermodynamic systems were encased in adiabatic, rigidly bounded enclosures, isolated (except for the ignored action of gravity) from the rest of the universe. Work developed in these systems can be modeled using fictitious ideal mechanisms that include pistons, deformable boundaries, and motor generators that are connected to a single weight. The processes inside these systems can be imagined as a series of internal constraints removed one by one until the system is at equilibrium. These processes can be mechanical—for example, the opening of a series of doors that separate a system into compartments. Imagine a box having four compartments with doors that open and close. One of these compartments holds 10,000 molecules of a gas; the other three compartments have no gas and are a vacuum. There is a large potential gradient between the sides of the doors. Upon opening the first door, or constraint, the system will come to its local equilibrium state of 5,000 molecules in each of two boxes, with no gradient between the two open boxes. The same process will occur as we open all the remaining doors, with 2,500 molecules in each of the four boxes at a local equilibrium state. Each time a constraint is removed, the system comes closer to equilibrium. After the whole system comes to equilibrium, one cannot determine the order in which the constraints or doors were opened. If the doors were pistons, work could be extracted from this system as the pistons come to equilibrium. These principles hold for a broad class of thermodynamic systems from chemical kinetic reactions to a cup of hot tea reaching room temperature.

In the mid-twentieth century, the thermodynamicists Hatsopoulos and Keenan in 1965 and George Kestin in 1966 formally proposed a principle that subsumes the zero, first, and second laws of thermodynamics (Hatsopoulos and Keenan, 1965; Kestin, 1966). "When an isolated system performs a process, after the removal of a series of internal constraints, it will always reach a unique state of equilibrium; this state of equilibrium is independent of the order in which the constraints are removed. The unique state of equilibrium is characterized by a maximum value of entropy" (Kestin, 1979, p. 2). This is called the Law of Stable Equilibrium by Hatsopoulos and Keenan and the Unified Principle of Thermodynamics by Kestin.

The unification and simplification of the second law by Hatsopoulos and Keenan and by Kestin are remarkable steps forward given little credit. Kestin's 1,342-page, two-volume text is filled with page after page of innovative explanations of thermodynamics and is a monument to the science (Kestin, 1979). This text is now in its seventh edition. Mathematically rich and accompanied by simple idealized models using pistons, membranes, adiabatic boxes, and such, the work sets forth a logical development of thermodynamics from the original special case of isolated systems to the general situation of open systems, including the complexity concentrators we see in whirling inanimate and genetically undergirded living systems.

The work of Hatsopoulos and Keenan, and of Kestin, describes the behaviors of a class of systems that are some distance from equilibrium, and with removal of constraints will come to a local equilibrium state. They present us with a formal statement of the second law of thermodynamics that encompasses all the attributes of the nature of entropy but sidesteps its use in its definition. We are thus on the cusp of a thermodynamics that describes systems some distance from equilibrium and the characteristics of their final equilibrium states. The importance of the work of Hatsopoulos, Keenan, and Kestin is that their statement imposes a direction and an equilibrium end point for all processes. Here is a thermodynamics that is telling us that the spontaneous degradation of gradients is a paramount part of thermodynamics. Any system that succeeds at getting to equilibrium has reduced constraining gradients.

Le Chatelier's principle in chemistry is an example of the equilibrium-seeking nature of the second law. Enrico Fermi, in his 1936 lectures on thermodynamics, noted that the effect of a change in external conditions on the equilibrium of a chemical reaction is prescribed by LeChatelier's principle (Fermi [1937], 1956). If the external conditions of an equilibrium thermodynamic system are altered, the system will tend to move in such a direction as to oppose the change in the external conditions. Fermi noted that if a chemical reaction were exothermal—that is, $(A + B = C + D + \text{heat})$—an increase in temperature in the reaction chamber will shift the chemical equilibrium in the reaction above to the left-hand side. Since the reaction from left to right is exothermal, the displacement of the equilibrium toward the left results in the absorption of heat and opposes the rise in temperature. Similarly, a change in pressure (at a constant temperature) results in a shift in the chemical equilibrium of reactions which opposes the pressure change. This thermodynamic behavior—very general and applicable to both open and closed thermodynamic systems—was discovered independently by Henry Le Chatelier in France and Karl Ferdinand Brauns in Germany. Kestin called this process the Principle of Spite, and it was called the Moderation Principle by Prigogine.

Lotka, Wicken, and the Thermodynamics of Life

Ludwig Boltzmann, the founder of statistical thermodynamics, emulated Charles Darwin. Like Darwin, he developed a natural law governing the behavior of change. In 1886 Boltzmann suggested that the energy gradient imposed on the Earth by the sun drives the living process. He postulated a Darwinian-style competition for entropy among energy-using living systems.

In 1922 Alfred Lotka wrote about energy flow in biological systems and suggested that the organism that can capture the most energy and turn it into seed, growth, and development will have the evolutionary advantage over an organism without these traits. He saw life as composed of interconnecting autocatalytic cycles that increase the flow of energy through biological systems. Others, including Harold Morowitz, Howard and Eugene Odum, Robert Ulanowicz, and Jeffrey Wicken have helped lead us toward a genuine "thermodynamics of life."

In the 1980s Wicken completed some of Lotka's and Erwin Schrödinger's unfinished thoughts on the thermodynamic nature of life. (Significantly, Schrödinger, best known in biology for his slim volume *What Is Life?* [Schrödinger, 1944], helped advance investigation into the genetic basis of life [and ultimately the discovery of DNA's role] and also, in the same volume, inquired into the thermodynamic basis of life.) Wicken laments that thermodynamics has been misunderstood as an obscure, dismal science whose attempt to illuminate life is just "one more attempt to reduce life to matter and motions."

Not so, Wicken retorts. "It attempts no such thing. Thermodynamics is, above all, the science of spontaneous process, the 'go' of things. Approaching evolution thermodynamically allows us to bring the "lifeness" of life into the legitimacy of physical process ... the emergence and the evolution of life are phenomena causally connected with the Second Law; and ... thermodynamics allows for the understanding of organic nature from organisms to ecosystems as relationally constituted processes bound everywhere by functional, part-whole relationships" (Wicken, 1987, p. 3).

Wicken persuasively argues that the second law is not just compatible with life; it is instrumental in its origins and evolution. His thesis is straightforward: thermodynamics infuses biology at all levels, and its second law, combined with imposed gradients, gives life—from its origin to ecosystem processes to the biosphere inclusive of human technical civilization—its direction. Needless to say, boundary conditions and stochastic events will affect the specific aspects, but the general direction will remain the same.

Boltzmann, Lotka, and Wicken have all proposed models of living systems that include thermodynamics at the core of their ideas. Recently Stuart Kauffman made a stab at describing the life–thermodynamics link in his book *Investigations* (Kauffman, 2000). Kauffman tentatively states at times that thermodynamics may be able to add to the discussion of living systems and physics. He suggests that "This coming to existence of self-constructing ecosystems must, somehow, be physics. Thus, it is important that we have no theories for these issues in current physics." He laments that "there can be no general law for open thermodynamics (2000, p. 82). As an answer to this conundrum Kauffman proposes a new law of thermodynamics, a "fourth law" which appears to hold for "self-constructing systems," including life. Kauffman proposes many candidate "fourth laws"; the following is a typical offering: "Indeed, there may be a general law for biospheres and perhaps even the universe as a whole along the following lines. A candidate fourth law: As an average trend biospheres and the universe create novelty and diversity as fast as they can manage to do so without destroying the accumulated propagating organization that is the basis and nexus from which novelty is discovered and incorporated into the propagating organization" (2000, p. 85). Later he becomes pessimistic about the role physics can play in biology: "for I will say that we cannot prestate the configuration space of a biosphere and, therefore, cannot deduce what will unfold" (2000, p. 113). Kauffman's caution probably results from a lack of knowledge of Lotka's, Prigogine's Morowitz's and Wicken's work. No new laws of thermodynamics are needed. All that is required is an expansion of the existing laws to the nonequilibrium region.

Instead of suggesting new laws for thermodynamics, let us propose a set of phenomenological observations or perhaps propositions that are common among nonequilibrium, nonliving, and living systems. These postulates frame the second law as an active agent in organizational processes with the ability to draw material and energy into structures and to delay their immediate journey to equilibrium. Instead of a second law portending the doom of life and the heat death of the universe, we propose an interpretation of the second law as an active participant in the emergence and evolution of dynamic dissipative systems. This list is preliminary and drawn from many sources. I am sure additional propositions will be forthcoming.

Properties of Open Thermodynamic Systems

1. In the linear near-equilibrium region the following properties exist.
   a. Reciprocity relationships exist in which forces and fluxes are coupled.
   b. Steady states exist at some distance from equilibrium in a state of minimum entropy production.
   c. Power is conserved in the system.
2. As systems are moved away from equilibrium by imposed gradients, they will use all avenues available to counter and degrade the applied gradients.
   a. As the applied gradients increase, so does the system's ability to oppose further movement from equilibrium.
   b. Systems moved away from equilibrium by greater gradients will be accompanied by increasing energy flows total system throughput (TST) and higher entropy production rates.
   c. The further a system is from equilibrium, the stronger the gradient that must be imposed and degraded to keep it there.
3. Open nonequilibrium systems reside at some distance from equilibrium, produce entropy that is exported out of the system, and maintain a low entropy level inside the system at the expense of disorder outside the system.
4. If a gradient is imposed on a system, and dynamic and or kinetic conditions permit, autocatalytic or self-reinforcing organizational processes, structures, and growth can be expected. These autocatalytic organizations draw material and energy into themselves (e.g., hurricanes and living systems). Although these processes are often described as self-organized, in fact they are organized by an accompanying gradient whose energy potential is drawn into cycling selves.
5. Autocatalytic organizations are nonlinear dynamic systems with flows of energy and material through them. These systems exhibit stable behavior and dynamic properties (e.g., attractor basins, bifurcations, and catastrophic behavior).
6. Biological systems optimally capture energy and degrade available energy gradients as completely as possible.
7. Biological processes delay the instantaneous dissipation of energy and give rise to energy and material storage, cycling, and structure.

Living systems delay or hold material and energy in states away from equilibrium, thus delaying their journey to equilibrium. To quote the early twentieth-century French biologist-philosopher Henri Bergson in a discussion of living systems, "[Life] has not the power to reverse the direction of physical changes, such as the principle of Carnot determines it. It does, however, behave absolutely as a force would behave which left to itself, would work in the inverse direction. Incapable of *stopping* the course of material changes downwards, it succeeds in *retarding* it" (Bergson, 1911, pp. 245–246). The capture and delay of photons, the transduction of light energy (photons) into chemical energy (the carbon–hydrogen and carbon–carbon bonds of organic compounds) by photosynthesis is the thermodynamic basis of most modern life. In subsurface environments like the deep sea vents and buried rocks there are methanogens which feed off carbon dioxide and hydrogen and are an alternative energetic pathway to life.

In short, dissipative autocatalytic structures degrade available gradients to maintain their gradient-degrading structures. Biological systems tend toward increasing energy capture, and by delay tactics like cycling and storage of biomass, they defer the instantaneous dissipation of energy. Complex material and energy cycles arise as dissipative structures bring some of the outside world into themselves. What appears to be a random walk of evolutionary change, a nonprogressive meander, is in fact a careful invisible hand, following to the letter an invisible thermodynamic message. Emily Dickinson's poetry—"The simple news/that Nature told/with tender majesty/Her message is/committed to hands I cannot see"—could be applied to the heretofore underappreciated thermodynamic shaping of nature. If evolution shows how all life is related by descent, and ecology shows how all life is spatially and temporally connected, then nonequilibrium thermodynamics links the complex behavior of life—so often explained by a vital principle, natural selection, or intelligent design—with the complex behavior of inanimate systems.

Successional Ecosystems

Ecology is the science of the interactions of living organisms with each other and their interplay with the physical and chemical components of the environment. Ecosystems can be viewed as the biotic, physical, and chemical components of nature acting together as nonequilibrium dissipative processes. They import high-quality energy from the sun and degrade it to lower-grade energy. Ecosystems recycle major chemical components required for life. As much processes as entities, ecosystems are part of the nested hierarchy of biology that scales from cell metabolism to organisms made of dividing and dying cells to communities of organisms to the biosphere as a global ecosystem. Each hierarchical level can be understood not only from the bottom up, as the result of its constituent subsystems, but also from the top down, thermodynamically.

When one looks at the development of ecosystems, one is struck by the ubiquitous process of ecosystem succession. Ecological successions have been discovered in all natural ecosystems, including terrestrial Amazon-like ecosystems, Arctic lakes, and tiny bacterial ecosystems; they all exhibit a characteristic pattern of development and have a thermodynamic rather than a genetic explanation. American ecosystem observation science seems to have started with Henry David Thoreau, who tracked the regrowth of grasses, shrubs, and trees in abandoned fields in Massachusetts. In 1899, the University of Chicago botanist Henry Chandler Cowles described plant communities in sand dunes on the shore of Lake Michigan. Near the lake, with the constant movement of wind, water, and sharp silicate sand, no rooted plants survive. Away from the shore, where there is less disturbance, succulent plants and grasses grow. The further one moves inland, the more trees such as juniper and pine take root. Proceeding from the lake across the dunes, an entire succession fans out horizontally in space rather than in time. There are spatial, temporal, informational, and energetic variants on the successional process, a pattern of rapid growth of monads followed by differentiation into a late successional phase of overall increasing entropy production.

In the 1940s the locus of American ecology was at Yale University, where, under the tutelage of the polymath G. Evelyn Hutchinson, a new branch of ecology blossomed. The Hutchinson school pioneered the study of material and energy flow through ecosystems. Hutchinson remarked that his most brilliant student, Raymond Lindeman, came to realize "that the most profitable method for [ecological] analysis lay in the reduction of all the interrelated biological events to energetic terms" (Lindeman, 1942, p. 417). Ecology, the study of the most complex systems we know, must be understood in terms of energy and energy flow, that is, thermodynamics.

One branch of the Hutchinson scientific family tree consists of Howard and Eugene Odum. Eugene, the dean of modern ecology, focused his scientific career on successional processes. In 1969 he published an influential paper, "The Strategy of Ecosystem Development," in which he synthesized the phenomenology of successional systems. He noted that early successional systems were inhabited by organisms with short lifetimes and high fecundity living in large niches. These systems recycled material more rapidly, had fewer cycles, and had cycles less adept at holding material and energy. More mature systems were more efficient, had higher energy flow through them, had longer residence time for materials, and were more complex with higher diversity. This carefully synthesized ecosystem phenomenology gives insights into other dissipative thermodynamic processes. One can see at this point that these ecosystem characteristics match many of the propositions presented above. Global humanity, for example, resembles a pioneer species colonizing a new niche; to achieve the global equivalent of successional maturity—to last in the biospheric long run—we will have to increase our connections with other species, and recycle our materials more adeptly through global biosystems of greater diversity and complexity. If the rapid growth of stem cells seen in embryos looks like the ghost of an ancient ecosystem, we should not be too surprised. Ancient cell colonies, including those that evolved into early animals and plants, were constrained by the same energetic verities that apply to ecosystems. Animal development exhibits successional processes. Indeed, the animal seems to be a spatiotemporally condensed, genetically "frozen" ecosystem. Traditional ecosystems are marvelous laboratories for study because they are often at the scale of and proximate to the humans who study them. But the thermodynamic lessons they hold go far beyond traditional ecology.

Beginning where Lindemen and Eugene Odum left off, Howard Odum developed his own schema to indicate storage and flow in energy-transforming ecosystems, machines, organisms, and economies. Like Lotka, he promoted the notion that the ability to access energy is directly related to survival (Odum and Pinkerton, 1955). Successful systems capture and store high-quality energy, use the stored energy, and recycle materials as needed.

Today only a handful of ecologists collect the data needed for complete analysis of energy flow in ecosystems. A modern practitioner of analyzing this data is the engineer-ecologist Robert Ulanowicz of the University of Maryland. Ulanowicz, a modern-day Lindeman, analyzes ecosystem energy flow, utilizing mathematical protocols used in economics (Ulanowicz, 1986). Ulanowicz has mastered the mapping of the material and energy flow through ecosystems and can calculate the total system throughput (TST) and ecological analogue of the gross national product as measures through and within the system. He has developed measures for cycling, tropic levels, and the interconnectivity between species. Ulanowicz's quantitative analysis confirms most of Eugene Odum's phenomenological syntheses.

Stress, Regression, and Recovery

Succession is not a one-way street; it can be reversed by environmental trauma. Stressed, deprived of the ability to use its ambient gradient, the ecosystem retreats to a less organized phase still within the identifiable trajectory of its own development. This exemplifies the thermodynamic basis of memory, which may also be operative in evolution and physiology.

A climax ecosystem is a mosaic of different stages of succession. Beyond the trees, grasses, and insects are the seeds, spores, and eggs. The community's potential extends far beyond what is visible at any given time. After a major setback like an extensive forest fire, community structure reverts to the equivalent of GO on a Monopoly board; the game is not over but set back to an initiation point, at which the progression begins again. Looking at an ecosystem's long-term accumulation of biomass or total system energy throughput, we see the characteristic pattern of ups and downs as it oscillates around a climax community of optimum energy use. Biomass builds, punctuated by occasional reversals. Organism-like ecosystems lack brains and central nervous systems, but their modularity demonstrates an ability to retrace thermodynamic paths from undamaged parts. Gradients reestablish and are reduced over wide areas of modular communities; they "remember" their previous states—older, more robust routes of gradient reduction. Ecosystems are dynamic processes that ebb and flow with the many variables that lead to an ecosystem state at any one time.

In the 1960s, at the Brookhaven Atomic Energy Laboratory on Long Island, New York, a forest was continuously irradiated with a highly radioactive gamma source (Woodwell, 1970). The experiment was set in a mature oak and pine forest ecosystem. The various species died out at different times in reverse succession, with pine trees dying first, followed by other tree species, then shrubs and small woody plants, and finally sedges, grasses, and herbs. Here we see a highly stressed ecosystem retreating along successional paths until the ground is bare.

Another pivotal experiment was carried out at the Hubbard Brook Experimental Forest in New Hampshire by Gene Likens, Frank Boorman, and their colleagues in the mid-1960s (Likens et al., 1970). In the fall and winter of 1965, a watershed in this experimental forest was clear-cut and then sprayed with herbicides. For several years afterward, researchers closely monitored the water and nutrient flow. They compared the drainage basin with uncut drainage basins in the surrounding ecosystem.

The results were astonishing. The stream runoff in the deforested system increased by 39 percent the first year and 28 percent the second year: the forest retreated to a very early successional system. It leaked its most valuable resource, water, and valuable nutrients at much greater rates than the other mature drainage basins. There were remarkable increases in ions running off the deforested system—417 percent for Ca^{++}, 408 percent for Mg^{++}, 1554 percent for K^{+}, and 177 percent for Na++−−. The nutrients—the "lifeblood"—were pouring out of the system. Eugene Odum had already pointed out that immature systems are relatively leaky, while more mature systems hold on to their materials, using energy to recycle their constituents in complex loops. The stressed, stripped-down system has been robbed of its ability to capture energy: no longer able to maintain the elaborate structures and processes that capture and retain energy, the ecosystem as a whole sickens. The thermodynamics of life gives us not metaphors of "raping" and "healing" the Earth, but real tools to measure ecosystem and environmental health.

In the 1970s and 1980s the world witnessed several large oil spills, including the *Amoco Cadiz* that ran ashore off the coast of Brittany and the *Exxon Valdez* that bled oil into Prince William Sound in Alaska. In both cases some bays and marshes were filled with over a foot of oil, followed by a chronic die-off of most species. Beaches were littered with millions of dead bivalves, and many benthic organisms died. Once the acute concentrations of the petroleum degraded or were washed out to sea, fast-reproducing phytoplankton and zooplankton species quickly moved in. These pioneer species—photosynthetic and amoeboid cells—filled the heretofore empty niches. Eighteen months after the spill succession was well under way.

In the 1970s Howard Odum and his students selected two adjacent tidal marsh creeks on the south Florida coast. As at Hubbard Brook, this study's value lies in its ecological rarity: a controlled experiment. One creek cooled the effluent of a large (2,400-megawatt) nuclear power plant and dumped hot water effluent into the marsh, raising its temperature by 6°C. The other creek maintained its natural character. The stress was the rising temperature as a result of the runoff. Analysis of the energy flows in these two ecosystems found the stressed system "starved," as attested to by a 34.7 percent drop in its biomass (Schneider and Kay, 1994). The total system throughput fell 21 percent in the stressed creek, and the number of cycles dropped by 51 percent. Again, the diminished energy flows led to characteristics of ecosystems in earlier developmental stages. Stress leading to decreased functional organization robs thermodynamic systems of the energy they need to maintain their complexity. They sicken, forget, and disintegrate. Energy, which William Blake rhapsodized is eternal delight, is behind the cyclic phenomena of beautiful diversity, ornate structure, and long memory we associate with healthy organisms and rich ecosystems.

Remote Sensing of Ecosystems

There is experimental and empirical proof not only that life is a gradient-reducing system, but also that the most complex and impressive ecosystems are the most effective at reducing the solar gradient. Only about 2 percent of the incoming solar energy is converted to biomass. Upon hitting Earth, solar energy is parsed into lower-grade energy driving a variety of processes. Some is reflected, some drives weather systems and runs the ocean currents. Some goes into evaporation and evapotranspiration, the process by which water is brought from roots to the stomata of leaves and evaporated. Incoming solar radiation is rich in high-quality, high-frequency quanta such as ultraviolet light, and has a greater exergy (usable energy content) than low-frequency quanta such as infrared radiation (heat). Solar energy that hits Earth and is reflected back into space is sensible molecular chaotic heat. Sensible heat is the sort of heat detected from a shopping mall, hot blacktop, or parking lot: the high-quality ultraviolet quanta convert to lower-grade vibrating molecular heat by bouncing off a dark surface. Another portion of the incoming solar energy is converted to latent heat—low-grade heat stored in atmospheric moisture. Evaporating water carries latent heat in the atmosphere, as does evapotranspiration. The evaporation of a gram of water requires 580 calories. The 580 calories per gram needed to evaporate water into the atmosphere is released when it rains or snows.

The radiative energy budgets of ecosystems should allow us to develop a test for our hypothesis that mature ecosystems degrade more energy than less mature systems. The sun, Earth, and surrounding space form a single thermodynamic system. The sun's energy hits the Earth, and some of that energy is radiated back into space. My general statement that "Nature abhors a gradient" in this case refers to the gradient between the warm Earth and the frigid temperatures of outer space. Thus the cooler our planet's surface, the less the gradient between Earth and the 2.7°K temperature of outer space. If the most effective degraders of energy are indeed the most mature successional ecosystems, then the energy content of incoming radiation should be sapped most by the most effective degrading system.

If this hypothesis pans out, one would expect mature ecosystems such as rain forests to have colder surface temperatures and lower reradiated temperatures than underdeveloped ecosystems such as, say, the steppes of Asia. One can use satellite data to look at surface temperatures on a global basis. Certain National Oceanic and Atmospheric Administration weather satellites routinely collect data on the outgoing clear sky longwave radiation from Earth's surface, which is a measure of the temperature of the surface of the Earth. The results match the theory. The surface over the rain forests is very cool, the steppes of Asia are warm, and deserts are hot—the temperature difference, the solar gradient, between the warm Earth and cold space seemed to be minimized by the mature rain forest ecosystems. But the temperature measured by satellites was the surface temperature of clouds over the rain forests rather than of the rain forests themselves. Once one realizes, however, that the tops of rain forests are represented by clouds produced via evapotranspiration by the trees beneath them, one can get a better appreciation of what is really going on. The overall rain forest with its reflective cloud cover is the true cooling system. The rain clouds must be regarded as part of the thermodynamic ecosystem. Once they are, the incoming gradient is degraded best by most diverse, late successional systems, as predicted.

By studying reradiated temperatures from beneath clouds, close to the ecosystem surface, one can identify specific ecosystems at their own temperatures. Field ecologists Jeffrey Luvall and H. R. Holbo of the NASA Marshall Space Flight Center in Huntsville, Alabama, helped develop a sophisticated device, a black body thermometer, carried on the underbelly of an airplane on flights below clouds (Luvall and Holbo, 1991). This permitted an excellent approximation of the incoming energy. The most interesting and tantalizing data were produced on short, low-level flights at noon on cloudless days. These allowed variations in ecosystem temperature at scales of 25 meters to be seen. The goal here was to compare the temperatures of ecosystems with a minimum of skewed variables such as varying rain moisture, soil type, and wind speed that might affect local temperatures. The data were unambiguous. More mature ecosystems degrade the incoming energy better than roads, clear-cuts, and young stands of trees.

Luvall also flew over a quarry, a clear-cut Douglas fir plantation with natural regrowth, and a 400-year-old Douglas fir forest (Luvall and Holbo, 1989). The quarry and clear-cut plantation degraded only 62 and 65 percent of the incoming energy, respectively, whereas the 400-year-old Douglas fir forest degraded 90 percent of the incoming solar radiation into lower-grade, latent heat. It thus seems incontrovertible that late succession, more diverse, and complex ecosystems are best at reducing the difference between the hot sun and cold space. Ecosystems seem to do this by generating coolness around them.

Evolution

As *The Ecological Theater and the Evolutionary Play*, the title of the book by the great Yale ecologist G. Evelyn Hutchinson, suggests, evolution and ecology are deeply related (Hutchinson, 1965). "Evolution," writes the Spanish ecologist Ramón Margalef, "cannot be understood except in the frame of ecosystems. By the natural process of succession, which is inherent in every ecosystem, the evolution of species is pushed —or sucked—in the direction taken by succession, in what has been called increasing maturity . . . evolution should conform to the same trend manifest in succession. Succession is in progress everywhere and evolution follows, encased in succession's frame" (Margalef, 1968, p. 81). Successional ecosystems go from a few fast-growing settler species to interwoven networks of many more species, including slower-growing ones. The climax ecosystem is larger, more complex, and more effective as a unit in degrading available energy. In our view, thermodynamics not only provides the underlying logic of ecosystem development, but informs evolutionary process as well. The biosphere, which is submitted to the same thermodynamic forces that organize local ecosystems, should, and does, show similar patterns of development.

Concern with the environment in the wake of humanity's innovative but disruptive spread points toward the future development of an energy-efficient climax global ecosystem in which humans are neither so isolated nor reproducing at rates incommensurate with the rest of the biota. Like our food monocrops, we are a potential gradient for viruses and microbes, as is made clear by increasing rates of sexually transmitted diseases and antibiotic-resistant strains of bacteria. The thermodynamic logic of ecosystems works to reduce bloated gradients, thereby producing climax biospheres with huge biogeochemical cycles and great amounts of biodiversity, species integration, and complexity. Indeed, one of Kauffman's generalizations—that biospheres maximize the average construction of the diversity of autonomous agents, and the ways those agents can make a living to propagate further—although not an example of the new fourth law he wants to apply to open systems, does contain some insight. Our biosphere and any conceivable biosphere will develop complexity cyclically while degrading energy in its cosmic milieu. Humanity's rapid spread via agriculture and technology has produced the entropic wastes familiar as pollution, wastes whose recycling demands more effective and biodiverse connections between our populations and those of our food plants and animals, fertilizing insects and birds, and especially recycling and soil-producing bacteria and fungi. Encapsulation, miniaturization, detoxification, and symbiotic integration are all ancient evolutionary themes being played out now via human technology in the global ecosystem. One can imagine technologically enabled recycling systems of humans and other organisms, protected by enclosures, in orbit, on Mars or other planets or moons, or even on Earth to protect against global pollution. Such systems would represent the reproduction of the global ecosystem. Even so, human engineering and design, themselves the result of genetic and thermodynamic processes, may still not have caught up with such things as total recycling, parallel processing computing, and nanotechnological construction—the original cyclical, energy-based high technology of life.

These views on natural selection owe much to Lotka and Wicken. Lotka held that natural selection works to increase both the mass of organic systems and the rates of circulation of matter through those systems. The capture, storage, and degrading of available energy is not like a drunk man's stagger across a sidewalk: it has a direction. Natural selection tends to bring the energy flux through a system to its optimum. Living systems strive—and we do not use this word lightly—to catch, store, and degrade gradients. These patterns, which mark ecosystems, apply to evolution, too. The direction in both ecosystems and evolution is toward increasing complexity (for example, in the number of species, toward more developed networks, toward increasing differentiation, increasing functional integration of thermodynamic flows, and the increasing abilities of organisms to adjust to changing resources). The development and evolution of Gaia is a thermodynamic as well as a genetic process that moves in a given direction but whose future outcomes are uncertain.

Acknowledgments

I would like to thank Dorion Sagan and Bob Ulanowicz, who provided help in the preparation of this manuscript.

References

Bergson, H. 1911. *Creative Evolution*, translated by A. Miller. New York: Henry Holt.

Boltzmann, L. 1886. The second law of thermodynamics. In B. McGinness, ed., *Ludwig Boltzmann: Theoretical Physics and Philosophical Problems*. New York: D. Reidel, 1974.

Fermi, E. 1956 [1937]. *Thermodynamics*. London: Dover Publications.

Glansdorff, P., and I. Prigogine. 1971. *Thermodynamic Theory of Structure, Stability, and Fluctuations*. New York: Wiley-Interscience.

Hatsopoulos, G., and J. H. Keenen. 1965. *Principles of General Thermodynamics*. New York: John Wiley.

Hutchinson, G. E. 1965. *The Ecological Theater and the Evolutionary Play*. New Haven, CT: Yale University Press.

Kauffman, S. 2000. *Investigations*. Oxford: Oxford University Press.

Kestin, J. A. 1979 [1966]. *A Course in Thermodynamics*, 7th ed. New York: Hemisphere.

Likens, G. E., et al. 1970. Effects of forest cutting and herbicide treatment on nutrient budgets in the Hubbard Brook Watershed-Ecosystem. *Ecological Monographs*, 40, 23–37.

Lindeman, R. L. 1942. The trophic-dynamic aspect of ecology. *Ecology*, 23, 399–418.

Lotka, A. 1922. Contribution to the energetics of evolution. *Proceedings of the National Academy of Sciences, USA*, 8, 148–154.

Luvall, J. C., and H. R. Holbo. 1989. Measurements of short-term thermal responses of coniferous forest canopies using thermal scanner data. *Remote Sensing of the Environment*, 27, 1–10.

Luvall, J. C., and H. R. Holbo. 1991. Thermal remote sensing methods in landscape ecology. In M. Turner, and R. H. Gardner, eds., *Quantitative Methods in Landscape Ecology*. New York: Springer-Verlag.

Margalef, R. 1968. *Perspectives in Ecological Theory*. Chicago: University of Chicago Press.

Morowitz, H. J. 1968. *Energy Flow in Biology*. New York: Academic Press.

Odum, E. P. 1969. The strategy of ecosystem development. *Science*, 164, 262–270.

Odum, H. T., and R. C. Pinkerton. 1955. Times speed regulator. *American Scientist*, 43, 321–343.

Onsager, L. 1931. Reciprocal relationships relations in irreversible relations. 1. *Physical Review*, 37, 405–426.

Prigogine, I. 1955. *Thermodynamics of Irreversible Processes*. New York: John Wiley.

Schneider, E. D., and J. J. Kay. 1994. Life as a manifestation of the second law of thermodynamics. *Mathematical and Computer Modeling*, 19(6–8), 25–48.

Schrödinger, E. 1944. *What Is Life?* London: Cambridge University Press.

Ulanowicz, R. E. 1986. *Growth and Development: Ecosystem Phenomenology*. New York: Springer-Verlag.

Wicken, J. S. 1987. *Evolution, Thermodynamics, and Information: Extending the Darwinian Program*. Oxford: Oxford University Press.

Woodwell, G. M. 1970. The energy cycle of the biosphere. In G. Piel, D. Lanagan, F. Bello, P. Morrison, J. Piel, J. Purcell, J. T. Rogers, A. Schwab, C. Strong, and J. Wisnovsky, eds. *The Biosphere*. San Francisco: W. H. Freeman.

Zhabotinski, A. M. 1974. *Self-oscillating Concentrations*. Moscow: Nauka.

5

Gaia, Extended Organisms, and Emergent Homeostasis

J. Scott Turner

Abstract

Gaia's most remarkable prediction is a biosphere-level physiology and all the organismal traits that implies, including global homeostasis. Its most formidable challenge is to explain how such properties can emerge from the welter of competing genetic interests which the biosphere comprises. This chapter explores the problem of "emergent homeostasis" in a model experimental system, the colonies of fungus-growing termites, in which a homeostasis of colony atmosphere emerges from a symbiotic assemblage between two heterotrophs, termites and fungi. The termite-fungus symbiosis is a coalition of genetically diverse organisms from which homeostasis emerges, driven by the symbionts' common physiological interests rather than their common genetic interests. This homeostatic system is therefore distinct in origin from the more common social homeostasis found among bees, ants, and wasps, in which the common genetic interest of the colony drives its evolution. I suggest that mechanisms of emergent homeostasis, in which common physiological interests are paramount, are a more appropriate model for understanding how Gaia's radical vision of an Earth physiology might work.

Introduction

Gaia is a goddess with two faces. On the one side there is, to paraphrase Daniel C. Dennett's (1995) description of natural selection, "Lovelock's dangerous idea," the radical proposition that the biosphere is, in some sense, alive, and not simply the harbor of life (Lovelock, 1987). In this conception, Gaia is an organism, with all the attributes that designation implies: self-sustenance, adaptability, homeostasis, even perhaps a sort of intentionality. Then there is Gaia's more prosaic face, unveiled as Gaia has evolved in recent years into Earth systems science (Schneider and Boston, 1991; Schneider, this volume). In this conception, Gaia is a research program concerned with the roles living organisms play in managing flows of matter and energy through the biosphere, and with identifying the feedbacks and potential control points that could give the Earth a semblance of life, if not life itself.

I am a physiologist who came fairly late to the concept of Gaia. I was initially attracted by Gaia's dangerous face, but not, I hasten to say, from any love of danger on my part. Rather, I was intrigued by the challenge "Lovelock's dangerous idea" presents to our understanding of physiology's central principle, homeostasis. To the physiologist, homeostasis is a phenomenon of organisms in the here-and-now, and its study is concerned primarily with mechanism: with controls and effectors, how they work and respond to various challenges. Yet homeostasis, like all other attributes of organisms, had to evolve somehow. Commonly, physiologists approach evolution as taxonomists do, taking what nature has provided us in the present—fully functioning organisms "engineered" by natural selection—and working backward through a process of "reverse engineering" to reconstruct how the organismal contrivances before them could have evolved. Rarely do physiologists ask the more fundamental question: How did homeostasis itself evolve? By and large, homeostasis is assumed simply to be axiomatic to the organismal condition: ultimate questions, such as how homeostasis itself might have evolved, are a troublesome complication. Gaia's radical appeal is that it no longer lets us safely ignore that ultimate question.

For the most part, our thinking on the question "How did homeostasis evolve?" has been shaped by the neo-Darwinist conception of homeostasis as a form of altruism among an organism's cells. Homeostasis requires an investment in physiological "machines" (organs and organ systems) which drive the flows of matter and energy through the organism, and in the control systems that manage those machines' operations. These machines arise through differentiation of the zygote's various cell lineages of the organism's somatic line. The somatic cells will never themselves reproduce: they are sacrificing their genetic interests to ensure the reproduction of the lucky few cells in the germ line. The sacrifice is redeemed for all by the assurance that organisms with well regulated internal environments—"good" homeostasis—will be more fecund than organisms with poorly regulated internal environments.

Gaia undercuts this tidy explanation because it posits the emergence of homeostasis in the absence of the common genetic interest that supposedly drives it in organisms. Now the question "How did homeostasis evolve?" becomes more problematic. Why should one organism with its own genetic interests make an investment in homeostasis so that another organism, with disparate genetic interests, might also benefit? This question has presented a major stumbling block for Gaia's winning even the grudging acceptance it now enjoys (Dawkins, 1982; Doolittle, 1981; Joseph, 1990; Kirchner, 1991; Williams, 1992). It is fair to say that acceptance has been won largely at the price of abandoning Gaia's most radical and (to me) appealing idea: the notion of the biosphere as a global organism.

My intent in this chapter is to argue that it is too soon to turn away from "Lovelock's dangerous idea." Gaia's proposition of a global homeostasis is in fact one of a large body of similar problems in what might be called emergent homeostasis. Emergent homeostasis asserts that the origin and evolution of homeostasis is driven by a sort of "physiological altruism," that is, pursuit of a common physiological interest by a genetically diverse assemblage of organisms. To the extent that physiological and genetic interests are congruent, as they are in organisms, homeostasis fits comfortably into conventional neo-Darwinism. However, common physiological interests can emerge independently of genetic interests, as is the case for most symbioses. And from these genetically disparate coalitions, homeostasis often emerges (Paracer and Ahmadjian, 2000). In these instances, the neo-Darwinist explanation of homeostasis is less robust (Margulis, 1997). The challenge is to explain how homeostasis nevertheless emerges.

Understanding emergent homeostasis would be aided by a model system, a sort of Gaia-in-microcosm, which could be studied experimentally. Below, I will outline such a model system: the colonial respiration of fungus-growing termites of the Macrotermitinae (subfamily of the Termitidae; Ruelle, 1970). The macrotermitines are marked by a sophisticated digestive symbiosis between termites, cellulolytic bacteria, and at least two types of cellulolytic fungi, which reaches its pinnacle in the mound-building genera *Macrotermes* and *Odontotermes*. Emerging from this symbiosis is a remarkable degree of homeostasis of the nest climate (Lüscher, 1961; Ruelle, 1964). The macrotermitines also are pivotal controllers of the physiology of the savanna ecosystems they inhabit (Dangerfield et al., 1998), and this ability is, in large measure, a consequence of the regulated environment in the termite nest. We have recently come to better understand this homeostasis and how it arises (Turner, 1994, 2000a, 2001; Korb and Linsenmair 1998a, 1998b, 2000). These findings may illuminate how homeostasis might emerge at any organizational level, ranging from cells and endosymbionts, to the colonies of termites, to the purported biosphere-level homeostasis posited by Gaia.

The Symbiosis

Termites are well known for their digestive symbioses with microorganisms (Wilson, 1971; Breznak, 1984; Wood, 1988; Wood and Thomas, 1989). The rationale for these associations is straightforward. The principal component of the termite diet is lignified cellulose, which termites have only limited ability to digest. Termites thus use microorganisms that can digest cellulose, commonly bacteria and protists composing an abundant gut flora. These intestinal symbionts express a rich array of cellulases and other enzymes that readily degrade cellulose to more easily digestible cellobiose, xylose, and oligosaccharides (Martin, 1987).

For most termites, the association with their digestive symbionts is obligatory and confined to the intestine. Termites sterilized of their gut flora macerate and pass cellulose through their intestines normally, but cannot extract useful energy from it, and thus starve to death. The gut flora are maintained by repeated inoculation of individuals' digestive tracts by other members of the colony, both upon hatching and following each molt. The macrotermitines have taken the digestive symbiosis a step further, however, "outsourcing" cellulose digestion to extracorporeal fungi (Martin, 1987; Batra and Batra, 1979). The fungi are cultivated on structures called fungus combs, built by the termites and maintained by them within the nest. This association probably arose from opportunistic invasions by soil fungi into termites' cached supplies of surplus food. This was gradually refined into the close symbiosis between the macrotermitine genera *Macrotermes* and *Odontotermes* and basidiomycete fungi of the genus *Termitomyces* (Mora and Rouland,

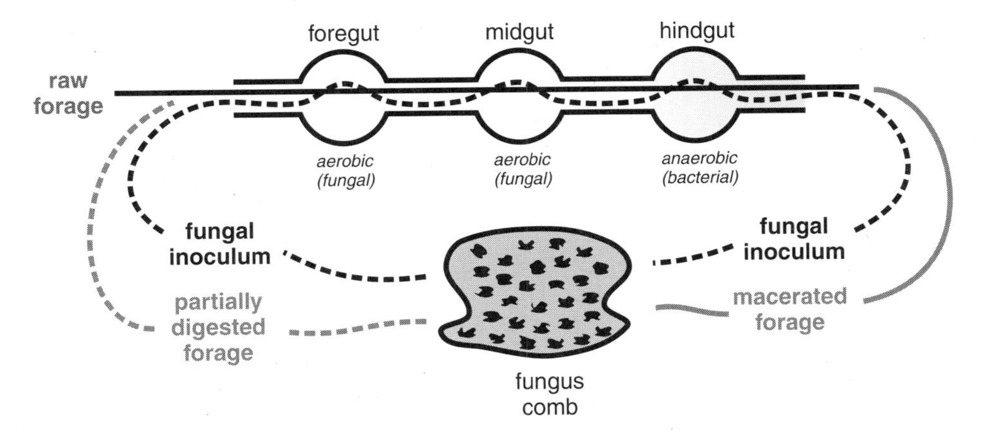

Figure 5.1
Scheme of the digestive symbiosis between *Macrotermes* and *Termitomyces*. Details in text

1995; Rouland et al., 1988, 1991; Thomas, 1987a, 1987b, 1987c).

The fungus combs are sites for conversion of low-quality lignified cellulose to a high-quality food of simpler sugars (Rouland et al., 1988a, 1991; Veivers et al., 1991). Foraging termites ingest food—mostly grass, but also bark, dead wood, and undigested material in fecal pats, and transfer it to minor workers and nurse workers upon return to the nest (figure 5.1). These pass the raw forage rapidly through the gut, which contains the usual culture of bacterial symbionts, but also is replete with fungal spores. These are mixed with the slurry of macerated raw forage as it passes through the gut. When defecated, the inoculated slurry is daubed by the workers onto the top of the fungus comb. The *Termitomyces* spores then germinate, and infiltrate their hyphae through the comb, digesting the raw forage (spores of other fungi remain dormant). Simultaneously, termites consume digested material from the bottom of the comb, and pass it again through the intestine, where it is digested by the termites' normal intestinal symbionts and by fungal enzymes which remain active in the intestine (Leuthold et al., 1989). Thus, the fungus comb is a sort of flow-through composter, with new material added continually to the top, and digested material continually being consumed from the bottom.

Energetics of Digestion Among the Macrotermitines

Most termites rely solely on intestinal digestion, which imposes upon them constraints in the extraction rate of energy from food. A termite intestine is analogous to a plug-flow digestive reactor, with the absorption rate optimized only at a certain feeding rate (Penry and Jumars, 1986, 1987). If the feeding rate of individual termites is already at this optimum, the colony can increase its energy intake rate only by increasing the number of workers in the colony—the number of individual plug-flow reactors. However, each worker itself represents an energy investment in development and maintenance, which increases exponentially by about the 0.4 power for each increment of termite biomass (Peakin and Josens, 1978). This must be repaid before the colony can accrue a net energy profit. These and other constraints conspire to limit both the body size and the colony population of termites that rely solely on intestinal digestion.

The macrotermitines have escaped this constraint by "outsourcing" cellulose digestion to the extracorporeal fungus combs, which are analogous to continuous-flow stirred tank reactors (Penry and Jumars, 1986, 1987). Expanding the colony's digestive capacity now involves the relatively cheap expansion of biomass in fungus combs. Predigestion by the fungi also enriches the diet, improving the digestive capacity of each worker termite. As a consequence, macrotermitine colonies are metabolically very active, consuming enormous quantities of food and releasing enormous amounts of energy for physiological work (Rohrmann, 1977; Rohrmann and Rossman, 1980; Darlington et al., 1997; Peakin and Josens, 1978). This has remarkable consequences for all aspects of these termites' biology. Macrotermitine bodies are typically two to three times larger than other termites' (Coaton and Sheasby, 1972), and their colony populations are, on average, 1 to 2 million workers, roughly an order of magnitude larger than the typical colony populations of species that rely solely on intestinal digestion (Wiegert, 1970; Darlington, 1990,

Table 5.1
Estimates of whole colony metabolic rates (in watts) of three nests of *Macrotermes jeanneli*

	Nest 1		Nest 2		Nest 3	
	MR (W)	M (kg)	MR (W)	M (kg)	MR (W)	M (kg)
Method A						
Worker total	75.61	61.25	49.54	34.31	24.93	13.39
Fungus combs	104.29	95.14	92.86	81.15	48.45	33.29
Nest total	179.90	200.79	142.40	145.76	73.38	58.78
Method B(1)						
Worker total	85.33	72.28	51.07	35.78	7.17	2.43
Fungus combs	122.50	118.61	72.45	57.76	8.34	2.99
Nest total	207.83	244.68	123.52	119.97	15.51	6.99
Method B(2)						
Worker total	73.33	58.73	58.51	43.11	23.90	12.65
Fungus combs	104.97	95.99	83.31	69.94	32.36	19.15
Nest total	178.30	198.35	141.82	144.96	56.26	40.85
Method C						
Nest total	227.95	277.70	373.51	551.47	75.23	61.06

Note: Method A estimates colony metabolic rate from biomass determinations in combination with estimates of mass-specific metabolic rates of each of the components. Methods B estimate colony metabolic rates from measurements of diameter of outflow air tunnels. Method C estimates colony metabolic rate from enrichment of carbon dioxide and volume flow rate of the outflow stream from the exhaust tunnels of intact nests. The latter includes the stimulation of soil respiration as well as respiration of the termites and fungi. Conversion of carbon dioxide enrichment to metabolic rate in watts was made assuming a respiratory quotient of 0.8. Total nest masses include workers, fungus combs, a lates, soldiers, royal couple, and eggs. After Darlington et al. (1997).

Table 5.2
Some estimates of the metabolic impact of macrotermitine termites on tropical ecosystems. After Dangerfield et al. (1997)

Measure	Location
Account for 40–65% of soil macrofauna biomass	African savanna
Standing biomass of 70–110 kg ha^{-1}	African savanna
Ungulate biomass of 10–80 kg ha^{-1}	African savanna
Annual turnover of termite biomass of 120 kg ha^{-1}	African savanna
Consume 1.0–1.5 t litter ha^{-1} y^{-1}	African savanna
Consume 23% of annual litter production	Nigerian forest

1994; Darlington et al., 1992). These factors combine to give a *Macrotermes* colony a high collective metabolic rate, similar to that of large ungulate herbivores (table 5.1). Finally, the presence of a colony stimulates soil respiration around it, which elevates the energy consumption rate of the "extended colony" further (Darlington et al., 1997). This metabolic effervescence makes these termites dominant components of the savanna ecosystems they inhabit (Dangerfield et al., 1998; tables 5.2 and 5.3). Their biomass exceeds that of all the other soil invertebrates combined; carbon flow through them is similar to that for all the ungulate herbivores combined (table 5.2); and they perturb soil at rates exceeding that of other mound-building termites in the environment (table 5.3).

The Nest and Its Infrastructure

The macrotermitines' most striking attribute is the nests and associated structures they build, most obviously the large aboveground mounds that are prominent features of the savannas they inhabit (Harris, 1956; Ruelle, 1964; figure 5.2). The mound is not the habitation for the colony. The queen, workers, fungus combs, and young reside in an underground nest, and are confined to a dense mass roughly 1.5 to 2 meters in diameter (figure 5.3). This, combined with a typical nest's high metabolic rate, results in the colony having a high metabolic power density (W m^{-3}). This substantially alters the concentrations of respiratory gases in the nest atmosphere. Carbon dioxide mole fractions commonly range from 0.5 percent to 1 percent, with concomitant reductions in oxygen concentrations (Korb and Linsenmair, 2000; Turner, 2001; Darlington et al., 1997). The nest atmosphere is also

Table 5.3
Rates of soil movement and standing crop of soil in mounds of various Ugandan termites

Site	Species	Soil standing crop (m^3 ha^{-1})	Upward soil transport rate (m^3 ha^{-1})	Downward soil transport rate (m^3 ha^{-1})
Natete (well-drained ridge)	*Macrotermes bellicosus* (Macrotermitinae)	8.08	0.90	1.15
	Pseudacanthotermes spp (Macrotermitinae)	0.28	0.10	0.12
Naluvule (poorly drained bottomland)	*Macrotermes bellicosus* (Macrotermitinae)	4.22	0.41	0.39
	Pseudacanthotermes spp (Macrotermitinae)	1.88	0.60	1.00
	Cubitermes spp (Termitinae)	0.02	0.02	0.00

Source: After Pomeroy (1976).

Figure 5.2
A mound of *Macrotermes michaelseni* in northern Namibia

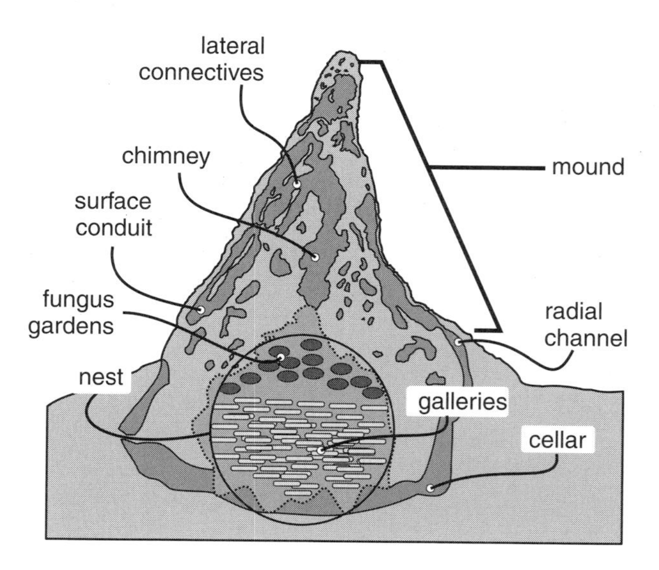

Figure 5.3
Cross section of a colony of *Macrotermes michaelseni*, showing locations of nest, galleries, fungus gardens, mound, and associated air passageways. From Turner (2000a).

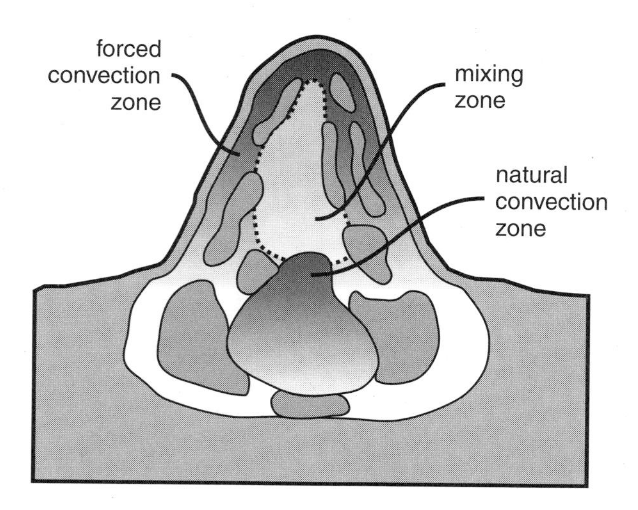

Figure 5.4
Postulated zones of gas exchange in a *Macrotermes michaelseni* mound and nest. Details in text. After Turner (2001).

rich in the gaseous products of the anaerobic and methanogenic bacteria that reside in termites' hindguts, and the volatile acids, alcohols, and other hydrocarbons produced by the fungi (Darlington et al., 1997). The nest atmosphere is also very humid, the water vapor supplied by a high production rate of metabolic water supplemented by soil water (Darlington et al., 1997; Weir, 1973; Turner, 2001).

The Mound and Nest Ventilation

The mound is the nest's physiological infrastructure. The colony's high metabolic power density requires that the nest be ventilated: if it were not, the colony would suffocate. The mound serves this function by capturing kinetic energy in winds. The ebb and flow of turbulent winds powers a tidal ventilation of the mound's peripheral air spaces, similar to the in-and-out movements of air during the respiratory cycle of the mammalian lung (Turner, 2000a). A similar pattern of ventilation has been observed in nests of the leaf-cutter ant *Atta vollenweideri* (Kleineidam and Roces, 2000; Kleineidam et al., 2001), and in nests of another macrotermitine, *Odontotermes transvaalensis* (Turner, 1994). The mound's elaborate tunnel network (figure 5.3) integrates wind-induced ventilation in the peripheral air spaces with metabolism-induced buoyant forces that loft spent air from the nest into the mound's chimney. The colony's respiratory gas exchange is therefore analogous to the three-phase gas exchange in the mammalian alveolus (figure 5.4): a forced convection phase in the bronchi and bronchioles, a diffusion phase in the alveoli, and a mixed convection-diffusion phase in the alveolar ducts and lower bronchioles. In the *Macrotermes* mound, the convection phase, driven by wind, is located primarily in the surface conduits and lateral connectives close to the surface. A natural convection phase, analogous to the diffusion phase in the alveolus, occurs in the nest and lower chimney, driven by relatively weak buoyant forces induced by colony metabolism. The upper chimney and inner parts of the lateral connectives form a mixed-regime phase, where both forced convection and natural convection are roughly equivalent (Turner, 2001).

Homeostasis in *Macrotermes* Colonies

Many social insect colonies exhibit social homeostasis, in which the collective activities of the inhabitants are coordinated to regulate the colony environment. For example, honeybee colonies regulate the temperature of the hive in ways that individual bees cannot (Southwick, 1983). As temperatures fall, for example, the hive's inhabitants cluster into a compact ball, conserving heat. Part of this involves behavioral differentiation among the members of the colony: certain bees in the cluster form an insulating layer, certain others generate the bulk of the heat to warm the cluster, and so forth. Social homeostasis is manifested in other ways, including matching of foraging rates to energy stores in the colony, control of reproduction, and so forth.

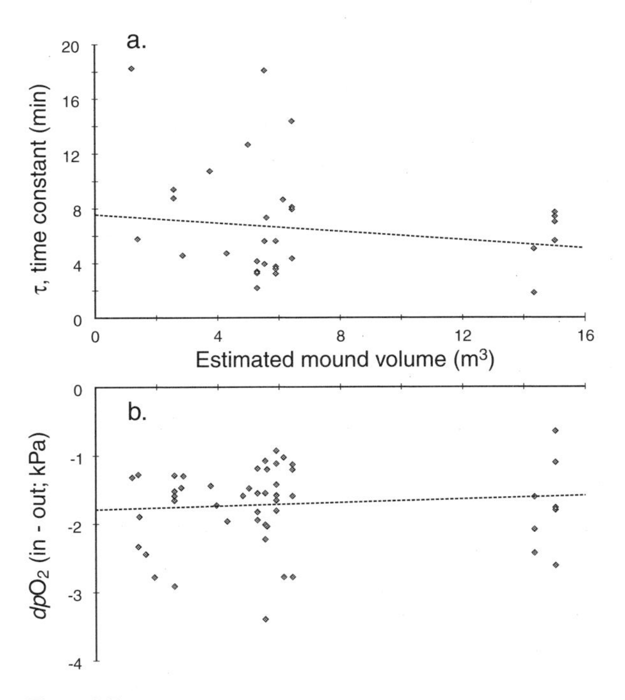

Figure 5.5
Homeostasis of nest atmosphere in mounds of *Macrotermes michaelseni*. (a) Time constant for clearance of tracer gas from *M. michaelseni* nests of various sizes. The time constant is the inverse of the rate constant for exponential clearance, and thus is inverse to the ventilation rate. Roughly 95 percent of the air in a space is replaced within a period of roughly three time constants. The time constant does not vary significantly with mound size. (b) Partial pressure differences for oxygen (dpO_2) between the nest atmosphere and external atmosphere. Despite large variation of colony metabolic rate with size, the depletion of oxygen within the mound is invariant with respect to colony size. Dotted lines indicate least-squares regressions for both panels.

Homeostasis in the *Macrotermes* colony appears directed largely to regulating the composition of the nest atmosphere (Turner, 2001). There are several lines of evidence for this. For example, concentration of oxygen in *Macrotermes michaelseni* nests' atmospheres do not vary appreciably with colony size, despite the large variation of colony metabolic rate that entails (Darlington et al., 1997). This constancy can occur only if ventilation keeps pace with the colony's growing demands for respiratory gas exchange (figure 5.5). Furthermore, metabolically active parts of the colony are ventilated more vigorously than relatively quiescent parts (figure 5.6). Finally, the nest atmosphere is steadier in composition and less susceptible to environmental perturbations during the active summer season than during the winter, when the termites are relatively inactive (Turner, 2001).

Homeostasis of the *Macrotermes* nest atmosphere arises from adaptive modification of mound architecture, which matches the capture of wind energy for ventilation to the rate of respiratory gas exchange in the nest. At its simplest, the matching is brought about through adjustment of mound height (Turner, 2001). The surface boundary layer presents a gradient in wind velocity. Consequently, the higher the mound, the more energetic the winds that the mound intercepts. If a colony's metabolic rate increases, as it might as the colony matures, demand rate for respiratory gases also grows. To regulate the composition of the nest atmosphere, the ventilation rate must increase commensurably. This is readily accomplished by building the mound upward through the surface

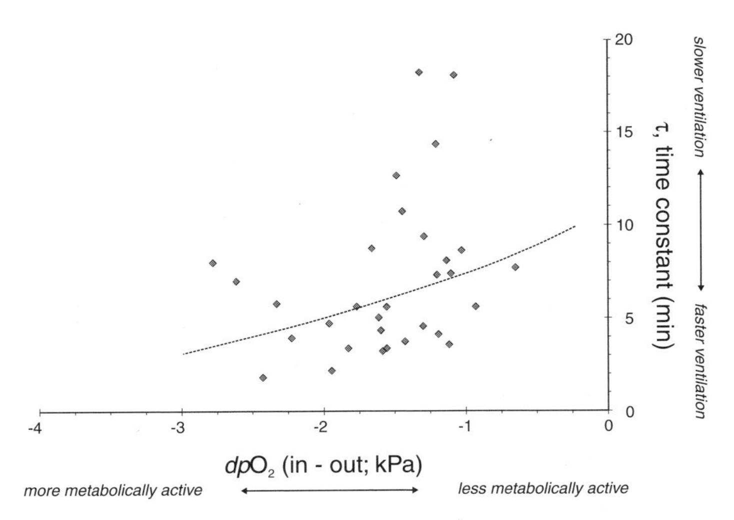

Figure 5.6
Relationship between time constant for clearance and oxygen partial pressure difference for mounds of *Macrotermes michaelseni*. Metabolically active areas (indicated by substantial oxygen depletion) are more intensely ventilated (indicated by short time constants) than metabolically more inert areas (indicated by little oxygen depletion). Dotted line indicates least-squares polynomial regression.

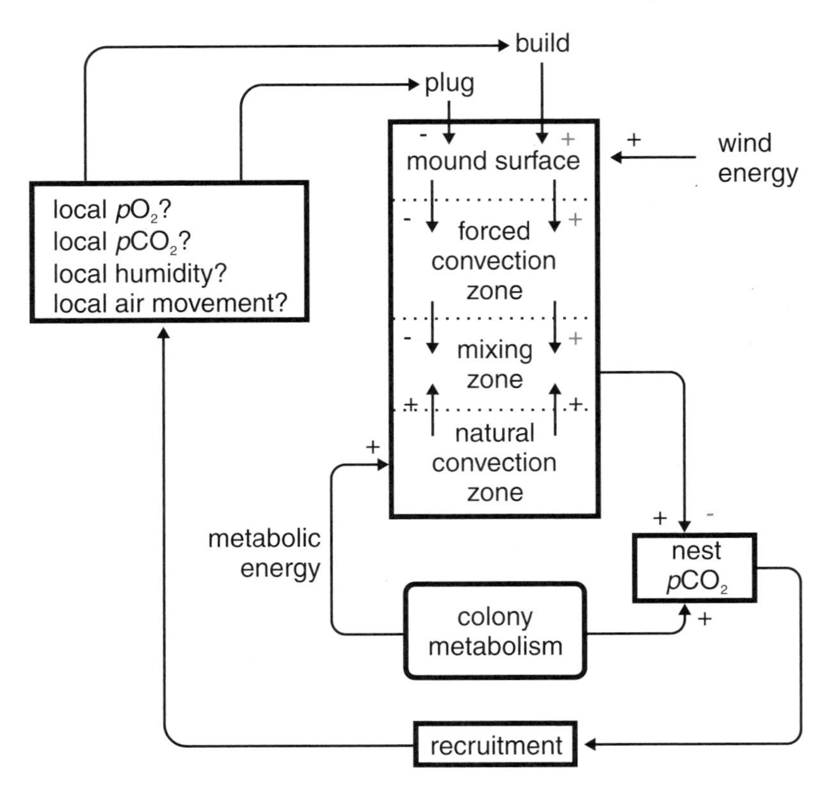

Figure 5.7
Simple operational scheme of homeostasis of nest atmosphere in *Macrotermes michaelseni*. For example, if concentration of carbon dioxide in the nest is too high, the ventilatory flux is too low to meet the colony's respiratory flux. This recruits workers from the nest, which build the mound upward, where it encounters more energetic winds, powering a more vigorous ventilation (gray symbols). Once ventilation flux and respiratory flux are brought back into balance (indicated by a return of nest carbon dioxide concentration to normal), worker recruitment ceases, as does the further upward extension of the mound. After Turner (2001).

boundary layer until it encounters winds sufficiently energetic to match ventilation rate with respiration rate (Turner 2000a, 2001).

The mound can be an effector for the regulation of the nest atmosphere because the mound's architecture is adjustable, and because mound morphogenesis is coupled to the nest's physiological state. Termites in the nest monitor local concentrations of CO_2, oxygen, and water vapor, and any disturbance of these properties indicates a mismatch between respiration and ventilation. When a disturbance is sensed, worker termites are recruited from the nest, where they normally reside, to the mound surface, where they build new surface, excavate vent holes, or seal porous layers of soil (figure 5.7; Turner, 2000a). Consequently, the termites' building activity alters the mound's capture of wind energy, which feeds back onto the stimulus (the perturbation in nest atmosphere) that initiates the building in the first place.

More generally, homeostasis in the *Macrotermes* nest arises through coupling of mechanisms of mound morphogenesis to large-scale gradients of potential energy in the mound (Turner, 2000a). Worker termites can be thought of as conveyors of soil along metabolism-induced gradients in concentration of respiratory gases. If the nest produces carbon dioxide at a certain rate, for example, this will establish gradients of pCO_2 within the mound that radiate away from the nest. How steep these gradients will be is determined both by the nest's metabolic rate and by the resistance to gas flux through the mound's air spaces. If worker termites convey soil down that pCO_2 gradient, the mound surface will grow outward and upward. This growth, in turn, alters the distribution of pCO_2 within the mound. Thus, the mound is both cause and effect of the gradients of carbon dioxide within itself. Homeostasis of the nest atmosphere follows when the intensity of soil transport is "tuned" properly to the changes of CO_2 distribution that result. Because the "tuning" involves adjusting the likelihood that individual termites will pick up or deposit grains of sand in response to a stimulus

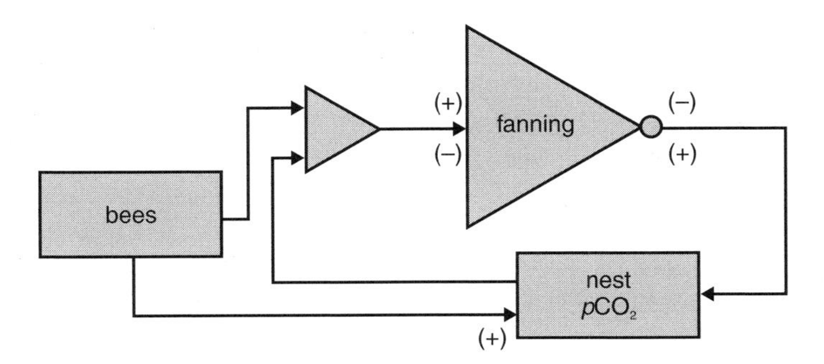

Figure 5.8
Social homeostasis of nest pCO_2 in a honeybee colony. Details in text. After Seeley (1974).

(say pCO_2), emergent homeostasis can arise through straightforward natural selection on variations in these likelihoods.

Why Homeostasis of the Nest Environment?

For homeostasis to be selected for, some selective benefit must result from more precise regulation of the nest atmosphere. The macrotermitines have the most sophisticated mechanisms for nest homeostasis of all the termites. However, what purpose the nest homeostasis serves is still an open question. Commonly, homeostasis was thought to be directed to regulation of nest temperature, which presumably allowed the macrotermitines to extend their ranges into thermal environments that were hostile to termites with less well-regulated nests. This is now doubtful, because manipulation of ventilation does not appreciably alter temperatures of underground nests of social insects (Turner, 1994; Kleineidam and Roces, 2000). More likely, the homeostasis arises to reinforce the symbiosis between *Macrotermes* and *Termitomyces*.

The fungal partner in the symbiosis seems to benefit the most from nest homeostasis. *Termitomyces* is in competition with another common soil fungus, *Xylaria*, for the rich trove of cellulose transported to the nest by the termites (Batra and Batra, 1979). *Xylaria* is fast-growing, presumably because its uptake of cellulose digestate is very rapid. In contrast, *Termitomyces* is slower-growing, perhaps because its uptake of cellulose digestate is slower. This may explain why *Termitomyces* rather than *Xylaria* is the favored fungal symbiont: its slower uptake of digestate leaves more for consumption by the termites. The exclusion of *Xylaria* appears to be through suppression of its germination and growth in the nest. Spores of both species of fungus are abundant in the nest soils, termites' intestines, and the fungus combs (Thomas, 1987a, 1987b, 1987c), yet only *Termitomyces* grows: germination and growth of all other fungal species is suppressed. Apparently, some aspect of the nest environment is responsible. Removing the comb from the nest invariably results in *Xylaria* spores germinating and aggressively taking over the comb, even if termites are given full access to it. Only within the nest is *Xylaria* germination and growth suppressed while *Termitomyces* growth is favored (Martin, 1987; Batra and Batra, 1979).

Precisely what aspect of the nest environment controls fungal growth is unknown at present, but the most intriguing hypothesis comes from Batra and Batra (1979), who suggest that the nest's high concentrations of carbon dioxide suppresses *Xylaria* growth, similar to the CO_2-induced suppression of fungal growth in nests of leaf-cutter ants (Kleineidam et al., 2001). If so, it points to why homeostasis of the nest atmosphere is crucial for the success of the symbiosis. If nest carbon dioxide concentrations are too low, the nest's culture of *Termitomyces* is threatened by enabling the runaway growth of *Xylaria*. In short, nest homeostasis may reinforce flow of energy through a particular association (*Macrotermes*/*Termitomyces*), while cutting out a competing association (*Macrotermes*/*Xylaria*).

Social Homeostasis versus Emergent Homeostasis

The macrotermitines exhibit what seems to be a unique type of social homeostasis, differentiable from the more familiar form found among, say, honeybees. In bee colonies, physiological and genetic interests are congruent: social homeostasis involves behavioral and physiological interactions only between genetically related members of the colony (figure 5.8). The honeybees are the principal source of carbon

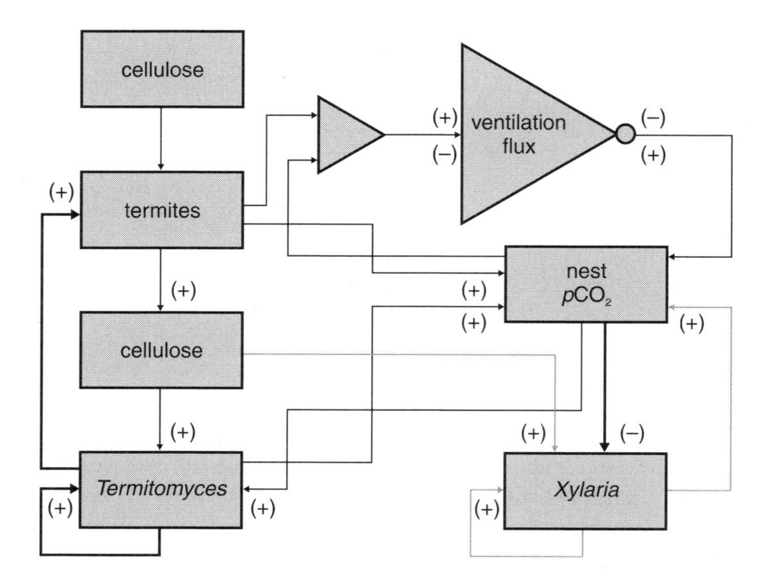

Figure 5.9
Emergent homeostasis of nest pCO_2 in a nest of *Macrotermes michaelseni.* Autocatalytic loops are represented by heavy arrows. Gray arrows represent potential avenues for flux of carbon that are not realized because of *Xylaria*'s sensitivity to the nest pCO_2.

dioxide production in the hive, and pCO_2 regulation is brought about through the activities of the worker bees (Seeley, 1974; Southwick and Moritz, 1987). In the *Macrotermes*/*Termitomyces* colony (figure 5.9), physiological and genetic interests are divergent. The nest's largest perturber of carbon dioxide concentration is not the termites, but the fungi. Yet it is the termites that do all the work of regulating the nest atmosphere. The termites clearly gain from the symbiosis, but so do the fungi, because the regulated nest atmosphere suppresses the growth of their principal competitor, *Xylaria.* In short, the homeostasis is maintained by a physiological interaction between genetically disparate partners in a symbiosis. This type of homeostasis we might call emergent, to distinguish it from the more commonly recognized social homeostasis where physiological and genetic interests converge.

Emergent Homeostasis and Gaia

Emergent homeostasis offers a different way of thinking about the evolution of homeostasis generally, and in particular for thinking about how the global homeostasis posited by Gaia might arise. The early and more radical conception of Gaia sought to paint the biosphere as a sort of superorganism writ large (Lovelock, 1987). This may have assured Gaia's initial controversial, even hostile, reception, because the superorganism concept has a long and troubled history (Wilson, 1971; Golley, 1993; Bowler, 1992). Since its heyday in the Clementsian ecology that spawned it, the superorganism idea has largely faded away, swept aside by the dominance of the philosophically incompatible doctrines of neo-Darwinism. It survives principally among students of the social insects, largely because it poses no essential challenge to neo-Darwinist conceptions of sociality, such as kin selection and inclusive fitness, which assume genetic and physiological self-interest to be congruent (Wilson, 1971). Yet if Gaia is to represent a biosphere-level homeostasis, it cannot involve any analogue of social homeostasis, because Gaia must involve assemblages of organisms in which genetic and physiological self-interests are dissociated. A Gaian biosphere-level homeostasis must therefore be an emergent homeostasis, similar in principle to that which arises among the genetically disparate organisms that enter into symbioses (Paracer and Ahmadjian, 2000). The challenge for Gaia is not to demonstrate whether or not there is a global homeostasis, but to explain how homeostasis arises from assemblages of genetically disparate partners. Understanding how emergent homeostasis works is the true research program of Gaia.

What lessons can be drawn from the "test case" of the *Macrotermes*/*Termitomyces* symbiosis? Certain features of this association stand out as conducive to emergent homeostasis, and these may be at the heart of Gaian homeostasis as well. Ranked in rough order of increasing unconventionality, these are the following.

• *Complementarity of metabolism:* In the *Macrotermes/Termitomyces* symbiosis, each partner brings a metabolic capability to the association that the other partner does not. The complementarity exists at many levels. On the one hand, the fungi contribute a variety of enzymes that digest woody material, lignases and endocellulases, while termites may contribute an exocellulase. The cocktail of enzymes from both breaks down woody material faster than would be possible for each partner alone. In addition, the termites bring capabilities for location, harvesting, and mechanical transport of food which the fungi obviously lack. With respect to Gaia, all agree that metabolic complementarity is the foundation upon which a purported homeostasis of the biosphere can be built (Volk, 1998). So far, much of the effort in study of Gaia has focused on microbes, because most of the biochemical diversity of the biosphere resides in them. Complementarity can exist at many levels of organization besides the biochemical, however, and Gaia will have to account for those.

• *Competition between loops and pathways for mass and energy flow:* The symbiosis between *Macrotermes* and *Termitomyces* persists in the face of a strong competitive challenge from *Xylaria*. The competition superficially involves the two fungal species, but it is really a competition between two supraorganismal pathways for energy flow. The termites convey cellulose to a focal point where it is available to both fungal species. Whether one fungal species or the other prospers depends not only on available food, but also upon how the termites bias the environmental conditions for each fungus's germination and growth. In an important sense, the real competition is not between two species of fungi for the same abundant resource, but between alternate pathways for carbon flow which involve plants, termites, and the fungi. One pathway channels carbon from plants to termite to *Xylaria*, leaving the fungus as the ultimate beneficiary. The other "closes the loop," channeling carbon from plant to termite to *Termitomyces* and then back to the termites. Closing the loop enhances biological work for both *Macrotermes* and *Termitomyces*, while the open loop only enhances growth of *Xylaria*. In the case of Gaia, all nutrients flow through the biosphere in closed loops (Barlow and Volk, 1990). The loops that persist and grow will be those that successfully retain nutrient flow within themselves, as in the high cycling ratios observed for rare nutrients like phosphorus or nitrogen (Volk, 1998).

• *Coordination of metabolism:* A successful symbiosis exhibits both complementarity and coordination of metabolism. In either an open pathway or a closed loop for energy flow, conservation of mass will dictate that flow through one partner in a symbiosis be matched to the other. A mismatch in rates will result in a "spillover" of material, either to competing loops or into energy sinks where it is hard to retrieve (Turner, 2000b). In either case the nutrient flow, and the capacity for physiological work that goes with it, are lost to the loop or pathway. In the *Macrotermes/Termitomyces* symbiosis, metabolic capacities are matched largely by adjustments in biomass of the respective partners. A high collective metabolism of the fungi makes energy available that can fuel increases in termite biomass, which can in turn increase the transport rate of carbon to the fungi. Conversely, an elevated transport rate of carbon to the colony is matched by an expansion of fungal biomass and metabolic capacity. This requirement for "tuning" the metabolism of the respective partners in a symbiosis acts as a natural check on one or the other of the partners pursuing its own selfish interests (Turner, 2000b). This is a somewhat controversial idea, because it asserts that the unbridled pursuit of selfish interests by individual organisms may be counterproductive, while restraint of selfish interest in favor of the partnership may enhance fitness of both.

• *Co-opting the physical environment into an "extended organism":* A symbiosis that joins the partners intimately, such as endosymbiosis, or the close associations found among lichens or mycorrhizae, facilitates the closing of loops of nutrient flow and the coordination of metabolism that is required for emergent homeostasis. A Gaian physiology must emerge from less intimate partnerships, in which the partners are separated to some degree by an unpredictable physical environment. In this circumstance, the controlled flow of nutrients between the partners could be disrupted, diminishing the likelihood that the metabolism of symbiotic partners could be attuned (Turner, 2000b). The disruption could be avoided, however, through adaptive modification of the physical environment, so that the flow of matter and energy between the partners could be controlled. The *Macrotermes/Termitomyces* association provides a dramatic example of this. The flow of carbon and energy through the association is subject to a variety of disruptive influences, like chaotic variations in the strength of prevailing winds. Disruption is prevented by the termites building an adaptive interface, the mound, between the outside environment and the environment of the nest, so that carbon can flow reliably between *Termitomyces* and *Macrotermes*. This extension of physiology beyond organisms' integumentary boundaries is a common feature of plants, animals, and microbes (Turner, 2000b). An emergent homeostasis for Gaia implies that the biosphere comprises a variety of such complementary and mutually coordinated extended organisms. The adaptive modification of the physical environment need not be something as tangible as a termite mound: it could include modifications of fluid density, wind speed, concentration of particular substances, oxidation state, and so forth.

• *Ecological inheritance:* The notion of a homeostatic biosphere has been most severely criticized for its supposed incompatibility with widely accepted principles of evolutionary biology, such as competition and differential reproduction, as well as for its purported failure to reconcile "selfish" genetic interests with the altruism that global homeostasis seems to demand. This is a less serious criticism now than it was when Gaia was first introduced: the collective pursuit of genetic self-interest among the partners in a symbiosis is no longer such a controversial idea. However, biosphere-scale homeostasis implies biosphere-scale physiology operating through the physiological outreach of extended organisms. This implies the perpetuation not only of the organisms that modify the environment but also of the modifications of the environment itself (Jones et al., 1997). Thus, Gaia may require a sort of ecological inheritance, in which the physical modifications of the environment take on a sort of extracorporeal genetic memory, shaping the selective milieu in which the biosphere's many extended organisms operate (Laland et al., 1996, 1999). In the *Macrotermes/Termitomyces* symbiosis, for example, the modifications of the soil environment associated with the colony outlast any of the individuals within the colony, and the success of future generations of workers and fungi depends in part upon the structural legacy left to them by previous generations. The structural legacy survives the death of the colony, enduring for centuries or even millennia, with substantial effects on the distribution and evolution of all the biota associated with it.

• *Telesymbiosis:* Finally, a homeostatic biosphere implies a level of symbiosis that extends biospherewide. This implies symbiosis between organisms that are vastly separated from one another in space and time, linked by an extended physiology controlled by the organisms that comprise it, a sort of symbiosis-at-a-distance, or telesymbiosis. There is evidence that telesymbiosis is present, as in the ecosystemwide coordination of metabolism implied by the high cycling ratios of certain nutrients (Volk, 1998). What is uncertain is how such telesymbioses could work. This is the real challenge faced by both supporters and critics of Gaia's radical conception of a homeostatic biosphere. For both, gathering evidence that the Earth's climate is, or is not, regulated by the biosphere is a dead end. Gaia will stand or fall on whether a convincing case can be made for how the telesymbiosis implied by Gaia could work.

Acknowledgments

I wish to thank the organizers of the Gaia 2000 conference for their generous invitation to attend the conference, and to contribute a chapter to this volume. Two anonymous reviewers provided thoughtful comments on the original manuscript, and I thank them for these. Original research presented here was supported by a grant from the Earthwatch Institute, and was carried out under a permit from the Ministry of Environment and Tourism, Republic of Namibia.

References

Barlow, C., and T. Volk. 1990. Open systems living in a closed biosphere: A new paradox for the Gaia debate. *BioSystems*, 23, 371–384.

Batra, L. R., and S. W. T. Batra. 1979. Termite-fungus mutualism. In *Insect–Fungus Symbiosis: Nutrition, Mutualism and Commensalism*, L. R. Batra, ed., pp. 117–163. New York: John Wiley and Sons.

Bowler, P. J. 1992. *The Norton History of the Environmental Sciences*. New York: W. W. Norton.

Breznak, J. A. 1984. Biochemical aspects of symbiosis between termites and their intestinal microbiota. In *Invertebrate–Microbial Interactions. Joint Symposium of the British Mycological Society and the British Ecological Society Held at the University of Exeter September 1982*, J. M. Anderson, A. D. M. Rayner, and D. W. H. Walton, eds., pp. 173–203. Cambridge: Cambridge University Press.

Coaton, W. G. H., and J. L. Sheasby. 1972. Preliminary report on a survey of the termites (Isoptera) of South West Africa. *Cimbebasia Memoir*, 2, 1–129.

Dangerfield, J. M., T. S. McCarthy, and W. N. Ellery. 1998. The mound-building termite *Macrotermes michaelseni* as an ecosystem engineer. *Journal of Tropical Ecology*, 14, 507–520.

Darlington, J. P. E. C. 1990. Populations in nests of the termite *Macrotermes subhyalinus* in Kenya. *Insectes Sociaux*, 37, 158–168.

Darlington, J. P. E. C. 1994. Mound structure and nest population of the termite, *Pseudacanthotermes spiniger* (Sjöstedt) in Kenya. *Insect Science and Its Application*, 15, 445–452.

Darlington, J. P. E. C., P. R. Zimmerman, J. Greenberg, C. Westberg, and P. Bakwin. 1997. Production of metabolic gases by nests of the termite *Macrotermes jeanelli* in Kenya. *Journal of Tropical Ecology*, 13, 491–510.

Darlington, J. P. E. C., P. R. Zimmerman, and S. O. Wandiga. 1992. Populations in nests of the termite *Macrotermes jeanneli* in Kenya. *Journal of Tropical Ecology*, 8, 73–85.

Dawkins, R. 1982. *The Extended Phenotype*. Oxford: W. H. Freeman.

Dennett, D. C. 1995. *Darwin's Dangerous Idea: Evolution and the Meanings of Life*. New York: Simon & Schuster.

Doolittle, W. F. 1981. Is nature really motherly? *The CoEvolution Quarterly*, Spring, 58–65.

Golley, F. B. 1993. *A History of the Ecosystem Concept in Ecology*. New Haven, CT: Yale University Press.

Grassé, P. P., and C. Noirot. 1961. Nouvelles recherches sur la systématique et l'éthologie des termites champignonnistes du genre *Bellicositermes* Emerson. *Insectes Sociaux*, 8, 311–359.

Harris, W. V. 1956. Termite mound building. *Insectes Sociaux*, 3, 261–268.

Jones, C. G., J. H. Lawton, and M. Shachak. 1997. Positive and negative effects of organisms as physical ecosystem engineers. *Ecology*, 78, 1946–1957.

Joseph, L. E. 1990. *Gaia: The Growth of an Idea*. New York: St. Martin's Press.

Kirchner, J. W. 1991. The Gaia hypotheses: Are they testable? Are they useful? In *Scientists on Gaia*, S. H. Schneider and P. J. Boston, eds., pp. 38–46. Cambridge, MA: MIT Press.

Kleineidam, C., R. Ernst, and F. Roces. 2001. Wind-induced ventilation of the giant nests of the leaf-cutting ant *Atta vollenweideri*. *Die Naturwissenschaften*, 88, 301–305.

Kleineidam, C., and F. Roces. 2000. Carbon dioxide concentrations and nest ventilation in nests of the leaf-cutting ant *Atta vollenweideri*. *Insectes Sociaux*, 47, 241–248.

Korb, J., and K. E. Linsenmair. 1998a. Experimental heating of *Macrotermes bellicosus* (Isoptera, Macrotermitinae) mounds: What role does microclimate play in influencing mound architecture? *Insectes Sociaux*, 45, 335–342.

Korb, J., and K. E. Linsenmair. 1998b. The effects of temperature on the architecture and distribution of *Macrotermes bellicosus* (Isoptera, Macrotermitinae) mounds in different habitats of a west African Guinea savanna. *Insectes Sociaux*, 45, 51–65.

Korb, J., and K. E. Linsenmair. 2000. Ventilation of termite mounds: New results require a new model. *Behavioral Ecology*, 11, 486–494.

Laland, K. N., F. J. Odling-Smee, and M. W. Feldman. 1996. The evolutionary consequences of niche construction. A theoretical investigation using two-locus theory. *Journal of Evolutionary Biology*, 9, 293–316.

Laland, K. N., F. J. Odling-Smee, and M. W. Feldman. 1999. Evolutionary consequences of niche construction and their implications for ecology. *Proceedings of the National Academy of Sciences USA*, 96, 10242–10247.

Leuthold, R. H., S. Badertscher, and H. Imboden. 1989. The inoculation of newly formed fungus comb with *Termitomyces* in *Macrotermes* colonies (Isoptera, Macrotermitinae). *Insectes Sociaux*, 36, 328–338.

Lovelock, J. E. 1987. *Gaia: A New Look at Life on Earth*. Oxford: Oxford University Press.

Lüscher, M. 1961. Air conditioned termite nests. *Scientific American*, 238(1), 138–145.

Margulis, L. 1997. Big trouble in biology: Physiological autopoiesis versus mechanistic neo-Darwinism. In *Slanted Truths: Essays on Gaia, Symbiosis and Evolution*, L. Margulis and D. Sagan, eds., pp. 265–282. New York: Copernicus.

Martin, M. M. 1987. *Invertebrate-Microbial Interactions: Ingested Enzymes in Arthropod Biology*. Ithaca, NY: Comstock Publishing Associates.

Mora, P., and C. Rouland. 1995. Comparison of hydrolytic enzymes produced during growth on carbohydrate substrates by *Termitomyces* associates of *Pseudacanthotermes spiniger* and *Microtermes subhyalinus* (Isoptera: Termitidae). *Sociobiology*, 26, 39–54.

Paracer, S., and V. Ahmadjian. 2000. *Symbiosis: An Introduction to Biological Associations*. Oxford: Oxford University Press.

Peakin, G. J., and G. Josens. 1978. Respiration and energy flow. In *Production Ecology of Ants and Termites*, M. V. Brian, ed. Cambridge: Cambridge University Press.

Penry, D. L., and P. A. Jumars. 1986. Chemical reactor analysis and optimal digestion. *BioScience*, 36, 310–315.

Penry, D. L., and P. A. Jumars. 1987. Modeling animal guts and chemical reactors. *American Naturalist*, 129, 69–96.

Pomeroy, D. E. 1976. Studies on a population of large termite mounds in Uganda. *Ecological Entomology*, 1, 49–61.

Rohrmann, G. F. 1977. Biomass, distribution, and respiration of colony components of *Macrotermes ukuzii* Fuller (Isoptera: Termitidae: Macrotermitidae). *Sociobiology*, 2, 283–295.

Rohrmann, G. F., and A. Y. Rossman. 1980. Nutrient strategies of *Macrotermes ukuzii*. *Pedobiologia*, 20, 61–73.

Rouland, C., A. Civas, J. Renoux, and F. Petek. 1988. Synergistic activities of the enzymes involved in cellulose degradation, purified from *Macrotermes mülleri* and from its symbiotic fungus *Termitomyces* sp. *Comparative Biochemistry and Physiology*, 91B, 459–465.

Rouland, C., F. Lenoir, and M. LePage. 1991. The role of the symbiotic fungus in the digestive metabolism of several species of fungus-growing termites. *Comparative Biochemistry and Physiology*, 99A, 657–663.

Ruelle, J. E. 1964. L'architecture du nid de *Macrotermes natalensis* et son sens fonctionnel. In *Etudes sur les termites africains*, A. Bouillon, ed., pp. 327–362. Paris: Maisson.

Ruelle, J. E. 1970. A revision of the termites of the genus *Macrotermes* from the Ethiopian region (Isoptera: Termitidae). *Bulletin of the British Museum of Natural History (Entomology)*, 24, 366–444.

Schneider, S. H., and P. J. Boston, eds. 1991. *Scientists on Gaia*. Cambridge, MA: MIT Press.

Seeley, T. D. 1974. Atmospheric carbon dioxide regulation in honey-bee (*Apis mellifera*) colonies. *Journal of Insect Physiology*, 20, 2301–2305.

Southwick, E. E. 1983. The honey bee cluster as a homeothermic superorganism. *Comparative Biochemistry and Physiology*, 75A, 641–645.

Southwick, E. E., and R. F. A. Moritz. 1987. Social control of ventilation in colonies of honey bees, *Apis mellifera*. *Journal of Insect Physiology*, 33, 623–626.

Thomas, R. J. 1987a. Distribution of *Termitomyces* and other fungi in the nests and major workers of *Macrotermes bellicosus* (Smeathman) in Nigeria. *Soil Biology and Biochemistry*, 19, 329–333.

Thomas, R. J. 1987b. Distribution of *Termitomyces* and other fungi in the nests and major workers of several Nigerian Macrotermitinae. *Soil Biology and Biochemistry*, 19, 335–341.

Thomas, R. J. 1987c. Factors affecting the distribution and activity of fungi in the nests of Macrotermitinae (Isoptera). *Soil Biology and Biochemistry*, 19, 343–349.

Turner, J. S. 1994. Ventilation and thermal constancy of a colony of a southern African termite (*Odontotermes transvaalensis:* Macrotermitinae). *Journal of Arid Environments*, 28, 231–248.

Turner, J. S. 2000a. Architecture and morphogenesis in the mound of *Macrotermes michaelseni* (Sjøstedt) (Isoptera: Termitidae, Macrotermitinae) in northern Namibia. *Cimbebasia*, 16, 143–175.

Turner, J. S. 2000b. *The Extended Organism: The Physiology of Animal-Built Structures*. Cambridge, MA: Harvard University Press.

Turner, J. S. 2001. On the mound of *Macrotermes michaelseni* as an organ of respiratory gas exchange. *Physiological and Biochemical Zoology*, 74, 798–822.

Veivers, P. C., R. Mühlemann, M. Slaytor, R. H. Leuthold, and D. E. Bignell. 1991. Digestion, diet and polyethism in two fungus-growing termites: *Macrotermes subhyalinus* Rambur and *M. michaelseni* Sjøstedt. *Journal of Insect Physiology*, 37, 675–682.

Volk, T. 1998. *Gaia's Body: Toward a Physiology of Earth*. New York: Copernicus and Springer-Verlag.

Weir, J. S. 1973. Air flow, evaporation and mineral accumulation in mounds of *Macrotermes subhyalinus*. *Journal of Animal Ecology*, 42, 509–520.

Wiegert, R. G. 1970. Energetics of the nest-building termite, *Nasutitermes costalis* (Holmgren), in a Puerto Rican forest. In *A Tropical Rain Forest: A Study of Irradiation and Ecology at El Verde, Puerto Rico*, H. T. Odum and R. F. Pigeon, eds. Washington, DC: Division of Technical Information, U.S. Atomic Energy Commission.

Williams, G. C. 1992. Gaia, nature worship and biocentric fallacies. *Quarterly Review of Biology*, 76, 479–486.

Wilson, E. O. 1971. *The Insect Societies*. Cambridge, MA: Belknap Press of Harvard University Press.

Wood, T. 1988. Termites and the soil environment. *Biology and Fertility of Soils*, 6, 228–236.

Wood, T. G., and R. J. Thomas. 1989. The mutualistic association between Macrotermitinae and *Termitomyces*. In *Insect–Fungus Interactions*, N. Wilding, N. M. Collins, P. M. Hammond, and J. F. Webber, eds., pp. 69–92. London: Academic Press.

6
Homeostatic Gaia: An Ecologist's Perspective on the Possibility of Regulation

David M. Wilkinson

Some ideas must be worked with and developed before they fit into the hypothetico-deductive context of contemporary science. Their validity as scientific constructs springs entirely from the interest that they generate among scientists. We fear that science can be trivialised by overriding emphasis on the narrow interaction between hypothesis and test.
—K. Vepsalainen and J. R. Spence (2000)

Abstract

A major question raised by Gaia theory is this: Is the long-term persistence of the biosphere entirely due to luck, or does it possess feedback loops that increase the possibility of long-term persistence? Such feedbacks are central to the idea of homeostatic Gaia. A common criticism of this idea is that these feedbacks could not evolve because they would require higher-level selection, and such group selection is too weak to overcome gene/individual level selection for self-interest. I argue that this need not be the case.

I point out that mutualisms are common in ecology and that Gaian-type interactions are likely to be "by-product" mutualisms. Such mutualisms do not pose a problem for evolutionary theory, because they are not susceptible to "cheats." This still leaves the question How could regulation develop without active selection for its presence? To approach this question, I draw analogies between population dynamics and Gaia. In population ecology the well-known process of density-dependent regulation is not usually considered to be a product of direct selection, but an emergent property of the behavior of individuals all trying to maximize their own fitness. This demonstrates that the idea that regulation requires active selection is incorrect—it can develop as an emergent property.

I conclude that biosphere regulation is theoretically possible, and as such is an important research question.

Introduction: Homeostasis in Physiology and Geophysiology

The definition of life has been a long-standing problem in biology (see, e.g., Schrödinger, 1948; Margulis and Sagan, 1995). The difficulty in providing a definition is not surprising from an evolutionary perspective. If life evolved from nonlife, then one would expect problems in establishing criteria that separate all life from all nonliving systems. However, one of the striking aspects of biological organisms is their highly ordered (information-rich) appearance; this is superficially surprising in a universe ruled by the second law of thermodynamics. Indeed, organisms have been described as "islands of order in an ocean of chaos" (Margulis and Sagan, 1995). Information (I) can be linked to entropy by the following equation (Lovelock and Margulis, 1974):

$$I = S_o - S, \tag{1}$$

where S_o is the entropy of the components of the system at thermodynamic equilibrium and S is the entropy of the system. Organisms maintain their high level of order (negative entropy) because they are not closed systems, but acquire energy from their environment and dump waste products back into their surroundings. This energy fuels complex biochemical reactions within the organisms; these reactions require feedback loops in order to maintain relatively stable conditions within the organism. A schematic representation of this situation for a simplified system with just two reaction paths is illustrated below (after R. J. P. Williams, 1980):

```
← feedback ←
↓           ↑
reaction 1 → products
                ↓
             feedback
                ↓
         → reaction 2 →
```

The ability of an organism to maintain a relatively constant internal environment is called homeostasis. Textbooks of Earth system science (e.g., Kump et al., 1999) also contain similar feedback diagrams. Indeed, early in the history of Gaia theory Lovelock and Margulis (1974) wrote of "atmospheric homeostasis by and for the biosphere"; this discussion included the use of equation (1) as an aid to thinking about the atmosphere. For homeostatic Gaia to exist, not only does life have to be linked through feedback loops to the nonliving environment, but these feedbacks must tend to stabilize the system, maintaining conditions suitable for life. Many biologists have claimed that such regulation could not have evolved. Such arguments are the main subject of this chapter. However, before addressing this question directly, it is important to consider the anthropic principle.

The Anthropic Principle

The anthropic principle has been most fully developed in cosmology (e.g., Carr and Rees, 1979; Rees, 1987; see Smolin, 1997 for an alternative view). Consider the following (from Hoyle, 1975):

$$\text{Universe}_{\text{scale}}/\text{Neutron star}_{\text{scale}} = \text{Universe}_{\text{number particles}}/\text{Neutron star}_{\text{number particles}} \quad (2)$$

If this equality did not hold, then stars with the number of neutrons found in our sun (10^{57}) would not be stable. Another example is that the nucleus of ^{16}O has an excited state that is slightly less than the sum of the rest mass energies of ^{12}C and an alpha particle. However, if the excited state in ^{16}O were slightly more than $^{12}C + \alpha$, then no carbon would have been formed, since the reaction $^{12}C + \alpha \rightarrow {}^{16}O$ would have removed carbon as fast as it was produced (Hoyle and Wickramasinghe, 1999). Many other examples were given in a classic paper in *Nature* by Carr and Rees (1979). There are two ways to view such coincidences: the strong anthropic principle suggests that the physical properties of matter are fixed to allow the existence of life, while the weak anthropic principle points out that there can only be observers (astronomers) in certain types of universe, so it is not surprising that we observe a universe of such a type! As Smolin (1997) pointed out, the strong form of the principle is more like a religious than a scientific idea.

Several authors have pointed out that the weak anthropic principle is relevant to Gaia (Lenton, 1998; Levin, 1998; Watson, 1999; Wilkinson, 1999). Any planet which is home to organisms as complex as James Lovelock or Lynn Margulis must have had a long period of time during which conditions were always suitable for life, and thus must give the impression of regulation for life-friendly conditions even if the persistence of life was purely a matter of chance (what Andrew Watson called "lucky Gaia" at the Valencia meeting; see also Lenton and Wilkinson, 2003). The key question to ask is Are the apparently regulating feedbacks we see in the Earth system a product of chance (lucky Gaia) or an expected outcome of a planet with abundant life (full Gaia, sometimes called probable Gaia or innate Gaia)?

Even if the Gaia of Lovelock (2000; see also Lenton, 1998) is real, there must still be an element of chance involved in the persistence of life on Earth. This is nicely illustrated by the extraterrestrial impact event that appears to have marked the Cretaceous-Tertiary (K-T) boundary (Alvarez, 1997). The K-T impact object could have been much larger than it actually was, as Watson (1999, 79) has pointed out: "It is presumably just luck that no such sterilising event has in fact occurred. One of the largest Earth crossing asteroids, Eros, has recently been shown to have about a 50 percent chance of colliding with the Earth in the next 100 Ma." It is surely luck, not Gaian regulation, which explains why life has not been wiped out by such a large impact; while many microbes would survive most impacts, even these would be killed by the largest collisions. There was also an element of "good luck" in the impact site of the K-T event. It hit rare sulfate-rich rock, which may have prevented major effects on the Earth's ozone layer (Cockell and Blaustein, 2000). There is clearly an element of chance involved in the persistence of Gaian systems.

The Criticism from Evolutionary Biology

Homeostasis in an organism is clearly to the advantage of that organism, and likely to increase its reproductive success. However, homeostasis at a planetary level cannot have a similar explanation, a point many biologists were quick to point out when Gaia theory was proposed (e.g., Dawkins, 1982; G. C. Williams, 1992). The main problem is that such a system would be wide open to cheats. Dawkins, in one of the best-known criticisms of Gaia, illustrated the point with the following example. He pointed out that if one considers oxygen production by plants for the good of Gaia, a mutant plant that did not produce oxygen would have fewer costs and as such would be at an advantage. This plant "would outreproduce its more public-spirited colleagues and genes for public-spiritedness would soon disappear" (Dawkins, 1982). This is a widely discussed problem in evolutionary ecology in the context of interspecific mutualisms (mutually beneficial relationships between members of different species; see Wilkinson, 2001 for discussion of terminology). Dawkins's criticisms of Gaia are really the same problem but at vastly larger spatial and temporal scales!

One way of classifying interspecific mutualisms is to split them into "investment mutualisms" and "by-product mutualisms" (Connor, 1995). In investment mutualisms both organisms provide some service to their partner at some cost to themselves, while in by-product mutualisms a waste product of one organism is used by its partner. Investment mutualisms are open to cheating (one partner could in theory reduce its investment while still taking the benefits), and much recent theoretical work has been devoted to trying to understand the mechanisms that allow such mutualisms to be stable (e.g., Hoekesema and Bruna, 2000; Wilkinson and Sherratt, 2001; Yu, 2001). However, many mutualisms are of the by-product type (Connor, 1995), in which there are no selective advantages to an organism's withholding its by-product. Indeed, if it were costly to prevent the partner from obtaining the by-product, then the subsequent fitness of a cheat would be lower than if it had cooperated in supplying the by-product (Wilkinson and Sherratt, 2001).

I have previously argued that Gaia is likely to be based on by-products (Wilkinson, 1999), as have several other authors (most notably Volk, 1998). This avoids the criticism of Dawkins, who, interestingly, used the example of oxygen production by plants, which is a by-product of oxygenic photosynthesis and thus not open to cheating. The point is nicely illustrated by the link between dimethyl sulfide (DMS) production, due to the activities of marine phytoplankton, and climatic regulation (Charlson et al., 1987; Simo, 2001). It now seems clear that plankton produce the precursors to DMS for their own use, its climatic effects are a by-product (Hamilton and Lenton, 1998; Simo, 2001; Sunda et al., 2002; Wilkinson, 1999), a possibility first suggested in the famous CLAW paper of 1987 (Charlson et al., 1987). The growing evidence that many phytoplankton blooms are not clonal undermines group selection explanations for DMS production (Thornton, 2002); these chemicals must benefit individual algal cells with their Gaian effects being a by-product.

To summarize this section, Gaian-type behavior by an organism is not a problem for evolutionary theory if that behavior is a by-product. However, this leaves a big problem: Can planetary homeostasis develop if there is no active selection for it? I address this question below by drawing analogies with population ecology.

Emergent Regulation: The Example of Population Dynamics

Consider a population of a particular species. Its numbers at a given point in time (N_{t+1}) will be a function of its numbers in the past (N_t). Formally;

$$N_{t+1} = f(N_t). \tag{3}$$

Different populations grow at different rates (r), so the rate of change in population size can be written;

$$dN/dt = rN. \tag{4}$$

However, this gives exponential growth, which clearly cannot last for very long. Darwin (1859, p. 64) appreciated the problem and considered the following thought experiment: "The elephant is reckoned to be the slowest breeder of all known animals, and I have taken some pains to estimate its probable minimum rate of natural increase: it will be under the mark to assume that it breeds when thirty years old, bringing forth three pair of young in this interval; if this be so, at the end of the fifth century there would be alive fifteen million elephants, descended from that first pair." While some of his details of elephant reproductive biology are now known to be wrong, the general point—of huge numbers of elephants in a relatively short space of time—still stands. Clearly there must be a leveling off of a population at higher numbers. The simplest equation with the properties of fast growth at low numbers leading to a steady population at higher values is the logistic equation

$$dN/dt = rN(K - N/N). \tag{5}$$

Note that if $K = N$, then

$$(K - N/N) = 0. \tag{6}$$

In this case $dN/dt = 0$. The value K, at which population growth is 0, is referred to as the carrying capacity, and is usually interpreted as the maximum number of individuals that the environment can support (for descriptions of the mathematics of population ecology see Case, 2000 or Krebs, 2001). K is the value around which the population is regulated. If the population is raised above K, then population growth should become negative and N should decrease until $N = K$; if $N < K$, the opposite should happen. Such regulation takes place through what ecologists call density-dependent processes. For example, at high population densities food may be in short supply or crowded conditions may favor disease outbreaks, whereas at lower population densities food may be plentiful and disease transmission lower, favoring higher reproductive rates and increased survival. From a Gaian perspective, the important thing to note is that there is no selection for regulation. All the individuals are attempting to maximize their reproductive success. The regulation around K emerges in the population's behavior because it is easier to survive and produce more offspring at lower densities than at higher densities.

The parallels with the debates about Gaia are strong. In the past some ecologists (e.g., Wynne-Edwards, 1962, who produced a detailed 653-page argument) assumed that population regulation was a product of selection at the species level to maintain a population size that maximized the survival chances of that species. By the mid-1970s most ecologists had realized that such an explanation suffers from the cheat problem and had abandoned such species selection explanations of population regulation (Maynard Smith, 1984). This may help to explain the reaction of biologists to talk of atmospheric regulation "by and for the biosphere" in the early work on Gaia in the mid-1970s. The phraseology seemed uncomfortably similar to Wynne-Edwards's unfashionable group selectionist arguments. In the population ecology case it is now clear that the regulation can emerge without selection for regulation.

For example, Wynne-Edwards (1962) famously argued that seabird colonies actively regulated their size to match food supplies in an attempt to avoid the overfishing that characterizes so many human fisheries. The idea was that at high population densities the birds would show restraint in the number of young reared to prevent the colony size exceeding available resources. One could envisage selection between colonies, with ones that didn't show restraint being more likely to suffer catastrophic decline. However, this mechanism runs into problems with the idea of cheats. A cheating genotype that didn't show restraint clearly would come to dominate the colony, although this would potentially reduce the long-term persistence of their colony. Some seabird colonies do show the sort of density-dependent regulation of colony size that Wynne-Edwards's ideas would predict. However, combined observational and modeling studies of the northern gannet (*Morus bassanus*) in Britain have shown how this appears as an emergent property of intraspecific competition for food between individual birds (Lewis et al., 2001).

Clearly the logistic equation (equation 5) is a great simplification of the complexities of population ecology. As such it is worth briefly considering a spatially explicit model that also leads to the same result, regulation without selection for regulation. All locations in an organism's range are seldom of equal quality; some areas will provide more suitable habitat than others. Consider a mobile population such as a bird species. At low densities most individuals will be able to breed in high-quality habitat and thus to rear a large number of young (therefore dN/dt will have a high positive value). As the population increases, space constraints will force an increasing number of birds to attempt to breed in poorer habitats, thus leading to a reduction in the mean growth rate of the population. This density-dependent mechanism is sometimes referred to as the "buffer effect," and has been documented for a range of species, including the European kestrel (*Falco tinnunculus*) (O'Connor, 1982) and overwintering black-tailed godwits (*Limosa limosa*) in Britain (Gill et al., 2001). Again, there is no selection for regulation—it is an emergent property of the population.

Population ecology can contribute another important insight to the Gaia debate. It is apparent from several of the contributions to this volume that the definition of regulation in a Gaian context is a controversial area. A naïve approach would be to consider that regulation of some factor (e.g., temperature or carbon dioxide) should always lead to stable values. However, consider the difference equation analogue of the differential logistic equation described above (equation 5). This can be written

$$N_{t+1} = N_t \exp[r(1 - N_t/K)]. \tag{7}$$

It is now well known that some values of r can give rise to chaotic behavior (May, 1974; Murray, 1993), although it is clear that this equation has "regulation" built into it. This cautions against expecting that regulated systems must always lead to constant values for the regulated variable.

Concluding Remarks

The key ideas of this chapter are that the criticisms of Gaia by biologists such as Dawkins are not valid if Gaian processes are by-products, and that regulation can emerge in a system without active selection for regulation. These points are important because some commentators are still criticizing Gaia because they cannot see how it can be a product of natural selection (e.g., Kirchner, 2002; Volk, 2002). The fact that regulation can be an emergent property of a system is well known in population ecology, although not normally described in such terms.

My use of analogies with population ecology is not the only possible route to these conclusions. The classic Daisyworld model of Watson and Lovelock (1983) was devised to illustrate the emergence of regulation in a simple biosphere. Similar conclusions about the emergence of regulation are arising in the area of mathematics called complexity theory (Levin, 1998; Stewart, 1998).

There is now widespread consensus that life can have a huge effect on the biosphere. This is illustrated by the many discussions of biological carbon sinks in the context of global warming. Within conventional ecology the idea of ecosystem engineering has recently stressed the importance of some organisms in modulating the availability of resources to other species by causing changes in the physical state of the environment (Jones et al., 1994, 1997; Wilby et al., 2001). This approach has similarities to Gaia inasmuch as it emphasizes two-way feedbacks between life and the abiotic environment.

The key question for homeostatic Gaia is "Does life tend to regulate the system in a way that prolongs the existence of life once it has developed on a planet?" Is it more than luck (the weak anthropic principle) that life has survived so long on Earth? Some recent modeling studies (e.g., Betts, 1999; Lenton and von Bloh, 2001) hint at the possibility that the answer is yes! It is at least now clear that such regulation cannot be ruled out by theoretical arguments that claim incompatibility with evolutionary theory. I conclude that biosphere regulation is theoretically possible, and as such is an important research question, worth "working with and developing" in the spirit of the chapter's epigraph.

Acknowledgments

My ideas on Gaia have benefited greatly from discussion with participants at the Gaia in Oxford (1999) and the Valencia 2000 meetings, especially Tim Lenton and Tyler Volk. Tom Sherratt has made important contributions to my related work on the evolutionary ecology of mutualisms. During the preparation of the final version of this chapter I was collaborating with Tim Lenton on another Gaian paper, so it is likely that some of Tim's ideas may have made their way into this chapter masquerading as my own. Scott Turner provided a detailed review of an earlier version of this paper. I thank Jim and Sandy Lovelock for encouragement, and dedicate this paper to Hannah O'Regan, with love.

References

Alvarez, W. 1997. *T. Rex and the Crater of Doom*. Princeton, NJ: Princeton University Press.

Betts, R. A. 1999. Self-beneficial effects of vegetation on climate in an ocean–atmosphere general circulation model. *Geophysical Research Letters*, 26, 1457–1460.

Carr, B. J., and M. J. Rees. 1979. The anthropic principle and the structure of the physical world. *Nature*, 278, 605–612.

Case, T. J. 2000. *An Illustrated Guide to Theoretical Ecology*. New York: Oxford University Press.

Charlson, R. J., J. E. Lovelock, M. O. Andreae, and S. G. Warren. 1987. Ocean phytoplankton, atmospheric sulphur, cloud albedo and climate. *Nature*, 326, 655–661.

Cockell, C. S., and A. R. Blaustein. 2000. "Ultraviolet spring" and the ecological consequences of catastrophic impacts. *Ecology Letters*, 3, 77–81.

Connor, R. C. 1995. The benefits of mutualism: A conceptual framework. *Biological Reviews*, 70, 427–457.

Darwin, C. 1859. *On the Origin of Species*. London: John Murray.

Dawkins, R. 1982. *The Extended Phenotype*. Oxford: Oxford University Press.

Gill, J. A., K. Norris, P. M. Potts, T. G. Gunnarson, P. W. Atkinson, and W. J. Sutherland. 2001. The buffer effect and large-scale population regulation in migrating birds. *Nature*, 412, 436–438.

Hamilton, W. D., and T. M. Lenton. 1998. Spora and Gaia: How microbes fly with their clouds. *Ethology, Ecology and Evolution*, 10, 1–16.

Hoeksema, J. D., and E. M. Bruna. 2000. Pursuing the big questions about interspecific mutualisms: A review of theoretical approaches. *Oecologia*, 125, 321–330.

Hoyle, F. 1975. *Astronomy and Cosmology: A Modern Course*. San Francisco: W. H. Freeman.

Hoyle, F., and N. C. Wickramasinghe. 1999. The universe and life: Deductions from the weak anthropic principle. *Astrophysics and Space Science*, 268, 89–102.

Jones, C. G., J. H. Lawton, and M. Shachak. 1994. Organisms as ecosystem engineers. *Oikos*, 69, 375–386.

Jones, C. G., J. H. Lawton, and M. Shachak. 1997. Positive and negative effects of organisms as physical ecosystem engineers. *Ecology*, 78, 1946–1957.

Krebs, C. J. 2001. *Ecology*, 5th ed. San Francisco: Benjamin Cummings.

Kirchner, J. W. 2002. The Gaia hypothesis: Fact, theory and wishful thinking. *Climatic Change*, 52, 391–408.

Kump, L. R., J. F. Kasting, and R. G. Crane. 1999. *The Earth System*. Englewood Cliffs, NJ: Prentice-Hall.

Lenton, T. M. 1998. Gaia and natural selection. *Nature*, 394, 439–447.

Lenton, T. M., and W. von Bloh. 2001. Biotic feedback extends the life span of the biosphere. *Geophysical Research Letters*, 28, 1715–1718.

Lenton, T. M., and D. M. Wilkinson. 2003. Developing the Gaia theory, a response to the criticisms of Kirchner and Volk. *Climatic Change*, 58, 1–12.

Levin, S. A. 1998. Ecosystems and the biosphere as complex adaptive systems. *Ecosystems*, 1, 431–436.

Lewis, S., T. N. Sherratt, K. C. Hamer, and S. Wanless. 2001. Evidence of intraspecific competition for food in a pelagic seabird. *Nature*, 412, 816–819.

Lovelock, J. 2000. *The Ages of Gaia*. 2nd ed. Oxford: Oxford University Press.

Lovelock, J. E., and L. Margulis. 1974. Atmospheric homeostasis by and for the biosphere: The Gaia hypothesis. *Tellus*, 26, 2–10.

Margulis, L., and D. Sagan. 1995. *What Is Life?* London: Weidenfeld and Nicolson.

May, R. M. 1974. Biological populations with nonoverlapping generations: Stable points, stable cycles and chaos. *Science*, 186, 645–647.

Maynard Smith, J. 1984. The population as a unit of selection. In B. Shorrock, ed., *Evolutionary Ecology*, pp. 195–202. Oxford: Blackwell.

Murray, J. D. 1993. *Mathematical Biology*, 2nd ed. New York: Springer-Verlag.

O'Connor, R. J. 1982. Habitat occupancy and regulation of clutch size in the European kestrel *Falco tinnunculus*. *Bird Study*, 29, 17–26.

Rees, M. 1987. The anthropic universe. *New Scientist*, 115(1572), 44–47.

Schrödinger, E. 1948. *What Is Life?* Cambridge: Cambridge University Press.

Simo, R. 2001. Production of atmospheric sulfur by oceanic plankton: Biogeochemical, ecological and evolutionary links. *Trends in Ecology and Evolution*, 16, 287–294.

Smolin, L. 1997. *The Life of the Cosmos*. London: Weidenfeld and Nicolson.

Stewart, I. 1998. *Life's Other Secret*. London: Allen Lane.

Sunda, W., D. J. Kieber, R. P. Kiene, and S. Huntsman. 2002. An antioxidant function for DMSP and DMS in micro algae. *Nature*, 418, 317–320.

Thornton, D. C. O. 2002. Individuals, clones or groups? Phytoplankton behaviour and units of selection. *Ethology, Ecology and Evolution*, 14, 165–173.

Vepsalainen, K., and J. R. Spence. 2000. Generalisation in ecology and evolutionary biology: From hypothesis to paradigm. *Biology and Philosophy*, 15, 211–238.

Volk, T. 1998. *Gaia's Body*. New York: Copernicus and Springer-Verlag.

Volk, T. 2002. Towards a future for Gaia theory. *Climatic Change*, 52, 423–430.

Watson, A. J. 1999. Coevolution of the Earth's environment and life: Goldilocks, Gaia and the anthropic principle. In G. Y. Craig and J. H. Hull, eds., *James Hutton—Present and Future*, pp. 75–88. Special Publication 150. London: Geological Society.

Watson, A. J., and J. E. Lovelock. 1983. Biological homeostasis of the global environment: The parable of Daisyworld. *Tellus*, 35B, 284–289.

Wilby, A., M. Shachak, and B. Boeken. 2001. Integration of ecosystem engineering and trophic effects of herbivores. *Oikos*, 92, 436–444.

Wilkinson, D. M. 1999. Is Gaia really conventional ecology? *Oikos*, 84, 533–536.

Wilkinson, D. M. 2001. At cross purposes. *Nature*, 412, 485.

Wilkinson, D. M., and T. N. Sherratt. 2001. Horizontally acquired mutualisms, an unsolved problem in ecology? *Oikos*, 92, 377–384.

Williams, G. C. 1992. Gaia, nature worship and biocentric fallacies. *Quarterly Review of Biology*, 67, 479–486.

Williams, R. J. P. 1980. On first looking into Nature's chemistry. Part I. The role of small molecules and ions: The transport of the elements. *Chemical Society Reviews*, 9, 281–324.

Wynne-Edwards, V. C. 1962. *Animal Dispersion in Relation to Social Behaviour*. Edinburgh: Oliver and Boyd.

Yu, D. W. 2001. Parasites of mutualisms. *Biological Journal of the Linnean Society*, 72, 529–546.

II
EARTH HISTORY AND CYCLES

7

Phosphorus, a Servant Faithful to Gaia? Biosphere Remediation Rather Than Regulation

Karl B. Föllmi, Federica Tamburini, Rachel Hosein,
Bas van de Schootbrugge, Kaspar Arn, and Claire Rambeau

Abstract

The global cycles of the biophile elements phosphorus and carbon are closely linked through their profound implication in two major biogeochemical processes, photosynthesis and biogeochemical weathering. In photosynthetic processes phosphorus may limit the transformation of atmospheric CO_2 into organic carbon, and in biogeochemical weathering processes atmospheric CO_2 may limit the mobilization of phosphorus. During environmental change, changes in both cycles are coupled and associated feedback mechanisms have important implications on the biosphere.

In this chapter, we study the character of the coupled changes in the phosphorus and carbon cycles during the last 160 million years and propose feedback mechanisms between the two cycles. We explore the effects of the proposed feedback systems on the biosphere and especially their capacity to regulate environmental conditions in a Gaian sense. For this purpose, we use marine phosphorus and carbon burial rates, a modeled atmospheric CO_2 curve, and stable carbon isotopes as proxies for temporal change in the global phosphorus and carbon cycles. Based on the temporal changes within these proxies, we postulate a period of fundamental change in feedback between weathering, productivity, and climate at around 32 million years ago, which is explained by the onset of major glaciation. This suggests that feedback mechanisms may not be uniform throughout Earth's history but may change during environmental change. We also observe evidence for complex interactions between the carbon and phosphorus cycles, which suggests that the two cycles are not necessarily coupled in a linear fashion.

Our general conclusion is that the phosphorus and carbon cycles are characterized by interactions and resulting feedback mechanisms, which show stabilizing effects only to a certain extent. The effects of global change extrinsic to the biosphere, such as volcanic events, changes in orbital parameters, and impacts, and intrinsic change related to biological evolution appear to overrule the tendency toward stable conditions, and the associated feedback mechanisms are considered to be remediative rather than regulatory.

Introduction

Phosphorus is an element essential to life. Within the biosphere, it drives plant growth and fosters the transformation of atmospheric CO_2 into organic carbon through photosynthesis. This vital process provides a close link between phosphorus and carbon and their respective cycles (figure 7.1).

The delivery of phosphorus to the biosphere is accomplished predominantly by biogeochemical weathering of continental rocks. This process is highly dependent on the availability of atmospheric CO_2, which—under natural conditions—forms a weak acid with water and dissolves rocks. For this reason, weathering provides a second interface with the carbon cycle, albeit with reversed signs. In photosynthetic processes, phosphorus provides a driving force for an important phase change within the carbon cycle, whereas in weathering processes carbon is instrumental in pushing an important phase change within the phosphorus cycle (figure 7.1). These two interfaces tightly couple both cycles, and therefore these cycles temporal and spatial changes occur interdependently.

Biogeochemical weathering is a process which involves biological participation, and photosynthesis is a biological process per se. There is thus an important biological momentum in linking the two cycles, and the question to be posed here is in how far this momentum shows Gaian properties—that is, regulative capacities in a homeostatic sense (Lovelock and Margulis, 1974; Lovelock, 1988). The approach taken here in trying to provide an answer is to reconstruct how temporal changes in the global phosphorus cycle have been linked to those in the carbon cycle during the last 160 million years (myr) of the history of the biosphere, and to evaluate whether these changes can

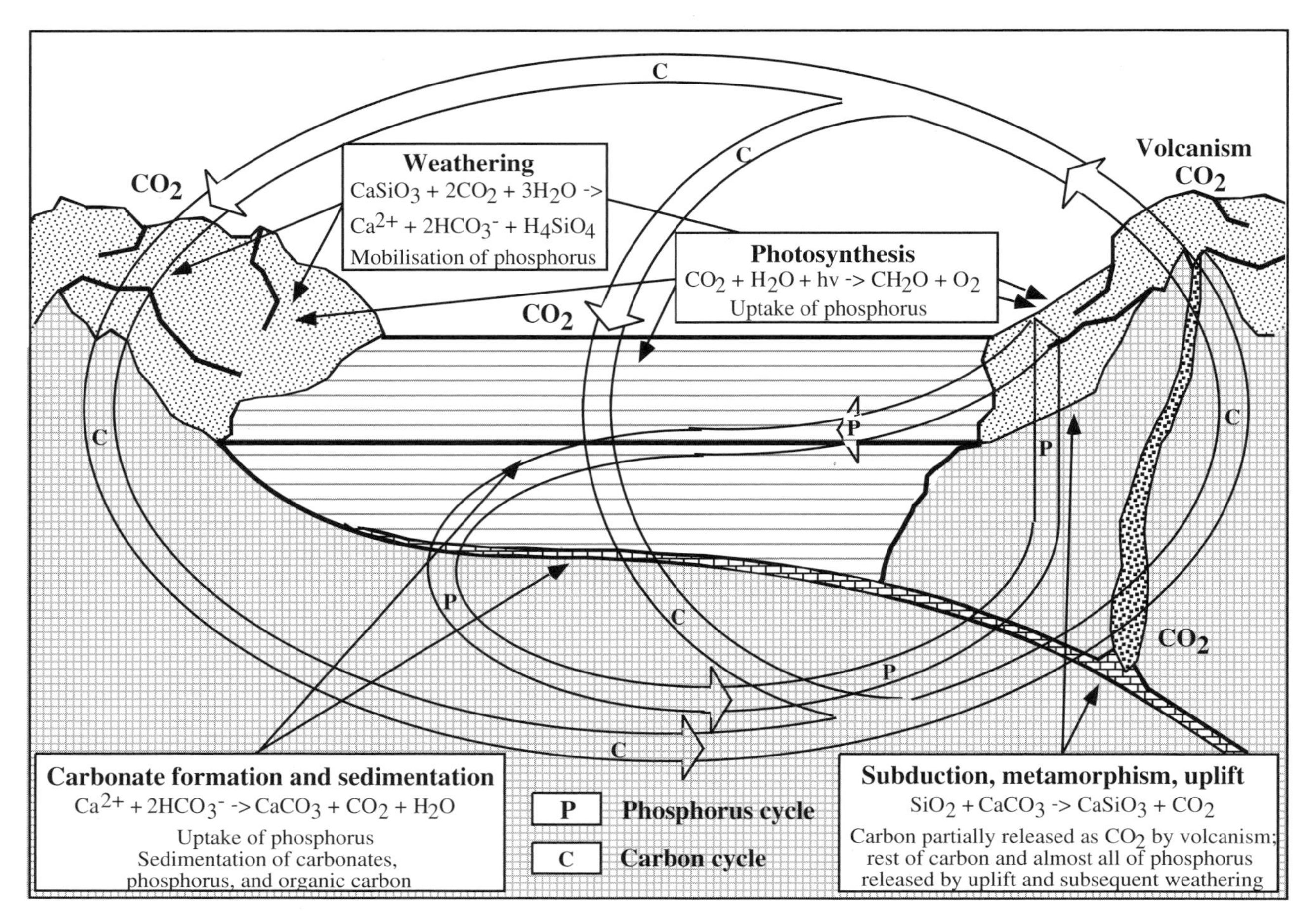

Figure 7.1
Qualitative and schematic overview of the long-term carbon and phosphorus cycles and the model reactions essential in the transfer and phase changes of carbon and phosphorus.

be interpreted as favorable to the biosphere (i.e., leading to a negative, and therefore stabilizing, feedback effect on environmental conditions). It will be shown here that—related to the nonlinear and complex character of change within the biosphere—no simple and uniform answer can be given.

Proxy for Temporal Change in the Global Phosphorus Cycle

The choice of a reliable proxy for temporal change in phosphorus input and sedimentary removal is limited to the oceanic sedimentary record. Such a record reflects changes in the phosphorus budget on local, regional, and global levels. This phenomenon is related to the observation that phosphorus is a reactive element and its residence time in oceanic systems is considered to be short (around 10,000 years; Colman and Holland, 2000). This means that its spatial distribution within the different ocean basins is not homogeneous, and is biased by its importance as an essential nutrient relative to other nutrients (Codispoti, 1989; Tyrell, 1999), by local to regional differences in phosphorus uptake through primary productivity and subsequent remineralization, and by the capacity of sediments to efficiently store phosphorus (Colman and Holland, 2000), among other factors. It is therefore crucial to use a record which is representative for large parts of the oceans, in order to extract a signal of wider significance.

The record we propose here is based on a compilation of phosphorus accumulation rates, which were calculated from systematically measured phosphorus contents in a great variety of Deep Sea Drilling Project (DSDP) and Ocean Drilling Program (ODP) cores (figure 7.2; Föllmi, 1995, 1996). This record is used as

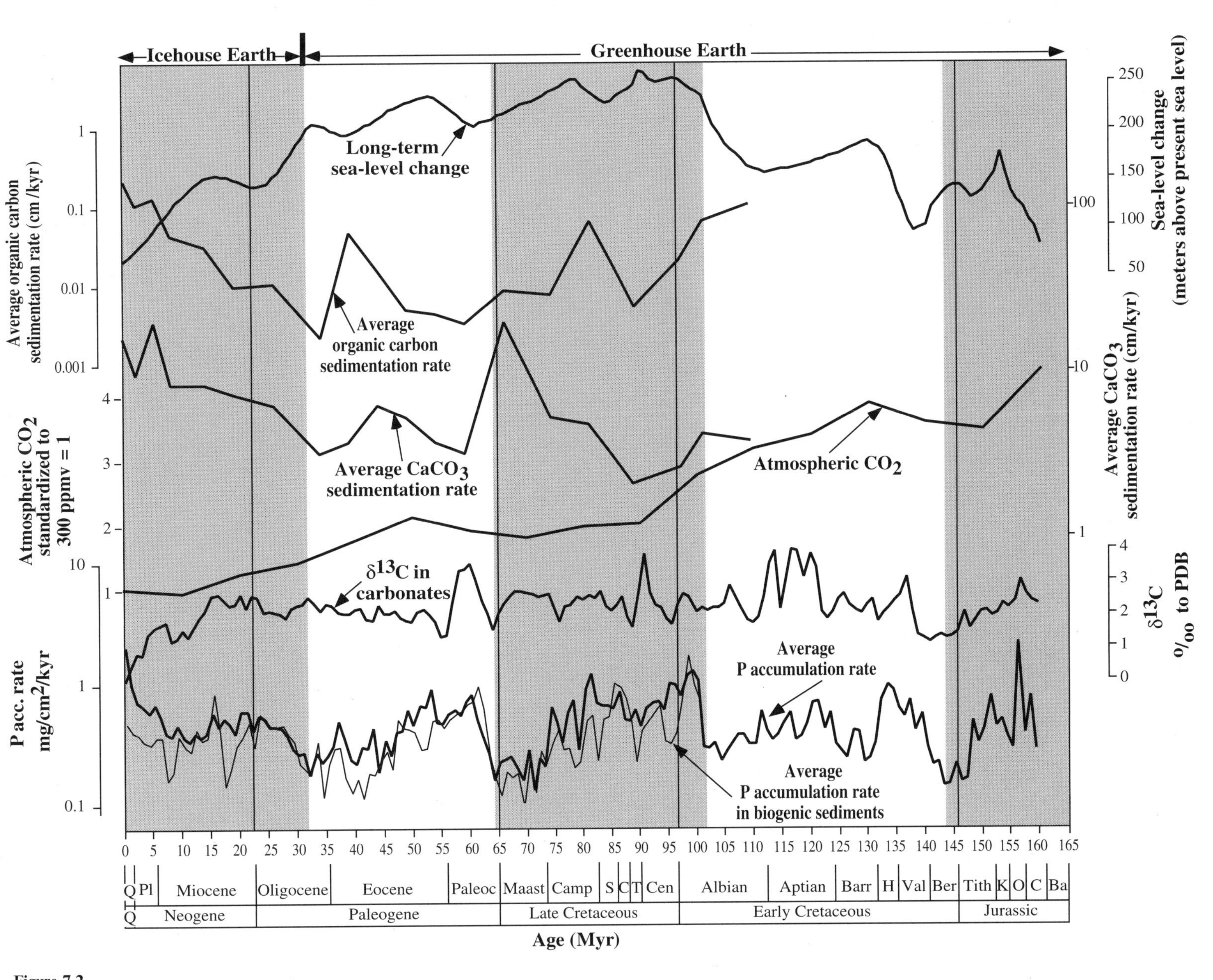

Figure 7.2
Evolution of marine phosphorus burial rates (for all types of sediments and separately for biogenic sediments) (Föllmi, 1995); the $\delta^{13}C$ whole-rock record of marine carbonates (Föllmi, 1996); atmospheric CO_2 contents (Berner, 1994); carbonate and organic carbon sedimentation rates (Southam and Hay, 1981); and long-term sea-level change (Haq et al., 1987).

a proxy for temporal changes in the global phosphorus cycle over the last 160 myr, which are coupled to changes in the intensity of continental weathering. It is suggestive of a close coupling between overall and chemical weathering rates, and between overall sediment and phosphorus accumulation rates. It shows also evidence for an important change in the coupling between climate, weathering, and productivity at around 32 myr. For the period prior to 32 myr, a positive correlation is observed, suggesting a link between warm climates, increased biogeochemical weathering rates, and phosphorus release and burial rates, whereas from 32 myr on, this correlation appears to be inverse, thereby suggesting a link between cooling periods, intensified weathering rates, and phosphorus mobilization and burial rates. The cause of this inversion is identified as the onset of widespread glaciation (Föllmi, 1995, 1996).

Since its first publication, we have received several criticisms challenging the assumption that this record is representative on a global scale. These concerns evolve around four observations: (1) older portions of the curve are constructed on fewer data relative to younger portions, and may not be as reliable; (2) glaciation is associated with mechanical weathering rather than biogeochemical weathering, and periods of increased glaciation should lead to lower mobilization rates of phosphorus rather than higher; (3) DSDP and ODP drill sites are for a large part located in the deeper waters of open oceans, and the shallower parts of shelf areas are generally underrepresented—this means that the effect of sea-level change and the associated shift in sedimentary depo-centers and phosphorus burial rates may influence phosphorus burial rates; and (4) ocean-inherent changes, such as the capacities to transform occluded phosphate phases into bio-available phases and to store phosphate, also may interfere (e.g., Van Cappellen and Ingall, 1994; Colman and Holland, 2000).

In order to explore the relevance of these observations, we decided to perform three tests. The first test considered observation (1) and consisted of a reexamination of an important part of its early Cretaceous portion (137 to 132 myr; Valanginian-Hauterivian), which is an older part within this compilation and based is on relatively few data in comparison to younger parts. For this test, 575 phosphorus concentrations were measured in eight continental sections in central and southern Europe for their Valanginian and Hauterivian portions (Van de Schootbrugge et al., in press). The resulting compilation correlates very well with the DSDP- and ODP-based data set, which suggests that the curve is robust for this time interval.

The second test considered observation (2) and consisted of a close-up study of the last full glaciation phase. Here different phosphorus phases were analyzed in a selection of eight ODP cores using a sequential extraction method (Ruttenberg, 1992; Anderson and Delaney, 2000; Tamburini, 2001). An important result was that during this last phase of glaciation, variations in phosphorus burial were coupled to climate change, albeit on a shorter time scale, in the range of the precession band frequency. In addition, glacial periods during this last glaciation show comparable to slightly higher phosphorus burial rates than interglacial stages (Tamburini, 2001; Tamburini et al., 2003).

The third test also considered observation (2) and included a detailed analysis of the importance of biogeochemical weathering processes in glaciated areas. Here we selected the Rhône and Oberaar glacier catchments—both situated within the crystalline basement of the Aare massif (central Switzerland)—and performed analyses on the geochemistry of the outlet waters, mineralogy of suspended material, and geochemistry and mineralogy of moraine material of different ages. One outcome was that glaciers have an important potential for increasing biogeochemical weathering rates during and especially immediately after glaciation phases (Hosein et al., in press).

With regard to observation (3), a positive correlation between sea-level change and phosphorus burial rates is given for a major part of the curve, indicating that the effect of sea-level rise in shifting depo-centers toward the continent is negligible relative to the increase in burial rates observed. Finally, with regard to observation (4), ocean-inherent changes in the capacity to transform and store phosphorus may considerably change the ocean residence time of phosphorus. For instance, it may very well be that during a period of increased anoxic conditions, such as the late Cenomanian to early Turonian, may have decelerated phosphorus burial rates on a wider scale (Van Cappellen and Ingall, 1994; see also below). However, the robustness of the curve displayed by the resemblance of the data extracted from biogenic sediments to the total curve (self-similarity) and by the meaningful correlation of the phosphorus burial curve with other paleo-environmental proxies such as sea-level change encourages us to consider the phosphorus burial curve as representative on a global scale.

Proxies for Temporal Change in the Global Carbon Cycle

There are several types of proxies for temporal change in the carbon cycle, all of which will be examined here. A first type is comparable with the one used for phosphorus, a compilation of carbonate and organic carbon burial rates, which is based on a series of DSDP drill sites (figure 7.2; Southam and Hay, 1981). The disadvantages this compilation is that it does not include the data sets of the more recent DSDP and ODP drill legs, and it does not show the same time resolution as the phosphorus curve. Unfortunately, more recent compilations are not available.

The organic carbon curve suggests high organic carbon sedimentation rates prior to approximately 100 myr, at around 80 and 40 myr, and from approximately 25 myr on. The carbonate curve shows maxima for the period around 65 and 45 myr and an increasing trend from approximately 25 myr onwards. The two curves show quite good correlation in the time intervals prior to 90 myr and from 65 myr on. The period between 90 and 65 myr is characterized by poor correlation between the two proxies.

A second type of proxy is based on the $\delta^{13}C$ whole-rock record of marine carbonates, which is interpreted as a measure of the ratio of carbonate carbon burial and organic carbon production and burial at a given time. Here sufficient data sets are available and the resolution is comparable with that of the phosphorus burial curve (figure 7.2). The $\delta^{13}C$ record correlates poorly to the sedimentation curves of carbonates and organic carbon, which is probably related to the differences in resolution and data density, in addition to original differences between production and burial rates.

A third type consists of modeled atmospheric CO_2 concentrations and temporal changes therein (figure 7.2; Berner, 1994). Here, too, resolution is poor in comparison with the phosphorus and $\delta^{13}C$ data sets. The curve shows a rather steady decrease toward the present, with relative maxima around 130 and 50 myr.

For the purpose of this chapter, we will concentrate on the $\delta^{13}C$ curve as a reference curve, and compare its evolution against the trends shown by the phosphorus curve. The other proxy curves will be used as additional sources of information.

Temporal Links Between the Carbon and Phosphorus Cycles and Their Interpretation

From a comparison of the phosphorus burial and the $\delta^{13}C$ curves, we intend to gain information on the character of coupling between the carbon and phosphorus cycles for the last 160 myr, as well as on the types of reactions and feedback mechanisms during periods of major environmental change. For this purpose, we divided this time period into five distinct intervals (figure 7.2), which we will examine separately.

Interval 160–143 myr

The oldest interval includes the last 160–143 myr (Bajocian–Berriasian), and is characterized by an increase in both $\delta^{13}C$ and phosphorus burial values up to 154 myr, followed by an unsteady decrease in values. With regard to the main trend, the two curves correlate rather well. This general trend is also seen in sea level (albeit with an offset between the two maxima). Atmospheric CO_2 values decrease until 150 myr and slowly increase thereafter. The rather good correlation between the $\delta^{13}C$ and phosphorus burial curve is explained by a period of warming and sea level rise until 154 myr, which led to intensified continental weathering and phosphorus mobilization (e.g., Bartolini and Cecca, 1999). The increased availability of phosphorus induced higher productivity and organic carbon burial rates (relative to carbonate carbon), which again is expressed by increasing $\delta^{13}C$ values. The period between 154 and 143 myr was one of sea-level fall, which probably was linked to a period of cooling (drop in atmospheric CO_2) and less intense weathering. Productivity levels were lowered and $\delta^{13}C$ values became lighter. This change is explained by a negative feedback reaction following the period of warming (figure 7.3).

The causes of the warming and sea-level rise have not been identified yet. Among the proposed mechanisms, an episode of intense volcanism and increased volcanic CO_2 outgassing is a likely candidate (Bartolini and Cecca, 1999).

Interval 143–101 myr

During the period between 143 and 101 myr (Berriasian to Albian), the general trends in phosphorus burial and $\delta^{13}C$ curve also correlate rather well. A pronounced increase in phosphorus burial at the beginning of this period was followed by an increase in $\delta^{13}C$ values. A relative minimum is seen for both curves at around 130–125 myr, whereas a relative

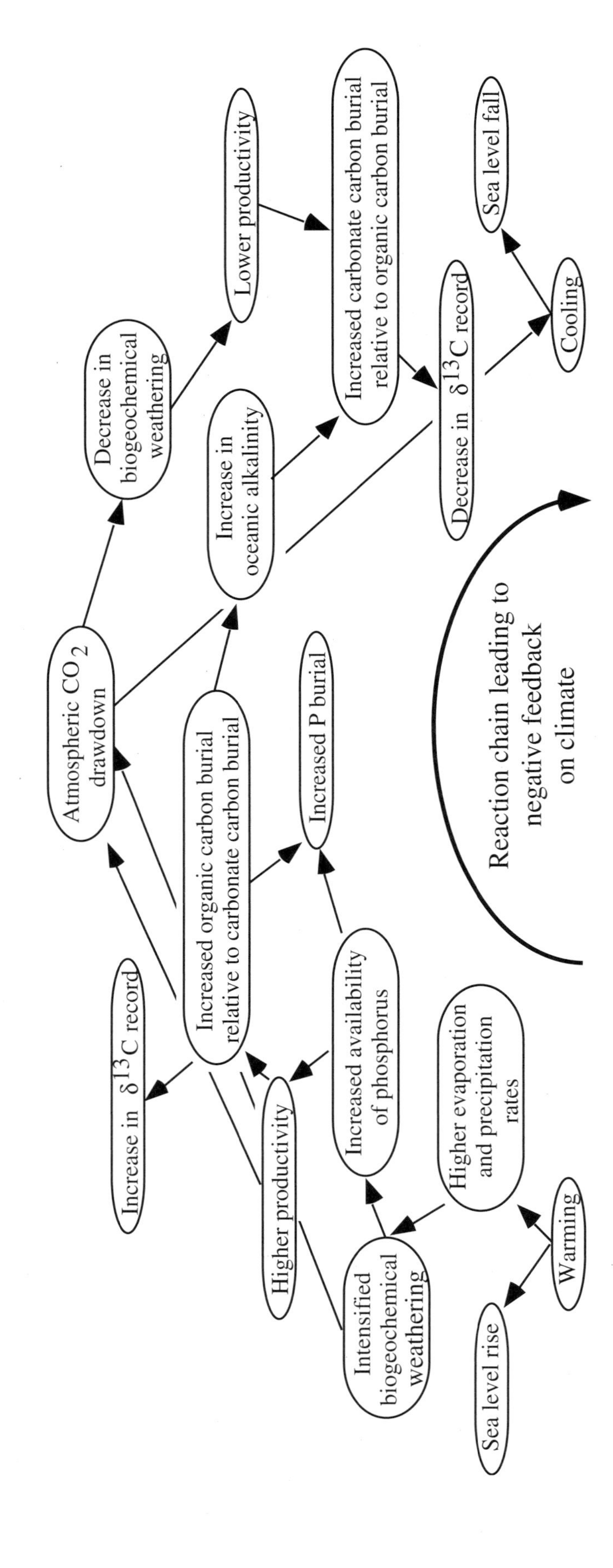

Figure 7.3
Network of reactions leading to negative feedback proposed for the period between 160 and 32 myr (after Föllmi et al., 1994).

maximum is present between 120 and 112 myr. During this period, the large increases in phosphorus burial led the increases in $\delta^{13}C$ by several million years. This may hint at a delay in response time by the $\delta^{13}C$ record to phosphorus-induced changes in productivity, due eventually to a difference in response time between phosphorus (short present-day oceanic residence time of around 10 kyr) and carbon (longer present-day oceanic residence time of several 100 kyr). It may also be noted here that the general shape of the phosphorus burial curve is mirrored by the long-term trend in sea-level change. This link may confirm the type of interaction between climate, continental weathering, and phosphorus mobilization, as was postulated for the previous period (e.g., Föllmi et al., 1994).

The period between 136 and 133 myr is characterized by a decoupling between the two signatures: The phosphorus burial record remains high until 133 myr, whereas the $\delta^{13}C$ record shows a decline in values at around 136 myr. This decoupling is interpreted as the consequence of the following chain of events (figure 7.3; Van de Schootbrugge, 2001; Van de Schootbrugge et al., in press): Analogous to the scenario invoked for the previous period, a warming period associated with a phase of pronounced sea-level rise led to intensified weathering on the continents, and therefore to increased phosphorus availability and higher productivity rates. One of the consequences of this environmental change was a widespread drowning episode of carbonate platforms (Schlager, 1981; Föllmi et al., 1994) and a presumed general increase in the ratio of buried organic carbon and carbonate carbon indicated by the positive excursion in $\delta^{13}C$ values (Lini et al., 1992). This increase and sustained high weathering rates led to a build up of oceanic alkalinity, which was then compensated for by a large increase in carbonate production and a second change in the ratio of buried organic carbon to carbonate carbon toward lower values (figure 7.3; Weissert et al., 1998; Van de Schootbrugge et al., in press).

Episodes of intensified volcanism and increased CO_2 output during the Valanginian, and Aptian and Albian (Larson, 1991; Föllmi et al., 1994; Van de Schootbrugge et al., in press) have been identified as the general cause of global warming during this time period.

Interval 101–65 myr

The time period of 101–65 myr (Albian–Maastrichtian) shows a rather positive correlation between phosphorus burial and long-term sea-level change, and the low-resolution organic carbon sedimentation curve appears to confirm both trends in such a way that the link between volcanism, warming, intensified weathering, mobilization of phosphorus, higher productivity, and higher organic carbon burial rates may also apply to this time period (figure 7.3).

In contrast to the two previous periods, the $\delta^{13}C$ curve shows almost no correlation with the phosphorus burial curve. An explanation may be found in the shape of the carbonate carbon sedimentation curve, which shows a decrease between 101 and 90 myr, followed by a marked increase in values from 90 myr on. This increase is correlated with an important phase of proliferation in planktonic carbonate-producing calcisphaerulides and Foraminifera (e.g., Scholle et al., 1983), and a corresponding increase in sedimentation rates of pelagic carbonates. This new mechanism of carbonate production leading to an efficient transfer of carbonate carbon to the oceanic sedimentary reservoir may have had an influence on the $\delta^{13}C$ signature in a way which has not yet been explored in detail. Possibly the formation of organic carbon associated with pelagic carbonate production was high as well, and this may have had a buffering effect on the global planktonic $\delta^{13}C$ signature. The burial rate of organic carbon, however, remained relatively low, which was eventually related to longer transfer ways (from 80 myr on).

A particular case is the "anoxic event" at around 90 myr (Cenomanian-Turonian boundary; e.g., Arthur and Schlanger, 1979; Arthur et al., 1985; Jenkyns, 1980), which is characterized by a pronounced positive excursion in $\delta^{13}C$ values, and relatively low corresponding values for phosphorus burial. Here the scenario developed by Van Cappellen and Ingall (1994) may help to explain this particular decoupling. The period around 90 myr is considered to be a time with particularly widespread dysaerobic conditions in oceanic bottom waters and a corresponding poor retention potential for phosphorus in anoxic oceanic sediments. The positive $\delta^{13}C$ excursion during this period was probably the expression of a shift in the ratio of carbonate carbon burial and organic carbon burial in favor of organic carbon, due in part to a widespread drowning event on carbonate platform systems (e.g., Jenkyns, 1991).

Interval 65–32 myr

The period between 65 and 32 myr (Paleocene–Oligocene) is characterized by a steep increase in phosphorus burial at its onset, followed by a long and irregular decrease in values. The general trend toward lower values was superimposed on by relative maxima around 52, 43, and 37 myr. The general trend in phosphorus burial is only partly mirrored by the $\delta^{13}C$ curve. Sea-level change appears to follow phosphorus burial, which may suggest that the feedback between climate, phosphorus mobilization, and productivity may be also applicable for this period. The trend in rising atmospheric CO_2 values until 50 myr, the initial marked increase in $\delta^{13}C$ values, and the initial decrease in the ratio of carbonate carbon and organic carbon burial are compatible with this scenario.

The increases in organic carbon and carbonate carbon burial from 60 myr on (with maxima at 45 and 40 myr, respectively) are not correlatable to any other trend, with the possible exception of the phosphorus burial curve. It would be good to have more detailed compilations of organic and carbonate carbon burial for this time period, in order to corroborate this lack of correlation.

Interval 32–0 myr

This last interval differs from the previous periods in a substantial way. Phosphorus burial and sea-level change are inversely correlated, which may indicate that continental weathering and phosphorus mobilization are favored in periods of long-term cooling (Föllmi, 1995). With regard to the phosphorus burial and the $\delta^{13}C$ curves, a similar inverse correlation is seen for the period between 32 and 28 myr, and between 6 myr and the present. The entire period is characterized by a long-term increase in carbonate carbon and organic carbon sedimentation rates, and by a long-term decrease in atmospheric CO_2 values. A first step of cooling from 32 to 25 myr was accompanied by an increase in phosphorus burial and in carbonate carbon and organic carbon sedimentation. This was followed by an interval of warming (25–14 myr), characterized by relatively high phosphorus burial rates and by a slower increase in carbonate and organic carbon sedimentation rates. The period from 14 myr on was again characterized by a period of cooling, which is correlated to further important increases in phosphorus burial rates, and in carbonate carbon and organic carbon sedimentation rates.

The inverse correlation between phosphorus burial rates and sea-level change suggests that the rates of phosphorus influx into the oceans covaried with climate change in such a way that phases of cooling were coupled with increased phosphorus mobilization rates. As already stated, the change in this coupling at 32 myr is explained by the onset of widespread glaciation, which may have acted as the driving force in continental weathering from 32 myr on (Föllmi, 1995, 1996). Glacial activity is very efficient in abrading and grinding bedrock, thereby providing fine-grained material which may subsequently be subjected to biogeochemical weathering. The change from a positive correlation between sea-level rise and phosphorus mobilization to an inverse one suggests that the effect of glaciation was powerful enough to change an entire chain of feedback mechanisms (figure 7.4). The uplift of the Himalaya and the Alps during this period may have been an additional factor which helped to raise weathering rates in general (e.g., Raymo, 1994).

The mechanisms invoked to slow down and stop this chain of positive feedback are not yet positively identified. The development of permafrost soils in periglacial areas and the change from warm-based to cold-based glaciers, which are frozen to the underlying bedrock, may have been important factors. Both phenomena may slow down physical weathering (and with that biogeochemical weathering) and transport of weathered material in such a way that the chain of positive feedback is halted.

In addition to the reversal in signs on the feedback chain, it appears that the rate of carbonate sedimentation was less hampered by the higher rates of phosphorus availability during cooling phases within this last period, whereas in the previous periods—during phases of warming—high phosphorus mobilization rates often led to drowning events of carbonate platforms and a general decrease in carbonate sedimentation. This may explain why the $\delta^{13}C$ signal becomes more negative in times of cooling and sea-level fall, in spite of the increased availability of phosphorus.

Relationships Between Phosphorus and Carbon in Changing Ecosystems

The coupling between phosphorus and carbon is also of interest in regionally confined ecosystems. An example is given here, in which we suggest a nonlinear coupling of the two elements, due to the dynamics of the ecosystem concerned.

The example concerns the temporal evolution of a shallow-water carbonate platform system during the latest Jurassic and early Cretaceous (approximately 147–112 myr). This platform developed along the former northern Tethyan margin, and large portions are presently preserved in the central Europe Jura Mountains and Helvetic Alps (Funk et al., 1993; Föllmi et al., 1994). Three phases have been distin-

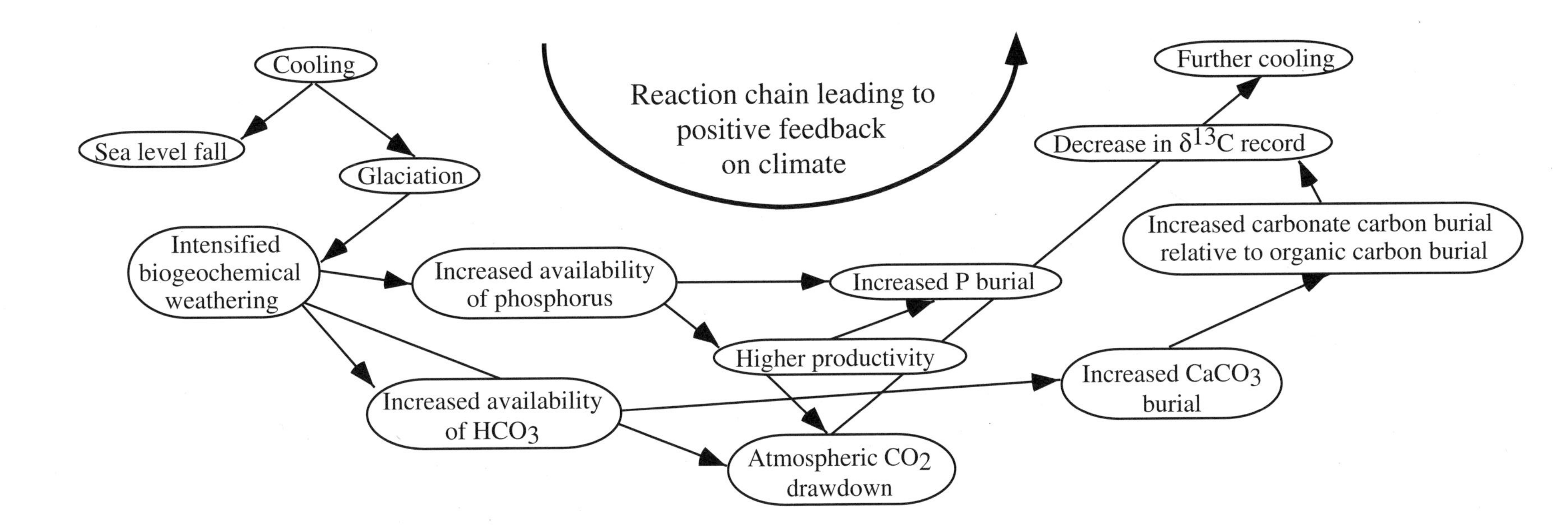

Figure 7.4
Network of reactions leading to positive feedback proposed for the period between 32 myr and present (after Föllmi, 1996).

guished in the evolution of this platform system: (1) platform growth in a "coral-oolith" mode, with carbonate production dominated by reef-type organisms (corals, chaetetids, rudists, stromatoporoids, calcareous sponges), as well as green algae and benthic Foraminifera. Another important mechanism is oolite formation ("chlorozoan mode," according to Lees and Buller, 1972); (2) platform growth in a "crinoid-bryozoan" mode, with carbonate production dominated by crinoids, bryozoans, bivalves, and brachiopods. Reef-related organisms are lacking ("foramol" mode, according to Lees and Buller, 1972); and (3) phases of platform drowning which are characterized by erosion and dissolution of preexisting carbonates, as well as the formation of thin and highly condensed, phosphate-rich horizons. The principal cause of change between the different modes was changes in nutrient levels, especially of phosphorus, as is shown by the changes in phosphorus burial rates in dependency of facies and the presence of macroscopic accumulations of phosphate-rich sediments during periods of platform drowning. Other factors such as changes in water temperature, related to sea-level change and the opening of gateways to the Boreal realm, were also of consequence (Van de Schootbrugge, 2001; Van de Schootbrugge et al., in press).

Sedimentation rates in the "coral-oolith" mode average around 2 cm/kyr and are maximally around 10 cm/kyr, whereas in the "crinoid-bryozoan" mode, they amount to average values around 5 cm/kyr and maximal values around 50 cm/kyr. The change from the first mode to the second was usually rather rapid and led to the buildup of well distinguishable lithostratigraphic formations. This suggests a link between phosphorus and carbonate carbon production (and sedimentation) in the following way: Carbonate production in the "coral-oolith" mode was under mostly oligotrophic conditions; an increase in phosphorus availability led to a rather rapid change in the composition of the platform ecosystem, including the loss of reef-related organisms. Carbonate production rates in the "crinoid-bryozoan" mode are significantly higher in comparison with the previous mode. Further increases in phosphorus availability led to a strong reduction in carbonate production and to the development of phosphate-rich drowning surfaces. In figure 7.5, we depict this development in a qualitative way to illustrate the development and the links observed between phosphorus availability and carbonate production.

Similar relationships are likely to occur in other ecosystems, such as coastal marine upwelling centers, where calcareous phytoplankton may be replaced by siliceous phytoplankton as a function of upwelling intensity, or in basins that become progressively eutrophic. In the analysis of population dynamics, Robert May pioneered the use of this type of "bifurcation" diagram (e.g., May and Oster, 1976), stating that dynamics are predictable until a threshold value is reached, from which point two states are possible (bifurcation). Within these two states, threshold values may be reached again, and four states are possible. Bifurcation may continue until a fully undeterminable situation is reached. From the research by May and many others, it appears that the relationships depicted in this example may be widespread in the biosphere.

Nonlinear Elements in the General Coupling of Phosphorus and Carbon and Its Importance to the Gaia Hypothesis

The relationships between phosphorus and carbon in the above-mentioned example are indicative of a nonlinear and complex system. This system has imminent properties which set it off from a linear and determinable system in a mathematic sense.

1. Figure 7.5 is characteristic of a nondetermined, complex system, in which different states are possible for the same set of conditions. In our example, production rates are dependent on the type of ecosystem available; once an ecosystem is not replaced by an ecosystem capable of coping with increased trophic conditions, productivity rates may break down, such as during phases of platform drowning.

2. The long-term evolution of phosphorus burial between 160 and 32 myr is characterized by the presence of an asymmetric pattern of development with main periodicities of around 33 and 18 myr (Tiwari and Rao, 1999). The asymmetry is expressed by short periods (3–5 myr) of rapid increase in phosphorus burial, followed by longer periods (12–30 myr) of irregular decrease. Superimposed on this long-term trend, shorter trends are visible which show a similar asymmetry (i.e., rapid increases in phosphorus burial followed by longer periods of decrease). This repetition of the same pattern on different time scales resembles a pattern of self-similarity, which is again indicative of a complex system. Comparable asymmetric patterns are known from other proxies, such as sea-level change (Haq et al., 1987), the $\delta^{18}O$ record in

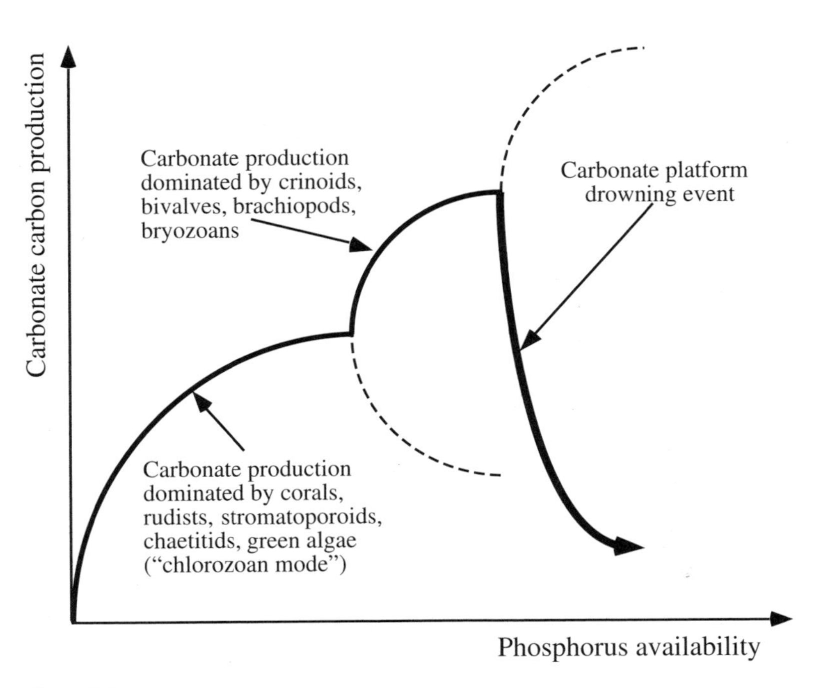

Figure 7.5
Qualitative bifurcation diagram depicting nonlinear relationships between phosphorus availability and carbonate carbon production.

ice cores, and the atmospheric CO_2 record measured in ice cores (e.g., Petit et al., 1999).

3. The asymmetric patterns in the phosphorus burial curve are characterized by the presence of minimal and maximal values between the periods of slow decrease and rapid increase. The extreme values are considered here to be threshold values, which mark a turning point within the chain of feedback reaction (i.e., the maximal values may mark the moment where negative feedback starts to have an effect on the environment (figure 7.3), whereas the minimal values appear to mark the effect of extrinsic events, which have a large impact on the entire system. The asymmetric shape appears, therefore, to be an expression of the interplay between extrinsic events (such as volcanic episodes; e.g., Courtillot, 1994), eventually helped by positive feedback (e.g., release of methane from clathrates during warming; Dickens et al., 1995; Kennett et al., 2000), and the effect of negative feedback, which tends to stabilize and prolong "stable" environmental conditions. The asymmetric shape is less pronounced in the phosphorus burial curve from the period from 32 myr to present, and this may be related to the change in sign on the postulated feedback mechanism (figure 7.4).

The presence of asymmetry in reaction, self-similarity in patterns, threshold values, nondeterminable behavior, and feedback reactions which may change signs according to environmental conditions—as suggested by the proxies used here to trace temporal change in the global phosphorus and carbon cycles—is characteristic of the complex, nonlinear dynamic character of the biosphere. It is the interaction of changes extrinsic to the biosphere, such as cyclic changes in astronomic parameters (Milankovitch cycles) and episodes of intense volcanism, and intrinsic to the biosphere, such as evolutionary patterns and degree of diversity in ecosystems, which renders this complexity to the biosphere system.

Within this pattern of complexity, we identify periods—such as those of slow and irregular decrease in phosphorus burial—in which negative feedback may have led to conditions of greater stability. In such periods, the pattern of change may be considered Gaian, albeit marked by great irregularity, as is shown by the repetitive character of change on different time scales. Of interest here is that during such periods, sea level was lowered, climate became cooler, weathering rates decreased, and phosphorus burial rates decreased. A feedback-guided convergence emerges from this pattern toward conditions of minimal energy use, toward a "saving mode," in which the transfer of energy and matter within the biosphere is minimalized. A similar trend is observed during

Pleistocene glaciation phases, where the trend goes toward colder temperatures in a comparable way (e.g., Petit et al., 1999).

The long-term trend toward a cooler climate was followed by a shorter phase of sea-level rise, warming, increased weathering rates, and increased phosphorus burial rates. Here the speed of change is related to the effect of extrinsic forcing—eventually helped by positive feedback; and it is during such periods that efficient negative feedback and regulation along Gaian pathways become important, in order to avoid protracted destabilization of the entire biosphere.

The period from 32 myr to the present is a different case: the role of phosphorus has changed in such a way that—in this last period—it has become an important driver in pushing positive feedback reactions, which seem to accelerate climate cooling (figure 7.4), whereas prior to 32 myr, it catalyzed negative feedback in times of warming (figure 7.3). This switch in feedback was forced by the change in the type of climate responsible for enhancing biogeochemical continental weathering rates. This change in the role of a biophile element in feedback processes has important implications for our understanding of the way the system Earth functions. The possibility of switching roles, and thereby changing signs on feedback mechanisms, adds an additional degree of complexity to environmental change and biosphere response.

In how far is the Gaia hypothesis compatible with the nonlinear dynamics described here? Signs of negative, stabilizing feedback are present in the records of the carbon and phosphorus cycles, and they allude to the regulative capacities of the Earth system. The shape of the patterns themselves, however, suggests that the nonlinear dynamics displayed during environmental change and induced by an interplay of intrinsic and extrinsic factors, prevent full-fledged stable conditions during longer time periods. During the last 160 myr, the feedback mechanisms described here appear to have remediated rather than regulated environmental conditions in response to the perils of global change, be it by extrinsic events, cyclic extrinsic change, intrinsic evolution and adaptation, or the unexpected consequences of complex dynamics.

Conclusions

Proxies used here in order to reconstruct temporal change in the global phosphorus and carbon cycles for the past 160 myr suggest the presence of feedback reactions which may change in character through time, the importance of threshold values, the possibility of self-similarity in patterns on different time scales, and the probability of nondetermined states of carbonate carbon and organic carbon production that depend on the trophic level. These phenomena are all indicative of the nonlinear and complex character of coupling between the elemental cycles of phosphorus and carbon, and associated change in environmental conditions in response to extrinsic and intrinsic change. Within the patterns of change, the effect of negative feedback is important; it plays a role in counteracting the effects of extrinsic change and stabilizes environmental conditions. The overall trend is toward a state of minimal transfer of energy and material within the biosphere. Even if the effect of negative feedback is important, it is not an element that permanently stabilizes environmental conditions. Global change related to extrinsic factors such as volcanic events, impacts, and cyclic change in astronomical parameters, as well as intrinsic factors such as the evolution and adaptation of ecosystems within the biosphere, overrules this tendency and imposes itself in such a way that the associated feedback mechanisms remediate rather than regulate environmental conditions.

The initial question posed here, if phosphorus is a servant faithful to Gaia, is negated. Due to the change in climate conditions at around 32 myr, continental biogeochemical weathering—the prime mechanism through which phosphorus is mobilized—is enhanced by glaciation rather than warm and humid climates. This change turns the sign on feedback mechanisms in which phosphorus is involved as an essential nutrient, and the servant appears to trim its sails to the wind (climate) rather than remaining trustworthy to Gaia.

Acknowledgments

We would like to express our thanks here to James Miller for his assistance in the revision of the manuscript and to Albert S. Colman and an anonymous reviewer for their constructive comments on an earlier version. We would also like to acknowledge the financial support of the Swiss National Science Foundation (projects 21-30611.91, 21-51616.97, 21-53997.98, 20-61485.00, and 21-65183.01) and of the University of Neuchâtel.

References

Anderson, L. D., and M. L. Delaney. 2000. Sequential extraction and analysis of phosphorus in marine sediments: Streamlining of the SEDEX procedure. *Limnology and Oceanography*, 45, 509–515.

Arthur, M. A., W. E. Dean, and S. O. Schlanger. 1985. Variations in the global carbon cycle during the Cretaceous related to climate, volcanism, and changes in atmospheric CO_2. In E. T. Sundquist and W. S. Broecker, eds., *The Carbon Cycle and Atmospheric CO_2: Natural Variations Archean to Present*, American Geophysical Union, Geophysical Monograph, 32, pp. 504–529. Washington, DC: American Geophysical Union.

Arthur, M. A., and S. O. Schlanger. 1979. Cretaceous "oceanic anoxic events" as causal factors in development of reef-reservoired giant oil fields. *American Association of Petroleum Geologists, Bulletin*, 63, 870–885.

Bartolini, A., and F. Cecca. 1999. 20 My hiatus in the Jurassic of Umbria–Marche Apeninnes (Italy): Carbonate crisis due to eutrophication. *Comptes Rendus, Académie des Sciences* (Paris), *Sciences de la Terre et des Planètes*, 329, 587–595.

Berner, R. A. 1994. Geocarb II: A revised model of atmospheric CO_2 over Phanerozoic time. *American Journal of Science*, 194, 56–91.

Codispoti, L. A. 1989. Phosphorus vs. nitrogen limitation of new and export production. In W. H. Berger, V. S. Smetacek, and G. Wefer, eds., *Productivity of the Ocean: Present and Past*, pp. 377–394. Chichester, UK: John Wiley.

Colman, A. S., and H. D. Holland. 2000. The global diagenetic flux of phosphorus from marine sediments to the oceans: Redox sensitivity and the control of atmospheric oxygen levels. In C. R. Glenn, L. Prévôt-Lucas, and J. Lucas, eds., *Marine Authigenesis: From Global to Microbial*, SEPM spec. vol. 66, 53–75. Tulsa, OK: Society of Sedimentary Geology.

Courtillot, V. 1994. Mass extinctions in the last 300 million years: One impact and seven flood basalts? *Israel Journal of Earth Sciences*, 43, 255–266.

Dickens, G. R., J. R. O'Neil, D. K. Rea, and R. M. Owen. 1995. Dissociation of oceanic methane hydrate as a cause of the carbon isotope excursion at the end of the Paleocene. *Paleoceanography*, 10, 965–971.

Föllmi, K. B. 1995. 160 m.y. record of marine sedimentary phosphorus burial: Coupling of climate and continental weathering under greenhouse and icehouse conditions. *Geology*, 23, 859–862.

Föllmi, K. B. 1996. The phosphorus cycle, phosphogenesis and marine phosphate-rich deposits. *Earth-Science Reviews*, 40, 55–124.

Föllmi, K. B., H. Weissert, M. Bisping, and H. Funk. 1994. Phosphogenesis, carbon-isotope stratigraphy, and carbonate-platform evolution along the Lower Cretaceous northern Tethyan margin. *Geological Society of America, Bulletin*, 106, 729–746.

Funk, H., K. B. Föllmi, and H. Mohr. 1993. Evolution of the Tithonian-Aptian carbonate platform along the northern Tethyan margin, eastern Helvetic Alps. In T. Simo, R. W. Scott, and J. P. Masse, eds., *Atlas of Cretaceous Carbonate Platforms*, American Association of Petroleum Geologists Memoir 56, pp. 387–408. Tulsa, OK: The Association.

Haq, B. U., J. Hardenbol, and P. R. Vail. 1987. Chronology of fluctuating sea levels since the Trias. *Science*, 235, 1156–1167.

Hosein, R., K. Arn, P. Steinmann, and K. B. Föllmi. In press. Calcite, the principal control on cation fluxes from the Rhône and Oberaar glaciers, Switzerland. *Geochimica et Cosmochimica Acta*.

Jenkyns, H. C. 1980. Cretaceous anoxic events: From continents to oceans. *Journal of the Geological Society of London*, 137, 171–188.

Jenkyns, H. C. 1991. Impact of Cretaceous sea-level rise and anoxic events on the Mesozoic carbonate platform of Yugoslavia. *American Association of Petroleum Geologists, Bulletin*, 75, 1007–1017.

Kennett, J. P., K. G. Cannariato, I. L. Hendy, and R. J. Behl. 2000. Carbon isotopic evidence for methane hydrate instability during Quaternary interstadials. *Science*, 288, 128–133.

Larson, R. L. 1991. The latest pulse of the Earth: Evidence for a mid-Cretaceous superplume. *Geology*, 19, 547–550.

Lees, A., and A. T. Buller. 1972. Modern temperate water and warm water shelf carbonate sediments contrasted. *Marine Geology*, 13, 1767–1773.

Lini, A., H. Weissert, and E. Erba. 1992. The Valanginian carbon isotope event: A first episode of greenhouse climate conditions during the Cretaceous. *Terra Nova*, 4, 374–384.

Lovelock, J. 1988. *The Ages of Gaia: A Biography of Our Living Earth*. New York: Norton.

Lovelock, J. E., and L. Margulis. 1974. Atmospheric homeostasis by and for the biosphere: The Gaia hypothesis. *Tellus*, 26, 2–10.

May, R. M., and G. F. Oster. 1976. Bifurcations and dynamics complexity in simple ecological models. *American Naturalist*, 110, 573–599.

Petit, J. R., J. Jouzel, N. I. Raynaud, N. I. Barkov, J. M. Barnola, I. Basile, M. Bender, J. Chappellaz, M. Davis, G. Delaygue, M. Delmotte, V. M. Kotlyakov, M. Legrand, V. Y. Lipenkov, C. Lorius, L. Pépin, C. Ritz, E. Saltzman, and M. Stievenard. 1999. Climate and atmosphric history of the past 420,000 years from the Vostok ice core, Antarctica. *Nature*, 399, 429–436.

Raymo, M. E. 1994. The Himalayas, organic carbon burial, and climate in the Miocene. *Paleoceanography*, 9, 399–404.

Ruttenberg, K. C. 1992. Development of a sequential extraction method for different forms of phosphorus in marine sediments. *Limnology and Oceanography*, 37, 1460–1482.

Schlager, W. 1981. The paradox of drowned reefs and carbonate platforms. *Geological Society of America, Bulletin*, 92, 197–211.

Scholle, P. A., M. A. Arthur, and A. A. Ekdale. 1983. Pelagic environment. In P. A. Scholle, D. G. Bebout, and C. H. Moore, eds., *Carbonate Depositional Environments*, American Association of Petroleum Geologists Memoir 33, pp. 620–691. Tulsa, OK: The Association.

Southam, J. R., and W. W. Hay. 1981. Global sedimentary mass balance and sea level changes. In C. Emiliani, ed., *The Sea*, vol. 7, pp. 1617–1684. New York: John Wiley and Sons.

Tamburini, F. 2001. Phosphorus in marine sediments during the last 150,000 years: Exploring relationships between continental weathering, productivity, and climate. Ph.D. thesis, University of Neuchâtel.

Tamburini, F., et al. 2003. Sedimentary phosphorus record from the Oman margin: New evidence of high productivity during glacial periods. *Paleoceanography*, 18.10.102g/2000PA000616.

Tiwari, R. K., and K. N. N. Rao. 1999. Periodicity in marine phosphorus burial records. *Nature*, 400, 31–32.

Tyrrell, T. 1999. The relative influences of nitrogen and phosphorus on oceanic primary production. *Nature*, 400, 525–531.

Van Cappellen, P., and E. D. Ingall. 1994. Benthic phosphorus regeneration, net primary production, and ocean anoxia: A model of the coupled marine biogeochemical cycles of carbon and phosphorus. *Paleoceanography*, 9, 677–692.

Van de Schootbrugge, B. 2001. Influence of paleo-environmental changes during the Hauterivian (early Cretaceous) on carbonate deposition along the northern margin of the Tethys: Evidence from geochemical records (C-, O-, Sr-isotopes, P, Fe, Mn. Ph.D. thesis, University of Neuchâtel.

Van de Schootbrugge, B., K. B. Föllmi, L. G. Bulot, and S. J. Burns. 2000. Paleoceanographic changes during the early Cretaceous (Valanginian–Hauterivian): Evidence from oxygen and carbon stable isotopes. *Earth and Planetary Science Letters*, 181, 15–31.

Van de Schootbrugge, B., O. Kuhn, T. Adatte, P. Steinmann, and K. B. Föllmi. In press. Mesotrophic carbonate production and high P-accumulation rate during the Hauterivian: Regulating the Valanginian global carbon cycle perturbation. *Palaeogeography, Palaeoclimatology, Palaeoecology*.

Weissert, H., A. Lini, K. B. Föllmi, and O. Kuhn. 1998. Correlation of early Cretaceous carbon isotope stratigraphy and platform drowning events: A possible link? *Palaeogeography, Palaeoclimatology, Palaeoecology*, 137, 189–203.

8
Self-Regulation of Ocean Composition by the Biosphere

Lee R. Kump

Abstract

The major element, nutrient, and trace metal concentrations of the ocean are regulated by a host of mechanisms involving biological and physical processes. This chapter explores those mechanisms in the context of the evolution of ocean chemistry on geologic time scales. Large fluctuations in the salinity of the ocean are unlikely to have occurred, although periods following the deposition of evaporite "giants" may have been anomalously low in salinity. The Ca concentration and alkalinity of the ocean are regulated by the sensitive response of the ocean's calcium carbonate compensation depth to variations in the oceanic $CaCO_3$ saturation state. Potential metal toxicity is prevented by the production of specific metal-binding ligands by the marine biota. Overall, the system is resilient, but human activity may be perturbing important state variables of the ocean to an extent that will be detrimental to marine life.

Introduction

What controls the composition of the ocean? Is it in fact regulated, or does it vary without constraint in response to variations in input and removal of its chemical constituents? These questions have for decades intrigued marine biogeochemists, including Sillén (1961), Mackenzie and Garrels (1966), Holland (1972), Berner et al. (1983), and Hardie (1996). Although the importance of biological processes was recognized in all of these works, the possibility that the biota was involved in a self-regulatory mechanism for ocean composition was not explicitly addressed. The potential for self-regulation emerges as a direct consequence of feedbacks in natural systems, and does not require teleological mechanisms, as demonstrated by Watson and Lovelock (1983) with the Daisyworld model.

The productivity of the marine biota depends on a number of physical factors, including temperature, light, nutrient availability and toxic metal activity. Through growth and metabolism, the biota can in turn influence each of these factors. For example, the settling of biogenic particles from surface waters and their remineralization in deep waters (the "biological pump") removes nutrients and trace metals from surface waters and enriches them in deep waters, creating a steady-state chemical stratification of the ocean. Moreover, because a small fraction of the elemental flux associated with the biological pump escapes remineralization and is removed to the sediments, biological productivity serves as a permanent sink for elements from the ocean, balancing the riverine, aerosol, and hydrothermal sources. Thus there is feedback between biological productivity and ocean composition on a variety of time scales, and the potential for self-regulation. This chapter explores these feedbacks and evaluates the possibility of self-regulation for the major element and trace element composition of the ocean. The issue of nutrient regulation has been reviewed by Falkowski et al. (1998), Tyrell (1999), and Lenton and Watson (2000), and thus will not be discussed here.

Major Elements

Salinity

The salinity of the ocean is determined by inputs from weathering of crustal rocks and output primarily through sedimentation. Among the dominant contributors to ocean salinity, Na^+, Ca^{2+}, Cl^-, and SO_4^{2-} have important sinks with evaporite minerals (NaCl, halite and $CaSO_4{\cdot}2H_2O$, gypsum). As seawater evaporates, calcite and dolomite, already supersaturated in seawater, precipitate. After the volume has been reduced by a factor of 75 percent or so, gypsum precipitates, and then halite at 90 percent evaporation. Further evaporation yields additional sodium, potassium, and magnesium salts of chloride, sulfate, and borate.

Evaporite formation requires special circumstances that allow evaporative concentration. These include a basin located in a region of net evaporation (e.g., the subtropics) with extremely limited exchange of seawater with the open ocean. Generally a sill separates these basins from the ocean. The restriction provided by the sill increases the water residence time in the basin, allowing for evaporative concentration, but also permits a slow replenishment of seawater. This replenishment, together with crustal subsidence of the basin, allows for the accumulation of thick deposits of evaporite minerals.

Such conditions are not common in the modern world. However, the geologic record indicates that at particular times in Earth history they have been, and evaporite "giants" were formed. Good examples include the Miocene "Messinian" evaporites of the Mediterranean region and the Permian "Castile" evaporites of Texas and "Zechstein" evaporites of Europe. The reason for this confluence of forcing factors is likely happenstance. However, in many cases the "sill" that restricts circulation is biogenic reef. This is somewhat puzzling, because many reef-building organisms today cannot survive in hypersaline conditions. Friedman and Sanders (1978, p. 382) present this as a paradox: "Organisms prefer not to be pickled in brine! Yet in the rock record, reefs and evaporites commonly are in lateral contact." The solution may be that the spatial association belies a temporal association; reefs form under normal salinity in arid areas of minimal riverine discharge, then sea-level fall or crustal uplift brings the reef above sea level, isolating the basin behind, killing the reef-makers, and stimulating evaporite deposition. Sea level rises again, reef organisms become reestablished, and the cycle repeats itself. Sea-level oscillations create an interfingering of carbonate and evaporite facies in the rock record (e.g., Sarg, 2001).

If in fact this is the extent of the relationship between reef building and evaporite deposition, then no true feedback involving reefs and evaporites exists that is capable of regulating ocean salinity. Indeed, no compelling feedback mechanism has been proposed. Based on this association between carbonate reef and evaporite, Lovelock (1991), citing unpublished work of Lynn Margulis and Greg Hinkle, has suggested that the biota may play a direct role in the regulation of salinity. However, many of the reefs that isolated evaporite basins in the geologic past were composed largely of cement. The classic Permian reefs of west Texas are largely devoid of framework-building taxa and are instead dominated by fine-grained material and marine cement (e.g., James, 1983). The Messinian reefs are also cement-rich. The precipitation of marine cements, in many cases aragonitic, implies high levels of supersaturation with $CaCO_3$ in surface waters, an expected consequence of evaporative concentration of seawater.

Might the salinity of the ocean be unregulated? The total mass of evaporites is approximately equal to 30 percent of the salt content of the ocean today (Holland, 1984). Thus, salinity would rise to only about $S = 45$ if evaporite deposition ceased but weathering continued for tens to hundreds of millions of years. (Even after all the evaporite deposits were weathered, salinity would continue to increase because volcanic HCl would continue to react with primary igneous rocks, albeit slowly.) Many marine organisms today are relatively stenohaline, with salinity limits of 34–36 (e.g., Sverdrup et al., 1942). However, the rate of change of salinity increase would be slow on adaptive and evolutionary time scales. Theoretically, the lower bound on salinity is 0 parts per thousand, but achieving such a low salinity would require a geologically long period of essentially no riverine input with continued evaporite deposition in highly evaporative basins. Both of these end members are highly unlikely. If one assumes that after correcting for the higher sedimentary recycling rate of evaporites (Garrels and Mackenzie, 1971; Veizer, 1988), the relative rate of evaporite deposition is recorded in the preserved evaporite record, one can calculate the resulting change in salinity over Phanerozoic time (e.g., Hay, 2000). The result is that salinity hasn't varied more than 30–40 percent over Phanerozoic time. Thus, the long (> 100 million years) residence time of salt in the ocean, together with the ability of organisms to adapt to a wide range of conditions, suggest that tight regulation is probably not required for the persistence of the marine biota. On the other hand, variations in salinity may have caused fundamental changes in the mode of ocean circulation (Hay, 2000).

Ca^{2+} and Alkalinity

Of all the elements, Ca^{2+} is perhaps the most tightly regulated element inside the cell and in the extracellular circulating fluids in multicellular organisms. Ca^{2+} regulation is also important to prokaryotes, which use calcium in the development of extracellular protection and enzymes (Frausto da Silva and Williams, 1991). In the ocean today, Ca^{2+} concentrations are approximately an order of magnitude larger than that which eukaryotes maintain in their circulating fluids and perhaps three orders of magnitude greater than the critical level for cytoplasm. Above the critical level cells "commit suicide by turning on internal proteases" (Frausto da Silva and Williams, 1991, p. 271). Precipitation of $CaCO_3$ and Ca phosphates and oxalates is a mechanism for maintaining the extracellular circulating fluid in homeostasis. The precipitation of a $CaCO_3$ shell buffers against fluctuations in Ca^{2+} activity, and when it is shed, it allows for the removal of potentially lethal buildups of the element. Thus a reasonable place to search for a feedback mechanism regulating Ca in the marine environment is in the cycling of $CaCO_3$.

The calcium cycle today is dominated by limestone weathering on land and deposition in the ocean. Secondary inputs of Ca^{2+} to the ocean come from the weathering of silicate minerals on land and the exchange of Mg^{2+} for Ca^{2+} during hydrothermal alteration of the seafloor. Imbalances between sources and sinks presumably have driven moderate fluctuations in the Ca^{2+} content of seawater over geologic time (Berner et al., 1983). Stanley and Hardie (1999) used a box model of the ocean's Ca^{2+} cycle to calculate the temporal evolution of seawater composition. They concluded that the Ca^{2+} concentration of the ocean today is at a low for the Phanerozoic (the last 540 million years). The concentration in the mid-Cretaceous (120 Ma) and mid-Cambrian (520 Ma) may have been three times the modern value of 10 mM. These results are generally consistent with other modeling efforts (e.g., Wallmann, 2001) and with the composition of fluid inclusions (Horita et al., 2002).

These observations, together with the tendency to maintain overall ocean saturation with respect to $CaCO_3$, suggests that the mid-Cretaceous and mid-Cambrian were times of high $pCO_2/(HCO_3)^2$ ratios, according to the following equilibrium (Kump and Arthur, 1997):

$$Ca^{2+} + 2\ HCO_3^- \Leftrightarrow CaCO_3 + CO_2 + H_2O$$

$$K = [Ca^{2+}][HCO_3^-]^2/pCO_2$$

Indeed, the mid-Cretaceous and mid-Cambrian are assumed to have been enriched in atmospheric pCO_2, either because of warm climates (mid-Cretaceous) or warmth together with reduced solar luminosity (mid-Cambrian). A consistent explanation would involve enhanced seafloor hydrothermal activity and volcanism that increased the Ca^{2+} concentration of seawater and the CO_2 content of the ocean and atmosphere directly, and the indirect conversion of bicarbonate to CO_2 to maintain the $CaCO_3$ saturation of the ocean.

In detail, the ocean is a well-designed system for maintaining saturation with respect to $CaCO_3$ and for balancing the alkalinity (largely HCO_3^-) inputs from rivers with its removal as $CaCO_3$. Surface waters are supersaturated because of photosynthetic drawdown of CO_2. Deep waters are undersaturated for two reasons: aerobic decomposition of the rain of organic detritus releases CO_2 into the deep waters, and the solubility of $CaCO_3$ is greater under the high pressure and low temperature of the deep sea (e.g., Broecker and Peng, 1982). The saturation horizon occurs at approximately 3–4 km depth in the ocean; above this depth, seafloor sediments are rich in $CaCO_3$, while below it, $CaCO_3$ content decreases rapidly with depth to essentially 0 below approximately 5 km (the $CaCO_3$ compensation depth or CCD).

The average depth of the saturation horizon and CCD are determined by the requirement for alkalinity balance in the ocean. To a first approximation, the flux of $CaCO_3$ from surface waters (the rain rate) is spatially uniform. The rate of accumulation of $CaCO_3$ on the seafloor thus depends on the area of seafloor above the CCD. If for some reason the alkalinity input from rivers were to decrease, the alkalinity of the ocean would fall and the CCD would shallow, ensuring that a smaller proportion of the seafloor accumulated $CaCO_3$ (Delaney and Boyle, 1988; figure 8.1). A more complicated response would accompany an increase in the pelagic flux of $CaCO_3$ from the surface waters resulting, for instance, from an ecological shift that favored Foraminifera over radiolarians or coccolithophorids over diatoms. The CCD would initially deepen because of the enhanced flux of alkalinity to the deep ocean. However, the resulting increase in $CaCO_3$ sedimentation (in excess of the riverine alkalinity input) would begin to deplete the ocean's alkalinity. The CCD would ultimately shallow as the ocean's alkalinity fell below its preperturbation level. Ultimately, the higher flux of alkalinity from the surface waters would be balanced by greater dissolution on the seafloor, and the removal rate of $CaCO_3$ would again match the input rate of alkalinity from rivers. If shallow-water benthic carbonate producers (corals, bivalves, etc.) were to increase their rate of $CaCO_3$ deposition, ocean alkalinity would decrease and the CCD would rise, reducing the pelagic carbonate deposition rate and restoring alkalinity balance (Opdyke and Wilkinson, 1988). Finally, if the atmospheric pCO_2 were to increase (e.g., from fossil-fuel burning), the CCD would immediately shallow (within centuries). However, on longer time scales (millennia) the consequences of enhanced continental weathering under warmer, wetter climates would be apparent, with the increased riverine alkalinity flux

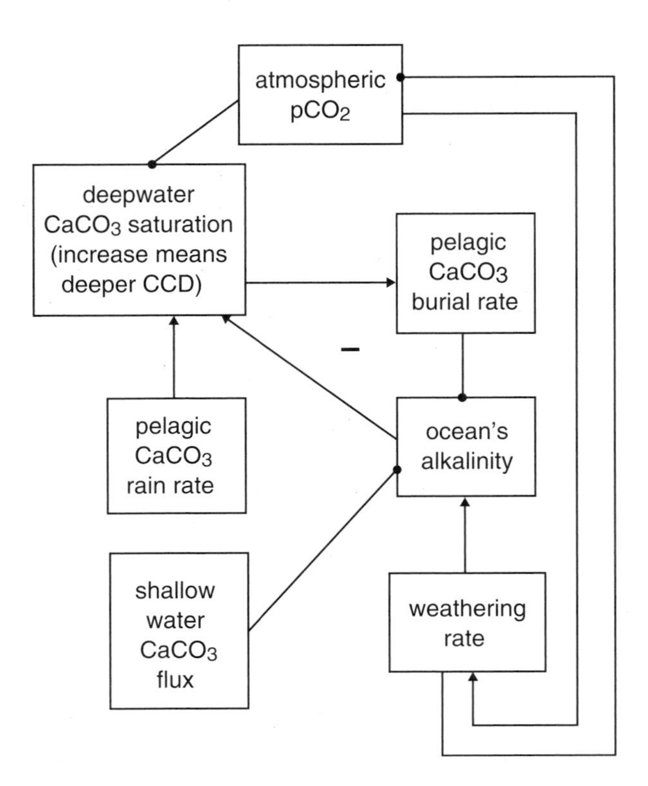

Figure 8.1
Proposed feedback loop for the regulation of the ocean's alkalinity (after Delaney and Boyle, 1988). Regular arrows are "positive couplings" (Kump et al., 1999) and indicate that the response of the system component at the arrow's head has the same sign as the sense of change of the component at the arrow's base (the stimulus). For example, an increase in atmospheric CO_2 causes an increase in weathering rates (via the greenhouse effect on climate and its effect on weathering). Circular arrowheads indicate a "negative coupling," whereby that the response has the opposite sign of the stimulus.

boosting the alkalinity of the ocean. The CCD would fall to a depth greater than that before the perturbation, and remain below this steady-state value until the excess CO_2 had been removed from the atmosphere and the excess alkalinity had been removed from the ocean (Walker and Kasting, 1992).

Did this mechanism operate before the evolution of planktonic calcifying organisms in the Mesozoic? Imagine, for example, the consequences of a sea-level fall that eliminated the shallow-water locus for carbonate deposition. Alkalinity inputs from rivers would increase the ocean's alkalinity and carbonate saturation state until surface waters became sufficiently supersaturated (~tenfold) that calcite or aragonite precipitated inorganically (or with facilitation from cyanobacteria). Thus, even an abiotic ocean would regulate its alkalinity.

On the other hand, the evolution of planktonic calcifying Foraminifera and coccolithophorids undoubtedly had significant consequences for the carbon cycle and ocean chemistry. The predominant locus of carbonate deposition shifted to the deep sea, aided at least in part by a secular decrease in tropical shelf area (Opdyke and Wilkinson, 1988). This shift has reduced the global rate of dolomitization, because deep-sea carbonates are not as readily dolomitized as are platform carbonates (Holland and Zimmerman, 2000). Because dolomitization is a major sink for Mg^{2+}, its reduction should have led to an increase in the Mg^{2+} concentration of seawater and a shift of the major sink for this element to seafloor hydrothermal alteration and the precipitation of authigenic clay minerals (Holland and Zimmerman, 2000). Finally, an increased delivery of $CaCO_3$ to seafloor sediments likely increased the delivery of these materials to subduction zones and their recycling through arc volcanoes (Volk, 1989).

Trace Metals

In marine and lake systems, biological productivity is responsive to changes in the concentrations of a host of trace metals, including Fe, Mn, Cu, Cd, and Zn (e.g., Bruland et al., 1991). At the same time, intense biological activity modifies the composition of surface and deep waters (Whitfield and Turner, 1987). Phytoplankton living in the photic zone strip carbon and nutrients, including a host of trace metals, from surface water, packaging them into organic compounds that provide nutrition for the bulk of the marine heterotrophic community. Much of this material is fine-grained and tends to remain in suspension; trace metals adsorb onto the surfaces of these particles. A portion of this organic matter is aggregated into larger particles that, with the incorporation of dense, biogenic $CaCO_3$ or lithogenic particles as ballast (Armstrong et al., 2002), sink into deeper waters and are largely decomposed, releasing nutrients and metals into solution. A fraction of trace metals is removed by sedimentation; at steady state, this sink balances the source of metals from riverine, groundwater, and atmospheric inputs. Trace metals and major nutrients in deep waters are returned to the surface by physical mixing, thus closing the recycling loop. The combination of these processes creates a chemically stratified system that is largely the result of biota-metal interactions.

The reciprocal effects of organisms on metal concentrations and vice versa create feedback loops that

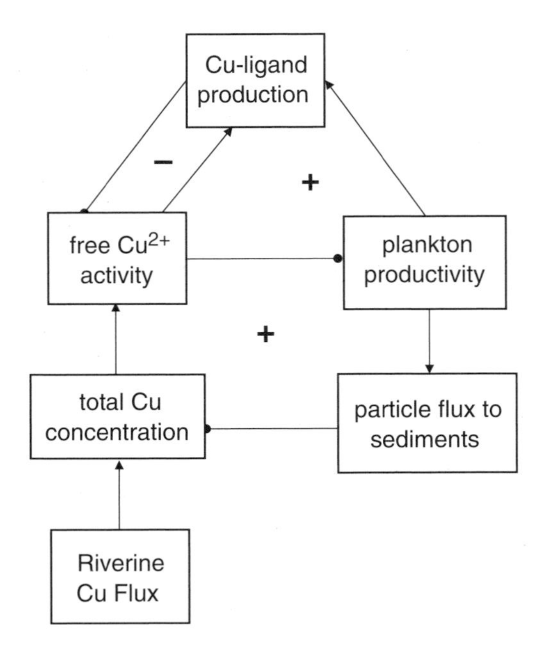

Figure 8.2
Proposed feedback loop for the regulation of the Cu^{2+} concentration of the surface ocean. See Figure 8.1 for explanation of symbols.

can be stabilizing or destabilizing. Stumm and Baccini (1978, 118), commenting on observations of metal scavenging in lakes, stated: "According to this observation, the biomass can significantly influence the lake internal Cu and Zn distribution. . . . Such a description implies a state of equilibrium for the subsystems of a lake. . . . It is obvious that the biomass is a main factor in any feed-back mechanism in controlling the heavy metal balance in lakes." They go on to point out that a potential positive feedback exists between metal input and biological productivity: an input of toxic metal reduces productivity, which also reduces the sink for the metal, allowing its concentration to build up even further (figure 8.2). This presents a conceptual puzzle: apparently the addition of even a small amount of additional toxic metal (e.g., Cu) to the oceans could eliminate the entire marine biota. Such an addition would reduce the production of biogenic particles that remove Cu. Rivers and other sources would continue to bring Cu to the oceans, and its concentration would rise. As the concentration increased, biological particle production would be reduced even further, and so would the rate of removal. This positive feedback loop would, theoretically, cause a total loss of the marine biota. Quantification of these effects is beyond the scope of this chapter. Clearly, though, there are other factors that provide stability.

One important component appears to be the production of copper-binding organic ligands by certain marine organisms, including cyanobacteria (e.g., Moffett et al., 1997). These ligands reduce the activity ratio of free to total metal in solution, and in doing so, reduce the toxicity of Cu (which is associated with the free form of the metal). In fact, at low concentrations Cu^{2+} can act as a limiting nutrient; like many substances, it can be a nutrient or a toxic substance, depending on its concentration. In the ocean, the Cu-binding ligands are abundant at the surface, where they account for orders of magnitude reduction in free-Cu^{2+} concentrations. The concentration of these ligands decreases dramatically with depth (Bruland et al., 1991), indicating both that their production rate is related to the overall marine productivity (which decreases exponentially with depth) and that they are decomposed at depth in the ocean.

The rate of production of these organic binding ligands is the product of the biological productivity of the organisms that produce them and the ratio of ligand to total productivity. If the Cu-binding ligand (L_1) production rates were simply proportional to overall biological productivity, then, counterintuitively the production of L_1 would destabilize the marine biota (figure 8.2). The argument goes that an addition of Cu to the ocean would reduce overall (and cyanobacterial) productivity, and thus the active uptake and passive scavenging of the metal into a particle sink. The free Cu^{2+} concentration of the ocean would thus increase, further reducing productivity and closing a positive feedback loop. However, laboratory and field studies (e.g., Gordon et al., 2000) indicate that increased free Cu^{2+} concentrations would trigger increased production of L_1 per unit cyanobacterial biomass; it is unclear without quantification whether this "per capita" increase in L_1 production rate would compensate for reduced overall productivity. A clue exists in the "conditioning" of seawater that takes place following upwelling (Barber and Ryther, 1969; Sunda et al., 1981). Upwelling waters have high total Cu and low complexing ligand concentrations; thus, they may not be able to support peak production until the water becomes conditioned (i.e., until ligand-producing organisms excrete sufficient complexing ligand to reduce the free Cu^{2+} concentration to "nutrient" concentrations). That these waters ultimately support high biological productivity suggests that the Cu-ligand feedback is negative (stabilizing; figure 8.2).

Another important component of the regulatory mechanism for marine trace metals is the so-called antagonistic effect among various metals. For example, it is well established that the toxic effects of Cu are reduced or eliminated in the presence of high concentrations of Zn, Mn, and Fe (Bruland et al., 1991). Apparently these metals can saturate uptake and adsorption sites on the cell membrane, preventing the uptake of Cu. Metal antagonism thus provides a possible solution to the puzzle posed above: an addition of Cu that causes reduced biological productivity, and thus reduced biogenic particle production, reduces the sink not only of the toxic metal Cu but also of the nutrient elements Zn, Mn, and Fe. If the buildup of these nutrient elements exceeds that of Cu, the Cu/(Zn + Mn + Fe) ratio decreases, and biogenic particle production rates are restored. A complete understanding of these effects requires consideration of the organic complexation and photochemistry of these other metals as well. For example, Fe^{3+} buildup in an oxygenated ocean is possible only as the result of organic complexation, Zn^{2+} is also strongly complexed by organic ligands, and Mn^{2+} buildup requires photoreduction of MnO_2 to Mn^{2+} (e.g., Whitfield, 2001).

The overall regulatory mechanism for trace metals in the ocean therefore involves the following components (figure 8.2): inputs of metals from a variety of sources; removal by active uptake and passive adsorption onto biogenic particles (organic tissues and inorganic skeletal materials; e.g., Clegg and Whitfield, 1990) and adsorption on nonbiogenic particles; oxidation and precipitation (e.g., Fe, Mn; not shown); diffusion into sediments followed by reduction (e.g., U, Mo; not shown); production of metal-binding organic ligands; and antagonistic interactions among various metals. The ability of these components to act together to regulate the ocean's composition is difficult to assess quantitatively. Instead, small-scale experiments (like those conducted recently to study the dependence of marine productivity on Fe availability; Behrenfeld et al., 1996; Coale et al., 1996; Turner and Hunter, 2001), observations of natural fluctuations, geologic investigations, and numerical modeling or larger-scale processes are needed.

Self-regulation should be apparent on a variety of time scales, from that of the seasonal phytoplankton bloom to the geologic (over millions of years). To my knowledge, the comprehensive studies of natural marine blooms that have been performed have failed to provide simultaneously collected data on primary productivity, organic ligand concentration and production rate, and the full suite of nutrient and trace-metal concentration data needed to adequately assess the model. Perhaps future field campaigns will be designed to provide data sufficient to convincingly support or refute the notion of marine trace metal regulation. Perturbations of the marine trace metal regulation system have likely ranged from relatively minor anthropogenic pollution events of modern times to the massive addition of metals after the asteroid or cometary impact associated with the Cretaceous-Tertiary (K-T) mass extinction, 65 million years ago (Erickson and Dickson, 1987). Recovery of the ocean's biological pump followed the K-T event, but it may have taken several hundred thousand years (Zachos et al., 1989). A possible explanation for this delay is the need to inorganically remove vast quantities of toxic metals from the ocean to jump-start the biological pump.

Conclusions

The marine self-regulatory mechanisms presented here are just a sampling of the many that could be conceived to be acting in the ocean today or in the distant past. Demonstrating their viability will require a focus of future field campaigns on the collection of relevant, comprehensive data. The iron-seeding experiments of the 1990s represent the type of approach to take, because it is only through manipulation that there is any hope of demonstrating homeostatic tendencies. However, deliberate manipulations of the major element chemistry of the oceans are unlikely to be possible (or well-advised), so a careful analysis of the geologic record is required. Unfortunately, sediments do not tend to retain unambiguous proxies of ancient ocean compositions. Nevertheless, great strides have been made in interpreting the isotopic and chemical compositions of limestones and shales, and of brine inclusions in evaporites (see Horita et al., 2002 for a current review of the literature).

Society, of course, has unwittingly taken this experimental approach, providing scientists with a variety of manipulative experiments at the global scale, including additions of greenhouse gases to the atmosphere, removal of vegetation from the terrestrial landscape, and addition of toxic metals to the ocean. Although they are not the explicit focus of the research done to date, self-regulatory mechanisms seem to be acting, removing half of the fossil-fuel flux of CO_2 from the atmosphere to terrestrial and oceanic sinks and stripping metals from the ocean onto sedimenting particles. Temperate forests are regrowing, although tropical rain forests may have existed in a metastable state and won't recover (Lovelock and

Kump, 1994). Overall, the Earth system appears to be resilient, with self-regulatory abilities that have sustained it as a living system for 3–4 billion years. However, global feedbacks may be fairly loose, and the fluctuations in state variables that result from human perturbation may be large enough to pose significant problems for advanced life on the planet in the centuries to come.

Acknowledgments

Support for this research was provided by the NASA Astrobiology Institute Cooperative Agreement (NCC2-1057) and the Geology and Paleontology Program of the National Science Foundation (EAR 0208119), and through a Shell UK Gaia Research fellowship. H. D. Holland and an anonymous reviewer provided helpful and constructive reviews of the manuscript.

References

Armstrong, R. A., C. Lee, J. I. Hedges, S. Honjo, and S. G. Wakeham. 2002. A new, mechanistic model for organic carbon fluxes in the ocean based on the quantitative association of POC with ballast minerals. *Deep-Sea Research Part II—Topical Studies in Oceanography*, 49, 219–236.

Barber, R. T., and J. H. Ryther. 1969. Organic chelators: Factors affecting primary production in the Cromwell Current upwelling. *Journal of Experimental Marine Biology and Ecology*, 3, 191–199.

Behrenfeld, M. J., A. J. Bale, Z. S. Kolber, J. Aiken, and P. Falkowski. 1996. Confirmation of iron limitation of phytoplankton photosynthesis in the equatorial Pacific Ocean. *Nature*, 383, 508–511.

Berner, R. A., A. C. Lasaga, and R. M. Garrels. 1983. The carbonate–silicate geochemical cycle and its effect on atmospheric carbon dioxide over the past 100 million years. *American Journal of Science*, 283, 641–683.

Broecker, W. S., and T.-H. Peng. 1982. *Tracers in the Sea*. New York: Eldigio Press.

Bruland, K. W., J. R. Donat, and D. A. Hutchins. 1991. Interactive influences of bioactive trace metals on biological production in oceanic waters. *Limnolology and Oceanography*, 36, 1555–1577.

Clegg, S. L., and M. Whitfield. 1990. A generalized model for the scavenging of trace metals in the open ocean—I. Particle cycling. *Deep-Sea Research*, 37, 809–832.

Coale, K. H., K. S. Johnson, S. E. Fitzwater, R. M. Gordon, S. Tanner, F. P. Chavez, L. Ferioli, C. Sakamoto, P. Rogers, F. Millero, P. Steinberg, P. Nightingale, D. Cooper, W. P. Cochlan, M. R. Landry, J. Constantinou, G. Rollwagen, A. Trasvina, and R. Kudela. 1996. A massive phytoplankton bloom induced by an ecosystem-scale iron fertilisation experiment in the equatorial Pacific Ocean. *Nature*, 383, 495–501.

Delaney, M. L., and E. A. Boyle. 1988. Tertiary paleoceanic chemical variability: Unintended consequences of simple geochemical models. *Paleoceanography*, 3, 137–156.

Erickson, D. J. III, and S. M. Dickson. 1987. Global trace-element biogeochemistry at the K/T boundary: Oceanic and biotic response to a hypothetical meteorite impact. *Geology*, 15, 1014–1017.

Falkowski, P. G., R. T. Barber, and V. Smetacek. 1998. Biogeochemical controls and feedbacks on ocean primary production. *Science*, 281, 200–206.

Frausto da Silva, J. J. R., and R. J. P. Williams. 1991. *The Biological Chemistry of the Elements*. Oxford: Oxford University Press.

Friedman, G. M., and J. E. Sanders. 1978. *Principles of Sedimentology*. New York: John Wiley and Sons.

Garrels, R. M., and F. T. Mackenzie. 1971. *Evolution of Sedimentary Rocks*. New York: W. W. Norton.

Gordon, A. S., J. R. Donat, R. A. Kango, B. J. Dyer, and L. M. Stuart. 2000. Dissolved copper-complexing ligands in cultures of marine bacteria and estuarine water. *Marine Chemistry*, 70, 149–160.

Hardie, L. A. 1996. Secular variation in seawater chemistry: An explanation for the coupled secular variation in the mineralogies of marine limestones and potash evaporites over the past 600 m.y. *Geology*, 24, 279–283.

Hay, W. W. 2000. Implications of higher ocean salinity for the Earth's climate system. *Abstracts with Programs, Geological Society of America*, 32, 67.

Holland, H. D. 1972. The geologic history of sea water—an attempt to solve the problem. *Geochimica et Cosmochimica Acta*, 36, 637–651.

Holland, H. D. 1984. *The Chemical Evolution of the Atmosphere and Oceans*. Princeton, NJ: Princeton University Press.

Holland, H. D., and H. Zimmermann. 2000. The dolomite problem revisited. *International Geology Review*, 42, 481–490.

Horita, J., H. Zimmermann, and H. D. Holland. 2002. Chemical evolution of seawater during the Phanerozoic: Implications from the record of marine evaporites. *Geochimica et Cosmochimica Acta*, 66, 3733–3756.

James, N. P. 1983. Reef. In *Carbonate Depositional Environments*, AAPG Memoir 33, P. A. Scholle, D. G. Bebout, and C. H. Moore, eds., pp. 345–440. Tulsa, OK: American Association of Petroleum Geologists.

Kump, L. R., and M. A. Arthur. 1997. Global chemical erosion during the Cenozoic: Weatherability balances the budget. In *Tectonics Uplift and Climate Change*, W. Ruddiman, ed., pp. 399–426. New York: Plenum.

Kump, L. R., J. F. Kasting, and R. C. Crane. 1999. *The Earth System*. Englewood Cliffs, NJ: Prentice-Hall.

Lenton, T. M., and A. J. Watson. 2000. Redfield revisited 1. Regulation of nitrate, phosphate, and oxygen in the ocean. *Global Biogeochemical Cycles*, 14, 225–248.

Lovelock, J. E. 1991. *Gaia: The Practical Science of Planetary Medicine*. London: Gaia Books.

Lovelock, J. E., and L. R. Kump. 1994. Failure of climate regulation in a geophysiological model. *Nature*, 369, 732–734.

Mackenzie, F. T., and R. M. Garrels. 1966. Chemical mass balance between rivers and oceans. *American Journal of Science*, 264, 507–525.

Moffett, J. W., L. E. Brand, P. L. Croot, and K. A. Barbeau. 1997. Cu speciation and cyanobacterial distribution in harbors subject to anthropogenic Cu inputs. *Limnology and Oceanography*, 42, 789–799.

Opdyke, B. N., and B. H. Wilkinson. 1988. Surface area control of shallow cratonic to deep marine carbonate accumulation. *Paleoceanography*, 3, 685–703.

Sarg, J. F. 2001. The sequence stratigraphy, sedimentology, and economic importance of evaporite–carbonate transitions: A review. *Sedimentary Geology*, 140, 9–42.

Sillén, L. G. 1961. The physical chemistry of sea water. In *Oceanography—Invited Lectures Presented at the International Oceanography Congress*, M. Sears, ed., pp. 549–581. Washington, DC, American Association for the Advancement of Science.

Stanley, S. M., and L. A. Hardie. 1999. Hypercalcification: Paleontology links plate tectonics and geochemistry to sedimentology. *GSA Today*, 9, 1–7.

Stumm, W., and P. Baccini. 1978. Man-made chemical perturbation of lakes. In *Lakes: Chemistry, Geology, Physics*, A. Lerman, ed., pp. 91–126. New York: Springer-Verlag.

Sunda, W. G., R. T. Barber, and S. A. Huntsman. 1981. Phytoplankton growth in nutrient rich seawater: Importance of copper–manganese cellular interactions. *Journal of Marine Research*, 39, 567–585.

Sverdrup, H. U., M. W. Johnson, and R. H. Fleming. 1942. *The Oceans: Their Physics, Chemistry, and General Biology*. Englewood Cliffs, NJ: Prentice-Hall.

Turner, D. R., and K. A. Hunter, eds. 2001. *The Biogeochemistry of Iron in Seawater*. New York: John Wiley.

Tyrrell, T. 1999. The relative influences of nitrogen and phosphorus on oceanic primary production. *Nature*, 400, 525–531.

Veizer, J. 1988. The evolving exogenic cycle. In C. B. Gregor, R. M. Garrels, F. T. Mackenzie, and J. B. Maynard, eds., *Chemical Cycles in the Evolution of the Earth*, pp. 175–220. New York: John Wiley and Sons.

Volk, T. 1989. Sensitivity of climate and atmospheric CO_2 to deep-ocean and shallow-ocean carbonate burial. *Nature*, 337, 637–640.

Walker, J. G., and J. F. Kasting. 1992. The effect of fuel and forest conservation on future levels of atmospheric carbon-dioxide. *Global and Planetary Change*, 97, 151–189.

Wallmann, K. 2001. Controls on the Cretaceous and Cenozoic evolution of seawater composition, atmospheric CO_2 and climate. *Geochimica et Cosmochimica Acta*, 65, 3005–3025.

Watson, A. J., and J. E. Lovelock. 1983. Biological homeostasis of the global environment: The parable of Daisyworld. *Tellus*, 35B, 284–289.

Whitfield, M. 2001. Interactions between phytoplankton and trace metals in the ocean. *Advances in Marine Biology*, 41, 3–128.

Whitfield, M., and D. R. Turner. 1987. The role of particles in regulating the composition of seawater. In *Aquatic Surface Chemistry*, W. Stumm, ed., pp. 457–493. New York: John Wiley and Sons.

Zachos, J. C., M. A. Arthur, and W. E. Dean. 1989. Geochemical evidence for suppression of pelagic marine productivity at the Cretaceous/Tertiary boundary. *Nature*, 337, 61–64.

A New Biogeochemical Earth System Model for the Phanerozoic Eon

Noam M. Bergman, Timothy M. Lenton, and Andrew J. Watson

Abstract

We present a new biogeochemical model of the Earth system that attempts to synthesize the different approaches of existing models. It combines the approach of Lenton and Watson's (2000a, 2000b) feedback-based "Redfield Revisited" papers, which model atmospheric oxygen levels and ocean nutrients with an interactive terrestrial biota, with elements of Berner's (1991, 1994) geological and tectonic-based Geocarb models, which model atmospheric pCO_2 throughout the Phanerozoic. The model includes ocean nutrients, interactive marine and terrestrial biota, atmospheric CO_2 and O_2, and mean global temperature, and predicts a coupled O_2 and CO_2 history. This is work in progress toward a "co-evolutionary" model of the Earth for geologic timescales.

Results demonstrate the major role played by evolution and feedback loops in controlling Earth system behavior. Life plays an important role through a combination of primary productivity, burial and weathering of organic matter, and amplification of silicate weathering. The evolution and spread of vascular land plants (from ca. 440–420 MyrBP) drives a fundamental transition in the state of the Earth system, including an order of magnitude drop in CO_2 levels accompanied by a significant rise in O_2 levels. The terrestrial biota's dependence on atmospheric O_2 and CO_2 levels, and its enhancement of chemical weathering, results in the two gases being strongly coupled thereafter. However, early Phanerozoic (before land plants) pO_2 is poorly constrained, and low Carboniferous pCO_2 ($<\sim 4$ PAL) requires unreasonably high pO_2 levels. Thus the model system is incomplete. Current work includes introducing the sulfur cycle into the model, which better constrains Phanerozoic predictions.

Introduction

Although concentrations of oxygen and carbon dioxide in the atmosphere throughout Earth history have been extensively studied, the two gases have generally been studied separately. A new biogeochemical model of the Earth system is presented which predicts a coupled O_2 and CO_2 history. The model incorporates geological forcings and biogeochemical processes, including ocean nutrients (NO_3 and PO_4), interactive and evolving marine and terrestrial biota, atmospheric CO_2 and O_2, and mean global temperature. The key carbon cycle processes and biogeochemical feedbacks controlling atmospheric O_2 and CO_2 concentrations are coupled, especially by land plants.

This work attempts to synthesize two different approaches of existing models. The first is that of the "Redfield Revisited" biological and biogeochemical-based feedback models of Lenton and Watson (2000a, 2000b), which model the ocean nutrient concentration and atmospheric oxygen content. These take a feedback-based approach with an interactive biota, changing ocean nutrient reservoirs, and an atmosphere-ocean O_2 reservoir. To this were added elements of Berner's (1991, 1994) Geocarb, tectonic and geological-based models of the global carbon cycle, which give predictions of atmospheric pCO_2 through Phanerozoic time that agree reasonably well with available geochemical proxies. Geocarb uses various geological forcings to calculate steady-state pCO_2 over time. Our work begins with these two different approaches, and builds toward a "co-evolutionary" model of the Earth for geologic timescales. This is work in progress, and the model is currently being expanded to include the sulfur cycle, and to predict sulfur and carbon isotope changes.

The Model

Ours is a box model made up of fluxes, reservoirs, and forcings, which are updated and recalculated with each time step. By varying parameters and flux functions, different feedback mechanisms can be incorporated. The response of the system to perturbations is tested, and when it is forced with proxy data, historical predictions are made. The chosen mechanisms are those which dominate the Earth system on the multimillion-year timescale. The historical period modeled is the Phanerozoic (the past 550 million years). Full model equations are in the appendix.

The model tracks the changes of four surficial biogeochemical reservoirs: the ocean nutrients phosphate (PO_4) and nitrate (NO_3), atmospheric oxygen (O_2) and atmospheric plus oceanic carbon dioxide (CO_2), as well as two larger and nearly constant reservoirs, carbonate (oxidized) carbon and organic (reduced) carbon in rock.

We include five external forcings: two geological, tectonic uplift (U) and metamorphic and volcanic degassing (D), based on geological and geochemical proxies; a time-dependent insolation forcing (I); and two terrestrial biota forcings, one representing evolution of vascular land plants and their colonization of the continents (E), and the other resulting biological enhancement of weathering (W). All of the forcings except insolation are normalized (i.e., at present $U, D, E, W = 1$). In the early Phanerozoic, before the evolution of land plants, $E = 0$ and therefore $W = 0$.[1] For a chosen set of functions and parameters, setting the five forcings to a constant value will yield a singular steady state. Although the model is not usually at such a steady state when run with changing forcings, when run forward through time it will always head toward the steady state defined by the forcings.

The geological forcings were patterned after Geocarb (Berner, 1991, 1994). U is calculated as Geocarb 2's uplift factor, using strontium isotope data from McArthur et al.'s (2001) Lowess fit and the accompanying computer-readable database for the past 509 million years, and from Burke et al. (1982) for the earlier Cambrian. D follows Geocarb 2's degassing rate in using data from Engebretson et al. (1992) for the past 150 million years and from Gaffin (1987) for the earlier Phanerozoic. The model uses the time-dependent insolation function of Caldeira and Kasting (1992).

The biological forcings E and W were quantified through rough estimates based on the Phanerozoic evolution of vascular land plants, as described by Stanley (1999). This includes the appearance of earliest land plants in the Late Ordovician, the evolution of seeds in the Devonian, trees and roots starting in the Devonian and into the Carboniferous, and the radiation of angiosperms in the Late Cretaceous. E starts at 0 in the Cambrian, and rises to 1 in the Late Carboniferous, when widespread forests covered the earth. W rises to 0.75 in the late Carboniferous, reaching 1 only in the late Cretaceous with the appearance of angiosperms. W is modeled after Geocarb's (Berner, 1991) weathering rate dependence on biological activity. It was found that the exact interpolations used for E and W as they rise during the Devonian and Carboniferous had little effect on model results.

Another important element of the model is the terrestrial biota parameter (V), representing land plant biomass and extent of continental cover. V is an interactive parameter set by feedbacks: it is a linear function of E, directly correlated with pCO_2, inversely correlated with pO_2, and has a weak temperature dependency; full V function description is in the appendix (section A2.3). Biological weathering is in turn a function not only of W but of V as well.

Several weathering and burial fluxes are included. These are weathering of silicate rock and carbonate rock, oxidative weathering of reduced carbon, and the phosphorus liberated in each; and terrestrial and marine burial of organic matter, and marine burial of inorganic carbonate carbon. Other model parameters relevant to O_2-CO_2 coupling include marine productivity and global temperature. Figure 9.1 shows a partial scheme of the model, focusing on CO_2-O_2 coupling feedbacks.

Although simple compared with the real Earth system, our model is complex enough to potentially exhibit a variety of responses and explore different proposed feedback mechanisms. Here we limit ourselves to describing initial results that seem robust and of general interest.

Land Plants Emerge—A Phase Change in the System

Here we explore the effects of perhaps the most dramatic evolutionary development of the Phanerozoic, the emergence of vascular land plants and their colonization of the continents. Plants amplified the weathering of rocks and became a new source of organic carbon for burial. This evolution changed the basic behavior of the Earth system by tightly coupling CO_2 and O_2 and changing the relations of the ocean nutrients. The rise of plants resulted in an order of magnitude drop in pCO_2 and an increase in oxygen.

Land primary productivity depends on pCO_2 and pO_2 levels, and plant productivity in turn controls the O_2 source via phosphorus weathering and organic carbon burial, and the CO_2 sink via silicate weathering. Thus plants amplify sensitivity of weathering to CO_2 and O_2, resulting in tighter oxygen controls and faster model responses due to the stronger feedback. The faster, tighter behavior increases oxygen stability.

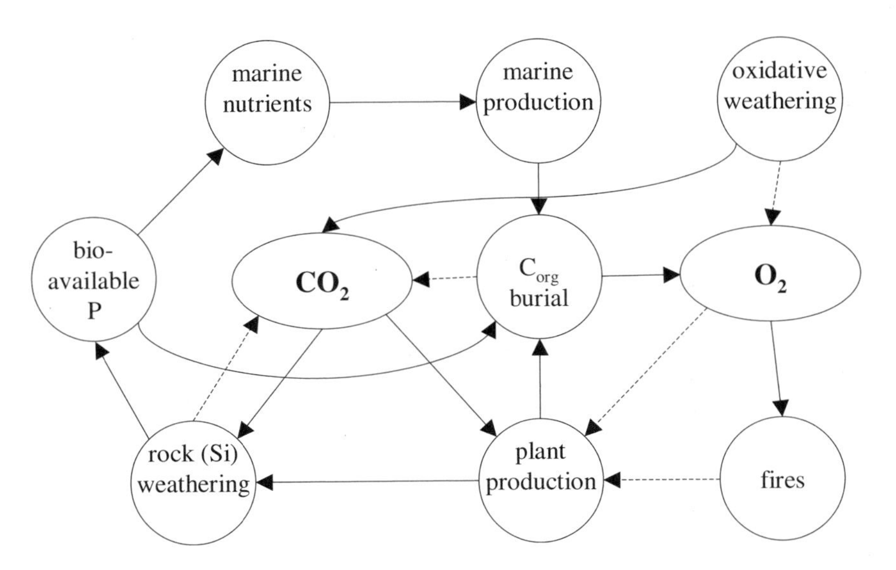

Figure 9.1
Model coupling of CO_2 and O_2 feedbacks. Ovals indicate reservoirs; circles, dependent variables. Arrows show a functional dependence of one component on another, with direction of causality. Solid arrows indicate a direct effect; dashed arrows, an inverse effect. A feedback loop exists if a closed circuit can be followed around the diagram. An even number of inverse relationships in the loop indicates positive feedback; an odd number, negative feedback. For example, an increase in atmospheric oxygen will reduce plant productivity; this reduces organic carbon burial, which in turn tends to decrease oxygen, acting as a negative feedback on the original change.

Model runs show that with land plants evolved as at present (i.e., with evolutionary colonization of land $E = 1$, and biological amplification of weathering $W = 1$), the system responds differently to perturbations and often achieves steady state faster than it does without plants ($E = 0$, $W = 0$). Oxygen is more tightly regulated, and changes only negligibly when tectonic uplift forcing (U) is changed. The ocean nutrients phosphate and nitrate react faster, and nitrate, coupled to 0.82–0.84 of its Redfield ratio to phosphate without plants, is coupled more tightly to phosphate, at 0.88–0.90 of the Redfield ratio. The coupling of the ocean nutrients is explained by Lenton and Watson (2000a, 2000b).

Figure 9.2 illustrates the system's response to three different perturbations, with and without plants. Increasing atmospheric oxygen by 50 percent (figure 9.2a) causes a similar oxygen response with and without plants, but the CO_2 response is dramatically different. Without land plants, pCO_2 shows a mild response to the oxygen increase, but when land plants are present, O_2-CO_2 coupling causes pCO_2 to more than double when oxygen is raised,[2] and the two decrease toward steady state together. Increasing the tectonic uplift forcing (U) by 50 percent (figure 9.2b) reveals a very different oxygen behavior when plants are present: pO_2 is more tightly constrained, and shows a slight decrease with time, unlike the increase in pO_2 when plants are absent. The CO_2 behavior is only slightly modified, but with a much faster response. Increasing the metamorphic and volcanic degassing forcing (D) shows oxygen slightly more constrained with plants present, but pCO_2 rising much higher (figure 9.2c).

What Goes Up Must Come Down

Steady States

It is convenient to analyze Earth system behavior over millions of years as steady states in which uplift of mountains and tectonic and metamorphic release of volatiles must be balanced by erosion and chemical weathering of rock. Silicate weathering[3] is the major geological sink for atmospheric CO_2, and reacts directly to uplift. Thus an increase in U increases silicate weathering and decreases pCO_2. The major geological CO_2 source is metamorphic and volcanic degassing. Thus, an increase in D increases pCO_2 and silicate weathering.

The model has two adjustive mechanisms in response to an increase in the CO_2 source (D): weathering rates can be increased chemically, by increasing pCO_2 and temperature (T), and biologically, by increasing terrestrial vegetation (V) or biological weathering enhancement (W). These responses are coupled, because the functional dependency of

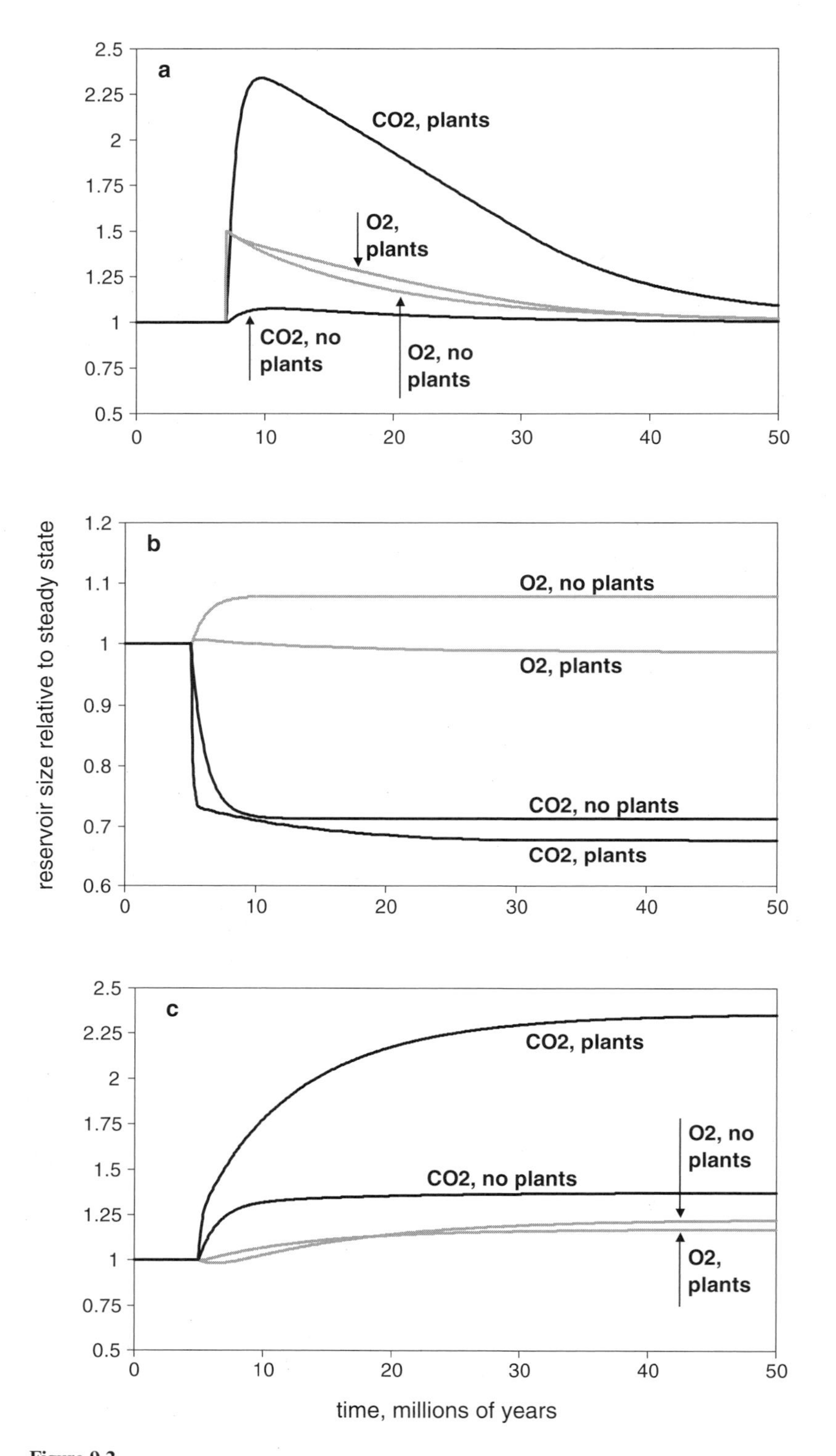

Figure 9.2
Model response to three perturbations, with and without plants: (a) increasing pO_2 by 50 percent; (b) increasing U, tectonic uplift forcing, by 50 percent; (c) increasing D, metamorphic and volcanic degassing forcing, by 50 percent. Responses shown are those of pCO_2 (dark lines) and pO_2 (light lines), with and without plants.

Table 9.1
Steady state values of Earth system biogeochemical parameters for four different model runs

Parameter	Run: a1 (S_0) Description: E = 1, W = 1	a2 E = 0, W = 0	a3 E = 1, W = 0	a4 E = 1, W = 1, C/P*2
Silicate weathering	1.00	1.00	1.00	1.00
Ocean phosphate	1.00	1.14	0.99	0.92
Ocean nitrate	1.00	1.09	0.96	0.95
Atmospheric pO_2	1.00	0.60	0.79	1.09
Atmospheric pCO_2	1.00	5.74	5.73	1.32
Temperature (°C)	15.0	24.2	24.2	16.2
Terrestrial vegetation	1.00	0	1.63	0.78
Land org. C burial	0.20	0	0.28	0.31
Marine org. C burial	1	1.19	0.92	0.90
Total org. C burial	1.20	1.19	1.19	1.20
Carbonate C burial	5.33	4.08	4.07	5.20
f_{org} (ΣOCB/ΣCB)	0.184	0.225	0.226	0.188

Temperature is in °C and f_{org} is a dimensionless fraction. Other parameters are relative to present steady state (S_0), with carbon burial parameters relative to S_0 marine organic carbon burial.

weathering on pCO_2 and T changes with V. pCO_2 (and T) levels for a given set of forcings are determined by the system heading toward a steady state where D is balanced by silicate weathering.

pCO_2 adjusts quickly to changed forcings. This implies that total atmospheric and ocean CO_2 (AOC), and specifically pCO_2, is more an indicator of the state of the system than a reservoir, unlike the much larger repositories of organic carbon and carbonate carbon in rock. For example, in model runs where the division of CO_2 between atmosphere and ocean was made dependent on temperature (CO_2 solubility is a function of T), predicted atmospheric pCO_2 history was *virtually identical* to predictions from runs without this dependency, but AOC predictions changed significantly.

The insolation function used determines T from pCO_2 and time. At any given time, T is therefore a function of pCO_2. V is a function of E, T, pCO_2 and pO_2. Thus, for a given set of forcings, a singular steady state is defined, with specific values for the parameters pCO_2, pO_2, and V, and the system will gravitate toward it. (Absence of plants largely decouples CO_2 and O_2, and changes their steady-state values greatly.)

Model Experiment

Demonstration of this can be seen in the results of four runs (a1–a4), shown in table 9.1. In each, the model was initiated with constant forcings, and allowed to run until a steady state was reached (tens of millions of years). The tectonic forcings uplift and degassing are left constant at their present values throughout (i.e., U = 1 and D = 1). The first run (a1) has normalized forcings equal to present values (i.e., E = 1, W = 1, and insolation as present). These were held constant until a steady state was reached, which hereafter will be referred to as S_0. In a2, no terrestrial vegetation evolves (E = 0, and therefore W = 0, V= 0). In a3, plants are assumed to evolve to current primary productivity rates (E = 1), without increasing weathering rates (W = 0).[4] In a4, plants evolve as in S_0, (E = 1, W = 1), but terrestrial carbon to phosphorus burial ratio (CPland) is doubled.

Run Results

In each of the four runs a different steady state was reached (table 9.1), but in all silicate weathering returned to its S_0 value, demonstrating the model's ability to adjust weathering rates through changes in pCO_2 and temperature, or through biological weathering enhancement (see section "Land Plants Emerge"). The results show further that the major effect of land plants on the carbon cycle is through weathering enhancement, not primary productivity or carbon burial (CB). When weathering enhancement is removed (a1 vs. a3), pCO_2 and temperature rise dramatically, while carbonate carbon burial (CCB) drops significantly, raising f_{org}, the organic fraction of buried carbon ($f_{org} = \Sigma OCB/\Sigma CB$). If primary productivity is then removed (i.e., no land plants at all, as in a2), there is no change in pCO_2, T, f_{org} or CCB.

Doubling CPland (a4) has a much smaller effect on the carbon cycle, slightly raising pCO_2, T and f_{org}, and slightly lowering CCB, from their S_0 values. At first it may seem surprising that doubling CPland raises pCO_2, but this is consistent with the system heading toward a new steady state in which fewer plants are needed to bury carbon. Fewer plants imply less biological weathering; therefore higher pCO_2 and T are needed to weather uplifted rock.

The dynamic process shows how this new steady state is achieved. When CPland is doubled, pCO_2 drops for the first ~400,000 years, while oxygen increases, both as a result of increased organic carbon burial. Land primary productivity then decreases due to decreasing atmospheric CO_2/O_2 ratio and increasing fire frequency. Terrestrial vegetation (V) is suppressed, thereby decreasing biological enhancement of weathering. Silicate and carbonate weathering decrease, resulting over millions of years in a decrease in CCB, thus reducing the CO_2 sink. Meanwhile, rising O_2 levels cause V (and land OCB) to drop further, shrinking this CO_2 sink. The result is a rise in pCO_2 levels until, after tens of millions of years, a new steady state is reached in which weathering equals uplift (table 9.1).

Oxygen levels show a departure of up to 40 percent from S_0 value in these runs. Oxidative weathering and degassing are kept constant in all four, and the only oxygen source or sink which changes is OCB. The effect on pO_2 is surprisingly large for small changes in OCB. Plants strengthen the regulation of oxygen: the largest departure from S_0 is found in a2, when $V = 0$. In a2 and a3, pO_2 decreases 20–40 percent although pCO_2 goes up. In a4 they both rise, and at a more comparable rate, once again demonstrating the tight O_2–CO_2 coupling when plants are fully evolved.

V is increased relative to S_0 when $W = 0$. This is because the lack of biologically enhanced weathering yields a high pCO_2, which increases primary productivity. It is plausible that plants have limited themselves through evolutionary competition: roots which weather rock and extract nutrients have yielded a steady state with lower pCO_2, and thus an environment which can support fewer plants. However, without efficient nutrient extraction, plants could not be so widespread.

The nutrients phosphate and nitrate show little change from S_0 in the four runs, and total OCB remains almost unchanged, merely partitioned differently between land and marine OCB. This may seem surprising, given plants' strong influence on phosphorus weathering and its partitioning between land and sea, but is consistent with the unchanged weathering rates: total organic carbon burial balances oxidative weathering. While the lack of biological weathering (a3) hardly changes ocean phosphate, when primary productivity is removed as well (a2), all organic phosphorus reaches the ocean, increasing ocean phosphate and marine OCB.

Phanerozoic Simulations

Four sample runs (b1–b4) for the entire Phanerozoic (figure 9.3) demonstrate the sensitivity of the atmospheric composition, predominantly oxygen history, to hypothesized feedbacks, and the change in the system as land plants evolve, including the coupling of oxygen and carbon dioxide.

The baseline run is b1. It includes land net primary productivity (NPP) as a function of pCO_2, pO_2 and temperature, based on the work of Farquhar et al. (1980); a fire feedback on land plants at high oxygen levels as introduced by Lenton and Watson (2000b) after the work of Watson (1978); a variable C/P burial ratio of marine organic matter dependent on bottom water anoxia, after Van Cappellen and Ingall (1994); and a doubling of C/P burial ratio of terrestrial organic matter during the Carboniferous, to represent increased terrestrial carbon burial at the time. In each of the three subsequent runs, one of these feedbacks is changed: in b2 terrestrial organic C/P burial ratio is constant; in b3 the fire feedback is turned off; and in b4 marine organic C/P burial ratio is independent of anoxia, as in Colman et al. (1997), and is set at 250.

Terrestrial vegetation and fire feedback calculations and land and marine C/P burial ratios are described in sections A2.3 and A2.1 of the appendix, respectively.

Early Phanerozoic Oxygen: As You Like It

Predicted oxygen levels in the first 200 million years of the Phanerozoic are poorly constrained (Lenton, 2003). Cambrian marine fauna supply evidence for $pO_2 > 0.02$ atm (~0.1 PAL) more than 500 MyrBP (million years before present). Early land plants appeared as long ago as 440–420 MyrBP and tended to increase atmospheric oxygen. By the Late Devonian (~370 MyrBP), the first trees and forests appeared, increasing weathering and reducing atmospheric pCO_2. Fossilized charcoal, evidence for pO_2 levels high enough to sustain fire (~0.15 atm or ~0.75 PAL),

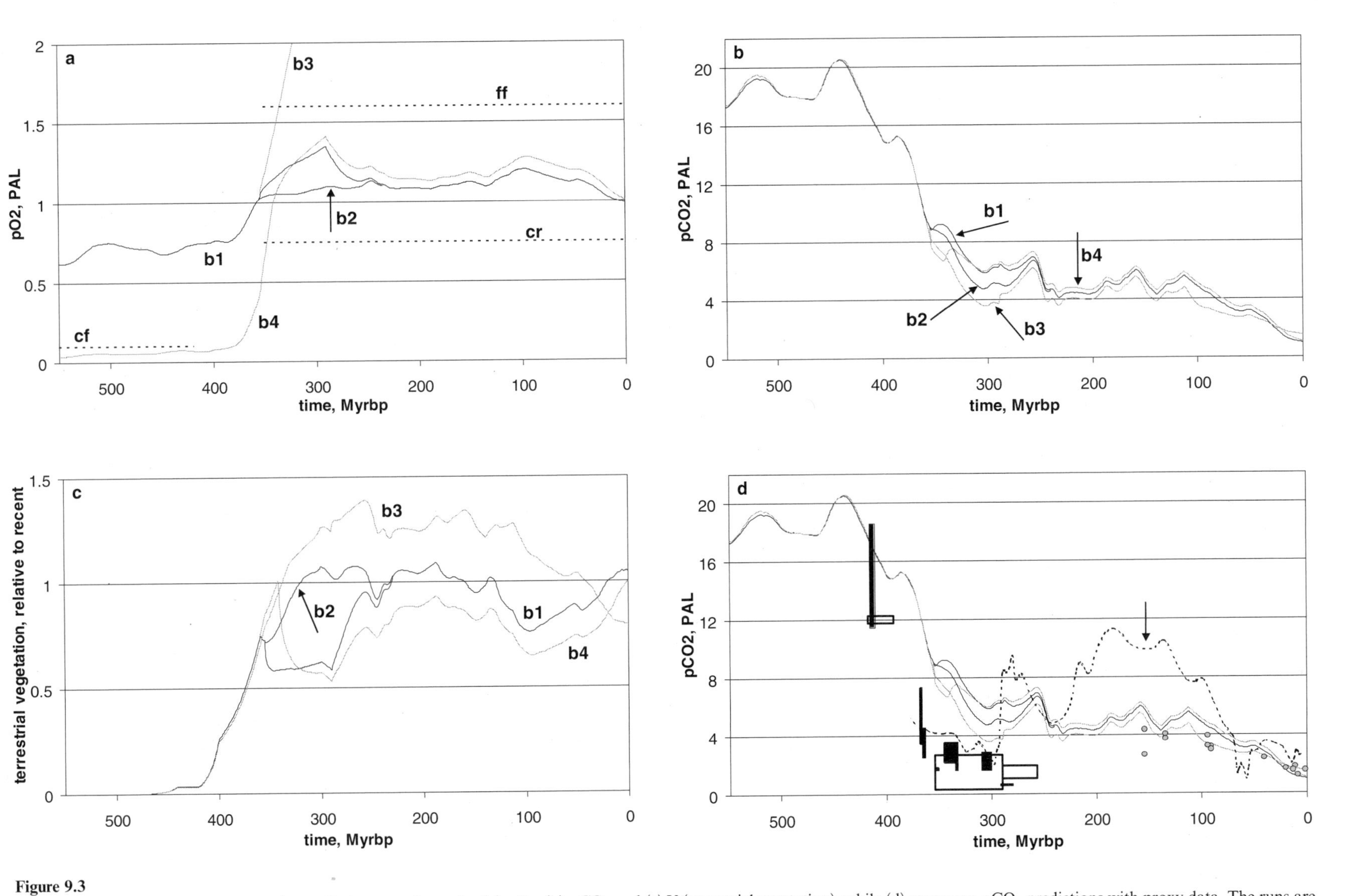

Figure 9.3
Sample model runs for the Phanerozoic: predictions are shown for (a) pO_2, (b) pCO_2 and (c) V (terrestrial vegetation), while (d) compares pCO_2 predictions with proxy data. The runs are (b1) baseline; (b2) constant C/P burial ratio of terrestrial organic matter; (b3) no fire feedback on terrestrial vegetation; and (b4) constant C/P burial ratio of marine organic matter. pO_2 constraints in (a) are shown as dashed lines: *cf*, estimated Cambrian fauna minimum (Holland, 1984); *cr*, charcoal record minimum (Lenton and Watson, 2000b); *ff*, forest fire estimated maximum. pCO_2 predictions are compared with proxies in (d): palaeosol carbonates, dashed line (Ekart et al., 1999) and solid rectangles (Mora et al., 1996); stomatal fossils, open rectangles (McElwain and Chaloner, 1995); and phytoplankton carbon isotope fractionation, circles (Freeman and Hayes, 1992).

appears before the end of the Devonian (~350 MyrBP).

Early Phanerozoic oxygen predictions of the model (figure 9.3a) change drastically in accordance with assumptions about the carbon cycle and oxygen sources and sinks, and are highly dependent on the phosphorus cycle. The baseline run (b1), in which the C/P ratio of buried marine organic matter is set to rise with bottom water anoxia, predicts pO_2 levels of ~0.7 PAL before the advent of land plants. However, b4 (having a constant marine C/P burial ratio) predicts pO_2 of <0.1 PAL in the early Phanerozoic, lower than the constraints for Cambrian fauna (Holland, 1984). Other possible feedbacks (not shown) yield intermediate values.

Before land plants appear, CO_2 and O_2 are largely decoupled: the order of magnitude difference in predicted oxygen levels makes no impact on the predicted pCO_2 history, with all runs nearly identical up to ~370 MyrBP (figure 9.3b). Only when land plants evolve sufficiently do the predictions part ways, and pCO_2 levels come to depend significantly on pO_2 as the latter rises to PAL and higher in the Carboniferous.

Plants, pCO_2 and Fire

The Geocarb models (Berner, 1991, 1994), as well as various geochemical and biological proxies such as palaeosol carbonates (Ekart et al., 1999; Mora et al., 1996) and stomatal fossils (McElwain and Chaloner, 1995), suggest that pCO_2 levels plunged from 15–20 PAL in the early Phanerozoic to a minimum of 1–4 PAL in the Permo-Carboniferous, accompanied by low global temperatures. Oxygen levels in this period are more disputed, with isotope-driven oxygen models (Berner et al., 2000) predicting a sharp rise in pO_2 in this period, reaching a maximum of up to ~2 PAL. However, forests have been widespread throughout the past 350 million years, and an atmospheric O_2 mixing ratio above ~25 percent (~1.25 PAL) is estimated to have made fires too frequent for forest regeneration (Watson, 1978; Lenton and Watson, 2000b).

This is not an absolute maximum, but O_2 mixing ratios above ~30 percent (~1.6 PAL) are hard to reconcile with continuous forest coverage. Model runs capture the pCO_2 plunge in the Devonian and Carboniferous with the rise of land plants (figure 9.3b, 9.3c), and a Permo-Carboniferous trough with a minimum of 3.5–6 PAL is predicted. Oxygen levels rise to present levels or higher as land plants emerge.

Comparing pCO_2 predictions to proxy data (figure 9.3d) shows that pCO_2 is not dropping early enough. This may indicate that land plants are introduced too late in the model, or that early land plants' amplification of weathering has been underestimated.

The Carboniferous pCO_2 trough may have been as low as 1 PAL (figure 9.3d). This translates to ~5.5°C below present-day global temperature, which is similar to the recent ice ages. Run b3 has a Carboniferous pCO_2 minimum of 3.5 PAL, and a temperature trough with a minimum of 0.2°C below the present. Runs b1 and b4 have Carboniferous pCO_2 no lower than ~6 PAL and a global temperature above the present temperature. This is inconsistent with evidence of a Carboniferous ice age, but many other factors influencing global temperature (including continental configuration) are not included in the model.

pCO_2 levels from 200 MyrBP on (figure 9.3d) are in good agreement with proxy data from phytoplankton carbon isotope fractionation (Freeman and Hayes, 1992), although palaeosol carbonates (Ekart et al., 1999) suggest higher levels.

The fire feedback has a major effect: When turned on (b1), oxygen reaches a Permo-Carboniferous peak of ~1.3 PAL, pCO_2 has a ~6 PAL minimum, and terrestrial vegetation hardly exceeds present levels. When turned off (b3), terrestrial vegetation increases uninhibited, exceeding current levels by up to 50 percent as oxygen levels soar to nearly 4 PAL after land plants emerge and stay high; the pCO_2 minimum drops to ~3.5 PAL. Runs with the fire feedback show a distinct maximum in V (figure 9.3c) as land plants emerge, when pO_2 exceeds PAL and fires limit terrestrial vegetation; this sharp maximum is absent in b3, and V declines slowly throughout the Mesozoic and Cenozoic as the pCO_2/pO_2 ratio declines.

The Carboniferous increase in the land C/P organic burial ratio also plays an important role. This was introduced to capture the increased organic carbon burial in Carboniferous coal swamps. Removing this forcing (b4) removes the Carboniferous-Permian oxygen maximum entirely (figure 9.3a), but allows for more land plants yielding a lower pCO_2 minimum. Thus an increase in C/P burial ratio increases pCO_2, as demonstrated in run a4 (see section "Run Results"). This contrasts with the Geocarb models, which do not track oxygen (and is a consequence of O_2–CO_2 coupling).

These results show an immensely varied behavior of oxygen depending on the feedbacks chosen, a significant variation in predicted Carboniferous pCO_2 minima, and a large variation in terrestrial vegetation predictions. Run b3 (no fire feedback) has the deepest predicted pCO_2 trough, which best fits the idea of a Carboniferous ice age, but unfortunately this combines with an unreasonably high pO_2 maximum.

Conclusions

The model results demonstrate the importance of evolution and feedback in determining Earth system behavior. The tendency of the system to gravitate toward a steady state is demonstrated: sources and sinks of carbon must be equal over millions of years, and atmospheric pCO_2 is not so much a reservoir as a parameter of the system, its value adjusting to allow weathering to balance uplift. In this the model agrees with earlier work, such as the Geocarb models (Berner, 1991, 1994).

Our work highlights the major role played by life in the Earth system, specifically how the emergence and spread of vascular land plants caused a shift, or "phase change," in the system. The plants' effect is mostly through enhancing rock weathering, bringing about steady states with much lower pCO_2, as reported by Berner (1991, 1994). Our model shows how this shift tightly couples O_2 and CO_2, affecting the CO_2 sink and the O_2 source. Thereafter, the dependency of land primary productivity on atmospheric partial pressures of CO_2 and O_2 exerts a strong control on O_2 levels, stabilizing oxygen.

We also move beyond the "Redfield Revisited" models (Lenton and Watson, 2000a, 2000b) in understanding the phosphorous cycle, through its coupling to the oxygen and carbon cycles: land plant-influenced phosphorus weathering patterns and organic matter C/P burial ratios, on both land and sea, strongly influence O_2 levels. Marine nitrate is coupled to just under its Redfield ratio with phosphate, as in the "Redfield Revisited" models, and we show here how this coupling is stronger after land plants arise.

Early Phanerozoic oxygen (before land plants) is poorly constrained, and prediction is made more difficult due to lack of coupling with CO_2. From the Carboniferous on, oxygen is better constrained, but still highly dependent on mechanisms chosen for different feedbacks. Early pCO_2 is predicted to have been 15–20 PAL, with a great decline in the Devonian and Carboniferous. In contrast to the Geocarb models, Carboniferous CO_2 levels are never predicted to fall as low as at present.

Predicted Phanerozoic pCO_2 levels are mostly in good agreement with proxy data, except during the Carboniferous. This carbon-oxygen puzzle indicates the necessity for further research into the long-term coupling of Earth system cycles. Current work focuses on the addition of the sulfur cycle to the model, and prediction of carbon and sulfur isotopes for comparison with available geological data. These additions to the model yield better-constrained Phanerozoic predictions, especially of O_2.

APPENDIX

A1. Model Constants, Modeled Reservoirs, and Fluxes

Table 9.2
Model constants

Name	Meaning	Baseline size
k_1	initial oxic fraction	0.86
k_2	marine organic carbon burial	$3.75 \cdot 10^{12}$ mol C yr^{-1}
k_3	nitrogen fixation	$8.7 \cdot 10^{12}$ mol N yr^{-1}
k_4	denitrification	$4.3 \cdot 10^{12}$ mol N yr^{-1}
k_5	ocean mass	$1.397 \cdot 10^{21}$ kg
k_6	Fe-sorbed P burial	$0.6 \cdot 10^{10}$ mol P yr^{-1}
k_7	Ca associated P burial	$1.5 \cdot 10^{10}$ mol P yr^{-1}
k_8	nutrient conversion: mol to μmol/kg	$7.16 \cdot 10^{-22}$ kg^{-1}
k_9	O_2 conversion: mol (atm) to μmol/kg (ocean)	$8.96 \cdot 10^{-24}$ kg^{-1}
k_{10}	(reactive) P weathering	$3.675 \cdot 10^{10}$ mol P yr^{-1}
k_{11}	fraction of P buried on land	0.02041
k_{12}	carbonate carbon degassing	$1.33 \cdot 10^{-9}$ mol C yr^{-1}
k_{13}	organic carbon degassing	$1.0 \cdot 10^{-9}$ mol C yr^{-1}
k_{14}	carbonate weathering	$2.67 \cdot 10^{-9}$ mol C yr^{-1}
k_{15}	normalized pre-plant weathering	0.15
k_{16}	O_2 conversion factor to atmospheric mixing ratio	3.762
k_{17}	oxidative weathering	$3.25 \cdot 10^{12}$ mol C yr^{-1}
k_{fire}	fire frequency parameter	20
$CPland_0$	initial organic C/P burial ratio, land	1,000
$CPsea_0$	initial organic C/P burial ratio, sea	250
$CNsea_0$	organic C/N burial ratio, sea	37.5
k_{oxic}	organic C/P burial ratio, oxic ocean	217.0
k_{anoxic}	organic C/P burial ratio, anoxic ocean	4,340.0

Table 9.3
Modeled reservoirs

Reservoir	Present-day size	Time step equation
Atmosphere and ocean O_2	$O_0 = 331.5\ \mu$mol kg^{-1}	$dO/dt = k_9 \cdot (mocb + locb - oxidw - ocdeg)$
Ocean phosphate	$P_0 = 2.2\ \mu$mol kg^{-1}	$dP/dt = k_8 \cdot (psea - mopb - fepb - capb)$
Ocean nitrate	$N_0 = 30.9\ \mu$mol kg^{-1}	$dN/dt = k_8 \cdot (nfix - denit - monb)$
Atmosphere and ocean CO_2	$A_0 = 3.19 \cdot 10^{21}$ mol C	$dA/dt = -dC/dt - dG/dt$
Carbonate carbon	$C_0 = 5.0 \cdot 10^{21}$ mol C	$dC/dt = mccb - carbw - ccdeg$
Organic carbon	$G_0 = 1.25 \cdot 10^{21}$ mol C	$dG/dt = locb + mocb - oxidw - ocdeg$

Table 9.4
Baseline fluxes

Flux	Calculation
Silicate weathering	$silw = k_{12} \cdot C_0 \cdot U \cdot f_C[k_{15} + (1 - k_{15}) \cdot W \cdot V]$
Carbonate weathering	$carbw = k_{14} \cdot C_0 \cdot U \cdot g_C(k_{15} + (1 - k_{15}) \cdot W \cdot V)$
Oxidative weathering	$oxidw = (G/G_0) \cdot U \cdot k_{17}$
Phosphorus weathering	$phosw = k_{10} \cdot [(2/12)(silw/silw_0) + (5/12)(carbw/carbw_0) + (5/12)(oxidw/oxidw_0)]$
Phosphorus to land	$pland = k_{11} \cdot V \cdot phosw$
Phosphorus to sea	$psea = (1 - k_{11}) \cdot V \cdot phosw$
Carbonate degassing	$ccdeg = k_{12} \cdot D \cdot C$
Organic degassing	$ocdeg = k_{13} \cdot D \cdot G$
Land organic C burial	$locb = \text{CPland} \cdot pland$
Marine new productivity	$newp = 117 \cdot \text{minimum}(P, N/16)$
Nitrogen fixation	$nfix = k_3[(P - N/16)/(P_0 - N_0/16)]^2$
Denitrification	$denit = k_4[1 + anox/(1 - k_1)]$
Ocean anoxia	$anox = 1 - k_1[(O/O_0)/(newp/newp_0)]$
Marine organic C burial	$mocb = k_2(newp/newp_0)^2$
Marine organic P burial	$mopb = mocb/\text{CPsea}$
Marine organic N burial	$monb = mocb/\text{CNsea}_0$
Marine carbonate C burial	$mccb = silw + carbw$
Iron-sorbed P burial	$fepb = k_6[(1 - anox)/k_1](P/P_0)$
Calcium-bound P burial	$capb = k_7(newp/newp_0)^2$

A2. Other Model Parameters

A2.1 C/P Burial Ratios of Organic Matter

In the baseline model, the C/P burial ratio of marine organic matter is determined by bottom water anoxia (Van Cappellen and Ingall, 1994; Lenton and Watson, 2000b):

$$\text{CPsea} = k_{\text{oxic}} \cdot k_{\text{anoxic}}/[(1 - anox)k_{\text{anoxic}} + anox\, k_{\text{oxic}}]. \tag{9.1}$$

An alternative is the constant burial ratio (Colman et al., 1997; Lenton and Watson, 2000b):

$$\text{CPsea} = \text{CPsea}_0. \tag{9.2}$$

The baseline C/P burial ratio of land organic matter assumes increased land carbon burial in the Carboniferous:

$$\text{CPland} = \text{CPland}_0, \quad 550 - 355 \text{ MyrBP}, \quad 290 \text{ MyrBP} - \text{present} \tag{9.3}$$

$$\text{CPland} = 2 \cdot \text{CPland}_0, \quad 355 - 290 \text{ MyrBP}. \tag{9.4}$$

An alternative is the constant value:

$$\text{CPland} = \text{CPland}_0, \quad \text{for the entire Phanerozoic.} \tag{9.5}$$

A2.2 Temperature

Temperature is calculated using the energy balance model of Caldeira and Kasting (1992). This uses time-dependent luminosity:

$$S(t) = (1 + 0.38t/\tau_0)^{-1} S_0, \tag{9.6}$$

where t is time in MyrBP; $\tau_0 = 4{,}550$ MyrBP, the Earth's age; and S_0 is current solar luminosity, 1,368 W m^{-2}. The blackbody energy balance is therefore

$$(1 - a)S/4 = \sigma \cdot T_{\text{eff}}^4, \tag{9.7}$$

where a is planetary albedo, σ is the Stefan–Boltzmann constant, and T_{eff} is the blackbody radiation temperature (K). The albedo is in turn temperature-dependent:

$$a = 1.4891 - 0.0065979T + (8.567 \cdot 10^{-6})T^2. \tag{9.8}$$

Planetary temperature includes T_{eff} and the greenhouse warming factor, ΔT:

$$T = T_{\text{eff}} + \Delta T. \tag{9.9}$$

ΔT is calculated by

$$\Delta T = 815.17 + (4.895 \cdot 10^7)T^{-2} - (3.9787 \cdot 10^5)T^{-1} - 6.7084\psi^{-2} + 73.221\psi^{-1} - 30{,}882T^{-1}\psi^{-1}, \tag{9.10}$$

where $\psi = \log(pCO_2)$, in bars. Thus, for a given t and pCO_2, T is calculated by iteration from equations (9.7)–(9.10).

A2.3 Terrestrial Vegetation

First, the relative net primary productivity (npp′) is calculated from pCO_2, pO_2 and T, using the biochemical leaf models of Farquhar et al. (1980) and Friend (1998):

$$npp' = V_{Cmax} \cdot (pCO_2 - \Gamma)/[pCO_2 + K_C(1 + pO_2/K_O)], \quad (9.11)$$

where Γ is the CO_2 compensation mixing ratio in the absence of day respiration, given by

$$\Gamma = [K_c/(2 \cdot K_o)] \cdot pO_2 \cdot (V_{Omax}/V_{Cmax}). \quad (9.12)$$

V_{Cmax} and V_{Omax} are the maximum velocities of carboxylation and oxidation, respectively. K_C and K_O are the Michaelis-Menten constants for CO_2 and O_2, respectively. All are temperature-dependent and are calculated from the Arrhenius law. Values for $T_0 = 15°C$ are $K_C = 88.7\ \mu$atm, $K_O = 349{,}950\ \mu$atm, $V_{Cmax} = 0.465\ \mu mol\ m^{-2}\ s^{-1}$, and $V_{Omax} = 0.396\ \mu mol\ m^{-2}\ s^{-1}$.

Next, this value is normalized for the model:

$$npp = E \cdot K_{normal} \cdot npp', \quad (9.13)$$

where E is evolutionary forcing and K_{normal} normalizes npp to 1 for present-day pCO_2, pO_2 and T.

Finally, terrestrial vegetation, V, is calculated from the primary productivity and the fire feedback of the "Redfield Revisited" model (Lenton and Watson, 2000b):

$$V = k_{fire} \cdot npp/(i + k_{fire} - 1), \quad (9.14)$$

where the ignition parameter, i, is given by

$$i = \text{maximum}(0, 586.2 \cdot (O/O_0)/[(O/O_0) + k_{16}] - 122.02). \quad (9.15)$$

When the fire feedback is turned off, this simplifies to

$$V = npp. \quad (9.16)$$

Notes

1. W represents enhancement of rock weathering by vascular land plants relative to the situation before their evolution (i.e., relative to weathering of rocks on which mosses and lichens grow, not relative to abiological conditions).

2. Without land plants, pCO_2 is significantly higher at steady state, ~5.7 PAL (see note 3), so it is not surprising that the relative change in pCO_2 is smaller. Simple calculation shows that the *absolute amount* of CO_2 added to the atmosphere as a result of the oxygen increase is greater when land plants are present: ~1.3 PAL with plants, versus ~0.5 PAL without.

3. "Silicate weathering" refers to the weathering of Ca and Mg silicates in the Urey reaction, involving uptake of atmospheric CO_2, followed by the precipitation of Ca and Mg carbonates.

4. This is a completely hypothetical situation, since terrestrial vegetation could not have become as extensive as it is without developing roots to extract nutrients from the soil, thereby increasing the erosion of rocks. Nonetheless, it serves a heuristic purpose.

References

Berner, R. A. 1991. A model for atmospheric CO_2 over Phanerozoic time. *American Journal of Science*, 291, 339–376.

Berner, R. A. 1994. Geocarb II: A revised model of atmospheric CO_2 over Phanerozoic time. *American Journal of Science*, 294, 56–91.

Berner, R. A., S. T. Petsch, J. A. Lake, D. J. Beerling, B. N. Popp, R. S. Lane, E. A. Laws, M. B. Westley, N. Cassar, F. I. Woodward, and W. P. Quick. 2000. Isotope fractionation and atmospheric oxygen: Implications for Phanerozoic O_2 evolution. *Science*, 287, 1630–1633.

Burke, W. H., R. E. Denison, E. A. Hetherington, R. B. Koepnick, H. F. Nelson, and J. B. Otto. 1982. Variation of seawater $^{87}Sr/^{86}Sr$ throughout Phanerozoic time. *Geology*, 10, 516–519.

Caldeira, K., and J. F. Kasting. 1992. The life span of the biosphere revisited. *Nature*, 360, 721–723.

Colman, A. S., F. T. Mackenzie, and H. D. Holland. 1997. Redox stabilisation of the atmosphere and oceans and marine productivity. *Science*, 275, 406–407.

Ekart, D. D., T. E. Cerling, I. P. Montañez, and N. J. Tabor. 1999. A 400 million year carbon isotope record of pedogenic carbonate: Implications for paleoatmospheric carbon dioxide. *American Journal of Science*, 299, 805–827.

Engebretson, D. C., K. P. Kelley, H. J. Cashman, and M. A. Richards. 1992. 180 million years of subduction. *GSA Today*, 2, 93–95, 100.

Farquhar, G. D., S. von Caemmerer, and J. A. Berry. 1980. A biochemical model of photosynthetic CO_2 assimilation in leaves of C_3 species. *Planta*, 149, 78–90.

Freeman, K. H., and J. M. Hayes. 1992. Fractionation of carbon isotopes by phytoplankton and estimates of ancient CO_2 levels. *Global Biogeochemical Cycles*, 6(2), 185–198.

Friend, A. D. 1998. Appendix: Biochemical modelling of leaf photosynthesis. In P. Jarvis, ed., *European Forests and Global Change*, pp. 335–346. Cambridge: Cambridge University Press.

Gaffin, S. 1987. Ridge volume dependence on seafloor generation rate and inversion using long term sealevel change. *American Journal of Science*, 287, 596–611.

Holland, H. D. 1984. *The Chemical Evolution of the Atmosphere and Oceans*. Princeton, NJ: Princeton University Press.

Lenton, T. M. 2003. The coupled evolution of life and atmospheric oxygen. In L. Rothschild and A. Lister, eds., *Evolution: The Impact of the Physical Environment*, pp. 35–53. London: Academic Press.

Lenton, T. M., and A. J. Watson. 2000a. Redfield revisited: 1. Regulation of nitrate, phosphate and oxygen in the ocean. *Global Biogeochemical Cycles*, 14(1), 225–248.

Lenton, T. M., and A. J. Watson. 2000b. Redfield revisited: 2. What regulates the oxygen content of the atmosphere? *Global Biogeochemical Cycles*, 14(1), 249–268.

McArthur, J. M., R. J. Howarth, and T. R. Bailey. 2001. Strontium isotope stratigraphy: LOWESS version 3. Best fit to the marine Sr-isotope curve for 0-509 Ma *and accompanying* look-up table for deriving numerical age. *Journal of Geology*, 109, 155–170.

McElwain, J. C., and W. G. Chaloner. 1995. Stomatal density and index of fossil plants track atmospheric carbon dioxide in the Palaeozoic. *Annals of Botany*, 76, 389–395.

Mora, C. I., S. G. Driese, and L. A. Colarusso. 1996. Middle to late Paleozoic atmospheric CO_2 levels from soil carbonate and organic matter. *Science*, 271, 1105–1107.

Stanley, S. M. 1999. *Earth System History*. New York: W. H. Freeman.

Van Cappellen, P., and E. D. Ingall. 1994. Benthic phosphorus regeneration, net primary production and ocean anoxia: A model of the coupled marine biogeochemical cycles of carbon and phosphorus. *Paleoceanography*, 9(5), 677–692.

Watson, A. J. 1978. Consequences for the biosphere of forest and grassland fires. Ph.D. thesis, University of Reading.

10
Gaia and Glaciation: Lipalian (Vendian) Environmental Crisis

Mark A. S. McMenamin

Abstract

The Lipalian or Vendian Period (600–541 million years ago) begins and ends with global environmental perturbations. It begins as the worst glaciation on record draws to a close. It ends with a sudden appearance of abundant skeletonized animals that mark the beginning of Cambrian ecology. Several key events in Earth history occur during the Lipalian, bracketed between severe glaciation (white or snowball Earth) and the initiation of modern marine ecosystems. The most notable of these events is the appearance of an unusual and conspicuous marine biota ("Garden of Ediacara"). This biota appears to have characteristics inherited from its sojourn beneath the ice. This chapter examines the role the cryophilic biota, consisting largely of cyanobacteria and chryosphyte and chlorophyte algae, may have played in ending the great ice age. It is hypothesized here that these microbes induced albedo reductions and other changes that rapidly improved global climate.

Introduction

The possibility of extreme glacial conditions followed by a rapid global warming poses a challenge for both Gaia theory and theories of climate change via abiotic mechanisms. There is convincing evidence that equatorial glaciation occurred during the Proterozoic. Assuming that the severe Proterozoic glaciation is correctly interpreted as a white or "snowball" Earth, Gaia theory (with its uniformitarian slant) is hard pressed to explain the intensity of the glaciation (with its catastrophic or even apocalyptic overtones). On the other hand, geochemistry alone is inadequate to explain the sudden deglaciation and extreme global warming that led to the deposition of the unusual cap carbonate deposits at the base of the Lipalian.

Proterozoic Glaciation and Climate Models

The fact that Earth has experienced severe glaciations poses a problem for Gaia theory. If the planet is maintained, in Daisyworld fashion, as an equable regime of climate over a wide range of solar luminosity values, then what do glaciations represent? James Lovelock compared the Pleistocene fluctuations between glacial and interglacial conditions to high-amplitude oscillations in simulated climate seen as one approaches the upper limit of solar luminosity in the Daisyworld model (Lovelock, 1990). Lovelock noted (1990: 137) that such a world "is inherently unstable." Fluctuations of this sort, perhaps involving a system at the threshold of chaotic behavior (Gleick, 1987: 170), might indeed be invoked to explain Pleistocene climate alternations.

One problem with modeling such fluctuations is keeping the system from going to extremes. In the early days of climate modeling, computer models of global climate tended to collapse into the deep freeze of a white Earth climate (Walsh and Sellers, 1993; Hyde et al., 2000). This collapse would no doubt be attributed to the inadequacies of the computer models if not for the fact that there appear to have been at least two white Earth climates in Earth history (Chumakov and Elston, 1989). The Sturtian Glaciation occurred about 760–700 million years ago, and the Marinoan Glaciation ended around 600 million years ago. Figure 10.1 shows glaciomarine diamictites from the Marinoan event.

Each glaciation may have lasted 10 million years, although estimates of ice duration range from 4 to 30 million years. Each of these two events is now referred to as a snowball Earth (a concept first proposed by Mikhail I. Budyko (1969, 1999)). The existence of these severe glaciations poses considerable challenges for Gaia theorists. If Earth can slip into a deep freeze, what does this say about the efficacy of biotic climate-regulating systems?

To answer this question, we must examine what is known about the Proterozoic snowball Earth. Research in this area may be divided into two camps: work directly involving geological evidence, and work using GCM (general circulation modeling). GCM utilizes computer models of the atmosphere and

Figure 10.1
Glaciomarine diamictite of the Marinoan glaciation interbedded with purple mudstones of unit 2 of the upper Tindir Group, Hard Luck Creek, Alaska. Coin (19 mm diameter) is for scale.

oceans to generate simulated climate and glacial ice behavior under a variety of conditions controlled by the input parameters influencing the model (Walsh and Sellers, 1993; Hyde et al., 2000; Donnadieu et al., 2003).

Conflicts have arisen between the results offered by the geologists and the general circulation modelers. Whereas early computer models of global climate slipped easily into white Earth conditions, the more sophisticated general circulation models have a difficult time generating what has been called "hard" snowball conditions, a catastrophic glaciation in which the oceans actually freeze over. This problem is due to the fact that oceanic heat transport in general circulation models is quite effective at preventing the oceans from forming an ice crust.

Geologists, on the other hand, have a different story to tell. Although there is currently a pitched controversy (Hoffman and Schrag, 2002; Lubick, 2002) concerning the extent of Proterozoic ice, the geological evidence (primarily evidence for equatorial glaciation) does seems to favor at least some version of the snowball Earth interpretation. Concerns that cold-based, low latitude glaciers would not be capable of generating the massive amounts of diamictite seen in the rock record have been allayed by modeling results suggesting that low-to-mid-latitude glaciers would have been wet-based (Donnadieu et al., 2003), and thus capable of forming the glacial tills.

Strange oxygen isotope ratios in carbonates within the glacial sequence of strata suggest the presence of at least a limited hydrological cycle. Most researchers now agree that ice sheets did indeed reach equatorial latitudes, but only on land (Hoffman and Schrag, 2002). Thus, we have a curious situation in which the computer models (formerly prone to extremes) are now generating only moderate freezes ("slushball Earth" or "soft snowball Earth"), whereas the geological interpretations (which until recently were beholden to a strong uniformitarian bias) suggest a Proterozoic global ice catastrophe followed immediately by catastrophic warming.

The key datum for the geological interpretation of ice catastrophe (hard snowball Earth) is compelling evidence for low-latitude glaciation at sea level (if not actually on the sea itself) during the Proterozoic (Embleton and Williams, 1986; Evans et al., 1997). This seemingly impossible situation (glacial ice at equatorial seashores) now seems to be an inescapable fact of Earth history. The new understanding that this can happen has been put together using lines of evidence as diverse as paleomagnetic evidence associated with glacial diamictite deposits, and periglacial features formed at low latitude, such as frost-heave structures and sand-wedge polygons (Williams and Tonkin, 1985; Hoffman, 1999; Berry et al., 2002).

Evidence of equatorial glaciation at sea level has induced a search for a special explanation. One pro-

Figure 10.2
Abrupt contact between the glacial deposits and the cap carbonate of the Ghaub diamictite, west Fransfontein, Namibia (P. F. Hoffman photo, used with permission).

posed solution to the problem is an interval of high obliquity of the Earth's rotational axis (Williams, 1993; Williams and Schmidt, 2000; Jenkins and Frakes, 1998; Jenkins and Smith, 1999; Jenkins, 2000a, 2000b). The difficulty with this extreme solution (the "big tilt hypothesis") is that it introduces a problem greater than the one it solves. For this hypothesis to be accepted, a mechanism must be found for tilting the Earth's rotational axis toward the sun and then moving it back again to restore normal climate. Although lack of a plausible mechanism should not invalidate the "big tilt hypothesis" (recall Alfred Wegener's difficulty with finding a convincing mechanism to drive continental drift), this absence of mechanism places a heavy burden of proof on the high obliquity theorists. Their case, nevertheless, has been strengthened by the lack of accord between the general circulation models and the geological evidence. It has even been suggested that the glaciations were caused by blockage of solar radiation by an Earth ring comparable to Saturn's ring (Sheldon, 1984). Such a ring might indeed be formed by a large impact event, but again one has the difficulty of removing the feature after millions of years of stability. Both the big tilt and earth ring hypotheses thus seem inadequate to the task of explaining the glaciations (Hoffman and Schrag, 2002).

Tajika (2001) presumed that all the major glaciations in the Proterozoic generated snowball Earths, and then he examined the possibility that reduced solar luminosity in the Proterozoic was the cause of the severity of the glaciations. He dismissed this idea, however, and blamed unspecified differences in carbon dioxide fluxes within the carbon cycle system during snowball conditions.

Cap Carbonates

A cap carbonate is a thin but very widespread limestone/dolostone rock unit that is usually interpreted as evidence for sudden and massive carbonate deposition under hot climate conditions. Compounding the difficulties associated with low-latitude glaciation is evidence for cap carbonate deposition immediately above the glacial deposits (figure 10.2). This juxtaposition is a major geological anomaly.

These unusual carbonates often represent the first deposits of the Lipalian system. They have three signal characteristics. First, they are depleted in heavy carbon (they have a $\delta^{13}C$ value of approximately -5, indicating a low $^{13}C/^{12}C$ ratio). A low $^{13}C/^{12}C$ ratio is usually interpreted as representing low marine biotic productivity (as indicated by a low ratio of buried organic carbon to carbonate carbon), since organisms preferentially take up the lighter isotope and leave seawater (from which the marine carbonates are precipitated) enriched in heavy carbon. Second, these deposits characteristically extend over vast areas, having been deposited in both shallow and relatively deep water. As such they provide a useful tool for lithostratigraphic correlation. Third, they contain primary dolostones, a highly unusual rock type owing to the rarity of primary precipitation of the mineral dolomite (dolomite usually forms as a secondary mineral). New cap carbonate occurrences are being described as more stratigraphic sections receive scrutiny. A newly discovered post-Sturtian cap carbonate of the Mina el Mesquite Formation in Sonora, Mexico, has been described (Corsetti et al., 2001).

Williams (1979: 385) noted in prescient fashion the climatic significance of the juxtaposition of cap carbonates above the glacial strata: "The sedimentological and oxygen-isotope data are consistent with relatively high formation-temperatures from the cap [carbonates]. Abrupt climatic warming—from cold glacial to at least seasonally high temperatures—at the close of the late Precambrian glacial epochs is implied."

Here again we see the dilemma that snowball Earth provides for Gaia theory. Not only does the geological evidence suggest a plummeting of global temperature into a white Earth climate (estimated by some to have a mean global temperature of −50°C), but the chill is ended by a dramatic fluctuation—global temperatures shoot up to hothouse conditions warm enough to precipitate carbonate in deep water. This appears to represent a convulsion of climate decoupled from any sort of Gaian regulation.

Ecology of a Snowball Earth

What would global ecology be like during a snowball Earth interval? The planet of any such time would have a surface ecology that was essentially a freshwater environment. The comment made by Ronald W. Hoham (Milius, 2000) with regard to snow ecosystems would equally apply to the ecology of snowball Earth: "[You have] got everything in it you would find in a freshwater ecosystem." McKay (2000) has argued that, in a hard snowball Earth scenario, "the thickness of the ice in tropical regions would be limited by the light penetrating into the ice cover and by the latent heat flux generated by freezing at the ice bottom." McKay calculates the tropical ice cover during white Earth conditions to be only ten meters or less (light can transmit through this thickness). With a transmissivity of ice of greater than 0.1 percent of incoming (and intense) solar radiation, pockets of photosynthesis could occur throughout the tropical ice layer. Pack ice elsewhere would have had a thickness of approximately 1 kilometer, but would not have been a continuous sheet because of ice fracture (forming open water areas called polynyas) and areas of ablation. Thus, a snowball Earth was far from being a lifeless Earth, especially if one considers the cryo-ecosystems that can develop as microbial mats on the surface of the ice or in brine channels within the ice (Vincent et al., 2000; Thomas and Dieckmann, 2002).

Bacteria in ice are able to survive for long stretches of geological time (Egorova, 1931; Kapterev, 1947; Anonymous, 1992). Actively growing bacteria have recently been found at the base of glacier ice, and are thought to be biogeochemically important in weathering via redox chemistry at the base of the glacier (Sharp et al., 1999; Skidmore et al., 1997; Wharton et al., 1995).

An algal microbiota is known to occur on the underside and even within ice found in marine and freshwater environments. These ice algae form a thick, brown layer at the ice–water interface. Some diatoms in this biota have growth optima at 2°C; growth ceases at 5°C (Sharp et al., 1999; Skidmore et al., 1997; Wharton et al., 1995). Other algal types grow at temperatures down to −4°C. The diatoms cause the formation of brine cells within the ice by withdrawing fresh water, and some species are able to withstand the frigid, hypersaline microenvironments that result (Vincent et al., 2000; Margulis et al., 1990; Hoham and Duval, 2001; Bowman et al., 1997). Such diatoms are able to withstand temperatures that freeze seawater, and they also occur in furrows left by ice wedges (Koivo and Seppala, 1994; Tazaki et al., 1994; Clasby et al., 1976; Clasby et al., 1972; Mosser et al., 1977; Bunt and Lee, 1972).

Melting the Snowball

Let us now consider the various mechanisms proposed for removing the Earth from snowball conditions. Kirschvink's (1992) original hypothesis for the mechanism of escape from snowball Earth involved the accumulation of carbon dioxide in the atmosphere as a result of volcanic input, the near cessation of photosynthesis, and the inhibition of silicate weathering via the Urey reaction (e.g., McMenamin, 2001a). Proterozoic marine iron formations were explained as buildup of dissolved iron in the water under the ice, followed by sudden oxidation and precipitation as the ice broke up and photosynthesizers could begin generating appreciable quantities of oxygen once again.

Hoffman and Schrag (2000, 2002) added that the cap carbonates would be an expected outcome of "extreme greenhouse conditions unique to the transient aftermath of a snowball earth." In this view the sudden leap from white Earth climate to cap carbonate climate is attributable to hothouse conditions resulting from unprecedented atmospheric CO_2 buildup.

A problem with this scenario is the speed with which Earth escapes the white Earth climate. In many cases the cap carbonates are deposited directly above the glacial strata (figure 10.2), implying a breathtakingly rapid transition from ice to heat. How could this transition occur so suddenly? Certainly as the ice began to melt, the planetary albedo would drop, and this process would contribute to surface warming and could even take on the characteristics of runaway feedback. However, it seems unlikely that all or even most land surfaces were covered by ice during peak white Earth conditions; thus the albedo changes may have been incremental rather than sudden and catastrophic. The same can be said for the accumulation of atmospheric carbon dioxide. The carbon dioxide sources would include volcanic emissions and respiratory release of oxidized carbon, the latter process presumably being inhibited as much as photosynthesis was by the frigid white Earth conditions. Enhancement of the atmospheric greenhouse could be expected to be incremental as well.

It would be possible to propose a nonlinear response of climate to greenhouse gas accumulation. In this case, once the CO_2 exceeded a certain threshold level, a catastrophic meltdown would occur. In fact, enhanced atmospheric carbon dioxide levels are indicated by the cap carbonate deposition. But how could the climate change so quickly, especially considering that the Urey reaction would kick in as a negative feedback as soon as temperatures began to rise on the continents (thus tending to limit the impact of a runaway greenhouse)? One might be able to account for the calcium in the cap carbonate deposits by rapid weathering of exposed carbonate rock at the end of the glaciation (Hoffman and Schrag, 2000), but this does not help with the net removal of carbon dioxide from the atmosphere. With the fragments of Rodinia forming a continental diaspora, the amount of relatively fresh continental material exposed to weathering along swollen rift margins may have been at an all-time high, tending to enhance the efficacy of the Urey reaction negative feedback. Such an effect might have contributed to the glaciation in the first place, since the most dramatic drop in $\delta^{13}C$ and the start of the Sturtian glaciation occur right at the end of the main pulse of the supercontinental breakup, shortly before 700 million years ago.

Unless we accept as plausible the idea that glaciation ended solely because of abiotic accumulation of volcanigenic carbon dioxide, the velocity and magnitude of global climate change at the end of the Proterozoic ice ages implies the action of a potent "climate change catalyst." Vladimir Vernadsky, who calculated the velocity of life and viewed life as the most important geological force (Vernadsky, 1998; Lapenis, 2002), also recognized the catalytic nature of living processes and the tendency for life to expand dramatically when conditions permit. The transition from snowball Earth to cap carbonate Earth is so sudden that it may bear a distinctly Vernadskian biotic signature. But what type of biology might lead to such a rapid climate switch? And if some type of organismic bloom is implicated in the climate change, might it not leave some type of fossil evidence (even if the transition episode was short-lived)? The cap carbonates will not help us here, for aside from some bacterial structures they appear to be devoid of fossils.

It is worth considering ways in which the biota might respond to end a glaciation. Or, to rephrase this in less anthropomorphic terms, it is worth considering how organisms might exploit a snowball Earth environment and how such modes of life might hasten the end of the glaciation. Recall that ice diatoms form a thick brown layer at the base of the ice in modern environments. Since light can pass through thick layers of ice, a dark algal layer at the bottom of a floating ice sheet could lead to considerable warming of the underside of the ice, and contribute to ice melt. The same could hold for the margins of an ice sheet on land, and indeed Donnadieu et al. (2003) note evidence for wet-based continental ice sheets.

Although diatoms were not present during the Lipalian to form the light-catching brown layer, other types of algae may have occupied extremophile niches similar to those now held by modern ice diatoms, and may have provided a Gaian feedback mechanism for melting the ice.

Tindir Group Microfossils: A Relict Ice Microbiota?

A microbial community of chrysophyte and chlorophyte algae and their bacterial neighbors (figures 10.3

Figure 10.3
Spirotindulus kryofili gen. nov. et sp. nov. from the upper Tindir Group unit 5, Yukon Territory, Canada. Diameter of fossil is 43 microns.

Figure 10.4
Cephalophytarion grande from chert of the Upper Tindir, Alaska and Canada. The species occurs in the Lipalian System of both North America and Australia. Algal filament is approximately 70 microns in length.

and 10.4), preserved in cherts of the Tindir Group, may be descended from a cryophilic microbiota. I propose here that this microbiota formed a dark-colored biofilm underneath Proterozoic ice sheets wherever they were thin enough to allow light penetration.

The Tindir Group of Alaska and the Yukon Territory is a sequence of strata that provides an outdoor laboratory for study of the events of the Lipalian period. It also provides a battery of potential tests of the hypotheses associated with snowball Earth. Unit 1 of the upper Tindir consists of volcanic rocks and pillow lavas associated with the final stages of the breakup of Rodinia. Units 2 and 3 consist of purple mudstones interbedded with Marinoan glaciomarine diamictites (figure 10.1). Unit 4 is a dolostone that may represent a cap carbonate. Unit 5 consists of resedimented limestones and clastic turbidites.

A curious biota of siliceous chrysophyte algae and bacteria was discovered in the late 1970s in cherts of the fetid (i.e., organic-rich) limestone in unit 5 of the Tindir Group in western Yukon Territory (Allison, 1981; Allison and Hilgert, 1986). Originally thought to belong to the Cambrian System, the fossils are now known to be Lipalian (McMenamin and Awramik, 1982; Young, 1982). An acid maceration sample (MAM-033) of a Tindir rock (10 of 7/22/79) produced a *Bavlinella*-like microfossil (McMenamin and Awramik, 1982). *Bavlinella* is considered by some to be diagnostic for the Lipalian (Lenk et al., 1982). Sponge spicules in the Tindir are among the earliest sponge spicules known (Allison and Awramik, 1989) and the report is now credible because Proterozoic sponge spicules are known elsewhere. The algal fossils occur in small chert nodules in limestone (Allison and Awramik, 1989).

The fossil chrysophytes are of great interest because they represent the first appearance of siliceous microfossils in the fossil record and the earliest known fossils of chrysophyte algae. The fossils consist of tiny, oval, opaline scales that in life formed an imbricate outer plating of each microbe's surface. The eukaryotic Tindir taxa are extinct, and they occur nowhere else in the rock record. The microbiota includes seventeen genera (among them *Archeoxybaphon*, *Hyaloxybaphon*, *Chilodictyon*, and *Altarmilla*). Several of the genera (Allison and Hilgert, 1986), particularly *Chilodictyon* and *Characodictyon*, have a superficially diatomlike aspect. Allison and Hilgert (1986: 979) noted that many of the Tindir chrysophytes most closely resemble modern species known "largely or exclusively from fresh water rather than marine environments." Interestingly, modern scaled chrysophytes are well known from arctic pond environments (Douglas and Smol, 1995).

A new fossil alga is described here as *Spirotindulus kryofili* gen. nov. et sp. nov. (figure 10.3). It represents a large, elongate resting zygospore of a chlorophyte green alga. Its grooved outer surface is quite similar to the flanges extending the full length of the cell in the elongate zygospores of the cryophilic chlorophyte algae *Chloromonas nivalis* and *Chloromonas polyptera*. *Spirotindulus kryofili* and *Chloromonas nivalis* zygospores have the same number of flange groove pairs (11 each).

These fossil chrysophyte and chlorophyte algae, plus their unique bacterial neighbors (including the cyanobacterial genera *Yukonosphaeridion*, *Microagglomeratus*, and *Phacelogeminus* and the species *Cephalophytarion grande* [figure 10.4]), are hypothesized here to be descended from the cryophilic microbiota that formed a brown biotic undercoat to snowball Earth ice wherever it was thin enough to allow light to penetrate through. Such microbes could have catalyzed melting of the ice, both by withdrawing fresh water from the ice (and forming ice-corroding brine pockets) and by trapping heat in the brown sub-ice mat or "hyposcum."

Recall that much of a white Earth glaciation would necessarily occur in low and even equatorial latitudes, where light levels are intense and light penetration through ice is greatest. The upper Tindir microbiota may largely represent a relict biota adapted for the tenebrous conditions beneath the ice. This biota, as a result of its albedo-altering characteristics, melted the ice and was subsequently thrust into moderate climate conditions after deposition of the dolostones of Upper Tindir unit 4. The marine environment, at last having returned to "normal," led to deposition of the fetid Tindir carbonate with its fossiliferous chert inclusions. The microbiota of the Tindir, with its preponderance of scaled chrysophytes, still carried with it the signature of its icy origin.

The spirally grooved zygospore *Spirotindulus kryofili* gen. nov. et sp. nov. (figure 10.3) is another member of this holdover community from glacial times. The spiral grooves on *Spirotindulus*, like those of the modern chlorophyte *Chloromonas*, were probably used to anchor these cells to frigid substrates (Bowman et al., 1997; Hoham and Duval, 2001).

The hypothesized ice microbiota may have existed as a stratified community at the base of the snowball Earth ice, analogous to modern ice algae (Milius, 2000), Antarctic rock-dwelling lichens or the farbstreifensandwatt of the Massachusetts and European coastlines. The similarities between the Tinder chrysophytes and modern ice diatoms and arctic lake chrysophytes are no coincidence; all these microbes evolved to occupy frigid habitats. It might even be argued that the unique ice biota ancestral to the biota of the Tindir fetid limestone evolved specifically to take advantage of the subice environment. One could also infer that the pre-Tindir brown layer consisted largely of euryhaline species that could also colonize periglacial freshwater environments.

At first the pre-Tindir brown layer may have been restricted in extent, but with continued ice thinning the habitat of the pre-Tindir microbes would have expanded rapidly. Special phytopigments and protective phenolic compounds, such as those found in the snow alga *Chlamydomonas nivalis*, could have developed to cope with increased amounts of incoming radiation (Duval et al., 2000) after thousands of years of faint light and UV exposure. Clay coatings may have played a similar role (Tazaki et al., 1994). Exceptional populations of *Chlamydomonas nivalis* show optimum photosynthesis at 0° to −3°C, although most populations do best at 10°–20°C.

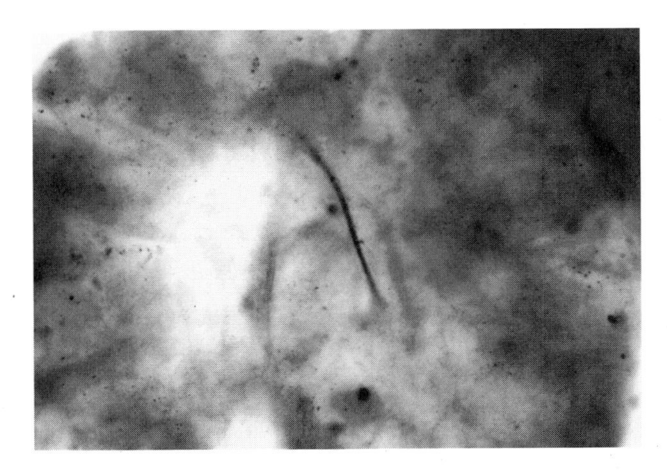

Figure 10.5
Eomycetopsis-like cyanobacterial filaments from the El Arpa Formation chert, Cerros El Arpa, Sonora, Mexico. Specimen MM-82-62; slide MM-82-62a; millimeters from reference $X = 26.3 \times 10.4$; width of filament is approximately 1.3 microns.

A Proterozoic Cryoconite

Cryoconite is a term applied to dark-colored material that forms on glacier ice, and consists primarily of wind-borne dust and microbial mat material. Cryoconite has been implicated in the acceleration of melting of Himalayan glaciers, and recent studies show that Alpine and Himalayan cryoconites consist of microbial mats formed by cyanobacteria and heterotrophic microbes (Sharp et al., 1999; Skidmore et al., 1997; Wharton et al., 1985; Takeuchi et al., 2001; Margesin et al., 2002). Cryoconite microbiotas play a role in reduction of glacier albedo, and are known to colonize open areas after the ice melts (Sharp et al., 1999; Skidmore et al., 1997; Wharton et al., 1985). Interestingly, the individual granules of granular cryoconites (Takeuchi et al., 2001: figures 4–6) bear a strong resemblance to Lipalian granules from the El Arpa Formation in Sonora, Mexico (McMenamin et al., 1983) both in granule morphology and in arrangement of the spherical microbial mat of filamentous cyanobacteria (figure 10.5). This resemblance may not be merely superficial. If the El Arpa Formation were to be correlated to the Canadian Tepee dolostone, as seems reasonable (McMenamin, 1996), it would qualify as a cap carbonate. The Tepee dolostone serves as the cap carbonate for the underlying, glaciogenic Ice Brook Formation.

Wind-blown sediments alone, volcanigenic or otherwise, do not greatly alter the albedo of ice surfaces because they are generally light tan or gray in color. But detrital cryoconites do begin to alter ice surface albedo when mat-forming microbes colonize them. Cryoconite microbial mats are known for forming dark humic acid accumulations that accelerate ice melting (Takeuchi et al., 2001; Margesin et al., 2002).

A Life-Induced Meltdown?

The ice of a snowball Earth would thus be biotically thawed from both above (cryoconite microbial mats) and below (subice algal/bacterial brown layer). A dark, humic acid-rich cryoconite layer would tend to block light from reaching any brown layer below, but this would be compensated for by cryoconite-induced thinning of the ice sheet above the brown layer. Furthermore, light entering the ice laterally from some distance away could become trapped in the ice between the upper and lower dark layers, further contributing to meltdown.

It would be to the immediate advantage of cryoconite microbes and brown layer microbes to expand their range and influence, thus melting more ice and expanding the areal extent of the band-shaped zone between the ice margin and the limit of light penetration through the ice. There would be few limits to growth, especially for a cold-adapted biota, because the chemical composition of the oceans of the time resembled "the nutrient media typically used for growing cyanobacteria in pure culture" (Duval et al., 2000) and a steep nutricline had not yet been reestablished in the oceans. In Vernadskian fashion, uncontrolled growth of the ancestors of the Tindir microorganisms may well have catalyzed rapid collapse of the ice sheets, heating them, as it were, from above, below, and within by passive solar gain. The organisms would also begin penetrating the ice itself by means of brine pockets and cryoconite holes (Sharp et al., 1999; Skidmore et al., 1997).

The oceans could thus become "prematurely" free of ice; that is, clear of ice before the normal operation of biogeochemical cycles could resume. The accumulated levels of atmospheric carbon dioxide would then represent a highly metastable situation, and carbon dioxide would literally collapse out of the atmosphere as the Urey reaction kicked in with full force in an anomalous hot climate. Carbon stripped from the atmosphere and sequestered as dissolved calcium carbonate would be delivered in massive amounts to the ocean basins, but this amount would be dwarfed by the amount of carbonate delivered by the weathering action of hot rain supercharged with carbonic acid falling onto exposed carbonate rock (Hoffman and Schrag, 2002).

The odd radiating crystal textures and unusual dolostones of the cap carbonates are evidence for precipitation as a result of what could be called oceanic supersaturation with respect to calcium carbonate. The depletion in heavy carbon (low $\delta^{13}C$ values) of the cap carbonates is also considered evidence for a chemical oceanographic origin (Kennedy et al., 2001a, 2001b; Pecher, 2002; Wood et al., 2002), although the source of this carbon is not entirely clear. Atmospheric carbon and carbon in methane from gas hydrates undergoing decomposition as global climate warmed have been suggested as sources. Newly described gas hydrate chimneys could have profound effects on the stability of gas hydrates (Wood et al., 2002).

Gaian Implications

Snowball Earth thus provides a challenge for both the Gaia theorists and those who prefer to emphasize abiotic geochemistry to explain climate change. For if Gaia is hard pressed to explain the intensity of the glaciation (Where were the biotic feedback controls on climate?), geochemistry alone is inadequate to explain the rapid deglaciation.

Perhaps we see here an emerging synthesis. Life processes, for Vernadsky (1998) *the* geological force, can indeed impart global climate regulation. Such influence, however, must always be considered in Vernadskian terms. The Vernadskian pressure of life will expand explosively at unexpected times and places, as with the Cambrian explosion that ended Lipalian times (McMenamin, 2003). The inferred growth of the hypothesized pre-Tindir microbe brown layer that led to the collapse of the white Earth ice sheets would serve as another example.

Some type of runaway biotic feedback must be responsible for the rapid transition from snowball conditions to tropical cap carbonate deposition. Conventional abiotic geochemical systems simply cannot respond with the requisite speed. Joseph Kirschvink (personal communication, 2001) has calculated that the atmosphere requires 0.6 bar of carbon dioxide to end hard snowball Earth conditions, and that it takes 70 million years to thaw the earth once this level of atmospheric carbon dioxide has been attained. This is more than seventy times the amount of time allowed by the rock record to accomplish the transition.

Runaway expansions of the biota eventually run out of space on our spherical planet. As the surge of life eventually plays itself out on the planetary surface, new and stable climate equilibrium is attained. The next generation of general circulation models must (in addition to improving their treatment of cloud cover) consider parameters designed to describe the surge aspect of biotic influence on global climate. Certainly our species could be considered responsible for this type of geologically sudden perturbation. Another good example of the phenomenon is the expansion of Hypersea in the Devonian (McMenamin and McMenamin, 1994) and associated climate instability (expressed as another glaciation; Saltzman, 2003) in the late Paleozoic.

The intensity of white Earth conditions during the Marinoan and Sturtian ice ages needs to be better understood if we hope to offer a more complete description of extreme climate dynamics. For example, carbonate sediments (peloids, oncolites, and marine cements) occurring within the glacial strata and presumably precipitated directly from glacial seawater of ancient Australia and Namibia have relatively high $\delta^{13}C$ values. These high values suggest to some that the global biological carbon pump was functioning normally, an interpretation that argues against ice-covered oceans (Hoffman and Schrag, 2002; Kennedy et al., 2001a, 2001b). Paul F. Hoffman and Adam C. Maloof (written communication, 2001) argue that these high values would be consistent with detrital origin for the intraglacial diamictite carbonates, in which case the carbonates would have a light carbon signature inherited from more ancient carbonates. This would also avoid the difficulty of having to deposit newly formed carbonate in mid-glaciation. Or perhaps the intraglacial sediment carbonates are indeed primary, but result from the unusual ecosystem at the base of the ice.

Strange and innovative biology seems to be associated with the subice biota. The hyposcum habitat would have been a suitable proving ground for a tenebrous proto-Garden of Ediacara ecology. Metacellular Ediacarans (McMenamin, 1998, 2001b) appear immediately after the end of the Marinoan glaciation, and may be descended from the elongate metacellular organisms *Archeomyces dimakeloides* and *Eophycomyces herkoides* of the Tindir Group cherts (Allison and Awramik, 1989). Allison and Awramik compare these genera to fungus-like chytrids. A chytrid affinity has been proposed as well for Ediacarans (McMenamin, 1998). Thumblike protrusions (Allison and Awramik, 1989) in *Eophycomyces* may be homologous to the "thumb structures" in the Ediacarans *Tribrachidium* and *Gehlingia* (McMenamin, 1998, 2001b). *Archeomyces* displays metacells that are rectangular in cross section, relatively rigid walls separating adjacent metacells, (branched?) tubular extensions coming from the larger metacells, and what appears to be metacellular budding—all characters seen in Ediacarans (McMenamin, 1998). The iterated and chainlike aspect of *Eophycomyces herkoides* completes the analogy between Ediacaran metacellularity and the modular architecture (particularly the ridgework) of the Spanish architect Antonio Gaudi (Collins, 1990).

The lack of large predators in the Lipalian may be a result of a relict cryophile ecosystem experiencing normal marine conditions for the first time. Indeed, ecological analogues to such an ecosystem are seen among modern protists. The giant branching foramineran *Notodendrodes antarctikos* lives under semipermanent sea ice in McMurdo Sound, Antarctica (Delaca, Lipps, and Hessler, 1980). It takes up dissolved organic matter largely as a function of its surface area, a mode of feeding that has been inferred for the Ediacarans (McMenamin, 1993). When large predators appeared at the end of the Lipalian or beginning of the Cambrian, the Garden of Ediacara ecosystem (an overgrown cryophilic microbiota?) was lost, although a few survivors apparently crossed the Cambrian boundary (McMenamin, 1998).

In conclusion, it seems reasonable to infer that a cryophilic biota could exert a major influence on white Earth conditions, and that white Earth conditions in turn could have a significant impact on the evolution of ecosystems. This inference could be tested by scalable experiments on modern ice microbes associated with floating, grounded, or anchored ice. Certainly the extreme fluctuation in carbon isotope values of Lipalian time implies substantive changes in the global and local biogeochemistry of microorganisms (Gaidos, 1999).

Ever since Mawson (1949), Harland, and Rudwick (Harland, 1965; Harland and Rudwick, 1964) drew attention to the severity of the Proterozoic glaciations, there have been concerns about the "bizarre potential, and possible past states of Earth's climate" (Fairchild, 2001). In particular, there are fears that alteration of the northward course of the Gulf Stream and the shutdown of North Atlantic bottom water generation could threaten the mild climate of northwestern Europe (Fairchild, 2001; Rahmstorf, 2000). We need to combine analysis of ocean dynamics (Poulsen et al., 2001) with analysis of biotic feedback mechanisms to arrive at an understanding of the constraints (or lack thereof) associated with extreme climatic fluctuations of the past. Such extreme fluctuations may not matter much for bacteria, since there is evidence suggesting that Proterozic preglacial bacterial communities in California were much the same as the synglacial communities (Fairchild, 2001), a fact not in accord with arguments that snowball Earth had catastrophic effects on the entire biosphere (Runnegar, 2000). Conversely, the modern biosphere in the guise of *Homo sapiens* appears to be having a major and perhaps soon to be catastrophic influence on the stability of polar ice sheets (Smith et al., 2002; Kaiser, 2002).

Acknowledgments

I would like to thank C. W. Allison, S. M. Awramik, P. E. Cloud, D. A. Evans, P. F. Hoffman, J. L. Kirschvink, L. Margulis, D. L. S. McMenamin, and G. Young for assistance with various aspects of this project.

APPENDIX: SYSTEMATIC PALEONTOLOGY

Kingdom Monera
Phylum Cyanobacteria
Class Cyanophyta
Order Nostocales
Family Oscillatoriaceae
Genus *Cephalophytarion* (Schopf, 1968)

Type species: *Cephalophytarion grande* (Schopf, 1968).

Geologic age: Late Sinian to Lipalian.

Cephalophytarion grande (Schopf, 1968) figure 10.4

1968: *Cephalophytarion grande* (Schopf, p. 669)

1989: *Cephalophytarion majesticum* (Allison, in Allison and Awramik, pp. 273–274).

Description: See Schopf (1968) and Allison and Awramik (1989).

Discussion: *Cephalophytarion majesticum* Allison is considered here to be synonymous with *Cephalophytarion grande* Schopf. The apical constriction in the original description of *C. grande* (Schopf, 1968) is surely a diagenetic/taphonomic feature (Allison and Awramik, 1989).

This species-level synonymy provides a potentially useful paleobiogeographic link between the Bitter Springs microbiota of Australia (Schopf, 1968) and the Tindir Group of North America. Australia and northwestern North America were probably juxtaposed as part of Rodinia (Hoffman, 1999; Donnadieu et al., 2003).

Geologic age: Lipalian period.

Kingdom Protoctista
Phylum Chlorophyta
Class Chlorophyceae
Order Chlamydomonadales
Family Chlamydomonadaceae
Genus *Spirotindulus* gen. nov.

Type species: *Spirotindulus kryofili* gen. et sp. nov.

Etymology: Named for the Tindir Group.

Diagnosis: A cylindrical fossil distinguished by ten relatively deep spiral grooves. Length of cylinder unknown. The cylinder is approximately 43 microns in diameter; the thickness of the tube wall is approximately 0.33 microns. No internal wall structures are known. The chert within the fossil is paler than the surrounding matrix. The grooves are V- to U-shaped, and are 3.5 microns wide and 3 microns deep.

Spirotindulus kryofili gen. et sp. nov. (figure 10.3)

1989: "Unnamed Spiral Grooved Form," Allison and Awramik, p. 284.

Holotype: University of California at Santa Barbara repository number SMA 524 at 30.9 × 11.7 (using reference "X" closest to center of slide).

Description: As for the genus.

Discussion: Allison and Awramik (1989) describe this fossil (their figure 10.6) as occurring between brown organic layers.

The flanges and grooves on its outer surface are this fossil's most distinctive feature. *Spirotindulus kryofili* gen. nov. et sp. nov. is most similar to the zygospores of *Chloromonas nivalis.* Each species has 11 flange/groove pairs along the length of the zygospore (see Hoham and Duval, 2001: figures 4.10–4.11). The zygospore of *Spirotindulus kryofili* gen. nov. et sp. nov., however, is 2.3 times larger in diameter, and its flanges are broader.

One of the flanges in *Spirotindulus kryofili* gen. nov. et sp. nov. is seen to branch (the broad flange in the lower right of figure 10.3 with the indentation on its outer part). This pattern of branching flanges is also recorded in *Chloromonas polyptera*, a species with zygospores bearing 19–20 flange/groove pairs (hence *polyptera*, meaning many wings or flanges; see Hoham and Duval, 2001: figures 4.20–4.21).

Geologic age: Lipalian period.

Locality: Field sample MM-80-118c (same as G4-80-182), Unit 5 of the upper Tindir Group, upper Tindir Creek, Yukon Territory, Canada. The sample site consists of a 6 cm thick bed of resedimented chert nodules. Shale occurs immediately below the chert bed and laminated limestone occurs immediately upsection.

References

Allison, C. W. 1981. Siliceous microfossils from the Lower Cambrian of northwest Canada: Possible source for biogenic chert. *Science*, 211, 53–55.

Allison, C. W., and S. M. Awramik. 1989. Organic-walled microfossils from earliest Cambrian or latest Proterozoic Tindir Group rocks, northwest Canada. *Precambrian Research*, 43, 253–294.

Allison, C. W., and J. W. Hilgert. 1986. Scale microfossils from the Early Cambrian of northwestern Canada. *Journal of Paleontology*, 60, 973–1015.

Anonymous. 1992. Lebende bakterien aus der eiszeit. *Naturwissenschaftliche Rundschau*, 45, 23.

Berry, R. F., G. A. Jenner, S. Meffre, and M. N. Tubrett. 2002. A North American provenance for Neoproterozoic to Cambrian sandstones in Tasmania? *Earth and Planetary Science Letters*, 192, 207–222.

Bowman, J. P., S. A. McCammon, M. V. Brown, D. S. Nichols, and T. A. McMeekin. 1997. Diversity and association of psychrophilic bacteria in Antarctic sea ice. *Applied and Environmental Microbiology*, 63, 3068–3078.

Budyko, M. 1969. The effect of solar radiation variations on the climate of the Earth. *Tellus*, 21, 611–619.

Budyko, M. 1999. Climate catastrophes. *Global and Planetary Change*, 20, 281–288.

Bunt, J. S., and C. C. Lee. 1972. Data on the composition and dark survival of four sea-ice microalgae. *Limnology and Oceanography*, 17, 458–461.

Chumakov, N. M., and D. P. Elston. 1989. The paradox of late Proterozoic glaciations at low latitudes. *Episodes*, 12, 115–120.

Clasby, R. C., V. Alexander, and R. Horner. 1976. Primary productivity of sea-ice algae. *Occasional Publications of the Institute of Marine Science*, 4, 289–304.

Clasby, R., R. Horner, and V. Alexander. 1972. Arctic ice algae studies. *Proceedings of the Alaska Science Conference*, 23, 54.

Collins, G. R. 1990. *Gaudi*. Barcelona: Editorial Escudo de Oro.

Corsetti, F., J. H. Stewart, and J. W. Hagadorn. 2001. Neoproterozoic diamictite-cap carbonate succession from eastern Sonora, Mexico. *Geological Society of America Abstracts*, 33, 21.

Delaca, T. E., J. H. Lipps, and Hessler, R. R. 1980. The morphology and ecology of a new large agglutinated Antarctic foraminifer (Textulariina: Notodendrodidae *nov.*). *Zoological Journal of the Linnean Society*, 69, 205–224.

Donnadieu, Y., F. Fluteau, G. Ramstein, C. Ritz, and J. Besse. 2003. Is there a conflict between the Neoproterozoic glacial deposits and the snowball Earth interpretation?: An improved understanding with numerical modeling. *Earth and Planetary Science Letters*, 208, 101–112.

Douglas, M. S. V., and J. P. Smol. 1995. Paleolimnological significance of observed distribution patterns of chrysophyte cysts in arctic pond environments. *Journal of Paleolimnology*, 13, 79–83.

Duval, B., K. Shetty, and W. H. Thomas. 2000. Phenolic compounds and antioxidant properties in the snow alga *Chlamydomonas nivalis* after exposure to UV light. *Journal of Applied Phycology*, 11, 559–566.

Egorova, A. A. 1931. Ueber bakterien im fossilen eis. *Arktis*, 4, 12–14.

Embleton, B. J. J., and G. E. Williams. 1986. Low paleolatitude of deposition for late Precambrian periglacial varvites in South Australia: Implications for palaeoclimatology. *Earth and Planetary Science Letters*, 79, 419–430.

Evans, D. A., N. J. Beukes, and J. L. Kirschvink. 1997. Low-latitude glaciation in the Palaeoproterozoic era. *Nature*, 386, 262–266.

Fairchild, I. J. 2001. Encapsulating climatic catastrophe: Snowball earth. *Geoscientist*, 11, 4–5.

Gaidos, E. J. 1999. Consequences of global glaciation for the microbial biosphere. *Geological Society of America Abstracts*, 31, 372.

Gleick, J. 1987. *Chaos*. New York: Viking.

Harland, W. B. 1965. Critical evidence for a great infra-Cambrian glaciation. *Geologische Rundschau*, 54, 45–61.

Harland, W. B., and M. J. S. Rudwick. 1964. The great infra-Cambrian ice age. *Scientific American*, 211, 28–36.

Hoffman, P. F. 1999. The break-up of Rodinia, birth of Gondwana, true polar wander and the snowball Earth. *Journal of African Earth Sciences*, 28, 17–33.

Hoffman, P. F., and D. P. Schrag. 2000. Quand le terre était gelée. *Pour la Science*, 268, 30–37.

Hoffman, P. F., and D. P. Schrag. 2002. The snowball Earth hypothesis: Testing the limits of global change. *Terra Nova*, 14, 129–155.

Hoham, R. W., and B. Duval. 2001. Microbial ecology of snow and freshwater ice with emphasis on snow algae. In H. G. Jones, J. W. Pomeroy, D. A. Walker, and R. W. Hoham, eds., *Snow Ecology*, pp. 168–169. Cambridge: Cambridge University Press.

Hyde, W. T., T. J. Crowley, S. K. Baum, and W. R. Peltier. 2000. Neoproterozoic "snowball Earth" simulations with a coupled climate/ice-sheet model. *Nature*, 405, 425–429.

Jenkins, G. S. 2000a. Global climate model high-obliquity solutions to the ancient climate puzzles of the faint-young sun paradox and low-altitude Proterozoic glaciation. *Journal of Geophysical Research*, 105, 7357–7370.

Jenkins, G. S. 2000b. Correction to "Global climate model high-obliquity solutions to the ancient climate puzzles of the faint-young sun paradox and low-latitude Proterozoic glaciation" by Gregory S. Jenkins. *Journal of Geophysical Research*, 105, 12519.

Jenkins, G. S., and L. A. Frakes. 1998. GCM sensitivity test using increased rotation rate, reduced solar forcing and orography to examine low latitude glaciation in the Neoproterozoic. *Geophysical Research Letters*, 25, 3525–3528.

Jenkins, G. S., and S. R. Smith. 1999. GCM simulations of Snowball Earth conditions during the late Proterozoic. *Geophysical Research Letters*, 26, 2263–2266.

Kaiser, J. 2002. Breaking up is far too easy. *Science*, 297, 1494–1496.

Kapterev, P. N. 1947. Anabiose im ewigen eise. *Deutsche Gesundheitswesen*, 2, 517.

Kennedy, M. J., N. Christie-Blick, and L. E. Sohl. 2001a. Carbon isotopic evidence that favors a gas hydrate versus atmospheric origin for Neoproterozoic cap carbonate formation. *Eos, Transactions, American Geophysical Union*, 82, S8.

Kennedy, M. J., N. Christie-Blick, and L. E. Sohl. 2001b. Are Proterozoic cap carbonates and isotopic excursions a record of gas hydrate destabilization following Earth's coldest intervals? *Geology*, 29, 443–446.

Kirschvink, J. L. 1992. Late Proterozoic low-latitude global glaciation: The snowball Earth. In J. W. Schopf and C. Klein, eds., *The Proterozoic Biosphere*, pp. 51–52. Cambridge: Cambridge University Press.

Koivo, L., and M. Seppala. 1994. Diatoms from an ice-wedge furrow, Ungava Peninsula, Quebec, Canada. *Polar Research*, 13, 237–241.

Lapenis, A. G. 2002. Directed evolution of the biosphere: Biogeochemical selection or Gaia? *The Professional Geographer*, 54, 379–391.

Lenk, C., P. K. Strother, C. A. Kaye, and E. S. Barghoorn. 1982. Precambrian age of the Boston Basin: New evidence from microfossils. *Science*, 216, 619–620.

Lovelock, J. 1990. *The Ages of Gaia*. New York: Bantam Books.

Lubick, N. 2002. Snowball fights. *Nature*, 417, 12–13.

Margesin, R., G. Zacke, and F. Schinner. 2002. Characterization of heterotrophic microorganisms in Alpine glacier cryoconite. *Arctic, Antarctic, and Alpine Research*, 34, 88–93.

Margulis, L., et al., eds. 1990. *Handbook of Protoctista*. Boston: Jones and Bartlett.

Mawson, D. 1949. The Late Precambrian ice-age and glacial record of the Bibliando dome. *Journal and Proceedings of the Royal Society of New South Wales*, 82, 150–174.

McKay, C. P. 2000. Thickness of tropical ice and photosynthesis on a snowball Earth. *Geophysical Research Letters*, 27, 2153–2156.

McMenamin, M. 1993. Osmotrophy in fossil protoctists and early animals. *Invertebrate Reproduction and Development*, 22, 301–304.

McMenamin, M. A. S. 1996. Ediacaran biota from Sonora, Mexico. *Proceedings of the National Academy of Sciences USA*, 93, 4990–4993.

McMenamin, M. A. S. 1998. *The Garden of Ediacara: Discovering the First Complex Life*. New York: Columbia University Press.

McMenamin, M. A. S. 2001a. *Dictionary of Earth and Environment*. South Hadley, MA: Meanma Press.

McMenamin, M. A. S. 2001b. *Evolution of the Noösphere*. Teilhard Studies no. 42. New York: American Teilhard Association for the Future of Man.

McMenamin, M. A. S. 2003. Origin and early evolution of predators: The ecotone model and early evidence for macropredation. In P. Kelley, M. Kowalewski, and T. Hansen, eds., *Predator–Prey Interactions in the Fossil Record*, pp. 379–400. Topics in Geobiology series 20. New York: Kluwer Academic/Plenum.

McMenamin, M. A. S., and S. M. Awramik. 1982. Acid resistant microfossils (acritarchs) from the Upper Tindir Group, Yukon Territory, Canada. *Geological Society of America Abstracts*, 14, 214.

McMenamin, M. A. S., S. M. Awramik, and J. H. Stewart. 1983. Precambrian–Cambrian transition problem in western North America: Part II. Early Cambrian skeletonized fauna and associated fossils from Sonora, Mexico. *Geology*, 11, 227–230.

McMenamin, M. A. S., and D. L. S. McMenamin. 1994. *Hypersea: Life on Land*. New York: Columbia University Press.

Milius, S. 2000. Red snow, green snow. *Science News*, 157, 328–330.

Mosser, J. L., A. G. Mosser, and T. D. Brock. 1977. Photosynthesis in the snow: The alga *Chlamydomonas nivalis* (Chlorophyceae). *Journal of Phycology*, 13, 22–27.

Pecher, I. A. 2002. Gas hydrates on the brink. *Nature*, 420, 622–623.

Poulsen, C. J., R. T. Pierrehumbert, and R. L. Jacob. 2001. Impact of ocean dynamics on the simulation of the Neoproterozoic "snowball Earth." *Geophysical Research Letters*, 28, 1575–1578.

Rahmstorf, S. 2000. The thermohaline ocean circulation: A system with dangerous thresholds? *Climatic Change*, 46, 247–256.

Runnegar, B. 2000. Loophole for snowball earth. *Nature*, 405, 403.

Saltzman, M. R. 2003. Late Paleozoic ice age: Oceanic gateway or pCO_2? *Geology*, 31, 151–154.

Schopf, J. W. 1968. Microflora of the Bitter Springs Formation, Late Precambrian, central Australia. *Journal of Paleontology*, 42, 651–688.

Sharp, M., J. Parkes, B. Cragg, I. J. Fairchild, H. Lamb, and M. Tranter. 1999. Widespread bacterial populations at glacier beds and their relationship to rock weathering and carbon cycling. *Geology*, 27, 107–110.

Sheldon, R. P. 1984. The Precambrian ice-ring model to account for changes in exogenic regimes from Proterozoic to Phanerozoic eras. In *Symposium of 5th International Field Workshop and Seminar on Phosphorite, Kunming, China, 1982*, vol. 2, pp. 227–243. Beijing: Geological Publishing House.

Skidmore, M. L., J. Foght, and M. J. Sharp. 1997. Microbially mediated weathering reactions in glacial environments. *Geological Society of America Abstracts*, 29, 362.

Smith, J., R. Stone, and J. Fahrenkamp-Uppenbrink. 2002. Trouble in polar paradise. *Science*, 297, 1489.

Tajika, E. 2001. Physical and geochemical conditions required for the initiation of snowball Earth. *Geological Society of America Abstracts*, 33, 451.

Takeuchi, N., S. Kohshima, and K. Seto. 2001. Structure, formation, and darkening process of albedo-reducing material (cryoconite) on a Himalayan glacier; a granular algal mat growing on the glacier. *Arctic, Antarctic, and Alpine Research*, 33, 115–122.

Tazaki, K., et al. 1994. Clay aerosols and Arctic ice algae. *Clays and Clay Minerals*, 42, 402–408.

Thomas, D. N., and G. S. Dieckmann. 2002. Antarctic sea ice—a habitat for extremophiles. *Science*, 295, 641–644.

Vernadsky, V. I. 1998. *The Biosphere*. New York: Copernicus.

Vincent, W. F., J. A. Gibson, R. Pienitz, V. Villeneuve, P. A. Broady, P. B. Hamilton, and C. Howard-Williams. 2000. Ice shelf microbial ecosystems in the high arctic and implications for life on snowball Earth. *Naturwissenschaften*, 87, 137–141.

Walsh, K. J., and W. D. Sellers. 1993. Response of a global climate model to a thirty percent reduction of the solar constant. *Global and Planetary Change*, 8, 219–230.

Wharton, R. A., C. P. McKay, G. M. Simmons, and B. C. Parker. 1985. Cryoconite holes on glaciers. *BioScience*, 35, 440–503.

Williams, G. E. 1979. Sedimentology, stable-isotope geochemistry and palaeoenvironment of dolostones capping late Precambrian glacial sequences in Australia. *Journal of the Geological Society of Australia*, 26, 377–386.

Williams, G. E. 1993. History of the Earth's obliquity. *Earth Science Reviews*, 34, 1–45.

Williams, G. E., and P. Schmidt. 2000. Proterozoic equatorial glaciation: Has "Snowball Earth" a snowball's chance? *Australian Geologist*, 117, 21–25.

Williams, G. E., and D. G. Tonkin. 1985. Periglacial structures and palaeoclimatic significance of a late Precambrian block field in the Cattle Grid copper mine, Mount Gunson, South Australia. *Australian Journal of Earth Sciences*, 32, 287–300.

Wood, W. T., J. F. Gettrust, N. R. Chapman, G. D. Spence, and R. D. Hyndman. 2002. Decreased stability of methane hydrates in marine sediments owing to phase-boundary roughness. *Nature*, 420, 656–660.

Young, G. M. 1982. The late Proterozoic Tindir Group, east-central Alaska: Evolution of a continental margin. *Geological Society of America Bulletin*, 93, 759–783.

11
Does Life Drive Disequilibrium in the Biosphere?

David W. Schwartzman and Tyler Volk

Abstract

Lovelock's Gaia hypothesis was born out of his insight that the atmosphere of a lifeless planet would be close to chemical equilibrium, while the robust presence of life would generate measurable disequilibrium. Others have expanded this view to postulate the growing disequilibrium of the Earth's surface system from life's influence over geologic time.

We show that the carbonate-silicate geochemical cycle (Urey reaction), the long-term control on the steady-state atmospheric carbon dioxide level, is far from equilibrium on the present Earth, approaching this state only on a billion-year timescale in the future as solar luminosity and surface temperature climb. Moreover, the progressive increase in the biotic enhancement of chemical weathering in the last 4 billion years, culminating in the weathering regime of the forest and grassland ecosystems, has brought the steady-state atmospheric carbon dioxide level closer to the Urey reaction equilibrium state. In contrast, the abiotic steady state is always further from this equilibrium state than the biotic, except near the origin of life and at their future convergence. These are counterintuitive results from a classical Gaian view.

Equilibrium is an apparent attractor state in biospheric evolution for the case of the Urey reaction and long-term atmospheric carbon dioxide levels, but apparently not for other atmospheric gases, especially oxygen.

Finally, an astrobiological flag: Lovelock's original insight may still be valid for some cases, but far-from-equilibrium abiotic steady states may arise on Earth and other planets, and should not be taken as a priori evidence for Gaian self-regulation.

Introduction

The origin of the Gaia concept is rooted in Lovelock's realization that the Martian atmospheric composition should indicate the presence or absence of an indigenous biota: "If the planet were lifeless, then it would be expected to have an atmosphere determined by physics and chemistry alone, and be close to the chemical equilibrium state. But if the planet bore life, organisms at the surface would be obliged to use the atmosphere as a source of raw materials and a depository for wastes. Such a use of the atmosphere would change its chemical composition. It would depart from equilibrium in a way that would show the presence of life" (Lovelock, 1990, 100). In particular, Lovelock pointed out the coexistence of both oxidizing (oxygen) and reducing (e.g., methane and nitrous oxide) gases in the present Earth's atmosphere.

Lenton (1998, 439) generalized this view: "Lovelock recognized that most organisms shift their physical environment away from equilibrium." Others have gone further, postulating the growing disequilibrium of the Earth's surface system as a result of life's influence over geologic time (see Guerzoni et al., 2001).

First, in considering the merits of these postulates, the departure from chemical equilibrium within the atmosphere should be distinguished from that of the crust/atmosphere interface. This chapter is mainly concerned with testing whether life has tended to bring the crust/atmosphere interface closer to equilibrium than an abiotic regime would. We will return to the proposal that an abiotic atmosphere should be close to equilibrium later. With respect to the crust/atmosphere interface, one should consider the possibility of a purely geochemical nonequilibrium condition of a planetary surface system resulting in steady-state properties, such as atmospheric carbon dioxide level, that are different from chemical equilibrium values.

The long-term carbon cycle (> 10^5 years) is controlled by the silicate-carbonate geochemical cycle. This cycle entails transfers of carbon to and from the crust and mantle. Walker et al.'s (1981) geochemical climatic stabilizer is a model of the operation of the silicate-carbonate geochemical cycle first described in the modern era by Urey (1952):

$CO_2 + CaSiO_3 = CaCO_3 + SiO_2$

The reaction to the right corresponds to chemical weathering of Ca silicates on land ($CaSiO_3$ is a simplified proxy for the diversity of rock-forming CaMg silicates such as plagioclase and pyroxene, which have more complicated formulas (e.g., Ca plagioclase: $CaAl_2Si_2O_8$ and diopside: $CaMg(SiO_3)_2$), while the reaction to the left corresponds to metamorphism (decarbonation) and degassing returning carbon dioxide to the atmosphere. The main aspects of chemical weathering, including the realization that plants are accelerators, and the long-term control mechanism on carbon in the atmosphere were first published over 140 years ago by the French mining engineer Jacques Ebelmen (Berner and Maasch, 1996). It is interesting that Hutchinson (1954), considering the Urey reaction, concluded that several factors on the Earth operate to preclude the likely attainment of an equilibrium level of atmospheric carbon dioxide, but argued that the equilibrium state must account for the general magnitude of the actually occurring level.

Computing the Urey Equilibrium over Geologic Time

We follow up here our previous discussion of this subject based on first approximation calculations of the equilibrium temperature and pCO_2 level, which assumed the classic Urey reaction and no dependence of K on temperature (Schwartzman et al., 1994; Schwartzman, 1999).

Calculation of Effective, Abiotic, and Equilibrium Temperatures

The effective blackbody radiation temperature, T_e, of the Earth was assumed to vary with age (t in Ga) as follows (see Kasting and Grinspoon, 1990):

$$T_e = 255/(1 + 0.087t)^{0.25}$$

This expression assumes a constant planetary albedo.

Hypothetical abiotic temperatures were computed assuming the biotic enhancement of weathering on the present Earth is 100, thereby subtracting out this biologically mediated cooling effect over the last 4 billion years; the factor of 100 is inferred from field and experimental studies. This model abiotic temperature history was computed from the inferred steady-state levels of atmospheric carbon dioxide over geologic time, controlled by the balance of the weathering sink and volcanic/metamorphic source of carbon dioxide with respect to the atmosphere/ocean reservoir (see Schwartzman and Volk, 1991; Schwartzman, 1999). Equilibrium temperatures and atmospheric pCO_2 levels were computed from thermodynamic data (Faure, 1991; units in kcal/mole) for the Urey-type reactions representing the commonest relevant minerals in the crust and most influential reactions:

(a) $CaCO_3 + SiO_2 = CaSiO_3 + CO_2$

$\Delta G^o = 9.8, \quad \Delta H^o = 21.28$

(b) $CaCO_3 + Al_2Si_2O_5(OH)_4$

$= CaAl_2Si_2O_8 + 2H_2O + CO_2$

$\Delta G^o = 8.6, \quad \Delta H^o = 31.2$

(c) $(CaMg)(CO_3)_2 + 2SiO_2$

$= CaMg(SiO_3)_2 + 2CO_2$

$\Delta G^o = 14.1, \quad \Delta H^o = 37.7$

K was computed as a function of temperature (T) between 0 and 100°C using the van't Hoff equation:

$$\ln(K_T/K_{298.15}) = -(\Delta H^o/R)[(1/T) - (1/298.15)],$$

$$\text{with } \Delta G^o = -RT \ln K$$

Coupled with a greenhouse function which takes into account the variation in solar luminosity, the temperature and carbon dioxide levels are solvable at any time (i.e., two equations, two unknowns). Only a carbon dioxide/water greenhouse is assumed for simplification, recognizing the possible importance of methane in the Archean/early Proterozoic prior to the rise of atmospheric oxygen (Kasting et al., 2001). We used the greenhouse function given in Caldeira and Kasting (1992) because it gives more plausible temperatures for the low pCO_2 range ($P_t < 0.03$ bar) than the Walker et al. (1981) function (Kasting, personal communication):

$T = T_e + \Delta T$, where T is the mean global surface temperature (°K).

$$\Delta T = 815.17 + (4.895 \times 10^7)T^{-2} - (3.9787 \times 10^5)T^{-1} - 6.7084y^{-2} + 73.221y^{-1} - 30{,}882T^{-1}y^{-1},$$

where $y = \log pCO_2$ (in bars).

For $P_t > 0.03$ bar, an updated version of Kasting and Ackerman's (1986) function, given in equation form by Caldeira (personal communication), with $T_t = f(P_t, S_t)$, where S_t is the relative solar flux at time t:

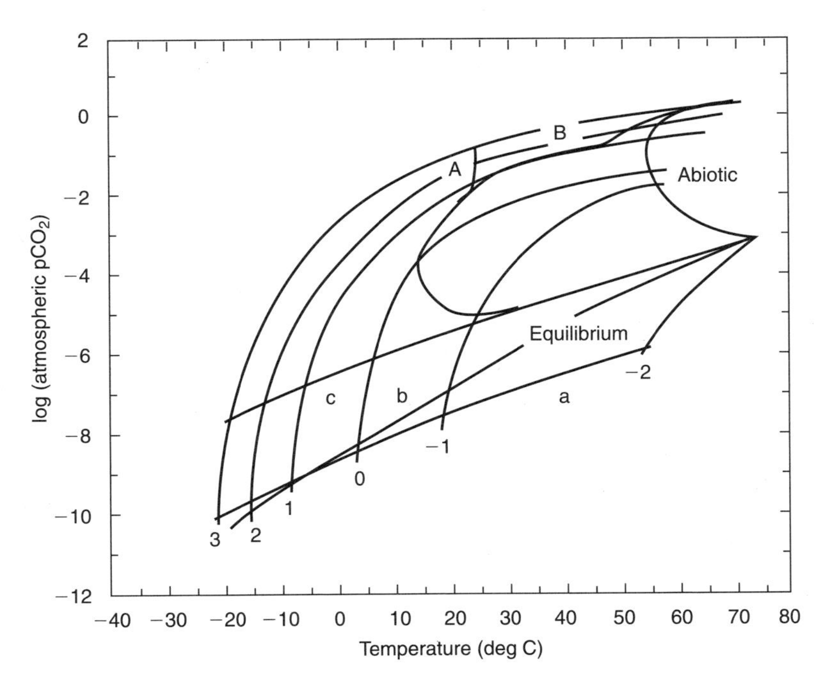

Figure 11.1
Log pCO_2 in atmosphere and surface temperature (deg C) as a function of time for Urey equilibria (labeled a, b, and c, corresponding to same reactions as in text); the surface biosphere (two histories, labeled A and B) and a hypothetical abiotic Earth surface (labeled Abiotic), assuming the present biotic enhancement of weathering is 100. Numbers on curves are in Ga (billion years): positive past, negative future, 0 now.

$$
\begin{aligned}
T_t = {} & 138.114 - 73.179(p) - 73.960(p^2) + 56.048(p^3) \\
& + 405.836(S_t) + 595.774(p)(S_t) \\
& + 385.004(p^2)(S_t) - 296.420(p^3)(S_t) \\
& - 316.907(S_t^2) - 839.205(p)(S_t^2) \\
& - 548.962(p^2)(S_t^2) + 461.125(p^3)(S_t^2) \\
& + 129.545(S_t^3) + 345.867(p)(S_t^3) \\
& + 251.4629(p^2)(S_t^3) - 216.438(p^3)(S_t^3),
\end{aligned}
$$

where $p = \log_{10}(P_t/(1 \text{ bar})$; the maximum error is 1.95 K; the r.m.s. error is 0.65 K).

The same greenhouse functions were used to compute atmospheric pCO_2 levels corresponding to the surface temperatures in the biotic and abiotic histories shown in figures 11.1 and 11.2.

Results and Discussion

The results are shown in figures 11.1 and 11.2. Two possible temperature and atmospheric pCO_2 histories for the surface biosphere are shown: curves labeled A and B, corresponding to an assumed conventional uniformitarian temperature history and a very warm Archean/early Proterozoic history respectively. Of course curves A and B refer to past and inferred future biotic Earth surfaces. The A curve corresponds to our approximation of the conventional uniformitarian view of past surface temperatures. The B temperature history is argued for in Schwartzman (1999, 2001), with the critical evidence being paleotemperatures derived from the climatic interpretation of the least altered chert oxygen isotopic record. Also plotted in figure 11.2 is the variation of the effective (no greenhouse) temperature as a reference curve. A series of curves labeled "equilibrium" (reactions a, b, and c) show abiotic equilibrium temperatures and corresponding atmospheric carbon dioxide levels, computed as previously outlined. These are hypothetical states which would be achieved only over long times at these relatively low temperatures because of the very slow kinetics of the abiotic solid/gas reaction under these conditions. Sources of the actual disequilibrium at the Earth's surface include the following:

- The steady-state level of atmospheric carbon dioxide is controlled by the balance of the surface weathering sink and the volcanic/metamorphic source from deep subsurface reactions at high temperature and pressure.

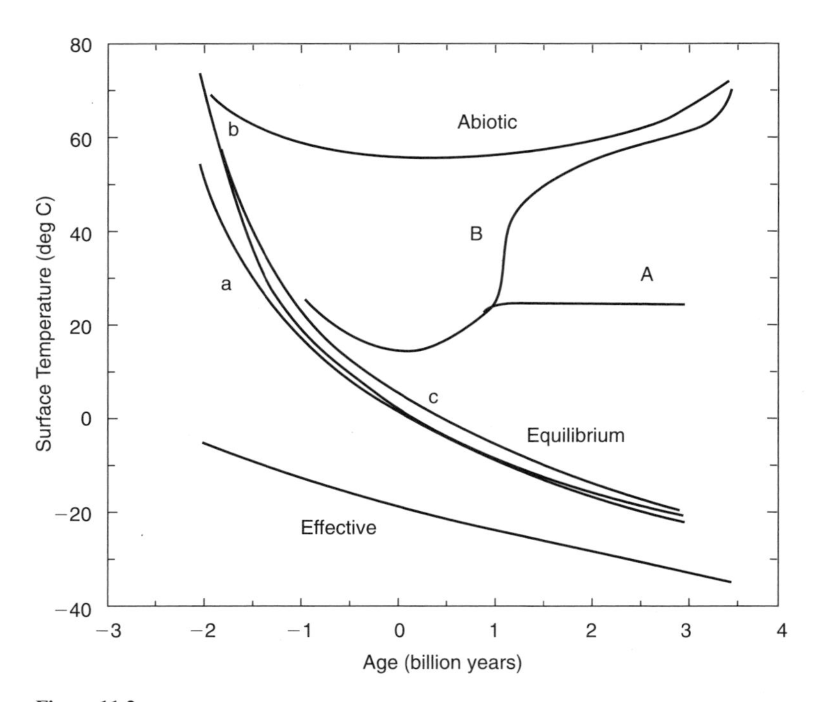

Figure 11.2
Same temperature trajectories as a funtion of age as in figure 11.1. Also plotted is the effective temperature.

• The volcanic/metamorphic source includes a juvenile component of carbon dioxide that is not derived from the reaction of calcium carbonate and silica in subduction zones.

The equilibrium state presumably would be more closely approached in a hypothetical isolated atmosphere/crust system with sufficient carbon dioxide to drive the Urey reaction to the left (reactions a, b, and c given above), exposed calcium silicate, water as a facilitator, and no source of carbon dioxide to the atmosphere other than from surface reaction. Given the slow kinetics of the Urey reaction, disequilibrium is reinforced by the resupply of carbon dioxide to the atmosphere by subduction/metamorphism by virtue of plate tectonics. On the other hand, tectonic uplift provides fresh rock for weathering, which tends to bring the system closer to equilibrium.

The computed results show that the carbonate–silicate geochemical cycle, the long-term control on the steady-state atmospheric carbon dioxide level, is far from equilibrium on the present Earth for both the abiotic and biotic cases, approaching equilibrium only on a billion-year timescale in the future as solar luminosity and surface temperature climb. Steady-state modeling of this geochemical cycle can ignore the equilibrium conditions until their future convergence.

The progressive increase in the biotic enhancement of chemical weathering in the last 4 billion years, culminating in the weathering regime of the forest and grassland ecosystems, has brought the steady-state atmospheric carbon dioxide level closer to the Urey reaction equilibrium state. At present, the equilibrium temperature is about 10°C lower than the actual mean surface temperature, while the corresponding atmospheric pCO_2 level is two to more than four orders of magnitude lower, depending on the equilibrium reaction. For the B scenario, the biotic enhancement of weathering has increased by nearly two orders of magnitude. In contrast, the abiotic steady state is always further from this equilibrium state than the biotic (for both possible temperature histories, A and B), except near the origin of life and at their future convergence. These are counterintuitive results from a classical Gaian view, which argues that life drives its environment away from equilibrium.

On a future Earth both the abiotic and the biotic temperatures will converge on the equilibrium temperature by about 2.5 billion years from now. The increase of solar luminosity overwhelms the regulatory capability of the biosphere, with the carbon dioxide greenhouse effect disappearing. At this point, temperature increases arise from a water greenhouse only, until the hydrosphere is lost. Self-organization of the biosphere here is defined as the increasing influence of life on the structure and physical properties of the Earth's surface. If the biotic enhancement of weathering is now high, and has been rising ever since the origin of life, the self-organization of the biosphere has been geophysiological. Biospheric self-organization has increased with the progressive colonization of the continents and evolutionary developments in the land biota, as a result of surface cooling arising from biotic enhancement of weathering, the equilibrium temperature/pCO_2 of the Urey reaction being the "attractor" state.

With future increase of solar luminosity, and ignoring the possibility of anthropogenic effects, the biospheric capacity for climatic regulation will decrease, leading to the ending of self-organization some 2 billion years from now. The Earth's surface will then approach chemical equilibrium with respect to the carbonate-silicate cycle. If the self-organization of the Earth's surface system is purely geochemical (inorganic), it will likewise end at the same time as the surface temperature converges on the equilibrium temperature. However, there is a possibility that the biotic enhancement of weathering may extend the life span of the biosphere with respect to complex life compared to an abiotic Earth, by delaying the loss of carbon dioxide from the atmosphere (Lenton and von Bloh, 2001).

The biota progressively speeds up weathering from the geologic past to the present, with the surface temperature approaching the equilibrium temperature while the biosphere "self-organizes." Self-organization in the sense previously defined is expressed by the achievement of lower steady-state atmospheric carbon dioxide levels (and therefore surface temperatures) obtaining under biotic conditions than under abiotic, at the same weathering flux at a given time under biotic conditions as abiotic. The key site for this self-organization is at the interface between land and atmosphere, the soil, where carbon is sequestered by its reaction (as carbonic and organic acids) with calcium magnesium silicates. The occurrence of differentiated soil goes back to the Archean, and is the critical material evidence for biospheric self-organization. This biotic invasion into the surface of the continental crust constitutes an ever-expanding front of high surface area/land area with progressive colonization of land and evolutionary developments culminating in the rhizosphere of higher plants. The microenvironments in soils in contact with mycorrhizae and plant roots (the rhizosphere) and microbial biofilms surrounding mineral grains can have significantly higher organic acid and chelating agent concentrations and lower pH than the soil waters commonly sampled (see Berner, 1995; Landeweert et al., 2001). Consider the implications of the estimated surface areas of fungal hyphae, plant roots, and bacteria being 6, 35, and 200 times the area of the Earth, respectively (Volk, 1998). Thus, these biologically created microenvironments probably play an important role in enhancing weathering rates of soil minerals.

Returning to the concept of equilibrium as an "attractor state" for the biosphere, the biotic catalytic role in weathering the crust brings the surface temperature closer to the equilibrium temperature of the Urey reaction. "A fundamental feature of life is its ability to catalyze reactions, and the presence of reactions proceeding at rates faster than predicted for abiotic processes is an indication that life might be present" (Conrad and Nealson, 2001, 20).

Biotic mediation and catalysis of inorganic reactions that are already thermodynamically favorable, such as in supersaturated solutions, brings the affected environment closer to equilibrium. Examples include chemical weathering (e.g., carbonation of silicates and oxidation of crustal Fe^{+2}, Mn^{+2}, and S^{-2}) as well the biologically mediated precipitation of calcium carbonate in the ocean, where surface waters are supersaturated with respect to both calcite and aragonite (Berner, 1971; Holmen, 2000). On the other hand, biologically mediated precipitation may move the environment further from equilibrium, as in the case of modern precipitation of silica in the ocean by diatoms, since surface waters are already undersaturated with respect to amorphous and opaline silica (Berner, 1971). Indeed, the supersaturation of $CaCO_3$ in oceanic surface waters is apparently a partial product of organic inhibition of nucleation (Westbroek and Marin, 2001). As a direct source, the biota may also bring affected environments away from equilibrium (e.g., oxygen rise in atmosphere, carbon dioxide rise in soils). While the rise of oxygen and its steady-state level in the atmosphere undoubtedly require a biotic source, the maintenance of the steady-state level far from equilibrium with respect to surface reduced carbon, especially living matter itself, is a partial outcome of the slow kinetics of surface oxidation (Butcher and Anthony, 2000), as well as of still uncertain biogeochemical feedbacks.

Thus we see that life shoves the environment around in various ways. As in the cases of precipitating $CaCO_3$ from the ocean and chemical weathering, life moves the environment closer to equilibrium. In the cases of precipitating of silica, creating marine

$CaCO_3$ supersaturation, and pumping high oxygen, life moves the environment away from equilibrium. The two cases involving $CaCO_3$ are particularly interesting because life is causing opposite effects within the same chemical system.

In our view, it cannot be said that life is moving the environment toward or away from equilibrium because it has been selected by evolution to do so. Hence we would expect a variety of kinds of effects by life on the environment. The effects on the Gaian (biosphere) scale are best characterized as by-products (Volk, 1998). Were some life to be evolved to actually change the large-scale environment in a certain direction, presumably benefiting from such changes (else why would some life be so evolved?), then cheats that do not perform the environmental manipulation would be at a higher reproductive advantage and soon take over. Oxygen production by plants is a waste by-product. Lowering of CO_2 by weathering is a by-product of mineral gathering, water retention, and microbial respiration, among other factors. In general, Gaia is likely built from environmental effects that are released as free by-products from organisms whose metabolisms evolved for direct benefit to their internal milieu.

Finally, an astrobiological flag: Lovelock's original insight may still be valid for some cases, but far-from-equilibrium abiotic steady states may arise on the surface of Earth and other planets and moons, and should not be taken as a priori evidence for Gaian self-regulation. The future search for alien biospheres will likely include attempts at spectral identification of ozone, indicating the presence of oxygen, and photosynthesis on extrasolar planets (Mariotti et al., 1997). However, the detection of ozone alone may not be conclusive evidence of a biosphere, since traces of oxygen can be generated by abiotic photodissociation of water (see discussion of detection of ozone on two of Saturn's satellites, Noll et al., 1997). One might imagine abiotic sources of methane (e.g., degassing from interior) in atmospheres that also contain such abiotically derived oxygen, thereby generating steady states far from equilibrium.

Thus we cannot simply state that life creates either equilibrium or nonequilibrium as a rule for Gaia or the biosphere. This subject needs more work. In general, we must be careful not to project Earth's example as we look into space for other inhabited worlds. The science of Gaia, applied to astrobiology, needs a new generation of models that can explore simulated planets which, without life, still show disequilibrium. And we need to ask whether certain kinds of life-created equilibria or disequilibria will be more or less probable.

References

Berner, R. A. 1971. *Principles of Chemical Sedimentology*. McGraw-Hill, New York.

Berner, R. A. 1995. Chemical weathering and its effect on atmospheric CO_2 and climate. In A. F. White and S. L. Brantley, eds., *Chemical Weathering Rates of Silicate Minerals*, pp. 565–583. Mineralogical Society of America, Washington, DC.

Berner, R. A., and K. A. Maasch. 1996. Chemical weathering and controls on atmospheric O_2 and CO_2: Fundamental principles were enunciated by J. J. Ebelmen in 1845. *Geochimica et Cosmochimica Acta*, 60, 1633–1637.

Butcher, S. S., and S. E. Anthony. 2000. Equilibrium, rate, and natural systems. In M. C. Jacobson, R. J. Charlson, H. Rodhe, and G. H. Orians, eds., *Earth System Science*, pp. 85–105. Academic Press, London.

Caldeira, K., and J. F. Kasting. 1992. The life span of the biosphere revisited. *Nature*, 360, 721–723.

Conrad, P. G., and K. H. Nealson. 2001. A non-earthcentric approach to life detection. *Astrobiology*, 1, 15–24.

Faure, G. 1991. *Principles and Applications of Inorganic Geochemistry*. Macmillan, New York.

Guerzoni, S., S. Harding, T. Lenton, and F. Ricci Lucchi. 2001. Introduction. In S. Guerzoni, S. Harding, T. Lenton, and F. Ricci Lucchi, eds., *Earth System Science—A New Subject for Study (Geophysiology) or a New Philosophy?*, pp. 1–8. Siena: International School of Earth and Planetary Sciences.

Holmen, K. 2000. The global carbon cycle. In M. C. Jacobson, R. J. Charlson, H. Rodhe, and G. H. Orians, eds., *Earth System Science*, pp. 282–321. Academic Press, London.

Hutchinson, G. E. 1954. The biochemistry of the terrestrial atmosphere. In J. P. Kuiper, ed., *The Earth as a Planet*, pp. 371–433. University of Chicago Press, Chicago.

Kasting, J. F., and T. P. Ackerman. 1986. Climatic consequences of very high CO_2 levels in Earth's early atmosphere. *Science*, 234, 1383–1385.

Kasting, J. F., and D. H. Grinspoon. 1990. The faint young sun problem. In C. P. Sonnett, M. S. Giampapa, and M. S. Matthews, eds., *The Sun in Time*, pp. 447–462. University of Arizona Press, Tucson.

Kasting, J. F., A. A. Pavlov, and J. L. Siefert. 2001. A coupled ecosytem–climate model for predicting the methane concentration in the Archean atmosphere. *Origins of Life and Evolution of the Biosphere*, 31, 271–285.

Landeweert, R., E. Hoffland, R. D. Finlay, T. W. Kuyper, and N. van Breemen. 2001. Linking plants to rocks: Ectomycorrhizal fungi mobilize nutrients from minerals. *Trends in Ecology and Evolution*, 16, 248–254.

Lenton, T. M. 1998. Gaia and natural selection. *Nature*, 394, 439–447.

Lenton, T. M., and W. von Bloh. 2001. Biotic feedback extends the life span of the biosphere. *Geophysical Research Letters*, 28, 1715–1718.

Lovelock, J. E. 1990. Hands up for the Gaia hypothesis. *Nature*, 344, 100–102.

Mariotti, J. M., A. Leger, B. Mennesson, and M. Ollivier. 1997. Detection and characterization of Earth-like planets. In C. B. Cosmovici, S. Bowyer, and D. Werthimer, eds., *Astronomical and Biochemical Origins and the Search for Life in the Universe*, pp. 299–311. Editrice Compositori, Bologna.

Noll, K. S., T. L. Roush, D. P. Cruikshank, R. E. Johnson, and Y. J. Pendleton. 1997. Detection of ozone on Saturn's satellites Rhea and Dione. *Nature*, 388, 45–47.

Schwartzman, D. 1999. *Life, Temperature, and the Earth: The Self-Organizing Biosphere*. Columbia University Press, New York.

Schwartzman, D., S. Shore, T. Volk, and M. McMenamin. 1994. Self-organization of the Earth's biosphere—geochemical or geophysiological? *Origins of Life and Evolution of the Biosphere*, 24, 435–450.

Schwartzman, D., and T. Volk. 1991. Biotic enhancement of weathering and surface temperatures on Earth since the origin of life. *Palaeogeography, Palaeoclimatololgy, Palaeoecology (Global Planet. Change Section)*, 90, 357–371.

Schwartzman, D. W. 2001. Playing the tape again, a deterministic theory of biosphere/biotic evolution, and Biotic enhancement of weathering: The geophysiology of climatic evolution. In S. Guerzoni, S. Harding, T. Lenton, and F. Ricci Lucchi, eds., *Earth System Science—A New Subject for Study (Geophysiology) or a New Philosophy?*, pp. 53–60 and 85–92. Siena: International School of Earth and Planetary Sciences.

Urey, H. C. 1952. *The Planets: Their Origin and Development*. Yale University Press, New Haven, CT.

Volk, T. 1998. *Gaia's Body: Toward a Physiology of Earth*. Copernicus and Springer-Verlag, New York.

Walker, J. C. G., P. B. Hays, and J. F. Kasting. 1981. A negative feedback mechanism for the long-term stabilization of Earth's surface temperature. *Journal of Geophysical Research*, 86, 9776–9782.

Westbroek, P., and F. Marin. 2001. The geologic history of calcification. In S. Guerzoni, S. Harding, T. Lenton, and F. Ricci Lucchi, eds., *Earth System Science—A New Subject for Study (Geophysiology) or a New Philosophy?*, pp. 93–108. Siena: International School of Earth and Planetary Sciences.

12
Biotic Plunder: Control of the Environment by Biological Exhaustion of Resources

Toby Tyrrell

Abstract

A major aim of Gaian research is to explain how large-scale environmental regulation can arise out of principles of natural selection. Here I describe how the tendency of all biological populations to proliferate when conditions are favorable frequently exerts a pressure on resources. Resources become exhausted (or nearly so), and stay that way. This ecological mechanism, biotic plunder, controls many features of the Earth's environment. Evidence is presented here showing how this mechanism is responsible for the "nutrient deserts" that cover the large majority of the ocean surface because phytoplankton plunder inorganic nutrients dissolved in seawater. Populations of herbivores, where there are no carnivores, can also execute biotic plunder on vegetation. The plunder of dead organic matter by decomposers (detritivores) ensures tight recycling of essential elements and restricts burial of organic matter on average to less than 1 percent of the total produced. Biotic plunder may also, possibly, be responsible in some way for the scarcity of carbon dioxide in the Earth's atmosphere.

Introduction

The Gaia hypothesis describes biological alteration of the Earth's atmosphere on a massive scale and also argues that stabilizing feedbacks have regulated some Earth environmental parameters (e.g., planetary temperature) over geological time. The phenomena pointed out by Lovelock have engaged the interest of a whole generation of environmental scientists. But the mechanism originally proposed by Lovelock (1979) involved sacrifice of individual genetic fitness for the global good and was biologically implausible; it ignored the requirement for organisms to pursue evolutionarily stable strategies (Maynard Smith, 1974)—that is, to behave so as to maximize expected genetic fitness in such a way that mutations toward different behaviors would not be more successful. The biological mechanism was rightly attacked (Dawkins, 1983, pp. 234–237; Doolittle, 1981).

Since that time the search has been on (e.g., Volk, 1998; Lenton, 1998) for mechanisms that can explain some of the properties described in the Gaia hypothesis without resorting to evolutionarily *un*stable strategies. According to Lovelock himself in his autobiography, "[Our critics] are right to insist that a large and still unanswered question remains: If the Earth is indeed self-regulating by biological feedback, how has this come about through natural selection?" (Lovelock, 2000, p. 263).

The first demonstration that regulation need not require altruism or conscious coordination was provided by the Daisyworld model (Watson and Lovelock, 1983). But while satisfying as a theoretical proof that regulation can arise automatically and unconsciously, Daisyworld is not a convincing depiction of how the world actually works. It contains several questionable assumptions (e.g., Robertson and Robinson, 1998), and temperature regulation on Earth has been primarily by alteration of greenhouse gas concentrations, not by alteration of albedo (Kump et al., 1999, p. 160).

Here I describe an environmentally stabilizing feedback that is derived directly from the "struggle for existence," one whose operation, certainly within the ocean, is directly observable.

Mechanism

Proliferation of Biological Populations

The ability to proliferate rapidly is both a fundamental and a feared property of all life. The "geometric" (i.e., exponential) increase of natural populations under favored circumstances was first brought to attention by Malthus (1803), more than 50 years before Darwin published his theory in *On the Origin of Species*. Both Darwin and Alfred Russel Wallace acknowledged the profound influence of Malthus as a catalyst for their ideas on natural selection. When the propensity for proliferation is unleashed in pathogens, it can lead to rampaging epidemics, such as influenza nowadays or bubonic plague in the past.

Even cancer can be considered a related phenomenon, whereby renegade cells "forget" the restrictions usually encoded in their genetic instructions and divide without restraint.

The ability to increase population size under favorable environmental conditions (e.g., warm and wet climate, plentiful food, lack of competition and predation) is shared by absolutely all life forms, including all plants, animals, fungi, and bacteria. It is easy to understand that this must be so in order for populations to recover from occasional natural catastrophes. Any population capable only of maintaining its numbers at a constant level, even when the population density is low and circumstances are otherwise suitable, would eventually die out due to occasional succumbing of individuals to accidents. Natural selection (selfish genes) leads to a scramble to contribute the maximum number of genes to subsequent generations. In many cases the best strategy in this scramble is high fecundity. The number of seeds produced by a healthy tree during its lifetime frequently amounts to millions, almost all of which will not succeed in propagating in turn. Almost all of the multitudinous seeds, spores, spawn, and eggs produced every year in the natural world fall "on stony ground" (meet with unfavorable environmental conditions, or get eaten) and do not contribute to future generations. It has to be so; otherwise, all populations would permanently resemble lemming outbreaks in their degree of proliferation, rather than the more stable population sizes that are normally seen.

No exponential population increase can be sustained for many generations because the numbers rapidly become ridiculously, unsustainably, large. For instance, the progeny of just one single bacterium (10^{-12} g), doubling in number every half-hour as they split by binary fission, could, at this extraordinary rate of increase ($P_t = P_0 * 2^{(t/0.5\,\mathrm{hrs})}$), attain the weight of the solid Earth (6×10^{27} g) within less than three days, if there were no limitation by predation and finite food supply. There is, however, a trade-off between "more" and "better" in the production of offspring, and many species have evolved toward producing fewer but more successful offspring. But even two elephants, to use Darwin's example of the slowest of breeders, will, if reproducing at the normal (slow) rate for elephants (but somehow all surviving to old age), have fanned out to about 20 million living descendants after 750 years.

Two types of human impact inadvertently reveal the power of this proliferation to lead to population explosions. The translocation of a species from one continent to another has often resulted in population explosions as the alien species thrives in the absence of its normal predators, or simply because it outcompetes the native species. For instance starlings introduced to New York had within 50 years colonised the whole of the United States (Elton, 1958). Reindeer brought from Lapland to Alaska in the 1890s had increased to over half a million by the 1930s (Elton, 1958). Hunting and culling (or, rather, their cessation) also reveal the tendency toward exponential rise in population, as a decimated population subsequently recovers at an exponential rate.

Exhaustion of Resources

What is it that reins in the intrinsic proliferation pressure of populations? What stops all populations from multiplying like lemmings or locusts ad infinitum?

When an inoculum of bacteria is placed on a petri dish (a small plastic dish smeared with a nutritious gel), the answer is clear—the bacteria carry on dividing and dividing until the gel is consumed. The same simple dynamic can be seen in a flask of sterilized lake water or seawater, especially after some nutrient salts have been stirred in. Each of the few algae initially added to the culture flask divides by binary fission into 2, then 4, then 8, 16, 32, 64, and so on, until the burgeoning number of descendants has drawn down to exhaustion the concentration of at least one of the nutrients in the flask, at which point growth can no longer continue.

Of course the situation in a real-life ecosystem need not be so simplistic. There are plenty of other candidates for density regulation of natural populations, including diseases which can be transmitted more efficiently and easily among more closely packed hosts. Predators may be individually more efficient at catching more abundant (more frequently encountered) prey. The numbers of a proliferating species can be prevented from rising too high either by increases in death rate or by decreases in birth rate at higher population density. If it is birth rate, then food (nutrients for plants and algae) is not the only requirement for reproduction that can run out as a result of escalating numbers of individuals. For instance, land plants can be prevented from growing more densely by shortages of water or light.

However, in this chapter the focus will be on the availability of resources (nutrients and food) as the natural regulator of population size. It is apparent that for many types of organisms, availability of resources outweighs any other consideration as the main factor in reining in proliferation pressure.

It is possible to look at this issue from two viewpoints. The viewpoint just considered is the endlessly fascinating question of what regulates animal or plant number. An alternative and interlinked viewpoint, even more intriguing from a biogeochemical (geophysiological) perspective, is the question of what regulates nutrient/resource level. As in petri dishes and algal culture flasks, the mechanism proposed here is that proliferation of consumers under conditions of abundant resources leads to rising consumer numbers until resources run out. Only at low (growth-limiting) resource availability will a steady state be achievable such that birth rate equals death rate for the consumers. Resources are kept at low levels because of consumer proliferation otherwise.

Predictions

Let the process by which the multiplication pressure of a consumer population keeps a resource scarce be called *biotic plunder*. For it to be true that a given resource is controlled by biotic plunder, I suggest that the following predictions must all hold true:

1. Concentrations of the resource must be low enough in the steady state to depress consumer replication rate to below its maximum.

2. If an intervention or event causes a reduction in the concentration of the resource to below its steady-state value, this should cause further depression of the consumer population growth rate.

3. Conversely, any increase in the resource availability should bring about an increase in the consumer population growth rate.

4. If some intervention or event causes a reduction in the standing stock of consumers to below its steady-state level, this should cause an ongoing accumulation of the resource in the environment (whatever processes supply new resources will now be opposed by a reduced uptake due to fewer consumers).

5. Likewise, any imposed increase in consumer numbers should lead to an increased consumption of the resource and a decrease in its already low concentration.

6. The feedback between consumer and resource must be sufficiently strong to dominate over all other processes and feedbacks potentially controlling either the resource or the consumer concentration.

Note that in predictions 2–5 the predicted responses all tend to drive the system back toward equilibrium. All are examples of negative feedback, whereby any perturbation to the system is opposed by a response that tends to minimize (and eventually remove) that perturbation. In prediction 2, for example, a scarcer resource should reduce consumer growth rate, which in turn should eventually reduce the consumer population size (assuming death rate stays the same), which in turn should reduce the uptake of the resource, which should finally lead to an increase in the resource, to oppose the initial perturbation.

Hypothesis: Biotic Plunder of Ocean Nutrients

While biotic plunder is discernible in other systems, I start by demonstrating that it occurs in the ocean. This section explains the hypothesis, and the next section gives evidence in support of it. The specific hypothesis in this particular case is:

> Phytoplankton proliferate until lack of inorganic nutrients prevents further proliferation. Biolimiting inorganic nutrients are thereby drawn down to low levels, and kept that way.

The consumers in this case are phytoplankton, which give rise to the large majority of primary production in the sea (seaweeds contributing only a small amount), and the resource is the inorganic nutrients they require for growth. Figure 12.1 gives an idea of the mechanism in this case. Regulation to low levels is here proposed only for bioavailable (reactive) nutrients—dissolved inorganic nutrients—and thus not for nutrients in dissolved organic matter or in particles, because these latter forms are not typically accessible by phytoplankton. Many chemical elements required by phytoplankton to create new cells (to be able to divide) are present in abundance in the ocean (carbon, for instance), whereas others are much less abundant relative to need and thus will run out earlier. This hypothesis applies only to biolimiting nutrients (i.e., those which sometimes run out). In the ocean this probably limits the hypothesis to dissolved nitrate, phosphate, silicate, and, in certain areas, iron.

This hypothesis assigns a lesser role for zooplankton (and viruses) in the control of phytoplankton proliferation, because evidence suggests that they are not generally able to prevent phytoplankton from increasing to numbers sufficient to consume and deplete nutrients if such nutrients are abundant. However, it should be appreciated that there is no common physiological characteristic of zooplankton (to this author's knowledge) that explains this lack of control over phytoplankton concentrations. One reason could be that the "evolutionary arms race" (Ridley,

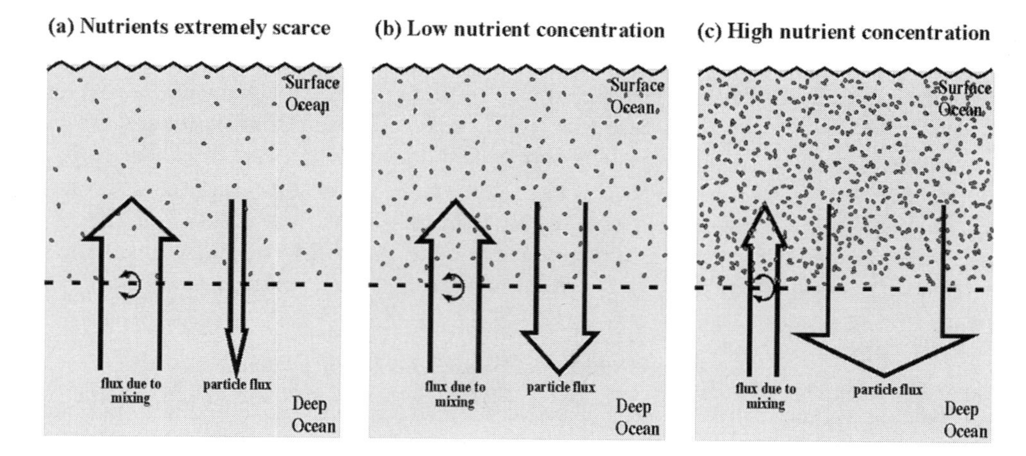

Figure 12.1
Schematic of the biotic plunder mechanism in the ocean: nutrient fluxes at different surface ocean nutrient concentrations: (a) very low nutrient concentration, (b) low nutrient concentration, (c) high nutrient concentration. Only major fluxes crossing the surface-deep boundary (thermocline) are shown. The shaded dots indicate phytoplankton cells, which increase in concentration in response to nutrient concentration. At high nutrient concentrations phytoplankton proliferate, leading to a large particulate flux of organic matter (principally in the form of zooplankton fecal pellets and "marine snow") out of the surface mixed layer. Mixing processes (diffusion, etc.) tend to equalize the surface and deep concentrations, and so oppose the strong vertical gradient in nutrients that biotic plunder brings about. Biotic plunder implements negative feedback because at high nutrient concentrations, output > input, whereas at very low nutrient concentrations, output < input. Input = output balance can be achieved only at low surface nutrient concentrations.

1994) between phytoplankton and zooplankton is fundamentally unequal, with phytoplankton having access to essentially infinite reserves of sunlight energy and the zooplankton, by contrast, having access only to that incorporated into phytoplankton (Colinvaux, 1993, chap. 10). A second possible reason is that there is an even more unequal arms race between phytoplankton and nutrients, since nutrients cannot evolve. However, speculation aside, all that can be said with any certainty is that phytoplankton proliferate when nutrients are abundant and that zooplankton, although thriving on the more abundant food, don't stop them from doing so.

This hypothesis goes further than the observation that primary production in the ocean is usually affected by shortage of nutrients, which is already well known. That observation is combined here with a population dynamic, proliferation, which tends to force lack of nutrients. This combination allows an explanation as to *why* the inhabitable (sunlit) surface is mostly stripped clear of limiting nutrients at the present time, why it must always have been so in the past (ever since phytoplankton evolved, except possibly during Snowball Earth or similar climate excursions), and why it will continue to be so in the future. Biotic plunder is key to our understanding of how the ocean has operated as a biogeochemical engine over geological time.

Evidence from the Ocean

Some, but not all of my earlier predictions can be shown to hold true for phytoplankton and nutrients.

Nutrient Scarcity in Non-Polar Waters

According to the first prediction, biolimiting inorganic nutrient concentrations should be low. This should hold for the sunlit surface (upper 100 m or so, depending on water clarity) but not for deeper water, where phytoplankton cannot get enough light to live and thus cannot plunder the nutrients. It also should apply wherever the surface ocean receives adequate sunlight, but not where it doesn't. And this, broadly speaking, is the pattern that is observed (figure 12.2). In the surface open ocean away from the poles the "macronutrients" nitrate (fixed nitrogen), phosphate, and silicate are continually stripped to low levels in most locations. Continuously favorable physical conditions mean phytoplankton never allow accumulations of nutrients to persist. They proliferate, or bloom, when nutrients are available, and continue to do so until they are no longer.

In those few non-polar areas that do contain high surface levels of macronutrients, such as the equatorial Pacific and the Southern Ocean, it has been demonstrated experimentally that a shortage of iron rather than of the macronutrients restrains proliferation there (Coale et al., 1996; Boyd et al., 2000; Wat-

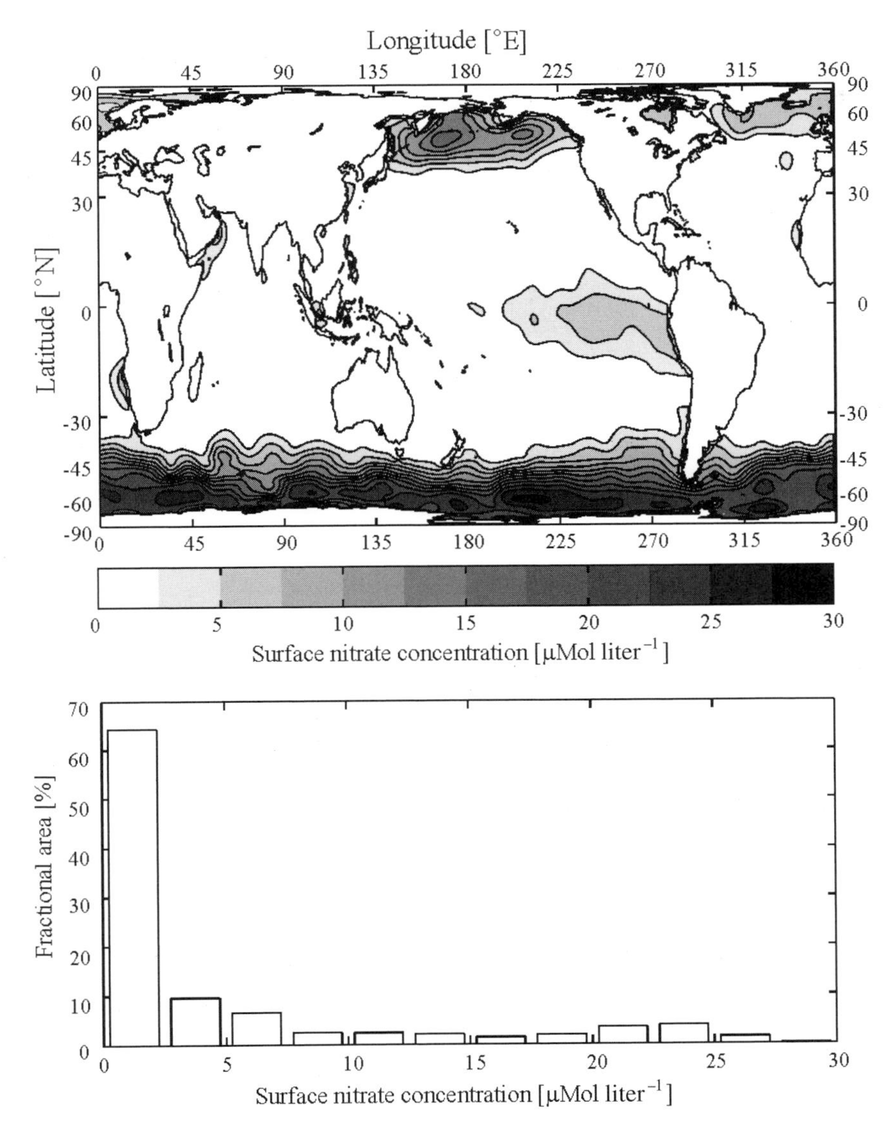

Figure 12.2
(a) Interpolated annual average surface nitrate concentration (μMol liter^{-1}) in the world's oceans (World Ocean Atlas '94 dataset; Conkright et al., 1994), shown on a Peters projection map; (b) fraction of total ocean area (expressed as a percentage) for which surface nitrate concentrations are within the ranges 0.0–2.5 μMol liter^{-1}, 2.5–5.0 μMol liter^{-1}, etc. The World Ocean Atlas is a very large archive of historical as well as more recent hydrographic data collected on ships of many different nations; some individual data points in the database will be inaccurate, but the gross patterns are correct. Annual averages are less meaningful in high latitude areas experiencing strong seasonal cycles. A Peters projection gives an accurate representation of the amount of area at different latitudes, in contrast to other projections, which overemphasize high latitude area. Plot courtesy of Andrew Yool.

son, 2001). A different nutrient becomes limiting first, and it is plundered instead of the macronutrients. In a graphic and beautiful demonstration of this, a recent iron fertilization experiment in the Southern Ocean made the ocean turn green over a large enough area that the patch was visible from space (Abraham et al., 2000). For only one area, the subarctic North Pacific, is there still an ongoing discussion about whether annual phytoplankton growth is primarily grazing-limited rather than iron-limited (Frost, 1991; see also Watson, 2001). All (or just possibly all but one) of the low- and mid-latitude oceans are limited by nutrients having run out.

The Response to Abundant Nutrients in Upwelling Areas

If it is true for a system, then the biotic plunder mechanism implies that a perturbation which introduces abundant nutrients to the system should be met with a negative feedback removing them (prediction 3). The operation of such a feedback is seen very clearly where nutrient-rich deep water upwells to the surface (Chavez and Smith, 1995). As deep water rises to the sunlit surface, and hence suddenly can be inhabited by photosynthetic life (there is insufficient sunlight in all waters below 100–200 m or so), a burst of growth occurs and the abundant nutrients are devoured by proliferating phytoplankton until they are removed. As expected, phytoplankton grow very rapidly in the high-nutrient water. The high nutrient content of water as it rises to the surface is almost completely removed by biological activity as the upwelled water is transported away horizontally.

The operation of biotic plunder (its dynamic response to the nutrient perturbation) is clearly seen in the many upwelling systems off the western coasts of continents (along the western coasts of Peru, southern Africa, Spain and Portugal, California, Mauritania, etc.); it is also seen when upwelling is driven by monsoonal winds, as in the Arabian Sea, or by divergent winds at the equator. In all these cases biotic plunder is not just a theoretical speculation but rather is an observable reality. High limiting nutrient concentrations are not a stable state for seawater receiving adequate sunlight.

Seasonal Cycles in Temperate Waters

Mid- to high-latitude environments oscillate between habitable and uninhabitable (for photosynthesizing phytoplankton) every year. In winter, growth is not possible. Examination of the contrast between summer and winter gives an indication of the effect that life (phytoplankton and the higher trophic levels they feed) has on its environment. Figure 12.3 shows one example, from close to the Isle of Man in the Irish Sea. Similar seasonal cycles occur in numerous other temperate aquatic environments, including the Baltic Sea and the NABE (North Atlantic Bloom

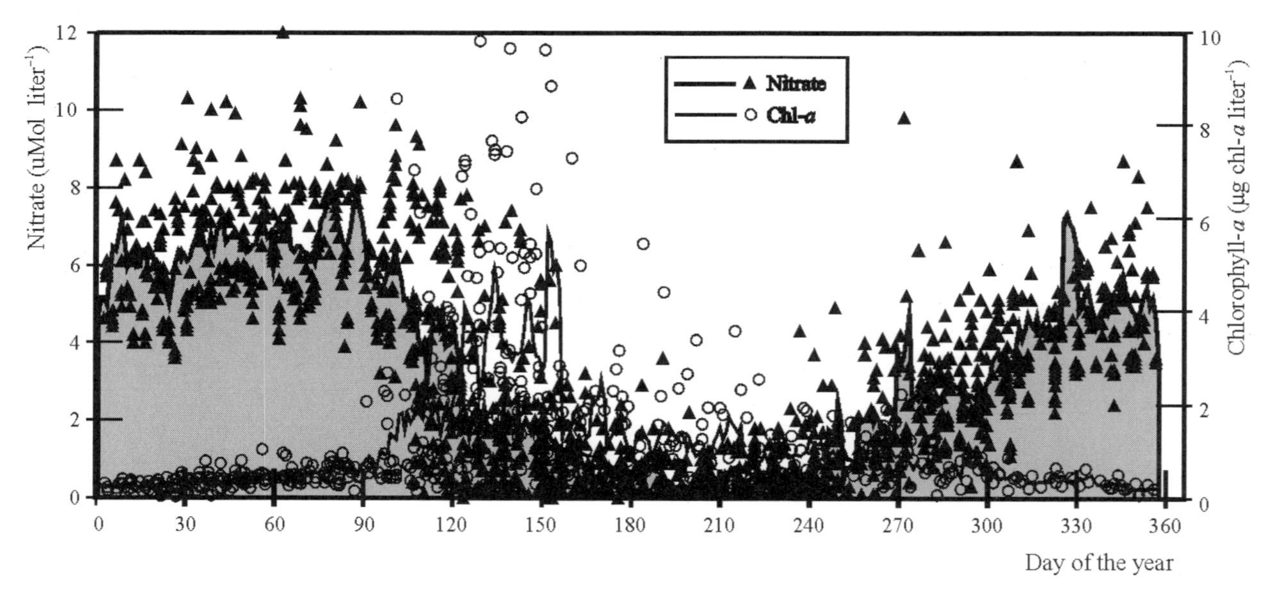

Figure 12.3
Seasonal pattern of phytoplankton (as measured by the concentration of chlorophyll-*a* pigment) and limiting nutrient in the Irish Sea close to the Isle of Man (CYPRIS time series), from all years between 1960 and 1994 (data from the NOWESP website, compiled by Russell Arnott). The circles and triangles on the plot are individual observations, the lines (and shaded area for nitrate ion concentration) are five-day averages. Some of the scatter in the data will be due to measurement inaccuracy, and some to inherent variability between years.

Experiment) site at (47°N, 20°W) in the North Atlantic.

Every year a similar pattern of events unfolds in these locations. Physical constraints (lack of light, low temperature, perhaps ice cover) prevent significant photosynthesis during the winter and allow nutrients stirred up by strong winds to persist and accumulate at the surface. In the late spring and early summer, as physical conditions become favorable, a bonanza ensues as the accumulated nutrients are rapidly consumed by multiplying phytoplankton. This is the classic "spring bloom." Loss of biomass to depth (for instance, in sinking marine snow or in zooplankton fecal pellets) depletes surface nutrient levels.

The spring glut is followed inevitably by the early summer crash, as nutrients run out. Low nutrients persist throughout the rest of the summer, the period of strongest biological control over the environment, with any new nutrient inputs being swiftly depleted by temporary resurgences of phytoplankton growth. Nutrients are so scarce in the summer as to starve many phytoplankton, or at least to prolong their acquisition of nutrients for so long that many of them are eaten before they can divide. The exact means by which low nutrients depress phytoplankton growth rate is unclear, but the most likely explanation is that phytoplankton cells delay binary fission (cell division) until they have been able to acquire sufficient nutrients to support two separate cells. However, see the subsection "Donkeys in the Desert" for a different explanation of how resource limitation can depress population increase rate. Regardless of the precise cause, low nutrients in summer restrain the proliferation pressure and keep phytoplankton numbers low.

As winter approaches, deterioration in weather and light levels prevents positive net growth of primary producers regardless of nutrient abundance, allowing nutrients to start accumulating near the surface once more. The scene is set for the same scenario to be repeated the following year.

The seasonal cycles of phytoplankton and nutrients in temperate latitudes again show the biotic plunder mechanism in action. In winter, conditions are inhospitable and biotic plunder cannot operate. Once phytoplankton can grow in spring, nutrients are brought down to very low levels, and are kept there throughout the summer and early autumn, the period of biological control. Biotic plunder keeps nutrients scarce.

Low Ocean Nutrients over Geological Timescales

The propensity of phytoplankton to divide without restriction when supplied with nutrients ensures that the only stable state is nutrient impoverishment. Although the evidence set out in the two preceding subsections describes a feedback responding on timescales of less than a year, this and related feedbacks have probably controlled ocean nutrients over multimillion-year timescales.

There are three different ways in which biotic plunder depletes nutrients: (1) nutrients dissolved in seawater are taken up into proliferating phytoplankton; (2) some of the phytoplankton organic matter then sinks out of the surface ocean (in the form of zooplankton fecal pellets, or in aggregates of miscellaneous material known as "marine snow"), taking its nutrients with it, most of which are subsequently remineralized into the deep ocean; (3) a small fraction of the sinking organic matter reaches the seafloor and is buried, eventually to be incorporated into marine sedimentary rocks (hence the rapid accumulation of thick, organic carbon-rich sediments beneath upwelling zones (e.g., Bremner, 1983). The first effect depletes the surface ocean stock of dissolved nutrients. The second depletes the surface ocean total amount of nutrients (sum of nutrients in organic matter and dissolved in the water), while at the same time raising the deep ocean total. The third runs down the whole ocean total amount of nutrients.

It is this last effect which is of greatest significance over geological timescales. If for some reason the ocean were to have a high inventory of nutrients, this would stimulate extra primary production and consequent enhanced rates of burial of organic matter. This negative feedback would cause ocean nutrient concentrations to fall until inputs (e.g., rivers) and outputs (e.g., burial) balance, as they must in steady state. A quantitative model of the ocean nitrogen and phosphorus cycles (Tyrrell, 1999) simulates all three of these feedbacks and demonstrates stable regulation of nitrate and phosphate at low surface concentrations. See also Yool and Tyrrell, (2003) for proliferation of diatoms leading to homeostasis toward scarce silicic acid. An intriguing geological calculation (Bjerrum and Canfield, 2002) suggests that dissolved phosphate was scarce in the oceans between 3,200 and 1,900 Mya (after which time the technique no longer works).

Biotic Plunder Elsewhere

There are many trophic interactions in different environments in which the higher trophic level does not plunder (keep scarce) the next one down. It is not suggested that biotic plunder is ubiquitous. However, there is evidence of varying quality suggesting that biotic plunder is much more widespread than just the phytoplankton-nutrient interaction in aquatic environments.

Herbivores Introduced to Remote Islands

Prior to the arrival of 29 reindeer on St. Matthew Island in the Bering Sea in 1944, there was abundant forage in the form of a thick and verdant blanket of lichens covering the island. This initial reindeer paradise fostered a burgeoning population which after 20 years had grown so large (6,000: a 200fold increase) as to graze the lichens down to almost nothing. Autopsies on a sample of the consequent >99 percent mortality revealed starvation as the main cause of reindeer death (Klein, 1968). A similar pattern was seen on one of the Pribilof Islands (St. Paul), also in the Bering Sea, but not on the other (St. George) (Scheffer, 1951). A similar but less dramatic outcome occurred to one of two reindeer herds on South Georgia, leading to stripping of the tussock grass vegetation (Leader-Williams, 1988, fig. 9.4b).

Moose on Isle Royale exhibited a similar pattern of proliferation following their arrival: increasing from 300 moose in 1918 to 3000 in 1930 (at which time all species of edible winter vegetation were observed to be overbrowsed), then plummeting to 500 in 1935 (Mech, 1966). Autopsies on 24 out of 400 or so dead moose showed malnutrition as the cause of death. A second moose irruption during 1980–1996 (when wolves which had arrived later were depleted by canine parvovirus) led to an unsustainable population of 2400 individuals in 1995; they stripped the available food, which led to starvation of all but 500 (80 percent mortality) by 1997 (Peterson, 1999).

Herbivores living in the absence of predators do not universally show such simple population trajectories. For instance, overpopulated cottontail rabbits on Fishers Island, New York, were decimated by a combination of tick infestation and browse exhaustion (Smith and Cheatum, 1944). Nevertheless, most island studies support the idea of herbivore populations expanding up to resource limits, and stripping those resources down to low levels in the process.

It may seem contradictory to demonstrate plunder of primary producers by herbivores in this section, having argued earlier against plunder of phytoplankton by zooplankton. However, the examples in this section and the next are all from terrestrial rather than marine environments, and also are from carnivore-free locations such as islands. Thus, there is no incompatibility. Plunder of primary producers by herbivores is not widespread in natural terrestrial ecosystems containing carnivores, perhaps because carnivores plunder herbivores (Hairston et al., 1960). Another theory argues that the impact of herbivores on plants is usually minor (failing condition (6) discussed earlier) because most parts of plants are not of sufficient food quality to sustain herbivores, being too dilute in nutrients (White, 1993).

Donkeys in the Desert

Biotic plunder by herbivores is, however, not restricted to islands, and can occur in a continental location if the herbivores there have no natural predators. Feral donkeys (*Equus asinus* (Equidae)) presumably were brought to Australia along with early settlers. But following escape from captivity, these wild asses are now numerous in the semidesert outback of the Northern Territories of Australia. It is thought that this feral population suffers no significant predation or parasitism; there is no evidence of dingoes killing or injuring donkeys, or of major equid diseases (Freeland and Choquenot, 1990). Over the century or more in which they have roamed free there, the population of donkeys seems to have come to an equilibrium with its food supply. Population is no longer increasing. Furthermore, it seems that this equilibrium conforms to expectations of biotic plunder.

In a single part of the range, more than 80 percent of the population was culled in 1982–1983. The fate of this culled part of the population was subsequently compared against that of the unmanipulated (presumably at carrying capacity) remainder of the population in the rest of the range. This type of perturbation (culling of consumers) should, according to the fourth prediction given earlier, increase resource availability (vegetation), which in turn should increase growth rate of the consumers (donkeys). Although vegetation density was not systematically measured in this study, it was nevertheless observed that there was less over-grazing, fewer bald patches, and less loss of soil where donkeys were less numerous three years after they had been culled (Freeland and Choquenot, 1990).

But in a subtle twist to the expected story, the number of donkey foals being born every year was not significantly different after the culling. The birthrate did not increase. But close monitoring of the population showed that the population left over after culling was recovering at the rapid rate of 20 percent or more per year, in contrast to the unculled population, which did not increase in size (Choquenot, 1991). If births did not increase, how could population do so? Further detailed investigations showed that the quality/quantity of vegetation made a significant difference to the healthiness and chances of survival to adulthood of the foals. A much smaller proportion of juveniles died in the area where culling had previously taken place (21 percent mortality per year) compared to the unculled population (62 percent mortality per year) (Choquenot, 1991).

The killing of so many of the donkeys in 1982–1983 set in motion a chain of events that will eventually return the system to its former equilibrium. Reduced grazing pressure allowed recovery of the grasses and herbs that, it is presumed, the donkeys normally keep heavily grazed down. Although the increase in quality/quantity of forage did not seem to affect the numbers of foals being born each year, it strongly improved each foal's chances of avoiding starvation to survive to adulthood. The resulting population spurt will eventually return the density of feral donkeys to carrying-capacity (steady-state) levels, at which point increased malnutrition (death of most juveniles) will once more prevent any further population increase.

Hairston, Smith, and Slobodkin (HSS): "We Aren't Making Much Coal"

It was previously proposed (Hairston, Smith, and Slobodkin, 1960) for terrestrial environments that decomposers, primary producers, and carnivores are all resource-limited (executing biotic plunder on their resources), whereas herbivores are not. The rationale for decomposers, for instance, is that only a very small fraction (typically less than 1 percent) of primary production survives decomposition to be permanently buried. To put it another way, "We aren't making much coal" (Hairston, 1981). If decomposers' proliferation were not resource-limited, then they would be unlikely to be so efficient in consuming their resource. The HSS theory has met with opposition (Ehrlich and Birch, 1967; Murdoch, 1966) but continues to be heavily cited (Hairston, 1981; Polis, 1999).

From the perspective of Gaia and an understanding of how organisms have shaped the Earth, biotic plunder therefore has the potential to help explain one of the great environmental mysteries. Many ecologists and writers have marveled at the wonderful efficiency of waste recycling on Earth. Ecosystems almost seem to be designed in such a way that whenever one organism produces some waste product, another organism is on hand to tidily clear it away and recycle it into a useful form for other organisms. Assuming HSS are right, then such efficient recycling can be understood in terms of waste as a resource, decomposers as consumers, and biotic plunder of the latter on the former.

Although perhaps difficult to prove right or wrong, the ramifications of such a process (on control of coal formation, organic carbon burial, long-term control on atmospheric O_2 and CO_2, sustenance of biological activity by minimizing the loss of essential elements in buried organic residues) are potentially of great significance.

Scarcity of Energy-containing Molecules

Organic matter is only one member of a class of energy-containing molecules. For instance, tube worms at hydrothermal vents (via symbiotic bacteria within them) catalyze the conversion of H_2S to SO_4, and from that energy they make a living. Likewise, there are whole guilds (*sensu* Volk, 1998) of microbes which have evolved to exploit the energy yield of certain chemical reactions. Nitrifying bacteria survive on the energy yield from converting ammonium ions to nitrite ions, and nitrite ions to nitrate ions. Yet others gain energy from reactions which convert hydrogen (H_2), ferrous iron (Fe^{2+}) and methane (CH_4) to other forms (Madigan et al., 2000, chap. 15).

From ecological theory there is no a priori prediction as to whether the propensity for proliferation of these microbes is resisted by increased grazing on them at higher microbe density, or instead by resource limitation—for instance, by running out of their chemical substrate. However, it can be observed that concentrations of these exploitable molecules are usually very low in the biosphere, lower than those of comparable but unprofitable molecules. For instance, of the various forms of inorganic nitrogen dissolved in the ocean, the chemically inert dinitrogen is by far the most abundant (as it is in the atmosphere) at $\sim$1000 μMol liter^{-1}, followed by nitrate at 0–40 μMol liter^{-1}, followed finally by the energy-yielding forms such as ammonium and nitrite, nearly always at less than 1 μMol liter^{-1} (Tyrrell, 1999). In forest soils or in streams and rivers, however, ammonium ions frequently are more abundant than nitrate ions, so this picture is by no means universal.

Nevertheless, the generally low concentrations of energy-yielding molecules in the biosphere may also be a result of biotic plunder keeping resource concentrations low. In the outflow channels of sulfur springs rich in H_2S, vast populations of sulfur bacteria develop. The vast populations are facilitated by the abundant H_2S resource and are not kept grazed down. These bacteria are present only in water containing H_2S. Further downstream, as the water flows away from the source and the H_2S is used up, the sulfur bacteria are no longer present (Madigan et al., 2000, chap. 15).

Atmospheric Carbon Dioxide

Although the proliferation potential of trees in rain forests is quite possibly held back by lack of light more strongly than by nutrient availability (Jordan and Herrera, 1981), there is potential for a form of biotic plunder on atmospheric CO_2, because the extent of forests is greater in high CO_2 (warm) climates than it is in low CO_2 (cold) climates (Olson, 1985). Rain forests may hold a similar density of trees per unit area in different climates, but the percentage of the land surface covered by dense forests increases dramatically during warm times (Olson, 1985). For forest abundance to feed back on climate, there would need to be an increase in long-term carbon sequestration (e.g., in organic carbon burial) as a result of the more widespread forests.

A major reason for considering the possibility of a form of biotic plunder regulating atmospheric CO_2 is the similarity between the trajectory of nitrate in temperate waters from spring to summer (figure 12.3) and that of atmospheric CO_2 from before to after the colonization of the land surface by trees. CO_2 decreased by about tenfold at the time of the first afforestation (Berner and Berner, 1997) and has remained at lower levels ever since, reminiscent of the decline in nitrate ion concentration during the spring–summer transition in temperate waters. Carbon dioxide affects plant growth at current (interglacial) levels (Prentice et al., 2001; Colinvaux, 1980, chap. 4) and must have been more strongly limiting during ice ages.

Thus biotic plunder may influence atmospheric CO_2 in at least three ways: (1) by biotic plunder of dead organic matter (by decomposers), leading to minimization of the accumulation and burial of organic matter within ecosystems; (2) by biotic plunder of nutrients in the ocean being a dominant control over the strength of the organic carbon pump there; and (3) by atmospheric CO_2 (and associated warmth and rainfall) controlling the "proliferation of forests" across the land surface.

Conclusions

A simple but hitherto largely unappreciated ecological and biogeochemical mechanism, biotic plunder, has been described. Whereas Daisyworld is a "parable," a theoretical construct, biotic plunder is a real-world process. It can be observed in action in different environments—supported by abundant evidence in some cases (e.g., spring blooms in temperate seas and oceans), by less in others. This mechanism is one means by which large-scale environmental regulation arises out of natural selection.

The mechanism of biotic plunder provides a robust regulation that can resist perturbations. It agrees with some of the predictions of Gaia, but not with others: (1) biotic plunder provides stable regulation of the environment; (2) that regulation is toward a habitable state; however, (3) that regulation is *not* toward an optimal state for the biota—far from it. As seen clearly for the Australian feral donkeys, regulation of the environment is necessarily toward resource scarcity, which is to say is toward malnutrition, even starvation of many.

References

Abraham, E. R., C. S. Law, P. W. Boyd, S. J. Lavender, M. T. Maldonado, and A. R. Bowie. 2000. The importance of stirring in the development of an iron-fertilized phytoplankton bloom. *Nature*, 407, 727–730. (See also the front cover of that issue.)

Berner, R. A., and E. K. Berner. 1997. Silicate weathering and climate. In *Tectonic Uplift and Climate Change*, ed. W. F. Ruddiman, pp. 353–365. Plenum Press, New York.

Bjerrum, C. J., and D. E. Canfield. 2002. Ocean productivity before about 1.9 Gyr ago limited by phosphorus absorption onto iron oxides. *Nature*, 417, 159–162.

Boyd, P. W., A. J. Watson, C. S. Law, et al. 2000. A mesoscale phytoplankton bloom in the polar Southern Ocean stimulated by iron fertilization. *Nature*, 407, 695–702.

Bremner, J. M. 1983. Biogenic sediments on the south west African (Namibian) continental margin. In *Coastal Upwelling: Its Sediment Record*, Part B, *Sedimentary Records of Ancient Coastal Upwelling*, ed. J. Thiede and E. Suess, pp. 73–103. Plenum Press, New York.

Chavez, F. P., and S. L. Smith. 1995. Biological and chemical consequences of open ocean upwelling. In *Upwelling in the Ocean: Modern Process and Ancient Records*, ed. C. P. Summerhayes et al., pp. 149–169. Wiley, Chichester, UK.

Choquenot, D. 1991. Density-dependent growth, body condition, and demography in feral donkeys: Testing the food hypothesis. *Ecology*, 72, 805–813.

Coale, K. H., K. S. Johnson, S. E. Fitzwater, et al. 1996. A massive phytoplankton bloom induced by an ecosystem-scale iron fertilization experiment in the equatorial Pacific Ocean. *Nature*, 383, 495–501.

Colinvaux, P. 1980. *Why Big Fierce Animals Are Rare*. George Allen & Unwin, London.

Colinvaux, P. 1993. *Ecology 2*. Wiley, New York.

Conkright, M. E., S. Levitus, and T. P. Boyer. 1994. *World Ocean Atlas 1994*, vol. 1, *Nutrients*. NOAA Technical Report. NOAA, Washington, DC.

Dawkins, R. 1983. *The Extended Phenotype*. Oxford University Press, Oxford.

Doolittle, W. F. 1981. Is Nature really motherly? *CoEvolution Quarterly*, Spring, 58–63.

Ehrlich, P. R., and L. C. Birch. 1967. The "balance of nature" and "population control." *American Naturalist*, 101, 97–107.

Elton, C. S. 1958. *The Ecology of Invasions by Plants and Animals*. Methuen, London.

Freeland, W. J., and D. Choquenot. 1990. Determinants of herbivore carrying capacity: Plants, nutrients, and *Equus asinus* in northern Australia. *Ecology*, 71, 589–597.

Frost, B. W. 1991. The role of grazing in nutrient-rich areas of the open sea. *Limnology and Oceanography*, 36, 1616–1630.

Hairston, N. G. 1981. *Current Contents AB&ES*, 12, 20.

Hairston, N. G., F. E. Smith, and L. B. Slobodkin. 1960. Community structure, population control, and competition. *American Naturalist*, 94, 421–425.

Jordan, C. F., and R. Herrera. 1981. Tropical rain forests—Are nutrients really critical? *American Naturalist*, 117, 167–180.

Klein, D. R. 1968. The introduction, increase, and crash of reindeer on St. Matthew Island. *Journal of Wildlife Management*, 32, 350–367.

Kump, L. R., J. F. Kasting, and R. G. Crane. 1999. *The Earth System*. Prentice-Hall, Englewood Cliffs, NJ.

Leader-Williams, N. 1988. *Reindeer on South Georgia—The Ecology of an Introduced Population*. Cambridge University Press, Cambridge.

Lenton, T. M. 1998. Gaia and natural selection. *Nature*, 394, 439–447.

Lovelock, J. E. 1979. *Gaia: A New Look at Life on Earth*. Oxford University Press, Oxford.

Lovelock, J. E. 2000. *Homage to Gaia: The Life of an Independent Scientist*. Oxford University Press, Oxford.

Madigan, M. T., J. M. Martinko, and J. Parker. 2000. *Brock Biology of Microorganisms*, 9th ed., Prentice-Hall, Englewood Cliffs, NJ.

Malthus, T. R. 1803. *An Essay on the Principle of Population; or, A View of Its Past and Present Effects on Human Happiness; with an Enquiry into Our Prospects Respecting the Future Removal or Mitigation of the Evils Which It Occasions*, new ed., J. Johnson, London.

Maynard Smith, J. 1974. The theory of games and the evolution of animal conflicts. *Journal of Theoretical Biology*, 47, 209–221.

Mech, L. D. 1966. *The Wolves of Isle Royale*. National Park Service Fauna Series, no. 7. U.S. Government Printing Office, Washington, DC.

Murdoch, W. W. 1966. Community structure, population control, and competition—a critique. *American Naturalist*, 100, 219–226.

Olson, J. S. 1985. Cenozoic fluctuations in biotic parts of the global carbon cycle. In *The Carbon Cycle and Atmospheric CO_2: Natural Variations Archaean to Present*, ed. E. T. Sundquist and W. S. Broecker, pp. 377–396. American Geophysical Union Geophysical Monograph 32. American Geophysical Union, Washington, DC.

Peterson, R. O. 1999. Wolf–moose interaction on Isle Royale: The end of natural regulation? *Ecological Applications*, 9, 10–16.

Polis, G. A. 1999. Why are parts of the world green? *Oikos*, 86, 3–15.

Prentice, I. C., et al. 2001. The carbon cycle and atmospheric carbon dioxide. In *Climate Change 2001: The Scientific Basis. Contribution of Working Group I to the Third Assessment Report of the Intergovernmental Panel on Climate Change*, ed. J. T. Houghton et al., sec. 3.2.2.4. Cambridge University Press, Cambridge.

Ridley, M. 1994. *The Red Queen*. Penguin, London.

Robertson, D., and J. Robinson. 1998. Darwinian Daisyworld. *Journal of Theoretical Biology*, 195, 129–134.

Scheffer, V. B. 1951. The rise and fall of a reindeer herd. *Scientific Monthly*, 73, 356–362.

Smith, R. H., and E. L. Cheatum. 1944. Role of ticks in decline of an insular cottontail population. *Journal of Wildlife Management*, 8, 311–317.

Tyrrell, T. 1999. The relative influences of nitrogen and phosphorus on oceanic primary production. *Nature*, 400, 525–531.

Volk, T. 1998. *Gaia's Body: Toward a Physiology of Earth*. Springer-Verlag, New York.

Watson, A. J. 2001. Iron limitation in the oceans. In *The Biogeochemistry of Iron in Seawater*, ed. D. R. Turner and K. A. Hunter, pp. 9–39. Wiley, New York.

Watson, A. J., and J. E. Lovelock. 1983. Biological homeostasis of the global environment: The parable of Daisyworld. *Tellus*, 35B, 284–289.

White, T. C. R. 1993. *The Inadequate Environment: Nitrogen and the Abundance of Animals*. Springer-Verlag, Berlin.

Yool, A., and T. Tyrrell. 2003. Role of diatoms in regulating the ocean's silicon cycle. *Global Biogeochemical Cycles*,17, doi:10.1029/2002GB002018.

III
PHILOSOPHY, HISTORY, AND HUMAN DIMENSIONS OF GAIA

13

Gaia: The Living Earth—2,500 Years of Precedents in Natural Science and Philosophy

Bruce Scofield

Abstract

The Gaia hypothesis orients us toward an understanding of the Earth as a living, dynamic, and self-sustaining system. Today, the various concepts organized under the name Gaia have been stated as a verifiable hypothesis, with postulates demonstrable by scientific methods. But Gaia is more than a testable scientific hypothesis—it offers a philosophical view of nature in the tradition of natural philosophy. Historically, important aspects of the Gaian view can be traced back at least 2,500 years. From the time of Thales and the birth of Greek philosophy in the sixth century B.C., through the seventeenth century, the concept of the "anima mundi" (living earth) can be found in the writings of many influential thinkers in the Western scientific tradition. Historical texts clearly indicate that many of these early scientists conceptualized the world as a living animal composed of innumerable subuniverses, each affecting all others. More specific precedents to the Gaia hypothesis are found in the philosophy of Stoicism, a widely accepted philosophy of Hellenistic and Roman times. Traces of Gaia are found in medieval alchemy, which certain historians argue strongly influenced the methods and social aims of modern science, and in Renaissance cosmology. We also find the idea of the Earth is a living being in the writings of Johannes Kepler (1571–1630), an astronomer/astrologer who lived during the transition period between ancient-medieval and modern science. In perspective, the mechanistic philosophy of Descartes and Newton emerges as an anomaly in the history of Western natural science.

The Gaia hypothesis is leading science toward an understanding of the Earth as a living, dynamic, and self-sustaining system. Originally proposed by James Lovelock in 1969 and supported by the (micro)-biologist Lynn Margulis, this hypothesis is not yet fully accepted within the larger scientific community. The key concepts of the Gaia hypothesis are the following:

1. The Earth is a single living ecosystem that is fueled mostly by solar energy.
2. Individual species and ecosystems function like the organs of a body.
3. Humans have no special place or role in Gaia.
4. Gaia is a system with many regulators.
5. The only goal is the tendency to perpetuate homeostasis.
6. There is no plan; Gaia is an outcome of evolution and operates opportunistically.
7. Organisms have produced and maintained the current composition of the reactive gases of the atmosphere, the temperature of the Earth's surface, and the acidity/alkalinity of the oceans.
8. Organisms incorporate and produce minerals in and around their bodies (biomineralization), a fact which in some ways blurs the distinctions between living and nonliving things.

The First Scientists

While this modern assembly of Gaian attributes is unquestionably unique, some interesting precedents to several of these key concepts are found in the history of Western philosophy and science. Greek philosophy and science was born in the lifework of the pre-Socratic Ionian philosophers, Thales (c. 630–545 B.C.), Anaximander, and Anaximenes during the sixth century B.C. Another influential contemporary was Pythagoras (572–497 B.C.). These protoscientists were philosophers of nature who used the intellect as a tool to investigate the world and offer explanations for the phenomena they observed. They pioneered some of the most basic principles of science as we understand it today, including the forwarding of rational, nonmythic interpretations of nature, the use of a single theory to explain a multiplicity of phenomena, and, in the case of Pythagoras, the use of mathematics to measure natural phenomena.

Certain elements of the Gaia hypothesis are apparent in the underlying assumptions of these early natural philosophers. Thales of Miletus regarded the world as an organism, specifically an animal. In his view, all organisms were complete in themselves, but they were also a part of a greater Earth organism. Anaximander, Anaximenes, Pythagoras, and others also took for granted that the Earth was a living organism, though their views differed in the details. Their animistic views of nature were not radical propositions for their time. Like scientists of all ages, they worked within a cultural matrix that shaped the direction and, consequently, the conclusions of their inquiries. The fundamental beliefs of ancient Greek culture were essentially animistic—Gaia was the Earth goddess, and she was therefore alive. So it follows that the concept of the Earth as an organism was unquestioned by them. Their interests lay more in determining the ultimate material the cosmos was made of and how it may have come into being. The concept of the "living Earth" was thus a tacit assumption of the earliest Greek scientists.

Over the centuries, philosophical terminology was created to better express the subtle distinctions that were made in regard to natural phenomena, terms that modern students of ancient Greek philosophy continue to analyze and interpret. While these definitions changed and evolved through centuries of philosophical argument, the short list below is an indication of how ancient natural philosophers were seeking to understand the nature of life itself. They recognized and grappled with the complex problem of the animation of matter and also the interactions between living things.

Animism: The world of nature is alive and intelligent.

Hylozoism: Matter is animated; that is, it has motion.

Psyche: The breath of life, ghost, vital principle, anima, soul—"psyche" is a term with many variations in definition, but is generally defined as an immaterial quality that has movement and intent. Psyche, or soul, has been defined as the vital principle of organic development, the form or field that gives rise to the characteristics of an organism.

Pneuma: Air, breath, spirit, spiritus; the animation factor.

Pantheism: God is equated with the forces and natural laws of the universe.

Sympatheia: Cosmic sympathy, the mutual interaction of the parts within the whole of the living Earth. All parts of the universe are interconnected.

Anima Mundi: The Earth as a living organism.

World Soul: The soul that resides in the living, organic cosmos that extends beyond the Earth itself.

Plato and Neoplatonism

Plato (429–347 B.C.) refined the early Greek notions of a living cosmos and established a long-lasting and influential tradition concerning the "world soul." In his dialogue *Timaeus*, Plato describes the world as having been created as a living being, endowed with soul and intelligence. Further, the world was not made in the likeness of any individual part of nature. The world is simply a whole (*Timaeus*, 30.d).

Plato may be best known for his theory of forms,[1] an idea that owes much to Pythagoreanism. The Pythagoreans believed that behind the appearances (i.e., empirical phenomena) were numbers—the ultimate reality. Plato's forms, like numbers, are not matter, which itself is "informed"; they are ideal images existing in a state of pure essence. They are the true reality, perceivable only by the intellect, not the senses (*Timaeus*, 30.d). As the thinking of Plato's followers evolved, form and matter drew further apart, a distinction that influenced the institutionalization of the mind–body, spirit–matter split in medieval and Renaissance thought.

Certain later followers of Plato, called Neoplatonists, also drew a sharp division between matter and a nonmaterial essence which animated and informed matter. Plotinus (c. 203–270) tells us that soul, which exists eternally, functions by bestowing life into matter. In the *Enneads* (V1.ii), a work assembled by his pupil Porphyry, he describes soul emanating from a separate, transcendent source, the All, and flowing into the Earth and giving it life. Since there is only one source for soul substance, every individual soul is one with the world soul (Gregory 1991:63).

For centuries, elements of Neoplatonism have been used by Christian thinkers to rationalize the separation of the (physical) body and the soul or spirit. Inherent in this view was the idea of separation between physical Nature and Man. Not only was Man set apart as special, he was placed in a hierarchically superior position and given God's blessing to dominate an "inferior" physical world. Through the Middle Ages and into the Renaissance, this way of thinking became so deeply embedded in the Western psyche that physical nature was looked upon as having negative qualities.

Stoicism

For about 500 years, from the third century B.C. to the second century A.D., Stoicism was probably the most widely embraced philosophical tradition in the ancient Greco-Roman world. Founded in 300 B.C. in Greece by Zeno of Citium (c. 336–263 B.C.), Stoicism evolved into a rigorous intellectual system that not only included ethics, for which it is most famous, but also produced a complete metaphysics. Historians of philosophy have argued that the Stoics brought very little that was completely new into Greek philosophy. However, they have also recognized that the power of Stoic philosophy lay in its ability to synthesize and extend existing doctrines. Stoicism was a syncretic body of ideas that evolved and stayed current with intellectual developments in the Greek and Roman worlds for hundreds of years. It appealed to all classes, and in many respects occupied a position in Roman society similar to our modern, scientific worldview.

Only fragments of the writings of the major Stoics have survived. One fact that is known for certain is that the Stoics believed the cosmos—the orderly, physical universe—was a living, intelligent being. Zeno taught that the cosmos is a material, biological organism, and its animating principle (which he called pneuma) is the element fire. Zeno's successor, Cleanthes, also taught that everything living is alive precisely because it has heat or fire, the vital force of the universe, within it. Later Stoics added that the cosmos grows continuously, gradually incorporating nonliving matter into itself. Stoic natural philosophy ignored Plato's ideal, transcendent forms and instead developed into a materialistic vitalism that some historians of philosophy have labeled a "cosmobiology" (Hahm 1977:136ff.).

Nature for the Stoics was not dead matter; it is a phenomenon driven by a creative, active force (fire) that caused growth in an intelligent, methodical manner. Nature advanced itself by virtue of this intelligent energy, and unlike the world soul of the Neoplatonists, nature was not separate from a transcendent, spiritual world of God.

According to the Stoics, the universe did not completely die. Periodic destructions, followed by a renewal of growth, did occur, and these were said to be structured by long-term astronomical cycles. In their view, there were four primary properties of nature that were symbolized by the four elements: fire, air, earth, and water. Since fire was the one element capable of transformation, the universe was thought to be imbued with fire and was explained as a pulsing, cycling organism that was eternally created and destroyed. This is a cosmological doctrine of both catastrophism and eternal recurrence.

In their synthesis of the Greek philosophical traditions, the Stoics arrived at a concept of nature that shares several ideas central to the modern Gaia hypothesis. First, there is the idea of the living Earth itself. The concept of nature as dynamic, craftsmanlike, and methodical foreshadows in some ways the concept of Gaia, a system of many regulators that operates opportunistically. Also, the recognition that organisms incorporate minerals into their bodies (biomineralization) is found in the Stoic idea that the living incorporates into itself the nonliving.

The Stoics brought together cosmology and biology, arriving at a position that described nature as more than simply the sum of many disconnected parts. Stoic natural philosophy strove for a single, unified, cosmobiological perspective. This view produced a unique set of concepts and explanations for phenomena that has been described as the most rational of their time. However, Stoic natural philosophy was not compatible with Christian views on nature, and considering the Church's power of censorship, few Stoic writings have survived.

Stoicism and Astrology

Western astrology originated in Mesopotamia at least as early as the second millennium B.C. Astrological reports recorded on Babylonian and Assyrian cuneiform tablets have revealed a long tradition of detailed astronomical observations far beyond what simple calendrics or orientations of buildings would require. Astronomical events such as planetary conjunctions and eclipses were observed and correlated with current data such as the weather, earthquakes, political events, wars, plagues, animal behaviors, and the price of wheat. Following the conquests of Alexander the Great in the fourth century B.C., Mesopotamian astrology entered the Greek world, where it became a more rigorous discipline.[2] The island of Rhodes and Alexandria in Egypt were among the places in the eastern Mediterranean where the mixing of Mesopotamian astrology with Greek mathematics took place.

Posidonius of Rhodes (circa 135–51 B.C.) was one of the most influential Stoic teachers. The few surviving fragments of his writings show that he was a powerful writer with a very wide-ranging and synthetic view of the world. His influence on many Roman intellectuals, including Seneca and Cicero, is well known. Galen called him the most scientific of the Stoics.

Like most of the Stoic teachers, Posidonius embraced astrology because astrological theory supported the Stoic philosophy of nature. Fundamental to astrological philosophy was the "doctrine of signatures," an ancient system of classification that assumed real correspondences among the various components of the living Earth, as well as between Earth and celestial phenomena. According to this cornerstone of ancient astrological theory, all phenomena were classifiable by a symbolic language of the planets. Each planet was thought to "rule" or resonate with a specific set of phenomena. An alignment or positioning of a specific planet produced a kind of vibration that was able to affect, through "sympathy," all the organisms and systems of the living Earth that were receptive to that specific planetary energy. Thus, for example, the organisms and systems that resonated with Mars would be affected by that planet's motions and configurations. Venus would affect other systems, and so on for the other planets and the sun and moon.

In this view, nature was composed of diverse parts, but these parts were interconnected in complex ways that were describable through the language of astrology. The empirical justification for this system of quasi-organic linkages was the astrological effect of the planets, which few contested. In general, no one questioned the branch of astrology called "universal" or "natural astrology," which concerned the effects of the planets on the physical Earth, such as weather, climate, earthquakes, and plagues. Those who argued against astrology were generally reacting to the branch of astrology called "judicial astrology," which concerned the effects of planets on individual humans. More specifically, their criticism centered on the issue of fate versus free will. In other words, astrology was not criticized on scientific grounds (as an explanation for the physical world), but on moral, religious ones.

In Stoic natural philosophy, astrology was a key to the validation of a living Earth. Posidonius sought to prove the theory that the living Earth responds to the planets by documenting a correlation between the moon and the tides, which he did in a scientific fashion by making observations in the field. He traveled to Gades (Cadiz) in Spain and for a time observed the Atlantic Ocean tides, which had far more extreme highs and lows than those in the Mediterranean. He regarded his observation that the sea responds to the movement of the moon as evidence of an interconnected cosmos (Sandbach 1975:131). Posidonius thought the sun to be much larger than the Earth and composed of pure fire, the animating principle of Stoic physics. He thought the sun was the cause of many earthly phenomena, including the generation of plants and animals, as well as gems. He was also interested in volcanoes and thought that the Earth itself contained fire and, consequently, was a stimulator of life.

The natural philosophy of Stoicism contains a number of interesting parallels to the philosophical view inherent in the science of Gaia. The general principles of nature according to the Stoics can be summarized as follows:

1. The world is a living creature, a rational animal, animated by a physical force that maintains the cyclic processes of nature. As Stoic cosmology evolved, the term "pneuma" (i.e., the breath of life, heat, fire, and the animating force) became synonymous with the term "soul."

2. Nature itself is a dynamic, living substance or matter, a craftsmanlike "fire" or energy that moves methodically toward genesis.

3. The sun, which is pure fire, enlivens the Earth and governs the world soul.

4. There is an unending series of world creations and destructions.

5. All parts of the universe obey the same laws.

6. All parts of the universe are connected and capable of influencing each other, directly or indirectly, due to sympathy. Astrology is the symbolic language through which these connections become apparent.

Alchemy

Alchemy has roots in the cultures of ancient Egypt and Mesopotamia. The earliest fragments of writings on the subject are found in the Hermetic corpus, which dates to the first and second centuries A.D. These writings describe a worldview based on ritual magic and alchemical and astrological principles, and are the basis of Hermetic philosophy. Over the centuries, the practice of alchemy evolved and stimulated the creation or development of numerous other activities and crafts, such as mining, dyeing, glass manufacturing, medicine preparation, and, eventually, the science that became chemistry. Like astrology, alchemy was not limited to Western culture. It also developed, more or less simultaneously and independently, in China and India.

The goal of alchemy was the acceleration of nature. The alchemist was a nurturer of nature, a conscious agent who assisted the natural processes of the Earth. In alchemical cosmology, the Earth was seen not only as alive but also as being a composite of lesser beings, each with its specific destiny and maturation process. The alchemist, who used fire as a transformative agent, intruded into nature to move these processes along. In the worldview of the alchemist, this was possible because humanity stood on the borderline between the earthly (and perishable) "sublunar" lower realm and the heavenly supralunar realm, which was eternal. Because he resided at the midpoint of this axis, the alchemist could serve as a transformer, an agent of change, or a midwife between the earthly and heavenly realms.

The tradition of alchemy absorbed many Hermetic, Neoplatonic, and Stoic ideas, including the living Earth, sympathetic action, and action at a distance. The doctrines of signatures and the four elements played a major role in alchemy. Also, distinctions between manifest and occult properties, as well as the mystical power of numbers (Pythagoreanism), were part of the intellectual structures of the subject that persisted well into the seventeenth century. Many respected founders of modern science, including Isaac Newton, practiced or were knowledgeable of alchemy.

The alchemical perspective on nature was to some extent resonant with the philosophy of the Gaia hypothesis. Alchemists regarded the Earth as a kind of mother. The parts of the Earth were seen as connected via sympathy and were explained by the doctrine of signatures. Astrology informed practitioners of alchemy in regard to the timing of alchemical operations and transformations. It was believed that changes in states of matter could occur only under certain astrological configurations which produced or coincided with fluctuations in the quality of the living Earth. Distinctions between the organic and the nonorganic were vague in alchemy. Minerals, believed to be alive, were subject to their own destiny and maturation process. Ores were believed to be in a slow process of change that would eventually lead to a ripening into gems or metals. Humans, through the alchemist intermediary, could intercede in nature and accelerate maturation and change.

Renaissance Cosmology

During the early Renaissance in western Europe, most philosophers of nature still regarded the world as a living organism that was animated by what had long been termed the world soul. They were influenced by a variety of doctrines, including alchemy, Hermeticism, astrology, Pythagorean number theory, Cabala, and Neoplatonism, all of which sustained the view of nature as complex and interconnected. John Dee (1527–1608), a leading scientific figure in the English Renaissance, practiced ritual magic and was therefore deeply influenced by Hermeticism. It was this philosophical foundation, however, that allowed him and others of his time to embrace the heliocentric hypothesis of Copernicus. To the Hermeticist, the sun's light was the underlying force and principle that unified and vitalized the universe. It deserved to be at the center of the cosmos.

Giordano Bruno (1548–1600) was a Hermeticist who thought that the infinite universe was a living being, and that God was totally within the manifest world. His pantheistic monism is reminiscent of the pre-Socratic philosophers who, as we have seen, were the founders of Western natural philosophy. These views were not considered particularly controversial at the time. What led to Bruno being burned at the stake was his belief that Nature could produce infinite worlds (beyond the Earth). His view of an infinite cosmos (in which the sun and the Earth had no privileged position) was in direct contradiction to Church doctrine. Bruno's notion, extremely radical in its time, anticipates in some ways the idea of Gaia as having no special purpose other than maintaining homeostasis, and certainly not functioning to serve the purposes of humans.

Johannes Kepler

Johannes Kepler (1571–1630) is best known for his solutions to the complex orbital problems of the planets that validated the heliocentric hypothesis of Copernicus. But Kepler's motivations were far more complex than those portrayed in secondary science history sources. His mathematical laws were designed to serve his larger attempt to construct a grand cosmological synthesis that accounted for the effects of the planets on the Earth, which consequently supported the notion that the Earth is alive. Kepler's vision was comprehensive; he strove to reconcile Pythagorean geometry, astrology, music, and the orbital mathematics of the planets. His first work, *Mysterium Cosmographicum* (Cosmic Mystery) laid out his grand scheme. His second major work, *Astronomia Nova* (The New Astronomy), was more strictly astronomical and offered concrete data on the orbits of the planets. His third major work, *Harmonice Mundi* (The Harmony of the World), continued and elaborated on the thesis of his first book. In this work he laid out his grand synthesis, now supported with mathematical proofs. Kepler was a brilliant mathematician who practiced a kind of Pythagorean–Platonic–Stoic astrology. In fact, he believed himself to be taking the next major step in this long tradition, and his arguments supported the notion of the world soul and several other key concepts that are similar to those of the modern Gaia hypothesis.

In *The Harmony of the World*, Kepler presented his ideas on the living Earth.[3] First, he commented on the Platonic doctrine of the world soul, arguing that if there is a soul of the universe (i.e., nonmaterial living essence), it probably resides in the sun. As for the Earth, which he called "sublunary," the effects of astrological aspects (angular configurations of the planets) on the weather were proof enough for him that the Earth is alive. This crucial part of Kepler's thesis has been overlooked by historians of science. For 30 years, Kepler conducted an empirical study—he kept a weather diary and compared the weather with the astrological aspects at the time.[4] This was not anything new in astrology; correlations between planetary aspects and weather had been a part of the astrological tradition since Babylonian times, and were the foundation on which weather forecasts in annual almanacs were based. However, Kepler was conducting his study in order to accumulate data to prove his thesis.[5] If the astrological aspects did coincide with weather on the Earth, then were they causing the weather through some kind of mechanism—or did the Earth simply respond to the geometry of the aspects? If it were the latter, as Kepler maintained, then the Earth must be sensing the existence of the planets, and sensing implies a living being. Kepler's ideas were not so far-fetched. Research from the twentieth century has produced evidence suggesting that heliocentric planetary alignments force solar conditions, which in turn cause variations in the external magnetic field of the Earth, with consequent effects on the Earth's weather (Seymour 1986:65–67; Seymour et al. 1992; Dewey 1968:278–281).

In several of Kepler's writings, he expresses his ideas about the Earth as a living entity.[6] Some of these suggest certain key concepts of the Gaia hypothesis. Kepler thought that the soul of the Earth was distributed throughout its sphere, both on the surface and below it. He also suggested that there was no essential difference between minerals and the various life-forms on the Earth; they were all part of a continuum. He further believed that the Earth breathed, as evidenced by the tides. Kepler recognized the apparent overall stability of natural phenomena, and he speculated on the causes of this balance. He considered the fact that the oceans of the Earth maintain a stable sea level; they do not overflow from the continuous inflow of rivers. He suggested that this stability was due to the Earth's absorption of seawater in a kind of feeding process. He thought that the evaporation/rain cycle served a balancing function in an organic cosmos, one that was necessary for the nourishment of plants according to season. These two ideas suggest that Kepler was aware of (though unable to fully express) the idea of homeostasis.

Kepler, who was perhaps the last great thinker in the Pythagorean-Platonic tradition and simultaneously one of the first modern scientists, regarded the Earth as a living being. He thought that the Earth's animate powers regulated the minerals, temperature, and water levels of the Earth, and that the Earth could "hear" the astrological aspects. He also wrote that the soul (life force) of the Earth came from the sun and that were are no essential differences between living things on Earth.

The Mechanistic Model

While Kepler strove for a balance between his mathematical modeling of astronomical phenomena and his organic conception of the cosmos, René Descartes (1596–1650) did not. Descartes rebelled against the idealistic philosophies of scholasticism, initiating one of the greatest revolutions in intellectual history. He argued that ancient learning was completely worthless, essentially invalidating anything Neoplatonic, Stoic, Hermetic, astrological, or that otherwise supported the notion of a living Earth. He further argued that numbers are the only test of certainty and that problems must be broken down into parts. Third, he argued that there were only two distinct realities—material substance (body/nature) and spiritual substance (mind/soul).

In Descartes's view, nature was an enormous machine, extended into space, that obeyed mathematical laws. Further, God merely created the machine and let it run according to its own, completely measurable logic. By establishing this distinction, Descartes opened up a mind-body problem of enormous proportions. His answer to the question of how the mind could actually interface in some way with the mechanical, external universe (i.e., the body) was far from scientific. He simply stated that God made it that way. The positivistic scientific currents of his time were reluctant to accept any theological explanations for this apparent dualism, and the problem was ultimately settled by defining the mind as something that existed in a part of the brain. Mind, or spirit, became defined as something that required physical substance for its existence. This was a major shift in thinking with powerful consequences in regard to humanity's relationship and interaction with nature.

Isaac Newton (1642–1727) brought together mathematical modeling and the Cartesian mechanistic perspective to form a new philosophy of nature, one that was powerfully imprinted on Western consciousness during the late seventeenth century. On close examination, Newton's thought actually embodied a mixture of scientific paradigms. He was deeply involved in alchemy; in fact, his treatment of gravity was an extension of his Hermetic understanding of cosmic sympathy. Although he was able to apply mathematics to gravitational phenomena, he could not say exactly what gravity was—it was truly an occult phenomenon. His public response to criticism (of his failure to explain gravity) was to argue that knowing what gravity was in essence was simply irrelevant—experimental philosophy (i.e., science) should concern itself only with measurement.

The seventeenth century was witness to a momentous shift in natural philosophy. The Earth was no longer perceived as living; it was dead and quantifiable, and this was very much in tune with the agenda of the rising forces of capitalism. The idea of matter as living and the cosmos as interconnected was not a view that supported this new capitalistic agenda. Dead nature, on the other hand, was more convenient; it presented no moral conflicts in regard to exploitation for profit.[7]

Since the second half of the seventeenth century, science and philosophy (for the most part) have viewed the world as a machine. Experimental science also has shifted away from an association with philosophy and become essentially an exercise in reductionism. One exception to this revolutionary trend was James Hutton (1726–1797), generally regarded as the founder of the science of geology. Hutton conceptualized the Earth not as a machine, which would have been more in line with the thinking of his age, but as an organized body, a superorganism. He saw the natural processes on Earth, both geological and biological, as interconnected and analogous to the circulation of the blood. Hutton wrote that machines wear out, but the Earth, with its constant cycling and recycling, repairs itself by virtue of its own productive powers (Hutton 1970:216).

Conclusions

The organic, holistic concept of a living Earth has a long history. Over 2,500 years ago, Greek natural science was born. Starting with only the logical tools of the critical mind, the ancient Greeks attempted to grasp the general principles of the world. They strove to understand the differences between living and nonliving matter and to discover the essence behind appearances. The doctrines of Stoicism, Platonism, and other ancient schools of thought envisioned a living Earth or living cosmos, composed of innumerable beings and living systems that together form a complete unit.

With the triumph of Christianity, the ideas of Plato and the Neoplatonists were appropriated to justify a doctrine of transcendence. By the late sixteenth and seventeenth centuries, the official religious doctrine that God was outside of nature, combined with the rise of a more practical-minded society, led to the Cartesian view of nature as a machine. The mechanical view of nature, promoted by Descartes, Newton, and others, combined with the Pythagorean concept of the application of mathematics to nature (as practiced by Kepler and Galileo) to bring about the birth of modern science.

History is written by the victors. Advocates of the current ruling paradigm, modern science, have generally regarded the thinking of previous ages as irrelevant or, at best, as quaint reminders of times we have long since outgrown. The sophisticated animism of the Greeks and Kepler has been considered as simply wrong. But Western science, while certainly experiencing a major revolution in the sixteenth and seventeenth centuries, can also be viewed as a long tradition whose roots extend to ancient times. On close inspection, the concept of a living Earth has long been a dominant theme in Western natural philosophy, and while it has been eclipsed since the seventeenth century, it has not been completely extinguished. In the twentieth century, from Vernadsky to Lovelock, a new and revolutionary perspective on nature has been emerging that treats the Earth holistically, but at the same time does not negate the methodologies of modern science.

As shown in this chapter, there has been a continuity between certain aspects of traditional Western natural philosophy and the modern Gaia hypothesis. The period since the seventeenth century, marked by the rise of modern reductionist science that claims to be objective, could be seen as a kind of aberration, or at least a discontinuity, in the history of Western natural science and philosophy.

Today, the mechanistic Cartesian legacy is still embraced for the most part by our leaders and has become increasingly ineffective in solving problems of a global nature. The Gaia hypothesis, in contrast, offers a way of understanding the Earth that is more conducive to dealing with the immense environmental problems of our time. It offers science a way of perceiving nature that is progressive and, as we have seen, oddly traditional. Gaian science is hard science working within a global perspective. There is no need to abandon scientific methodologies in order to embrace the larger worldview that guides the science of Gaia.

Notes

1. Elements of Plato's theory of forms resurfaced in the twentieth century in the philosophy of Alfred North Whitehead, in the vitalistic theory of Hans Driesch, and in the morphic fields of organicism.

2. By the time of Ptolemy (c. A.D. 150) astrology had become organized into two main branches. Universal astrology (later called natural astrology) concerned itself with weather, climate, plagues, and cultural history, and was intellectually respectable. Judicial astrology was a body of rules for interpreting the positions of the planets at the time and place of a birth or of an event. This branch has always been at odds with religion. See Ptolemy (1940), book II.1.

3. See Kepler, *The Harmony of the World*, chap. VII, "Epilogue on Sublunary Nature and on the Inferior Faculties of the Soul, Especially Those on Which Astrology Depends."

4. The planetary aspects are determined by specific vectors extending from the Earth to the planets and measured on the ecliptic. The traditional aspects, called the Ptolemaic aspects after the Greek astronomer and astrologer Ptolemy, are angles of separation of 0, 60, 90, 120, and 180 degrees. Kepler perceived these aspects as harmonic intervals and added several of his own invention, including ones based on division of the circle by 5 or by multiples of 72 degrees.

5. John Goad (1616–1687) followed Kepler's example and kept a detailed weather diary in England for thirty years. He analyzed these records and those of Kepler in terms of the astrological aspects, a study that was documented in his book *The Astro-Metorologica*. Interestingly, one of Goad's observations, that moisture (i.e., rain or snow) occurs with more frequency when the moon was either 60 or 240 degrees from the sun, was replicated to some extent in several twentieth-century studies, including one by Bradley and Woodbury (*Science* 137 [1962]:748–750).

6. Arguments are found throughout the following *De Fundamentis*, *Harmonice Mundi* (book VII), and *Tertius Interveniens*.

7. Historians of science have argued that the Scientific Revolution is properly understood in a social context, and that the shift from Hermeticism to mechanism was favorable to capitalism. See Berman (1981:115–132).

References

Armstrong, A. H. 1940. *The Architecture of the Intelligible Universe in the Philosophy of Plotinus*. Cambridge: Cambridge University Press.

Berman, Morris. 1981. *The Reenchantment of the World*. Ithaca, NY: Cornell University Press.

Burtt, E. A. 1954. *The Metaphysical Foundations of Modern Science*. Garden City, NY: Doubleday.

Butterfield, Herbert. 1957. *The Origins of Modern Science 1300–1800*. New York: The Free Press.

Collingwood, R. G. 1960. *The Idea of Nature*. New York: Oxford University Press.

Copleston, Frederick, S. J. 1962a. *A History of Philosophy*, vol. 1, parts I and II, *Greece and Rome*. New York: Image Books.

Copleston, Frederick, S. J. 1962b. *A History of Philosophy*, vol. 2, *Mediaeval Philosophy*, part I. New York: Image Books.

Crombie, A. C. 1959. *Medieval and Early Modern Science*, vol. 2. Garden City, NY: Doubleday.

Dewey, Edward R. 1968. Forecasting radio weather. *Cycles*, 278–281. December.

Eliade, Mircea. 1956. *The Forge and the Crucible*. Chicago: University of Chicago Press.

French, Peter J. 1972. *John Dee: The World of an Elizabethan Magus*. London: Routledge & Kegan Paul.

Gregory, John. 1991. *The Neoplatonists: A Reader*. London and New York: Routledge.

Hahm, David E. 1977. *The Origins of Stoic Cosmology*. Columbus: Ohio State University Press.

Hamilton, Edith, and Huntington Cairns, eds. 1961. *The Collected Dialogues of Plato*. Princeton, NJ: Princeton University Press.

Hunt, H. A. K. 1976. *A Physical Interpretation of the Universe: The Doctrines of Zeno the Stoic*. Melbourne, Australia: Melbourne University Press.

Hutton, James. 1970 [1788]. *Theory of the Earth*. Reprinted New York: Hafner Press.

Kepler, Johannes. 1942 [1602]. *De Fundamentis Astrologiae Certioribus or Concerning the More Certain Fundamentals of Astrology*. Reprinted New York: Clancy.

Kepler, Johannes. 1997. *The Harmony of the World*, translated by E. J. Aiton, A. M. Duncan, and J. V. Field. Philadelphia: American Philosophical Society.

Lovelock, J. E. 1995. *The Ages of Gaia: A Biography of Our Living Earth*. New York: Norton.

Peters, F. E. 1967. *Greek Philosophical Terms: A Historical Lexicon*. New York: New York University Press.

Ptolemy. 1940. *Tetrabiblos*, translated by F. E. Robbins. Cambridge, MA: Harvard University Press.

Read, John. 1966 [1936]. *Prelude to Chemistry*. Reprinted Cambridge, MA: MIT Press.

Rist, J. M. 1969. *Stoic Philosophy*. Cambridge: Cambridge University Press.

Sambursky, S. 1956. *The Physical World of the Greeks*. New York: Collier Books.

Sandbach, F. H. 1975. *The Stoics*. Bristol, UK: Bristol Press.

Seymour, P. A. H., M. Willmott, and A. Turner. 1992. Sunspots, planetary alignments and solar magnetism: A Progess Review. *Vistas in Astronomy*, 35, 39–71.

Seymour, Percy. 1986. *Cosmic Magnetism*. Boston: Adam Hilger.

Schrodinger, Erwin. 1996 [1954]. *Nature and the Greeks and Science and Humanism*. Reprinted Cambridge: Cambridge University Press.

Scott, Alan. 1991. *Origen and the Life of the Stars*. Oxford: Clarendon Press.

Tester, Jim. 1987. *A History of Western Astrology*. New York: Ballantine Books.

Timothy, H. B. 1973. *The Tenets of Stoicism*. Amsterdam: Adolf M. Hakkert.

Yates, Frances A. 1964. *Giordano Bruno and the Hermetic Tradition*. New York: Vintage Books.

14

Concerned with Trifles? A Geophysiological Reading of Charles Darwin's Last Book

Eileen Crist

I take shame to myself for not having earlier thanked you for the Diet of Worms, which I have read through with great interest. I must own I had always looked on the worms as amongst the most helpless and unintelligent members of the creation; and am amazed to find that they have a domestic life and public duties! I shall now respect them, even in our Garden Pots; and regard them as something better than food for fishes.
—Sir Joseph Dalton Hooker, Botanist (Letter to Charles Darwin)[1]

Abstract

The last book that Charles Darwin wrote was published one year before his death. It has been celebrated in the fields of soil biology and earthworm ecology as a landmark contribution. Yet outside this specialized literature, there is surprisingly little commentary on what is often referred to as Darwin's "worm book." I argue that the relative neglect of Darwin's worm book stems from his investigating an unorthodox (until recently) topic: the formative impact of organisms on their physical, chemical, and biotic surroundings. I present Darwin's understanding of the global effects of earthworms, and then show that his conceptual and empirical framework is consonant with geophysiology—the science of the Earth as a living system. I conclude with a discussion of the implications of Darwin's argument in the contemporary context of global environmental degradation.

Introduction

The last book that Charles Darwin wrote, *The Formation of Vegetable Mould, Through the Action of Worms, with Observations on Their Habits*, was published in 1881, one year before his death. Often referred to as his "worm book," it has been celebrated in the fields of soil biology and earthworm ecology as a landmark contribution.[2] Darwin introduced its subject matter as "the share that worms have taken in the formation of vegetable mould which covers the whole surface of the land in every moderately humid country" (1985:1). This low-key introduction suggests that the work examines a natural phenomenon of interest only to specialists. Yet the author's insinuation that the topic is narrow and specialized is profoundly misleading. His study examines the enormous impact of earthworm species on the Earth's physical, chemical, geological, and biological environment. Darwin's investigation into how organisms shape their surroundings marks his worm book as pioneering science. Its uniqueness, however, has been somewhat obscured by a misperception of "mould formation by worms" as a phenomenon of limited interest.

Barring its importance in soil biology and earthworm ecology, Darwin's last book has been relatively neglected.[3] The author himself may be partly to blame for this—he was terribly self-conscious that a study of earthworms might be deemed a triviality, and began his manuscript somewhat apologetically: "The subject may appear an insignificant one, but we shall see that it possesses some interest; and the maxim '*de minimis lex non curat*' [the law is not concerned with trifles], does not apply to science" (1985:2). In his typically modest style, Darwin might be paraphrased as saying "I studied the formation of mould by earthworms because trifles are not beneath science." The author's modesty notwithstanding, his last work is far from "concerned with trifles": its main theme—the ways organisms shape, in his words, "the history of the world"—has been a scarce topic of investigation, at least until recently.

Ecologist Paul Ehrlich has remarked that he would not be surprised if a search of "Darwin's work from end to end found that Darwin was prescient about Gaia" (1991:19). In this chapter, I do not contend that Darwin anticipated what James Lovelock would call the "Gaia hypothesis," and yet Ehrlich did hit something of a mark: the argument of Darwin's last work can be read as Gaian or geophysiological. Geophysiology investigates living and nonliving processes on Earth as a coevolving system, within which organisms not only adapt to their environment but also modify it in ways that support their survival and proliferation. I argue that Darwin's presentation of the formidable effects of earthworms on their abiotic and biotic surroundings constitutes a conceptual and empirical approach consonant with present-day geophysiology.

The Formation of Vegetable Mould is a scientific study about the reciprocal influence of life and environment. In investigating this relationship, however, Darwin focused less on earthworms' adaptation to their surroundings, and far more on their significant transformation of the environment. He showed that worms alter the appearance of the landscape; they change the physical texture and chemical composition of the soil; they contribute to disintegration and denudation with subsequent geological-level effects; and their presence or absence is critical with respect to the livelihood of plants. Darwin thus documented the biogeochemical significance of animals that, at first blush, might be dismissed as trivial.

I begin by summarizing Darwin's thesis of the environmental impact of worms; next, I show the affinity between his argument and a geophysiological perspective; I conclude with some thoughts about the environmental implications of his worm book some 120 years after its publication.

The Impact of Worms on the Environment

In constantly upturning and swallowing the soil, earthworms' "chief work," according to Darwin, "is to sift the finer from the coarser particles, to mingle the whole with vegetable debris, and to saturate it with their intestinal secretions" (1985:174–175). Worms turn over, blend, chemically modify, and internally triturate the soil, thereby contributing significantly to its makeup. For nutritive purposes and in the process of burrowing, earthworms swallow earth and then expel it from their bodies as viscid, tower-shaped "castings"—otherwise known as "vegetable mold." (Darwin noted that "the term 'animal mould' would be in some respects more appropriate" [1985:4].) Vegetable mold has passed through worms' bodies countless times, tends to be darker in color from the subsoil, and contains no fragments of stone larger than those that can pass through a worm's alimentary canal (1985:236). Writing to naturalist informants around the world, Darwin inquired whether the formation of vegetable mold is a widespread phenomenon; on the basis of their responses, he deduced that "earthworms are found in all parts of the world, and some of the genera have an enormous range" (1985:120).[4] The author was thus not simply concerned with the work of worms in his own pots, garden, and countryside, but with whether their impact on the soil was a planet-wide phenomenon (1985:128–129).

After establishing the baseline that worms "do much work," Darwin proceeded to measure how much earth they bring up by determining the rate at which surface objects are buried beneath vegetable mold. He discussed numerous cases of natural and man-made objects being eventually buried beneath worm castings, slowly but ceaselessly produced. One description depicts changes in a field near Darwin's house:

> [T]he field was always called by my sons "the stony field." When they ran down the slope the stones clattered together. I remember doubting whether I should live to see these larger flints covered with vegetable mould and turf. But the smaller stones disappeared before many years had elapsed, as did every one of the larger ones after a time; so that after thirty years (1871) a horse could gallop over the compact turf from one end of the field to the other, and not strike a single stone with his shoes. To anyone who remembered the field in 1842, the transformation was wonderful. This was certainly the work of the worms, for though castings were not frequent for several years, yet some were thrown up month after month, and these gradually increased in numbers as the pasture improved. (1985:143–144)

Darwin calculated that the average rate of accumulation was one inch of mold added every dozen years—a steady increase that completely transformed the "now miscalled 'stony field'" (1985:145). All his observations indicated the same trend: the slow and uniform sinking of small and large stones through the action of worms, at rates on average somewhat greater than those determined for the stony field (see 1985:171–172).

Darwin complemented observations of worm-caused changes in landscape appearance with quantitative measurements. He calculated the amount of earth brought up by worms by selecting different terrain types, collecting castings, and weighing them.[5] Different measurements are provided, and four summarized cases yield an average of 14 tons of earth annually ejected per acre (1985:168–169). This impressive amount of earth brought to the surface by worms, and subsequently spread over the land by wind, gravity, and rain, appears less astounding when one considers the "vast number of worms [that] live unseen by us beneath our feet" (1985:158). Citing the calculations of Von Hensen, a German physiologist, Darwin estimated an approximate population of 53,767 worms per acre.[6]

Darwin also documented the impact of earthworms on the landscape by investigating their role in the burial of antiquities (1985:176–229). He examined a number of Roman sites, adducing evidence that ancient objects as well as entire buildings are eventually interred through the slow but relentless work of worms. Through the study of archaeological sites Darwin confirmed how worms change the appearance of the land, and he was also able to determine the depths they inhabit. Excavated areas revealed countless burrows beneath the floors of ruins, indicating that worms can live up to six feet or more inside the earth. Darwin deduced that the burial of ancient buildings is instigated by worms' undermining work below and ultimately is completed by their becoming covered with worm-generated mold above (1985:189–190).

Darwin's study of earthworms' modification of their surroundings culminates in the presentation of geological consequences: worms contribute to "denudation," defined as the removal of "disintegrated matter to a lower level" (1985:231). His argument that worms are a geological force unfolds with an almost deliberate slowness, beginning with a summary of geological findings regarding the disintegration of rocks and denudation. Darwin noted that geologists at first regarded sea waves as the chief agency driving these phenomena, but they eventually included other powerful forces such as rain, streams and rivers, frost, volcanic eruptions, and wind-driven sand (1985:232–235). He reviewed geologists' findings on disintegration and denudation, while also gently exposing how they had been compelled to recognize a larger set of forces at work. He thus set the scene for expanding the list yet again—this time adding the biogeochemical agency of global earthworm activity.

According to Darwin's detailed studies, worms contribute to disintegration in two ways: (1) by virtue of the chemical action of their intestinal secretions and (2) by virtue of the mechanical action of their gizzards on swallowed earth. Soil composition is chemically modified by means of processing organic and inorganic matter in their mildly acidic intestines (1985:240–243). The chemical composition of castings has consistent and cumulative effects, since "the entire mass of the mould over every field passes, in the course of a few years, through their alimentary canals" (1985:243). Castings affect the acidity/alkalinity levels of surface soils as well as deeper layers, contributing to the disintegration of coarse materials, organic debris, and rock fragments. The consequence of these disintegration effects is that the amount and thickness of soil tend to increase (1985:244). Darwin went on to note that "not only do worms aid indirectly in the chemical disintegration of rocks, but there is good reason to believe that they likewise act in a direct and mechanical manner on the smaller particles" (1985:246). Worms have gizzards lined with a coarse membrane and surrounded by muscles that contract forcefully, grinding together, and thus levigating, small particles of earth; the attrition of swallowed materials was supported by Darwin's observation that castings resemble "paint" (1985:249). He concluded this section on the following note: "The trituration of small particles of stone in the gizzards of worms is of more importance *under a geological point of view* than may at first appear to be the case" (1985:257; emphasis added).[7]

After discussing the part worms play in disintegration, Darwin considered how this contributes to "denudation." He found that with rain, worm castings flow down even mildly inclined slopes, while in dry weather they disintegrate into pellets and roll, or are blown, to lower levels. Numerous observations and measurements are given to show that long-term denudation effects are notable. On the basis of eleven sets of observations of the downward flow of castings, coupled with the calculated amount of castings annually brought to the surface, Darwin estimated that each year approximately 2.4 cubic inches of earth cross a horizontal line one yard in length (1985:268). He argued that this amount—a few handfuls of earth—is far from negligible when the inference is drawn with respect to large-scale and long-term effects:

> This amount is small; but we should bear in mind how many branching valleys intersect most countries, the whole length of which must be very great; and that the earth is steadily traveling down both turf-covered sides of each valley. For every 100 yards in length in a valley with sides sloping as in the foregoing cases, 480 cubic inches of damp earth, weighing above 23 pounds, will annually reach the bottom. Here a thick bed of alluvium will accumulate, ready to be washed away in the course of centuries, as the stream in the middle meanders side to side. (1985:269–270)

Darwin's reasoning is at once simple and profound. As Stephen Jay Gould and other scholars have noted, his genius resided in the ability to discern momentous consequences as the cumulative import of small local changes (Gould 1985; Ghilarov 1983). Like his worldview-shattering argument for evolution—as the cumulative upshot of (often) minor variations in organisms over geological time—his panoramic vision of the gradual transformation of a valley over the course of centuries due to earthworm activity reflects his far-reaching insight.

The processing of soil by earthworms is not only significant "under a geological point of view." Throughout the manuscript there are allusions to the vigorous growth of plants and turf in mold-rich regions and, conversely, to the scarcity of plant life where worms are absent. In his conclusion, Darwin tackled the relationship between worms and plants directly. He argued that worms are key for the flourishing of plants, for they "prepare the ground in an excellent manner for the growth of fibrous-rooted plants and for seedlings of all kinds" (1985:309). (He noted that seedlings sometimes germinate when covered with castings—a finding confirmed by recent studies [1985:311].) By burrowing through the earth, bringing up castings and thus replenishing topsoil, and contributing to the disintegration of organic and inorganic materials, worms ceaselessly aerate, blend, and thicken the soil. A soil texture is thereby created that retains and diffuses moisture, and nutritional elements are uniformly distributed and brought closer to the roots of plants.

Darwin captured the intimate affiliation of plants and worms in a beautiful and ecologically astute passage:

> [Worms] mingle the whole intimately together, like a gardener who prepares fine soil for his choicest plants. In this state it is well fitted to retain moisture and to absorb all soluble substances, as well as for the process of nitrification. The bones of dead animals, the harder parts of insects, the shells of land-mollusks, leaves, twigs, &c., are before long all buried beneath the accumulated castings of worms, and are thus brought in a more or less decayed state within reach of the roots of plants. (1985:310)

His ecological view was markedly in contrast to beliefs of his day, when worms were widely believed to be injurious to plant life, and the agricultural and horticultural literature recommended methods to exterminate them (Graff 1983:7). Darwin went on to argue that by benefiting plants, the "gardeners" benefit themselves: as plants flourish, more food for worms is produced—for their main sustenance is leaves and petioles that fall to the ground. In such a biologically rich environment, worms proliferate, thereby increasingly thickening and blending the soil, upon which plants will continue to thrive. Darwin essentially argued that earthworms and plants are connected in a mutually sustaining partnership.

In sum, in his last book Darwin showed that worms have a significant impact on the appearance, chemical constitution, physical structure, geological shaping, and biological organization of the land. By bringing up their castings, earth from deeper layers is continually conducted to the surface; by ceaselessly burrowing, worms mix and aerate the earth; by passing soil through their intestinal tracts, they contribute to the chemical decomposition of earth materials; and by the action of their gizzards, worms directly triturate swallowed matter—including tiny rocks. The compounded consequence of these ongoing worm activities is disintegration of organic matter and small rocks, such that the soil becomes finer and thicker. This contributes to denudation, for there is a tendency for the castings (or "fine earth") to move downward, as well as leeward, through the action of gravity, rain, and wind. Over the course of long time spans, disintegration and denudation facilitated by worms lead to geological-level consequences. Further, as a result of blending and thickening of soil, earthworms support the flourishing of all manner of plants. A thriving plant life, in turn, has significant consequences for both soil and worms: plants counteract the tendency toward denudation by anchoring the soil; and they favor the proliferation of earthworms, for leaves and other plant parts constitute a significant part of worms' diet.

Darwin introduced his inquiry as "the share that worms have taken in the formation of vegetable mould," yet he achieved far more than this low-key statement would suggest: his "worm book" is a pioneering and sophisticated study of the ways that a group of invertebrates interact with, and shape, a broad range of nonliving and living processes of the Earth's land surface.

The Affinity of Darwin's Argument with Geophysiology

Through an interdisciplinary study that is both qualitative and quantitative, that weds geological and biological knowledge, that is empirically rigorous in local inquiry and imaginatively inferential for long-term effects, Darwin demonstrated that something as ostensibly "trite" as vegetable mold is a massive, ceaselessly produced, life-generated, life-enhancing, near-global phenomenon. He painstakingly documented that, far from being inconsequential, earthworms are a significant shaping force of their nonliving and living environment. Nor did he forever shy away from affirming this as the major finding of his study. In the last chapter, the author expressed the scope of his findings with less modesty but far greater accuracy: "Worms," he wrote, "have played a more important part in *the history of the world* than most persons would at first suppose" (1985:305; emphasis added).[8] In this section, I review Darwin's analysis in a geophysiological light. First, a few words about the science of geophysiology.

Geophysiology, or Gaian science, refers to the contemporary scientific inquiry into the biogeochemical dynamics of the Earth.[9] Geophysiology studies the ways that organisms affect and alter their surroundings, and the subsequent repercussions of such life-driven effects for the organisms themselves and for life as a whole. Central to geophysiological thinking are the following working ideas or broadly shared hypotheses: living and nonliving processes are tightly knit—whether systemically coupled or seamlessly continuous; life powerfully modifies (and perhaps even regulates) local, regional, and ultimately global environmental conditions; the adaptive modulation of environmental parameters by living processes constitutes a highly plausible explanation for the resilience of life on Earth for well over 3 billion years; and such a "planetary takeover" by the biota would be favored by (and is consistent with) natural selection, if indeed it has led to the maintenance of key environmental conditions within ranges viable for life as a whole.[10]

Life understood as a planetary phenomenon implicates interdependency and cooperation between organisms in the creation of life-sustaining atmospheric composition, hydrological regimes, and land surface (especially soil) constitution. For geophysiology, air, water, and land environments constitute, metaphorically speaking, the "commons" of the biota: the biota does an impeccable job of preserving those commons—not as static environmental settings but as ever-fluctuating yet always viable surroundings. The commons are not thought to be deliberately maintained, but consist in emergent biogeochemical phenomena such as element cycling, waste use, recycling, decomposition, and regulatory feedback loops—in all of which living organisms, as interdependent players, have the leading role. In studying emergent phenomena that involve intricate connections between life and environment, geophysiology weds the Earth and life sciences. The aim of such interdisciplinary inquiry is to understand the dynamics of the Earth as a living whole.

I now turn to argue that Darwin's understanding of the formidable impact of earthworms on environment is essentially geophysiological. To show this, I rely on a discursive methodology that social scientists have called "rational reconstruction" (see Habermas 1981:197). This methodological approach allows me to explicitly disclaim imputing to Darwin a geophysiological perspective: Gaia was not in Darwin's repertoire, and he might well have disagreed with the present-day geophysiological conception of planet Earth had he encountered it. At the same time, the virtue of reconstructive analysis is that it enables me to demonstrate that the argument of his last book can be conceptualized in geophysiological terms, without distorting or overinterpreting Darwin's own presentation of the impact of earthworms on the land. Using the methodology of rational reconstruction is critical, for I am not claiming that Darwin anticipated the Gaia hypothesis; rather, I am arguing that the thesis of his last book can be read, without strain, as a geophysiological thesis.

Geophysiology emphasizes that organisms play an important role in the creation of their environments. Organisms thereby play an important role in creating themselves, for the environmental conditions they contribute to forming subsequently exert selective pressures on them and their descendants: on straight evolutionary reasoning, it follows that organisms which create a favorable environment for themselves will tend to be selected for through consequent feedback effects from the environment. Darwin's understanding of the shaping force of earthworms on their surroundings resonates strongly with this view: he did not take the land as the given background to which worms adapt, but saw it as a medium actively created and maintained, in large part, by these animals themselves.

In exploring the mutual shaping of life and environment, it has been noted that Darwin's last book is an ecological analysis (see Carson 1994). However, the commentator M. S. Ghilarov has observed that a (subsequently) neglected aspect of ecology was brought to the fore in Darwin's study of earthworms: "Up to a short time ago, ecologists only studied dependence of organisms on their environment." "Darwin," he continues, "has shown brilliantly the other side of the medal—the influence of organisms on their environment, i.e. the dependence of the milieu, of the environment, on their activity" (1983:3–4). Indeed, the other side of the ecological coin—how organisms shape their surroundings—is the main subject matter of geophysiologists, who insist that life does not adjust to "an inert world determined by the dead hand of chemistry and physics" (Lovelock 1988:33). In resonance with this perspective, Darwin showed that worms do more than simply adapt to their surroundings. By producing and tilling the soil, they partake in forming an environment that favors their livelihood. His thesis was thus stronger than the observation that earthworms simply affect the land: he argued that worms transform their environment in ways that contribute to creating a favorable habitat for themselves.

The idea that life and environment shape each other may be regarded as trivially true. However, the geophysiological understanding of the relationship between life and environment as a system (life ↔ environment) is neither trivial nor obviously true. In Darwin's last work, as well, the connection between worms and soil is systemically conceptualized. The ecologist J. E. Satchell, editor of the 1983 collection *Earthworm Ecology: From Darwin to Vermiculture*, writes that "it has always been difficult to formulate a balanced judgement on how far earthworm activity creates fertile soils and how far fertile soils create a favorable environment for earthworm activity" (1983:xi). One hundred years earlier, Darwin was not waylaid by the "chicken-egg" appearance of this matter. He showed that earthworms contribute to the formation and thickening of the soil and simultaneously emphasized that worms prosper in thicker soil. Darwin thus did not conceptualize the relationship between "earthworm activity" ↔ "fertile soils" as a circular formulation in need of resolution, but understood it as a precise description of the phenomenon—which is circular itself.

Indeed, the interaction between earthworms and plants—which is what creates and sustains fertile soils—was essentially described by Darwin as a positive feedback loop: the proliferation of worms and the proliferation of plants are mutually causal, and this two-way causation results in the acceleration of the proliferation of both. The feature of acceleration in positive feedback loops has been a central insight of systems theory—developed decades after Darwin's death. Yet this feature is explicitly put forward in Darwin's observations about the "stony field" (cited above), which, he remarked, became a misnomer after being covered over by mold and turf within a few decades. He observed that "this was certainly the work of the worms, for though castings were not frequent for many years, yet some were thrown up month after month, and *these gradually increased in numbers as the pasture improved*" (1985:143–144; emphasis added). While he provided the average rate of mold increase per year, he emphasized that the actual rate was not constant, but "must have been much slower at first, and afterwards considerably quicker" (1985: 144). The more castings the worms brought up, the more plants grew on the field, the more favorable the environment for worms to thrive: such a feedback mechanism translated into an accelerated rate of mold formation—which increases soil fertility. Fertile soils were thus understood by Darwin as an emergent effect of the tight coupling of earthworms and plant life.

Interestingly, there is also implicit reference in Darwin's work to what would be called a "negative feedback loop" (or homeostasis). Specifically, Darwin noted that through the interplay of living and nonliving processes a relatively stable thickness of vegetable mold is maintained. His reasoning went as follows. Worms tend to increase the thickness of the soil, but this tendency toward ever-increasing thickness is countered by denudation due (especially) to rain and wind; as a result of these inanimate forces, "the superficial mould is prevented from accumulating to a great thickness" (1985:307). Thus, while the layer of vegetable mold that "covers the whole surface of the land" is continually thickened and resifted, it also tends to remain fairly constant. In support of this view, Darwin cited a passage from John Playfair's *Illustrations of the Huttonian Theory of the Earth*: "'In the permanence of a coat of vegetable mould on the surface of the earth,' wrote Playfair in 1802, "we gain a demonstrative proof of the continued destruction of the rocks'" (quoted in Darwin 1985:290).

The remainder of Playfair's passage (not quoted by Darwin) continues without pause: "And cannot but admire the skill, with which the powers of the many chemical and mechanical agents employed in this complicated work, are so adjusted, as to make the supply and the waste of the soil exactly equal to one another" (Playfair 1956:107). Playfair maintained that soil thickness is kept constant by means of ceaseless flux; the "complicated work" of nonliving forces, ever at play, results in "adjustment" or stability. But to the stability of soil conditions achieved somewhat inexplicably, according to Playfair, by "chemical and mechanical agents," Darwin added a truly explanatory biological factor: earthworms. For Darwin, the disintegration effects of worms are integral to the relatively steady constituency and density of the upper layer of the land. Thus he essentially described the attainment of dynamic homeostasis:[11] the incessant activities of worms, combined with the ever-present erosion effects of wind and water, create a steady state of topsoil (see also Gould 1985).

In geophysiological terms, when life plays a key role in creating and sustaining environmental conditions, the process is referred to as a "biological system of regulation" (Lavelle 1996:211). Darwin presented earthworm activity as a biological system of regulation of soil conditions—especially fertility and thickness. Ultimately, worms, organic and inorganic debris, soil, plants, and nonliving forces were understood by Darwin as dynamically integrated—an understanding consonant with a geophysiological perspective. This perspective is expressly interdisciplinary, crossing the boundaries of Earth and life sciences to understand the ways biotic and abiotic processes are enmeshed. "Why run the Earth and life sciences together?" inquires James Lovelock, rhetorically. He responds: "I would ask, why have they been torn apart by the ruthless dissection of science into separate and blinkered disciplines?" (1988:11). Darwin's investigation into the impact of worms on the land draws extensively on Earth and life sciences, producing knowledge that is a contribution from both, but which each alone could never have yielded.

The originality of Darwin's synthesis of geological and biological knowledge hinged especially on his approach to time: by considering the effects of earthworms in extended time frames, he contextualized their living activities in ways that yielded geological insight. Worms were shown to shape the landscape, and ultimately to be a geological force, only by considering their impact over time. The passage of time was variously gauged: in Darwin's observations of the "stony field," for example, the impact of worms was assessed within a human lifetime; in the burial of antiquities, their impact was surmised in terms of centuries; the downward and leeward movement of worm castings was judged considerable "in the course of thousands of years" (1985:289); and consideration of greater time periods—for example, 1 million years—gave insight into the "not insignificant" effects of earthworms:

> Nor should we forget, in considering the power which worms exert in triturating particles of rock, that there is good evidence that on each acre of land ... [that] worms inhabit, a weight of more than ten tons of earth annually passes through their bodies and is brought to the surface. The result for a country the size of Great Britain, within a period not very long in a geological sense, such as a million years, is not insignificant; for the ten tons of earth has to be multiplied first by the above number of years, and then by the numbers of acres fully stocked with worms; and in England, together with Scotland, the land which is cultivated and is well fitted for these animals, has been estimated at 32 million acres. And the product is 320 million tons of earth (1985:258).

The time frame of organisms' life cycles and the time frame of geological events diverge by orders of magnitude. In order, therefore, to discern the geological impact of earthworms, their cumulative effects had to be deductively and inductively extrapolated by Darwin for the long run.

Darwin was impatient with those who could not grasp the significance of the cumulative effects of "a continually recurrent cause" in the course of time. He quoted one critic who remarked about Darwin's conclusions that "considering [worms'] weakness[12] and their size, the work they are represented to accomplish is stupendous" (Darwin 1985:6). Darwin responded with uncharacteristic brusqueness: "Here we have an instance of that inability to sum up the effects of a continually recurrent cause, which has so often retarded the progress of science, as formerly in the case of geology, and more recently in that of the principle of evolution" (1985:6). The influence of a single worm on the environment is obviously negligible, Darwin conceded, but the additive effects, over time, of the widely distributed genera of earthworms—often numbering in thousands of individuals per acre—is, indeed, "stupendous."

It is intriguing that the great evolutionist focused on how earthworms transform the global environment rather than solely exploring the selective pressures that shaped their evolution—as seen in their anatomy, physiology, habits, and adaptive radiation. Natural selection has clearly forged the physical and behavioral characteristics of these animals, and their wide geographic distribution (and 600-million-year presence [Lee 1985]) attests to their success. In his worm book, however, Darwin chose to investigate the other side of the relationship between environment and life—the ways earthworms transform the land and their important role in the history of the world: locally, regionally, and globally, and on timescales from one human lifetime to 1 million years.

Darwin's analysis may be deemed geophysiological on the following counts: (1) the portrayal of the tight coupling of earthworms and land, with particular emphasis on how worms modify their environment; (2) the argument that worms change chemical and physical conditions in ways that are beneficial—rather than neutral or haphazard—to themselves; (3) the implicit description of positive and negative feedback loops that depict systemic connections between worms and their biotic and abiotic environment; (4) the emphasis on earthworms' impact as nearly planetary in scope; and (5) the interdisciplinary character of his study, which joined biological and geological knowledge by extrapolating from local biotic activities to cumulative abiotic effects. In consonance with a geophysiological perspective, Darwin's assessment of the impact of worms essentially agrees with Lovelock's assessment: "The Earth's crust ... [is] either directly the product of living things or else massively modified by their presence" (1988:33).

Gaian literature sometimes cites Darwin's evolutionary perspective as incomplete for missing a geophysiological angle (for example, Lovelock 1988:63). While the geophysiological theme of Darwin's worm book—in which the effects of organisms on their surroundings are emphasized more than their adaptive traits—may not be typical of his work as whole, it is no less significant for this reason.[13] Its significance for Darwin is evident in that his research on earthworms spanned his entire career; as his biographers Adrian Desmond and James Moore write, the book was "forty years in the making" (1992:654).[14] Stephen Jay Gould commented that Charles Darwin is the greatest ally any perspective can claim. On the basis of his last work, I believe geophysiology can rightfully claim Darwin as an antecedent thinker.

Environmental Implications of the Worm Book

The pioneering character of Darwin's last work may be underscored by comparison with the perspective of another great scientist, evolutionary and conservation biologist, E. O. Wilson. In his celebrated essay "The Little Things That Run the World," Wilson presents a generalized version of Darwin's analysis—with the significant addition of implications for a conservation ethic. He draws attention to the role of invertebrates in the preservation of biodiversity by means of a compelling scenario:

> The truth is that we need invertebrates but they don't need us. If human beings were to disappear tomorrow, the world would go on with little change. Gaia, the totality of life on earth, would set about healing itself and return to the rich environmental states of a few thousand years ago. But if invertebrates were to disappear, I doubt that the human species would last more than a few months. Most of the fishes, amphibians, birds, and mammals would crash to extinction about the same time. Next would go the bulk of flowering plants and with them the physical structure of the majority of forests and other terrestrial habitats of the world. The earth would rot. As dead vegetation piled up and dried out, narrowing and closing the channels of nutrient cycles, other complex forms of vegetation would die off, and with them the last remnants of the vertebrates. The remaining fungi, after enjoying a population explosion of stupendous proportions, would also perish. Within a few decades the world would return to the state of a billion years ago, composed primarily of bacteria, algae, and a few other very simple multicellular plants. (1987:345)

Wilson sums up the ecological role of invertebrates in the terse report of what would transpire without them: the earth would rot. In the same vein, a hundred years earlier Darwin averred that "long before [man] existed the land was in fact regularly ploughed, and still continues to be thus ploughed by earthworms" (1985:313).

In her work *Silent Spring*, Rachel Carson was the first to note the significant environmental implications of the worm book. After summarizing Darwin's thesis of the importance of earthworms for the soil—and adding to his observations the crucial role of microorganisms and their connection of earthworms—she wrote: "This soil community, then, consists of a web of interwoven lives, each in some way related to the others—the living creatures depending on the soil, but the soil in turn a vital element of the earth *only so long as this community within it flourishes*" (1994:55–56; emphasis added). More recently, the science journalist Yvonne Baskin referred to the underground community that sustains fertile soils "as a work force, a global service corps of rot and renewal" (1997:108).

Carson observed that "the very nature of the world of soil"—in which the living and nonliving components are inseparably coupled—"has been largely ignored" (1994:57). "Chemical control of insects," she noted, "seems to have proceeded on the assumption that the soil could and would sustain any amount of insult via the introduction of poisons without striking back" (1994:57). As Carson forecast, the soil has struck back: the United Nations Environmental Program reports that since 1945, 300 million hectares of land have been so degraded as to be rendered useless for agriculture (Baskin 1998:107). Agrochemicals are not the sole cause of the global problem of soil ruin; but in killing micro- and macroorganisms that are integral to it, chemicals contribute to dismantling the ecology and undermining the health of the soil. Many agrochemicals are directly toxic to earthworms.[15] The earthworm ecologist K. E. Lee provides an extensive inventory of the effects of various biocides (insecticides, herbicides, fungicides, and fumigants) on earthworms, most of which range from "slightly toxic to very toxic" (1985:292–314).[16]

Biologists have corroborated Darwin's finding that earthworms boost soil fertility, in both arable and wild lands, by maintaining soil structure, aeration, and drainage, and by breaking down organic matter and incorporating it into the earth (e.g., Syers and Springett 1983; Blair et al. 1995). The entomologist C. A. Edwards notes that if soil organic content is maintained, and harmful chemicals are avoided, earthworm populations thrive. "A better understanding of the ecology of earthworms," he adds, "could enable their activities to be manipulated to improve soil fertility" (1983:134–135). Such a biological, age-tested means for creating, enhancing, and tilling the soil—a 600-million-year-old geophysiological technology—was pinpointed with peerless precision by Darwin; yet in the interest of short-term benefits and sky-high profits, soils continue to be treated with abrasive machinery, artificial fertilizers, and toxic chemicals that sooner or later degrade arable lands (see Ehrlich and Ehrlich 1998:243).[17] The impoverishment of soil biodiversity through harsh methods of modern agriculture has been implicated in eventual declines of soil fertility and crop yields (Baskin 1998:113).

Besides their significance for arable lands, earthworms play a critical role in forest soils, especially by preventing the accumulation of leaf litter on forest grounds. In some forest environments, worms consume up to 20 percent of the annual leaf fall—removing organic matter from the surface, breaking it down, enhancing microbial decomposing activity, and thus contributing to the production of humus and the cycling of elements, such as carbon, nitrogen, and phosphorus (see Lee 1985:200–228; Satchell 1983:166–168). Satchell maintains that "in forest soils in which they flourish, earthworms are fundamental to the dynamics of the ecosystem" (1983:168).

Various economic activities damage forests by undermining the soil communities, including the earthworm populations, that support them. Acid rain due to air pollution has had dire effects on forestlands throughout the world—especially in Europe, China, and North America. Acidification kills soil organisms, and is particularly detrimental to earthworms: as a result, forest litter accumulates on the ground, nutrient cycling is retarded, and the aboveground ecosystem eventually suffers (Ehrlich and Ehrlich 1998:148).[18] Moreover, clear-cutting is not only (obviously) detrimental to the aboveground forest flora and fauna, but also to the underground ecosystem vital for a forest's sustenance. Attempts to get new seedlings started on clear-cuts often fail for decades, because the ecological infrastructure of the soil has been undermined beyond viability thresholds—whether by overly efficient vegetation removal, logging skids, or application of fumigants and herbicides (ironically) to assist replanting (Baskin 1998:117).

The chemist M. H. B. Hayes notes that to this day there is insufficient awareness of the importance of earthworms, even though research clearly indicates that soil structure is invariably good when an adequate earthworm population is present (1983:28–29).[19] Darwin's prescience is all the more remarkable when one considers that the constructive role of earthworms for the land continues to be ignored or underestimated—even in the present, when topsoil erosion, soil degradation, and forest declines are widely acknowledged as critical ecological problems.

In the roughly 120 years that have elapsed since its publication, Darwin's last work has proven to be both pathbreaking and current. Contemporary soil and earthworm ecologists have confirmed the accuracy of Darwin's findings regarding the important contribution of worms to the nature of the land. His understanding embedded an ecological perspective and systems thinking decades before the respective fields emerged. If the analysis of this chapter is correct, in his last work Darwin also anticipated key elements of geophysiological science almost a century before its inauguration with the work of James Lovelock and Lynn Margulis. Finally, Darwin's understanding of the ecological role of earthworms speaks to the plight of cultivated and forest lands, and is profoundly current in its conservation implications. As Stephen Kellert has so aptly noted, to reverse the "trend toward the increasing impoverishment of the planet's biological diversity, we will need to acquire a more appreciative attitude toward the biological matrix of so-called 'lower' life forms represented by the invertebrates" (1993:852).

Conclusion

I suggest that the significance of Darwin's last work has been largely overlooked because its interdisciplinary, arguably geophysiological perspective has been incongruent with disciplinary and specialization trends in science. Add to this that in his study of their habits, Darwin argued that earthworms show "some degree of intelligence"—an apparently embarrassing idea for much subsequent behavioral science (see Crist 2002). The fate of the worm book was further sealed by a common human propensity to belittle and even despise invertebrates (Kellert 1993). In these ways, *The Formation of Vegetable Mould, Through the Action of Worms with Observations on Their Habits* has been, to use a colloquialism, "outside the box."

Perhaps it is not hyperbolic to say that the force of this work drives a punch that has made it difficult to acknowledge without disturbing scientific and commonsense orthodoxies. The author of the worm book was not known to abide by either.

Charles Darwin did not study worms because he was an inveterate naturalist who enjoyed tinkering with the soil—although he was that, too. Darwin was not above studying simple organisms like earthworms because his scientific vision was unusually lucid: he was able to unearth the deep knowledge that life safeguards in the most unsuspected recesses of its universe.

Acknowledgments

I would like to thank Lynn Margulis, Duncan Porter, Michael Ghiselin, and an anonymous reviewer for reading earlier drafts of this chapter and providing valuable commentary and criticism.

Notes

1. Cited in Porter 1988:8.

2. It has been key in the understanding of "pedogenesis"—the study of the formation and development of soil. In an edited collection of papers on contemporary developments in earthworm ecology, J. E. Satchell notes that "Darwin's views on earthworm pedogenesis have been fully vindicated" (1983:xi–xii). See also K. E. Lee (1985). On the overall positive reception of Darwin's last book, including reviews in his own day, see Graff (1983). For a review of the 1985 edition of the work, see Porter (1988).

3. Exceptions are Yerkes (1912), Ghiselin (1969), Graff (1983), and Gould (1983, 1985).

4. While Darwin's claim that earthworms "have an enormous range" has proven correct, his wording that they are "found in all parts of the world" is somewhat overstated. Elsewhere in the book, he gives a more accurate portrayal of worms found in "all humid, even moderately humid, countries" (1985:235–236). According to the soil biologist K. E. Lee, "in tropical and temperate regions alike earthworms are among the most widespread of invertebrate animals and are found mainly in the soils of forests, woodlands, shrublands and grasslands, which together cover ... *ca.* 54 percent of the land surface of the earth (1985:179). The ecological function that Darwin attributed to earthworms—that they "plough the land" (1985:313)—has since been both corroborated and made more accurate, by generalizing this role to include all invertebrates and associated microbial groups of the Earth's surface (see Wilson 1987; Lavelle 1996).

5. For these measurements, Darwin relied in part on the help of "a lady on whose accuracy I can implicitly rely" (1985:165).

6. Commenting on this estimate, he wrote that "the above result, astonishing though it may be, seems to me credible, judging from the number of worms that I have sometimes seen, and the number daily destroyed by birds without the species being exterminated" (1985:159). On the close collaboration via correspondence between Darwin and Hensen, see Graff (1983).

7. Worms' triturating effects are mostly significant, Darwin emphasized, at very fine levels of particles—no bigger than can pass through their alimentary canals. Their impact on a diminutive scale is all the more crucial for that very reason: on this scale, agencies like running water or ocean waves have negligible effects, acting "with less and less power on fragments of rock the smaller they are" (1985:257).

8. In the last paragraph of the book, Darwin reiterated his conclusion that worms have played an "important part in the history of the world" (313); by using the expression "history of the world" twice, he indicated that he meant it seriously and not as an offhand turn of phrase.

9. As a terminological caveat, I note parenthetically that I am partial to the term "geophysiology" over "Gaian science." "Gaia" (the "G word," as Lynn Margulis puts it) produces strong reactions among different constituencies in ways that often divert from the substance of the science. On the other hand, the concept "geophysiology" has the virtues of carrying no distracting extrascientific implications, being general enough to include a longer history, and precise enough to capture the idea of the Earth as a living system.

10. On the Gaia hypothesis and Gaian science, see Lovelock (1979, 1987a, 1987b, 1988, 2002); Margulis and Lovelock (1974); Margulis (1981, 1987, 1996, 1998); Margulis and Sagan (1997); Volk (1998). Edited collections on Gaia include Schneider and Boston (1991); Bunyard (1996); Barlow (1991); and Thompson (1987). A good introduction to the history and idea of Gaia is Joseph (1990).

11. The idea of homeostasis has been central in present-day geophysiology: it is defined as the maintenance of conditions by active biotic control, indicating the Earth system's capacity to sustain a range of chemical and physical parameters viable for life as a whole. Darwin's citation of Playfair is interesting for forming an indirect link to contemporary geophysiology. Playfair was a proponent of James Hutton's geological views, and Hutton has been cited as a precursor of the geophysiological perspective. At the 1785 meeting of the Royal Society of Edinburgh, Hutton maintained "that the Earth was a living organism and that its proper study should be geophysiology" (quoted in Lovelock 1988:10).

12. From his observations of worm burrowing, and of the depths they are capable of penetrating, Darwin deduced that worms possess great "muscular power" (1985:188). This is corroborated by recent findings regarding the high protein constitution of earthworms.

13. Darwin's first book, *The Structure & Distribution of Coral Reefs*, in which he describes the different types of coral reefs and explains the origin of their peculiar forms, was also concerned with the geological-level effects of life—in this case, of marine organisms. Darwin was aware of the connection between the themes of his first and last books, and ended *The Formation of Vegetable Mould* as follows: "Some other animals ... still more lowly organized [than earthworms], namely corals, have done far more conspicuous work in having constructed innumerable reefs and islands in the great oceans; but these are almost confined to the tropical zones" (1985:313). See Ghiselin (1984).

14. Prior to the 1881 publication of *The Formation of Vegetable Mould,* Darwin published three papers on the formation mold by worms (in 1837, 1844, and 1869). These papers are reprinted in *The Collected Papers of Charles Darwin*, edited by Paul H. Barrett.

15. Toxic chemicals also adversely affect wildlife by entering the food chain through the worms. Earthworms are composed mostly of protein and therefore are highly nutritious. They are prey for a number of animals, including some amphibians, reptiles, birds (even raptors such as owls and kestrels), and mammals (for example, moles, raccoons, foxes, and badgers) (Macdonald 1983).

16. On the other hand, Edwards et al. note that earthworm populations rise as agriculture moves away from fertilizers and biocides, relying instead on organic methods of rotation and biological sources of fertility and pest control. They predict that the role of earthworms for productivity will become increasingly prominent as trends favoring organic farming continue (1995:186).

17. Plowing destroys worms by digging them up and making them available to predators—especially birds.

18. Environmental analyst Charles E. Little notes that "on the forest floor in the Middle West, and presumably elsewhere, the earthworms are dying" (1997:229). Citing the ecologist Orie Loucks, he writes that parts of Ohio and Indiana subject to acid rain show a 97 percent decline in the density of earthworms (1997: 229).

19. Organic farming, or agroecology, constitutes the exception: the utilization of worms to enhance the nutritional value, and structural efficacy, of soils is known as vermiculture.

References

Barlow, Connie, ed. 1991. *From Gaia to Selfish Genes: Selected Writings in the Life Sciences*. Cambridge, MA: MIT Press.

Baskin, Yvonne. 1998. *The Work of Nature: How the Diversity of Life Sustains Us*. Washington DC: Island Press.

Blair, J. M., R. W. Parmelee, and P. Lavelle. 1995. Influences of earthworms on biogeochemistry. In *Earthworm Ecology and Biogeography in North America*, Paul F. Hendrix, ed. Boca Raton, FL: Lewis Publishers.

Bunyard, Peter, ed. 1996. *Gaia in Action*. Edinburgh: Floris Books.

Carson, Rachel. 1994 [1962]. *Silent Spring*. Boston: Houghton Mifflin.

Crist, Eileen. 2002. The inner life of earthworms: Darwin's argument and its implications. In *The Cognitive Animal*, Marc Bekoff, Colin Allen, and Gordon Burghardt, eds., pp. 3–8. Cambridge, MA: MIT Press.

Darwin, Charles. 1837. On the formation of mould. Reprinted in *The Collected Papers of Charles Darwin*, vol. 1, Paul H. Barrett, ed., pp. 49–53. Chicago: University of Chicago Press.

Darwin, Charles. 1844. On the origin of mould. Reprinted in *The Collected Papers of Charles Darwin*, vol. 1, Paul H. Barrett, ed., pp. 195–196. Chicago: University of Chicago Press.

Darwin, Charles. 1869. The formation of mould in worms. Reprinted in *The Collected Papers of Charles Darwin*, vol. 2, Paul H. Barrett, ed., pp. 135–136. Chicago: University of Chicago Press.

Darwin, Charles. 1984 [1844]. *The Structure and Distribution of Coral Reefs*. Reprinted Tuscon: University of Arizona Press.

Darwin, Charles. 1985 [1881]. *The Formation of Vegetable Mould, Through the Action of Worms with Observations on Their Habits*. Reprinted Chicago: University of Chicago Press.

Desmond, Adrian, and James Moore. 1991. *Darwin*. London: Penguin Books.

Edwards, C. A., P. J. Bohlen, D. R. Linden, and S. Subler. 1995. Earthworms in agroecosystems. In *Earthworm Ecology and Biogeography in North America*, Paul F. Hendrix, ed. Boca Raton, FL: Lewis Publishers.

Ehrlich, Paul. 1991. Coevolution and its applicability to the Gaia hypothesis. In *Scientists on Gaia*, Stephen Schneider and Penelope Boston, eds. Cambridge, MA: MIT Press.

Ehrlich, Paul, and Anne Ehrlich. 1998. *Betrayal of Science and Reason: How Anti-Environmental Rhetoric Threatens Our Future*. Washington DC: Island Press.

Ghilarov, M. S. 1983. Darwin's *Formation of Vegetable Mould*—Its philosophical basis. In *Earthworm Ecology*, J. E. Satchell, ed. London: Chapman and Hall.

Ghiselin, Michael. 1969. *The Triumph of the Darwinian Method*. Berkeley: University of California Press.

Ghiselin, Michael. 1984. Foreword. In *The Structure and Distribution of Coral Reefs*, by Charles Darwin, pp. vii–xii. Tuscon: University of Arizona Press.

Gould, S. J. 1983. Worm for a century and all seasons. In *his Hen's Teeth and Horse's Toes*. New York: W. W. Norton.

Gould, S. J. 1985. Forward. In *The Formation of Vegetable Mould*, by Charles Darwin. Chicago: University of Chicago Press.

Graff, O. 1983. Darwin on earthworms—the contemporary background and what critics thought. In *Earthworm Ecology*, J. E. Satchell, ed. London: Chapman and Hall.

Habermas, Jürgen. 1981. *The Theory of Communicative Action*. Vol. 1, *Reason and Rationalization of Society*. Boston: Beacon Press.

Hayes, M. H. B. 1983. Darwin's "vegetable mould" and some modern concepts of humus structure and soil aggregation. In *Earthworm Ecology*, J. E. Satchell, ed. London: Chapman and Hall.

Joseph, Lawrence E. 1990. *Gaia: The Growth of an Idea*. New York: St. Martin's Press.

Kellert, Stephen. 1993. Values and perceptions of invertebrates. *Conservation Biology*, 7(4), 845–855.

Krumbein, W. E., and A. V. Lapo. 1996. Vernadsky's biosphere as a basis of geophysiology. In *Gaia in Action*, Peter Bunyard, ed. Edinburgh: Floris Books.

Lavelle, Patrick. 1996. Mutualism and soil processes: A Gaian outlook. In *Gaia in Action*, Peter Bunyard, ed. Edinburgh: Floris Books.

Lee, K. E. 1985. *Earthworms: Their Ecology and Relationships with Soil and Land Use*. Sydney: Academic Press (Harcourt Brace Jovanovich).

Little, Charles E. 1997. *The Dying of the Trees.* New York: Penguin Books.

Lovelock, James. 1979. *Gaia: A New Look at Life on Earth.* Oxford: Oxford University Press.

Lovelock, James. 1987a. Geophysiology: A new look at Earth science. In *The Geophysiology of Amazonia: Vegetation and Climate Interactions,* Robert E. Dickinson ed. New York: John Wiley and Sons.

Lovelock, James. 1987b. Gaia: A model for planetary and cellular dynamics. In *Gaia, a Way of Knowing,* William Irwin Thompson, ed. Great Barrington, MA: Lindisfarne Press.

Lovelock, James. 1988. *The Ages of Gaia: A Biography of Our Living Earth.* New York: W. W. Norton.

Lovelock, James. 2002. What is Gaia? *Resurgence,* 211, 6–8.

Macdonald, D. W. 1983. Predation on earthworms by terrestrial vertebrates. In *Earthworm Ecology,* J. E. Satchell, ed. London: Chapman and Hall.

Margulis, Lynn. 1981. *Symbiosis in Cell Evolution: Life and Its Environment on the Early Earth.* San Francisco: W. H. Freeman.

Margulis, Lynn. 1987. Early life: The microbes have priority. In *Gaia, a Way of Knowing,* William Irwin Thompson, ed. Great Barrington, MA: Lindisfarne Press.

Margulis, Lynn. 1996. Jim Lovelock's Gaia. In *Gaia in Action,* Peter Bunyard, ed. Edinburgh, Floris Books.

Margulis, Lynn. 1998. *Symbiotic Planet: A New Look at Evolution.* New York: Basic Books.

Margulis, Lynn, and James Lovelock. 1974. Biological modulation of the Earth's atmosphere. *Icarus,* 21, 471–489.

Margulis, Lynn, and Dorion Sagan. 1997 [1986]. *Microcosmos: Four Billion Years of Evolution from Our Microbial Ancestors.* Berkeley: University of California Press.

Playfair, John. 1956 [1802]. *Illustrations of the Huttonian Theory of the Earth.* Urbana: University of Illinois Press. Facsimile of the original.

Porter, Duncan. 1988. Review of *The Formation of Vegetable Mould, Through the Action of Worms with Observations on Their Habits.* Chicago: Chicago University Press 1985 [1881]. *ASC Newsletter,* 16, 8–9.

Satchell, J. E. 1983. Earthworm ecology in forest soil. In *Earthworm Ecology,* J. E. Satchell, ed. London: Chapman and Hall.

Satchell, J. E., ed. 1983. *Earthworm Ecology: From Darwin to Vermiculture.* London: Chapman and Hall.

Schneider, Stephen, and Penelope Boston, eds. 1991. *Scientists on Gaia.* Cambridge, MA: MIT Press.

Syers, J. K., and J. A. Springett. 1983. Earthworm ecology in grassland soils. In *Earthworm Ecology,* J. E. Satchell, ed. London: Chapman and Hall.

Thompson, William Irwin, ed. 1987. *Gaia, a Way of Knowing: Political Implications of the New Biology.* Great Barrington, MA: Lindisfarne Press.

Volk, Tyler. 1998. *Gaia's Body: Toward a Physiology of Earth.* New York: Copernicus and Springer-Verlag.

Wilson, E. O. 1987. The little things that run the world. *Conservation Biology,* 1(4), 344–346.

Yerkes, R. M. 1912. The intelligence of earthworms. *Journal of Animal Behavior,* 2, 322–352.

15
Gradient Reduction Theory: Thermodynamics and the Purpose of Life

Dorion Sagan and Jessica Hope Whiteside

Abstract

We argue that nonequilibrium thermodynamics, informed by a general version of the second law that applies equally to open and closed systems, connects flow processes of life to nonlife in illuminating ways. Energy delocalizes or, in the words of Eric D. Schneider (see chapter 4 in this volume), "nature abhors a gradient." Investigating how this version of the second law informs the natural organization of complex flow systems, we rethink the teleological status of systems, such as organisms and Gaia, usually covertly or overtly endowed with "purpose" or "mind." Our conclusion is that teleology as found in purposeful organisms, including humans, derives from inanimate flow systems thermodynamically organized "to" (this is their function, their prebiotic physiology, and their materialistic purpose) reduce ambient gradients.

A Fourth Copernican Deconstruction

Although scientifically anchored, this chapter's aim is to sample the breadth of gradient-reduction (nonequilibrium thermodynamic) theory, especially as it applies to Gaia—that is, to Earth's surface considered as a global physiological system. Although lack of space prevents us from being able to present either a detailed description of the gradient-based or open-system thermodynamics on which this chapter is based, or the full range of philosophical implications of such a scientific perspective, we hope to hit on the major points (Schneider and Kay, 1994b; Sagan and Schneider, 2000).[1]

One can provisionally identify four scientific displacements or deconstructions of human's special place at the center of the cosmos;[2] the first three of these are (1) Copernicus's decentering of Earth, (2) Darwin's destruction of humanity's special place above the rest of the animals (a corollary of which is our microbial ancestry), and (3) the disproof of the vitalistic conceit that life is composed of any special stuff, associated both with Fredreich Woeller's 1828 synthesis of urea (an "organic compound") from ammonium cyanate (an "inorganic compund") and the details of nucleosynthesis—the production of the chemical elements of which life is made (carbon, oxygen, hydrogen, nitrogen, sulfur, phosphorus, etc.) in nuclear reactions inside exploding supernovae (Gribbin and Gribbin, 2000).

A fourth deconstruction, we argue, is that life shares its basic process—of being end-directed toward gradient reduction in regions of energy flow—with other naturally complex, materially cycling systems. This fourth deconstruction is based on the philosophical repercussions of the realization that the second law of thermodynamics can be restated for open systems. The most famous descriptions of the second law are the classical, quantitative descriptions of the nineteenth century. They involve the tendency of entropy (originally heat divided by temperature, and later given a statistical formulation) to increase in isolated systems.

However, modern thermodynamics has realized that not all systems (life being the most spectacular example) head inevitably toward equilibrium. The most complex thermodynamic systems in the universe are open to their energetic surroundings. The second law absolutely does not contradict life's tendency to become more organized and regulated over evolutionary time because life is an open system dispersing waste as heat, entropy, and reacted gases into its surroundings. In fact, life, like other natural complex systems, seems not only to have as its most basic function (its natural "purpose") the laying to waste of ambient gradients. It also seems, by forming natural gradient-reducing "machines," positively to accelerate the efficiency with which organized environments are rendered disorganized in accord with the second law. Complex, growing, cyclical (and in the case of life, at least, self-regulating and reproducing), thermodynamic systems tend to be much better (quicker and more efficient) at reducing the gradients that sustain them than random particle interactions.

Thus, despite the quantitative efficacy of entropy as a measure of energy's tendency to become lost in sealed systems, the second law must be stated in a more general and qualitative way that includes open systems. Restated to include open systems, the second law says that energy delocalizes—or, as Eric D. Schneider puts it, nature abhors a gradient. From this point of view life, not just in its matter but also in the essence of its evolutionary process, is a particular historically developed thermodynamic system whose trends from planetary expansion and prokaryotic metabolic innovation to increasingly efficient energy use and even the rise of animal intelligence are all in harmonious keeping with the second law mandate to reduce gradients.

This deconstruction, showing that life is an energetic process deeply related to certain other complexity-building processes, links life to nonlife, matter to mind, and purpose (in its manifold forms, ranging from function and physiology to human intentionality) to the unconscious thermodynamic "computations" of equilibrium-seeking systems, of which humanity (we make the materialistic assumption) is a long-evolved example. Although others have come to similar philosophical conclusions, they have done so on the basis of an entropy maximization law that is demonstrably incorrect.[3] Our aim thus is to show roughly some major philosophical implications of a nonequilibrium thermodynamics which displays purpose (a stumbling block for acceptance of geophysiology, or Gaia, criticized for being teleological) and yet is based on the energetics of nonliving, nonconscious systems. As Lovelock's Daisyworld models have shown (in rebutting such criticism), coordinated thermoregulatory behavior—mistaken for mind, purpose, or complexity requiring extended periods of natural selection—is a natural consequence of organismic growth within constraints (Watson and Lovelock, 1983).

In our view, this is a specific example of a general, thermodynamic phenomenon:[4] complex thermodynamic behavior can be taken (especially out of context) for purposeful behavior or consciousness because nature's thermodynamic systems, engulfed in the genetic systems of life, are at the root of the complex, intelligent-acting behaviors of life. Gaia arose in the recognition of thermodynamic atmospheric disequilibria; any "purposeful" or physiological behaviors displayed by the biosphere, a population of one, must be understood in terms of thermodynamics, not natural selection among competing variants. In short, our proposed fourth Copernican deconstruction goes beyond the stuff of life to the process of life, linking mind and matter, human purpose and purposeful-seeming planetary behavior to physiology and pre-physiological gradient-reducing behaviors.

Nature's Abhorrence of Gradients

Classical and statistical thermodynamics suffer from their characterization of the special (but experimentally easier to observe) case of isolated (energetically and materially sealed) systems as universal. Real complex systems, however, including those which potentially exhibit intelligence (e.g., humans) or a simulacrum of it (e.g., computers), inevitably tend to be open to, and to a large extent defined by, their material and energetic flows. Thus the tendency of improbable matter and energy distributions to settle to equilibrium in isolated containers—prematurely and perhaps egregiously universalized into the notion that the universe is inevitably headed toward cosmic standstill, or "heat death"—continues to impede understanding of the crucial ways in which energy flows organize complex systems. This situation is alleviated by Schneider's rephrasing of the second law (Schneider and Kay, 1989). Significantly, the notion that "nature abhors a gradient," as opposed to "entropy inevitably increases in isolated systems," applies to open systems—and focuses our attention on the flows that sustain and help organize them.

Complex systems (approaching and, in the case of the origins of life, apparently achieving selfhood) tend to appear spontaneously in nature under the influence of appropriate gradients when and where dynamic conditions permit. A gradient is a measurable difference across a distance of temperature (the classic thermodynamic gradient which runs heat engines), pressure, chemical concentration, or other variables. At the limit the difference may not be graded much, as in the clichéd difference between the something (air molecules) and nothing (their lack) of a vacuum, which nature is correctly said to abhor. There are also exploitable economic and mathematical gradients—for example, between rich and poor people, and between actual and probable distributions of playing cards (see below). The value of gradient reduction theory can be seen in its ability to illuminate many widely disparate energy-based systems, including those involving intelligent perceivers actively looking for gradients from which they can profit.[5]

From primordial differences gravitationally manifesting into the major distinction between stars and space (Chaisson, 2001) to temperature and pressure gradients within the protosolar nebula organizing the distribution of chemical elements and compounds (Harder, 2002), differences are exploited by complex systems, generating further differences; although we are acutely aware that a theory which explains everything explains nothing, the notion of difference seems pivotal not only scientifically in terms of measurement, but also in the philosophical realms of ontology and epistemology (e.g., Derrida, 1967; Bateson, 1979).[6]

That nature abhors a gradient restates thermodynamics' second law (Schneider and Kay, 1989; Sagan and Schneider, 2000). Differences in barometric pressure, for example, lead to hurricanes and tornadoes, complex and cyclical processes that dissipate such gradients and vanish when done. Although such gradient-driven, nonrandom, cyclically complex processes may be chemical, biological, and economic (as well as purely physical), their appearance is not assured merely by the existence of a gradient. Kinetic, chemical, and thermodynamic constraints—an appropriate infrastructure—must also be in place before they "pop into existence," exhibiting selflike recursive and teleomatic behaviors. Moreover, the pressure, temperature, electron potential, semiotic, or mathematical gradient must be "just right": if it is too steep, or not steep enough, no complex system will form. Finally, that complex behaviors "eat" gradients of a certain steepness, temporarily and cyclically reducing them (and thus the source of organization for the complex systems themselves), may be at the root of the cyclical (and fundamentally thermodynamic) processes of physiology.[7]

The Goldilocks Paradox and Perception

A tendency to misattribute mystifyingly "mindlike" factors to the emergence of complex systems sensitively reacting to gradients of only certain steepnesses and under certain constraints can be considered a kind of "Goldilocks Paradox"—the complex systems behave as if they recognize or "know" their external gradients (Sagan, 2000).[8] A similar situation, which Harold Morowitz refers to under the general rubric of immanent natural rules of "informatic" matter, and specifically as the nondynamical "noetic" character of the Pauli exclusion principle in quantum mechanics, leads to the know-how of particles to arrange themselves nonrandomly in the elements of the periodic table (Morowitz, 2002). Leaving aside questions of immanence versus transcendence and divinity, it is clear that complex systems in their sensitivity to external conditions may be confused with living and intelligent systems, both directly and by default in cases where complexity seems irreducible and thus explicable only by design. From electrically generated computer algorithms and Belousov-Zhabotinsky chemical clocks kept going by chemical gradients to globally regulating Gaia, kept alive by the solar gradient and its offshoot, the atmospheric redox gradient, complex systems are inevitably "fed" by physical potentials in their surroundings; they do not appear ex nihilo.

Although they span a huge range of complexity, the would-be mysterious systems that elicit awe and a rush toward overly complex explanations are not always particularly complex. If exposed to constraints, nature will fluidly "attempt" to reach equilibrium, sometimes giving the appearance of "choice" or "will," as in the case of a dusted streamer of warm air "seeking" the way out of a leaky house "in order to" come into equilibrium with the cooler air outdoors. Whether near equilibirum or far from equilibrium, such complex behaviors with their ability to mimic "mindfulness" inevitably occur within an environment demarcated by a gradient—a gradient whose existence may go undetected because, although necessary (if not sufficient) for the complex process, it lies just beyond the observer's frame of reference or focus of attention. Epistemologically, ontologically, and eschatologically, the gradient represents preexisting organization, a potential for energy and future activity which must be unlocked by intelligence or its reasonable simulacrum.

James Clerk Maxwell defined dissipated energy as "energy which we cannot lay hold of and direct at pleasure, such as the energy of the confused agitation of molecules which we call heat. Now, confusion, like the correlative term order, is not a property of material things in themselves, but only in relation to the mind which perceives them.... It is only to a being in the intermediate stage, who can lay hold of some forms of energy while others elude his grasp, that energy appears to be passing inevitably from the available to the dissipated state" (Nørrestranders, 1991, 21–22). We would be reluctant to interpret this comment by the inventor of Maxwell's demon to be a lapse into some sort of idealism or non-Cartesian mysticism; it rather seems to us to augur the ultimate necessity of dismantling, on the basis of gradients and their recognition, the wall separating mind (teleology) from body (mechanism).

Life, Gradients, and Information

Gaia, defined as the biosphere (sum of flora, fauna, and microbiota) with physiological properties of thermoregulation, modulation of atmospheric gas composition away from chemical equilibrium, maintenance of pH and salinity of oceans, and possible homeorrhesis of other global variables such as plate tectonics and global persistence of water, does not exist in a vacuum (Sagan and Margulis, 1993; Lovelock, 1979, 1991). Rather, Gaia is a system organized by the gradient formed by 5800 Kelvin incoming solar radiation and 2.7 Kelvin temperature of outer space. The biosphere keeps itself cool. Gaian (and/or abiotic or relatively abiotic) thermoregulation (Schwartzman, 1999) reduces this solar gradient; this physiology-like gradient reduction, moreover, takes place over different temporal and spatial scales—for example, both over the history of the biosphere (probably mainly via eons-long secular increases in the deposition of carbon, reducing atmospheric quantities of carbon dioxide) *and* via evapotranspiration, in which mature tropical ecosystems, producing reflective rainclouds, keep Earth's surface much cooler than it would otherwise be.

The astronomical improbability of stars, an offshoot of our universe's vastly improbable set of physical constants (Smolin, 1997), is "condensed" into the improbability of living and other systems organized by stellar gradients. The biggest gradient on Earth's surface is that created by cyanobacteria, whose photosynthetic growth, using sunlight to liberate oxygen in water, created some 1800 million years ago an energy-rich oxidizing atmosphere (Schopf, 1992). Biospherically, the solar gradient has become a redox gradient supporting all aerobic prokaryotes and eukaryotic cells, including most protoctists and all plants, fungi, and animals. Primarily nonsolar gradient-supported ecosystems such as those at vents deep in the ocean depend on sufide-oxidizing bacteria degrading a redox gradient. Such symbiotic sulfide-oxidizing bacteria permit pogonophorans (giant red tube worms) to thrive without mouths or any sort of feeding and digestion. Such dark gardens, it has been speculated, run on gradients similar to those which fueled life's earthly origins (Corliss, 2001; Russel et al., 1998; Baross and Hoffman, 1985; Shock, 1990).

Gradients are involved not only in Gaian-style biospheric complexity and the origin of life, but also in chemical interactions among organisms, which are open thermodynamic systems, inducing them under certain conditions to organize into societies or "superorganisms." Over evolutionary time gradients help generate genuine new organisms (e.g., eukaryotes from once-free-living prokaryotes; animals from collections of eukaryotic cells). The use of chemical gradients in evolution even represents an archetypal form of referentiality or "language." Cyclic AMP (acrasin), a chemical secreted by ameobas of the genus *Dictyostelium* when they are individually unable to find sufficient food, causes them to aggregate into a throbbing slime mold slug: here the *lack* of a food gradient used by nucleated cells causes individual amoebas to produce a semiochemical that leads formerly independent amoebas to reorganize as a collectivized individual.

More common, perhaps, are gradients signaling the *presence* of food—for example the pheromone trails laid down and then followed by ants to a lump of sugar. Bacteria effectively recognize sugar gradients and swim up them (although the biomechanics may involve sampling and alternating relatively random with relatively directed movements); magnetotactic bacteria even orient themselves according to the electromagnetic gradient of Earth: they swim toward the north magnetic pole. Chemical secretions can stand in for a gradient unperceivable on its own. Various species of bees signal the presence of pollen in distinct ways; "primitive" signalers return to hive with flower fragrance on them, inducing hive mates to search; bumblebees perform elaborate dances to convey the exact position of nectar-rich blossoms (Von Frisch, 1967; W. H. Kirchner, 1993): here the chemical gradient becomes an informational one.

Male silk moths can recognize a female from a single molecule a mile away, and then head in the direction of the concentrated source of such molecules, along an olfactory and sexual gradient. Flying swarms of locusts seem to form from singly feeding cricketlike individuals when neurons on the backs of their legs transform a mechanical signal into a chemical one, perhaps via proteins sensitive to precise mechanical stimulation which lead them to add (or subtract) a phosphate group: phosphorylation (or dephosphorylation).

Human language, too, occurs by material change, such as writing on scraps of bark or by pieces of charcoal—bright colors, whether artificial dye or skin of fruit, we associate with sweet citrus—the fructose gradient of our primate ancestors. The energy of naturally dwindling gradients selects in life for recognition of increasingly subtle cofactors, leading perhaps to referentiality, if not language, long before humans.

Measurable as quantities by science and recognizable as qualities by the living processes which depend on them, gradients as a concept have also, although not under that name, generated thought in philosophy. In a section of the essay "White Mythology" titled "Flowers of Rhetoric," Jacques Derrida, using the polysemy of the word "heliotrope" (both a stone and a flower and, literally, "movement toward the sun") shows that there is no ultimate anchor for the literalness of language (as distinguished from tropes or figures of speech)—except perhaps the sun, which generates subsequent linguistic and symbolic distinctions. The focus on the sun as a master metaphor or primordial signifier itself beyond signification seems to have been lifted by Derrida from Georges Bataille, who in the late 1920s read a book by the essentially thermodynamic thinker Vladimir Ivanovich Vernadsky, translated into French as *La Biosphère*. Significantly, Vernadsky conceived of the biosphere and its living processes as part of a continuum with solar energy fixed and redeployed on Earth's surface (Sagan, 1992).

Gaia and Gaia Theory

Gaia is the good "four-letter word" suggested to James Lovelock by William Golding, his neighbor in Cornwall, Great Britain, to refer to a suite of observations centering on the thermodynamically nonequilibrium status of Earth's atmosphere (Lovelock, 1979). Retained by NASA to explore possible ways to detect life on Mars, Lovelock used the thought experiment of attempting to detect life on Earth from space. It occurred to him that spectroscopic remote gas detection techniques would reveal Earth to have a chemically, and thus thermodynamically, unstable atmosphere. For example, at 1 to 2 parts per million in the air, methane exists at concentrations 100000000000000000000000000000000000 (10^{35}) times higher than would be expected in a closed system with one-fifth reactive free oxygen. This is equivalent to a miracle—similar to receiving several straight flushes in a row in a game of poker dealt from a fairly shuffled pack of cards. (The chances of getting a royal flush are about 1.5 in a million; the chances of shuffling the cards and having them all wind up in factory order are 1 in 10^{68}—beyond merely unlikely.)

Understanding where this "miracle" comes from goes to the core of Gaia and the problems with Gaia theory. Notice that the original scientific interest of the problem is thermodynamic. The nonequilibrium atmosphere is unexpected and mysterious. It is quite far from random. And it is connected to life, which also is unusual in the universe. Lovelock compares the atmospheric chemistry of our planet to a sandcastle on a beach—it may not be life, but it looks created. In fact, both sandcastles and the atmosphere were built by life. Discussions with the American microbiologist Lynn Margulis revealed to Lovelock that methanogens, bacteria teeming in marshes (they are the source of marsh gas) and the rumens of cows, continuously release methane into the atmosphere. As in the disappointment of spectators who discover the simple means of an initially astounding magic trick, the "miracle" in this case was accomplished by nothing more rarefied than farting cattle.

The juxtaposition of grand effect and humble means remains a theme not only of Gaia but of science generally. Gaia, shorthand for a physiological Earth, represents a biosphere whose environment is as nonrandom and teleological as a mammal's body. "She" (for the term Gaia, which shares the Greek root "ge," meaning Earth, with geology and geography, is the ancient Greek Earth mother of the Titans) is a global body able somehow to regulate not only "her" atmospheric chemistry but also global mean temperature, ocean pH and salinity, and other variables that should theoretically—at least according to classical equilibrium thermodynamics—be randomly "shuffled."

On the one hand, the naming of Gaia after a mythological figure, let alone the observed phenomena of biospheric nonrandomness, tends to exacerbate both (scientifically productive) mystery and (scientifically unproductive) mystification, leading to all sorts of cultural ramifications, not the least of which is an archaifying and blending of the geosciences into phenomenology and Earth mother religion (seen as an alternative to transcendental monotheism).

On the other hand, the mystery of Gaia as organismlike can be explained by mundane factors such as microbes whose organizing activities "bleed" or "leak" into the environment, thereby organizing it. Much of the mystery or not of Gaia (liberally and perhaps most fruitfully understood as a physiological, "bodylike" Earth) has to do, again as in a magic trick, with one's context: focused on in isolation, without regard to the thermodynamic cosmic context, the biosphere's properties of "*self*-regulation" and "*self*-organization" (we italicize this word to question it) excite more discomfort in scientific circles—and more satisfaction among relatively uncritical New Age minds—than they otherwise might.

Objections to and Defenses of the Gaia Hypothesis

The objections to and defenses of Gaia theory are instructive, as much for what they say about the research program of a physiological Earth as for the light they shed on the cultural history of ineluctably human science. Among the complexities to be reckoned with are that "the Gaia hypothesis," even in its strict scientific formulation, has undergone changes. Thus, for example, J. W. Kirchner (1991) identifies multiple versions of Gaia, from a "strong" version that "Earth is a living organism," which the analytic philosopher maintains is untestable (and likens to the Shakespearean line that "all the world is a stage") to weak versions which amount to a claim of coevolution of life and the environment, which he dismisses as true but already known and relatively trivial. This "divide and conquer" rhetorical ploy is reminiscent of Arthur Clarke's Law of Revolutionary New Ideas: "All revolutionary new ideas ... pass through three stages, which may be summed up by these reactions: 1. 'It's crazy—don't waste my time'; 2. 'It's possible, but it's not worth doing'; and 3. 'I always said it was a good idea' "(Clarke, 1972). In other words, Gaia in its strongest form—that Earth is an organism—is crazy, while in its weaker forms of environmental-organismic coupling it is trivial. If and when a stronger form becomes accepted, according to this trajectory it will be portrayed as unrevolutionary.

Biospheres and Closed (Ecosystemic) Versus Open (Organismic) Systems

However, even at this relatively superficial level of analytical philosophical rhetoric, an interesting defense can be, and was, made for the strong form. Assuming reproduction is the signal trait of organismhood, it was argued in the form of a thought experiment that the development of closed ecosystems would represent de facto reproduction of Gaia (Sagan, 1990). Since it is easy to imagine technologically enclosed biospheres—more advanced versions of Biosphere II, the failed human experiment in creating a giant enclosed ecosystem near Tucson, Arizona—separated from Earth on Mars, in orbit, or even in spacecraft, the organismic status of Gaia, taken as the planetary biota and its environment, was considered proven. Moreover, this thought experiment illuminated the "deep ecological" perspective that the global life-form was transhuman, since in any currently imaginable technological scenario, the only way for global life to reproduce would be by carting into space recycling systems including edible plants, waste-recycling bacteria and fungi, and living green matter from cyanobacteria to vegetables to produce oxygen and take in carbon dioxide. Humans and technology, in other words, while necessary for present global reproduction, are only part of the alleged "Gaian superorganism" (Kelly, 1995; Margulis and Sagan, 1997).

However, we now find this defense interesting more for its implication of humans as unknowing participants in a biology-like extraterrestrial expansion of Earth life—the "budding" or "sporulation" of Gaia, and our human involvement in this "strange brood"—than in the technical claim of Gaia as an organism. Thermodynamically, Gaia does not qualify as an organism because the global ecosystem is largely a closed rather than an isolated or open system. Open systems in thermodynamics, whether construed as near or far from equilibrium, include all actively living cells and organisms made of cells, and always enjoy an influx of materials; organisms require incoming sources of carbon, hydrogen, nitrogen, sulfur, phosphorus, and oxygen to maintain, grow, and reproduce their bodies.

Closed systems, of which Gaia is a good example, are (again, in thermodynamics) closed to material but open to energetic influx. In the case of the biosphere, including the deep, hot biosphere, energy is provided from above the terrestrial envelope in the form of sunlight and from below in the form of chemical gradients, such as the energetic difference between sulfide- and oxygen-feeding bacteria. As in ecosystems, or their desktop simulations—glass-enclosed ecospheres containing shrimp and algae once sold as novelties—the matter of the living Earth (except for occasional meteorites coming in and astronautic stuff going out) forms basically a closed system. At present, as technological humanity is finding out, there is limited room to grow: if there were room, we would not be faced with the ethical quandaries of territorial violence, eating other intelligent animals, and so on. Perhaps only if the biosphere did evolve sufficient ecotechnoscience to actively funnel matter into growth of viable miniature biospheres would we be justified in considering it an organism, *sensu stricto*. Until then it remains a superecosystem, a biosystem or biosphere with physiological properties whose origins must still be addressed.

Doolittle's Objection

Other interesting objections to and defenses of the Gaia hypothesis include those of the neo-Darwinists Richard Dawkins and W. Ford Doolittle. Doolittle, a

Canadian molecular biologist, in a paper titled "Is Nature Really Motherly?," ridiculed the notion that there might be a concerted network of interactions among diverse life-forms, thereby somehow ensuring global regulation of environmental variables (Doolittle, 1981). Note that, as in early objections to continental drift by plate tectonics, the phenomenon (global physiology, like continental drift) is dismissed because there seems to be no reasonable explanatory mechanism (Lovelock's cybernetic links among organisms being as unpalatable as plate tectonics). Doolittle spoke dismissively of a "secret consensus" among organisms, suggesting the absurdity of late-night committees of organisms coming together to discuss their common interests. Perhaps in tacit rebuttal Lovelock remarked that Gaia is no doting nanny but has all the sympathy for humanity of a microprocessor in the warhead of an intercontinental nuclear missile.

Cybernetics, the computer-based discipline inaugurated by Norbert Weiner (1948) to study the nature of control in organisms and machines, came to the fore in Lovelock's attempts to provide an acceptable scientific mechanism for the seeming miracle of global physiology. Cybernetic mechanisms, through positive and negative feedbacks, can amplify or attenuate trends automatically. Thus, what appeared to be intelligence and unified organismhood could accrue without any secret committees or personified collusions among presumably mindless organisms. There was neither an implicit motherliness to nature nor anything more mindful than could be found in computers.

Despite the expediency and versatility of cybernetics as the Gaian mechanism of choice, however, it may have been flawed. A thermodynamic analysis, for example, would distinguish sharply between the ideal case of a machine as an isolated system, inevitably coming to equilibrium in accord with the second law, and the real status of organisms as open systems, indefinitely postponing their tendency to return to an equilibrium state by making more of themselves. In other words, the machines and organisms conflated by cybernetics in its rush to understand control, contrasts rather dramatically with distinctions between classical thermodynamics based on a study of steam engines and nonequilibrium thermodynamics attempting to understand life.

The Cybernetic Turn and Abiotic Thermoregulation

We would also suggest that the recent turn in Gaia science away from cybernetics to natural selection (e.g., Lenton, 1998; Harding and Lovelock, 1996) represents a turn both toward a more orthodox (and thus scientifically and politically acceptable) explanatory principle and a turn away from cybernetics (spawned by control mechanisms in ballistic missiles), away from a too machine-focused science. Nonetheless—and we are not intelligent enough to say precisely how—we sense that cybernetic feedback behaviors, insofar as they exist in the natural world beyond the realm of human engineering, stem from autocatalytic networks feeding on gradients.

For example, Bénard-Rayleigh convection cells—hexagonal structures that appear on the surfaces of substances such as spermaceti (a waxy solid from whales), silicone, and sulfur hexafluoride gas that are exposed to temperature gradients within a certain steepness range (and that range only)—may exhibit thermoregulatory behaviors (Koschmieder, 1993). The phase transition from disordered conduction of heat to organized convection occurs at certain nondimensional numbers. At this critical point, associated with the difference in temperatures between the top and bottom of the liquids, heat transfer suddenly becomes more efficient: the system's convectional complexity, accelerating the rate of heat loss, appears to readjust itself to dissipate the gradient more effectively. The appearance of more efficient convection when the temperature below is raised, indicates an ability of the inanimate system "to cool itself." Do we see here a thermoregulatory mechanism that owes nothing to life (let alone natural selection)—a thermoregulatory system homologous to Gaia's alleged temperature control of the planet, which shows multiple signs of cooling itself for hundreds of millions of years in the face of increasing luminosity from the sun?

Dawkins's Objection and Daisyworld

The British zoologist Richard Dawkins, a staunch neo-Darwinist and defender of evolutionary theory against what he sees as religious or pseudoscientific threats, objected to Gaia on the grounds that a physiological Earth might be plausibly postulated only if it could have evolved, like animals, by natural selection (Dawkins, 1982). But since, Dawkins reasoned, there is only one living planet, it could not in principle have evolved by natural selection, which by definition requires competition among variants. If, Dawkins further suggested, Earth was but one of many living planets, some of which had not survived, competing in our solar system—if this solar system were "littered" with imperfectly physiological planets—then, he allowed, there would be the possibility of a physiological planet.

As with Doolittle, and earlier objectors to now-accepted plate tectonics, we see the failure of logic and seeming common sense when confronted with a phenomenon that has no obvious mechanism. The situation is analogous to a spectator witnessing a magician make a coin completely disappear and disavowing the mystery because he is not privy to the method. Of all the objections to Gaia, Dawkins's is at once the most interesting and the most cavalierly symptomatic of the limitations of reductionist science.

If the suggestion above of a protophysiological thermoregulatory Bénard cell in the total absence of life and complex chemistry—let alone reproduction, genetics, or natural selection—is not enough, many more examples can be adduced. The putative ancestor of life, if considered the first single cell, also cannot, any more than the present biosphere, be explained cogently by natural selection: in both cases the complex phenomenon is selflike, even a self, yet a population of one. Other cyclical selflike systems appear in the neighborhood of gradients, increasing our suspicion that all selves may not owe their existence to natural selection.

For example, hurricanes (often given first names) appear from gradients; their complexity and cyclicity have nothing to do with natural selection and everything to do with the formation of locally improbable gradients whose "job," in thermodynamic terms, is to destroy a preexisting improbability. "Whirlpool"—that is its name—is a permanent cycling eddy downstream of Niagara Falls. And chemical clocks, such as Belousov-Zhabotinski reactions, show intricate and unexpected patterns that grope toward individuality and selfhood as they reduce electron potential gradients. Although they do not reproduce, and therefore do not produce variants which can be naturally selected, they do grow and they do show complexity which, if observed out of context, would no doubt seem mysterious and unexpected.

Lovelock's response to Dawkins was to show how a model of a planet, consisting only of daisies of light and dark hue, would, with very simple biological assumption of growth (no natural selection!), thermoregulate a planet exposed to increasing luminosity from its sun (Watson and Lovelock, 1983). The albedo of light daisies growing in clumps tended to reflect light, thereby cooling the planet when it got too hot; the albedo of the dark daisies tended to absorb heat when the sun was proportionately less luminous. Together the daisies raised planetary temperature by absorbing more radiation in the sun's early years (stars are thought to become more luminous as they age) and reflecting more when the star might have overheated the planet. Because the clumps of daisies died when it got too hot or too cold, the thermoregulatory effect was not perfect, but operated only within a certain temperature range.

One can glean how the perfectly credible biological assumption of growth within a certain temperature range translates into thermoregulation, homeostasis, or homeorrhesis at a planetary scale. Ultimately, we would argue, it is the growth properties of the daisies—analogous if not homologous to Bénard-style complexity, appearing only within the window of a certain steepness of gradient—that confers the complex "physiological" phenomenon of thermoregulation on the planet. It is a phenomenon that Dawkins must dismiss because it seems to him, as it does to Doolittle, too mysterious to be explained by natural selection, which cannot be operating either on the lone planet or on the nonreproducing daisies. Parenthetically, one might argue, because such concerted planetary behavior, in the absence of natural selection, seems to call forth references to mystical directing powers, that Daisy World satisfies the Turing Test—a computer program that, in retrospect anyway, mimics the behavior of a teleological entity, either conscious or physiological, whose behavior is in fact a simple extrapolation of the growth properties of daisies—at least as far as Dawkins is concerned.

In real life, Gaian global cooling has been postulated to involve coccolith algae that grow in sunlight and emit sulfur gases that serve as condensation nuclei for raindrops depriving the algae of sunlight—a negative means of planetary "air-conditioning." Perhaps more obvious, if less studied, is the role of evapotranspiration: clouds appear regularly over rain forests exposed to high incident radiation. In this way areas such as Amazonia, with highly evolved ecosystems, cool themselves and reduce the gradient between hot sun and 2.7 Kelvin space. Life, as Lovelock has repeated, likes it cool—and cooling at the planetary surface via tree-produced cloud cover necessarily entails dissipation of heat farther out. The situation is symmetrical to a room heated up by a refrigerator. The sun is like the plug: if we were unaware of it, the cooling of the magic icebox would indeed be miraculous.

The Biological Anthropic Principle

A final objection to the Gaia hypothesis of which we are aware was made in passing by Stephen Jay Gould at a colloquium. Asked about the peculiar habitability of the Earth going by the name of the Gaia hypothesis, Gould replied that, were we not gifted with a supportive environment, we would not be around to marvel at the question of the fine tuning of the environment. He added that if the colloquium were attended not by humans but by octopi, the question might be raised as to the basis of our near-miraculous possession of eight lovely arms.[9] Gould's response was consistent with his view of contingency in evolution: that if it were "replayed"—if the genetic deck of speciation were reshuffled, one might say—the chances of humans reevolving, or of intelligence reappearing, would be nil. Gould's dismissal of the Gaia hypothesis can also be seen as a biological version of the anthropic principle in physics, the weak version of which says that were the universe not so perfectly suited for the evolution and emergence of conscious life, conscious life would not be present to marvel at it.

Again, looking at Gaia as a magic trick, this seems to be the equivalent of accepting as trivially not in need of explanation a feat of surpassing improbability. Doolittle agrees: "If the fitness of the terrestrial environment is accidental, then is Lovelock not right in saying that for life to have survived to reach the stage of self-awareness 'is as unlikely as to survive unscathed a drive blindfold through rush-hour traffic?' I think he is right; the prolonged survival of life is an event of extraordinary low probability. It is however an event which is a prerequisite for the existence of Jim Lovelock and thus for the formation of the Gaia hypothesis.... Surely if a large enough number of blindfold drivers launched themselves into rush-hour traffic, one would survive, and surely he, unaware of the existence of his less fortunate colleagues, would suggest that something other than good luck was on his side" (Barlow, 1992, p. 33).

What we wish to stress, however, is not so much the improbability of Gaia as its natural appearance as a gradient-breaking structure from the improbable gradients of space. Gaia need not be one of many failed systems but, rather, a low-entropy gradient reducer fomenting external chaos in tune with the second law as it builds up internal complexity and history, "concentrating" improbability.

Nonequilbrium Thermodynamics and Extending the Second Law

A better way of understanding global physiology (and other examples of apparently inexplicable complexity or intelligent design) may be to return Gaia to its roots in nonequilibrium thermodynamics. Thermodynamics is notoriously confounding, in part because its conclusions of a universal tendency toward equilibration seem to contradict complexity and evolution, and in part because the mathematical equations for thermodynamic and information theory entropy, in addition to using the same term, are formally similar. There are many formulations of the second law, but the basic idea was formalized by Sadi Carnot in his miliatrily motivated attempts to improve steam power to battle the British navy and industry. Carnot pointed out that it was not simply the temperature of the steam-producing boiler that made pistons pump hard and fast in an engine, but rather the difference between the temperatures of its hot boiler and cooler radiator. "The production of heat is not sufficient to give birth to the impelling power. It is necessary that there should be cold; without it, the heat would be useless" (Guillen, 1995, p. 179).

The second law of thermodynamics, later understood in Boltzmann's statistical mechanics as matter's tendency to drift into states of increasing probability—there are more ways, for example, for cream particles in your coffee to be mixed with coffee than there are for them to be separated—linked Newtonian mechanics to the phenomenological observations of inexorable loss, decay, forgetting—thus producing, in Eddington's words, "the arrow of time" (Blum, 1968, pp. 5–6). Although this derivation is itself problematic—in infinite time even very unusual arrangements would be repeated an infinite number of times—it preceded evolutionary theory's equal, if opposite, linear time-based view of the cosmos. However, in thermodynamics the projected end of the cosmos was one of spent embers, with no energy left available for work, the "heat death" of the universe.

Thermodynamic Biology and the Purpose of Life

The rectification of the second law's degradation with life's complex maintenance and evolution was broached by Schrödinger, and major contributions to the physics of biology were made by Lotka, Vernadsky, Prigogine, Odum, Lovelock, Morowitz, Wicken, and Schneider. Morowitz, for example, in what is sometimes described as "a fourth law" of thermodynamics, argues that "In the steady state systems, the flow of energy through the system from a source to a sink will lead to at least one cycle in the system" (1979, p. 33). It is crucial to realize that the second law generalized the move to equilibrium in isolated systems, taking a very contrived and artificial experimental condition and applying it far beyond its ken.

In fact, we now see that stars, Bénard cells, Taylor vortices (which occur in counterrotating pairs as a result of rotational pressure gradients), whirlpools, dust devils, hurricanes, chemical clocks, and other nonliving selflike systems crop up spontaneously and grow (like the daisies of Daisy World) in response to ambient gradients.

Earth is cooler than a simple interpolation between Mars and Venus (whose atmospheres are "reacted-out" mixtures of mostly carbon dioxide) would suggest: Gaia itself is a giant gradient-reducing system. Wicken argues that "Thermodynamics is, above all, the science of spontaneous process, the 'go' of things. Approaching evolution thermodynamically allows us to bring the 'lifeness' of life into the legitimacy of physical process ... the emergence and the evolution of life are phenomena causally connected with the Second Law" (Wicken, 1987, p. 5). The genetic mechanisms of replication and reproduction provide, in Wicken's view, "stable vehicles of degradation" for ambient gradients to be reduced. At the same time, the randomizing effects of the second law inevitably disturb the copying process of the chemical Rube Goldberg machines which are living things, taking available energy from their environment and using it (up) not only to maintain and grow their structure, but also to seek out new gradients upon which their existence as selves, as forms depending on a whirling flux of materials, inevitably depends. Evolution, in this view, is second not only to selfhood but also to thermodynamic processes conferring metastability in the coherent areas of matter-degrading ambient gradients. These open systems are low-entropy and highly organized—indeed, organisms—within their frame because they are helping to randomize the surroundings outside their open bodies.

Which brings us to the precipice, or rather foothill, of the great and scientifically frightening edifice of teleology. Why is life? What is its purpose? The reader familiar with the literature of Gaia, or research funding for geophysiological studies within biology, will discern that the link to teleology has long been a thorn in the side of Gaia studies. But thermodynamics suggests a way around this impasse. Organisms as cells and bodies have a natural physiology: they exist "to" break down gradients in much the same way that lungs exist "to" take in air or the heart exists "to" pump blood. Indeed, the future orientation of beings whose genetic makeup presumably evolved piggyback on imperfectly reproducing vehicles of gradient degradation becomes naturalistic in a Gaian-thermodynamic view.

Organisms are, in Kantian language, "natural purposes" (Wicken, 1987) whose means are wrapped up with their ends in functional closure, autocatalytic chemical organization, and energetic and material openness to the environment. Humans have proliferated relative to other primates in large part due to a combination of neural plasticity (Skoyles and Sagan, 2002) and complex social relations (mediated by language) whose net result is a much enhanced ability to identify and deploy the food and other gradients necessary to move agricultural and technical civilization into the material evolutionary form which is humanity. Despite civilization and classical music, a disproportionate amount of waking human life is devoted to thoughts and activities revolving around the procuring of food, the finding of mates, and the making of money—activities necessary to maintain and perpetuate a particular form of genetically undergirded gradient-reducing organization.

Other species obey the same thermodynamic imperative arguably behind the processes of life's origination, growth (increase in biomass), reproduction, increase in respiration, energy efficiency, number and types of taxa (biodiversity), rates of circulation of elements, numbers of elements involved in biological circulation, and increase in intelligence (which identifies new gradients to be exploited and means of escaping the pollution which inevitably and thermodynamically accompanies rapid growth). The thermodynamic imperative or arrow thus points ahead, if not specifically in the direction of humanity; the teleology exists, but is prosaic. We thus are partially in accord with Ernst Mayr (1982), who distinguishes between the second law as teleomatic, the evolved physiology of animals as teleonomic, and conscious awareness as teleological. We see the teleological (so defined) as an outgrowth of the teleomatic; we would disagree, however, with the notion that the second law's status as law with regard to life is no different from the law of gravity.

The second law's character appears (at least proximately) to be more foundational to living teleology, and indeed provides the impetus for resisting the effects of gravitation in flight and motility by gathering, via biochemistry, energies from the environmental surround. The material purpose of life is to degrade the solar gradient (and perhaps this is connected with any "higher" purpose it may have). The tendency to retreat into ideational realms of mathematical or religious "ultimate reality" (e.g., ideas of heaven) during hard times may also reflect a thermodynamic tendency—the panbiological tendency to "shut down" (thus preserving a given gradient-reducing material form); in neurally plastic humans, this tendency may manifest itself in a relative foregrounding of previously imaginary realms and a correlated willingness to die for the social collective. Thus a thermodynamically based teleology, while at first glance seemingly allied to scientifically taboo thoughts of religious purpose, in fact maintains a materialism more uncompromising than Cartesian dualism—which would, perhaps for ultimately practical reasons, bracket all purpose and free will, fencing it off with divine authority in a "humans-only" realm.

This fourth Copernican deconstruction, refusing to consider special human mindlike computational and perceptive processes, allies the conscious teleology we perceive in ourselves to the nonliving realm of complex thermodynamic processes. Here we can discern that part of the sociopolitical problem with science's reception of Gaia has been the perceived vitalism of the hypothesis. But in a thoroughgoing thermodynamic worldview, inanimate matter already displays teleological behavior, precisely that of "seeking" gradients to come to equilibrium: the "teleological" behavior of a biosphere, acting as if it "knew" its surrounding environment by sensing and reacting to it, thus becomes a moot argument against the existence of Gaia or Gaia-like processes.

The Processes Themselves

And yet, our consciousness, our perception, may be at least in part an elaboration of such equilibrium-seeking, distorted by our need to feed on available gradients to maintain ourselves (or our relatives, associates, or children) as stable vehicles of degradation. Here we would have to disagree with the "candidate fourth laws" put forth by Stuart Kauffman (2000), who argues the need for a thermodynamic explanation of biology and technology. But the complexity of biology and technology, so dependent on energy and so productive of waste and pollution, is directly related to their status as nonequilibrium vehicles of degradation. Why invent a complex fourth law (and, moreover, one which applies disproportionately to life and technology) when the second law—a law which, as we stress, was originally based on the special case of isolated rather than the general case of open systems—can simply be extended?

Here we must accede both to Occam's razor and the connecting spirit of Darwinian evolution to chose a simpler, more general principle over a more complex and ad hoc one. As Nobel laureate Stephen Weinberg said, science rests on the discovery "of simple but impersonal principles."[10] We nominate Schneider's extension of the second law into a gradient-destroying tendency as such a principle. We see no reason why "explanations" of complexity should, instead of simplifying, apprentice themselves to the complexity of the phenomena they purport to explain. On the other hand, the iteration of the entire universe from simple, mindlike (computer algorithmic) rules by Stephen Wolfram (2002) seems to us to err in the opposite direction, and to commit the original thermodynamic sin of generalizing a highly specific situation (now computers, then the behavior of energetic systems in closed adiabatic containers) and applying it precipitously to the entire universe. The mathematician-philosopher Edmond Husserl, in founding phenomenology, advocated a "return to the things themselves." Similarly, those who study complex processes should return to "the processes themselves"—only a subset of which appear on computer screens.

Life and Nonlife

Gradient-based thermodynamics links life to nonlife, and linguistic, conscious human teleology to inanimate purpose in nature. Because complex material processes arise and persist to degrade gradients, and because thinking organisms represent a genetic example of such a process, there is a natural link between the behavior of matter (gradient reduction) and of mind (gradient perception). Gradient reduction theory thus would seem to further science's historical trajectory of linking us to the rest of the physical universe. If we are not at the center of the universe, if the atoms of our bodies are not special but common star stuff, our information- and energy-handling abilities also have a cosmic context. Taylor vortices jump to new states dependent upon their past history—they show a fledgling memory.

Parsimony suggests our animate purpose has roots in thermodynamic teleology. If this is the case, then the prosaic purpose of life can be understood. Our desires for food, sex, power, and money reflect us as open selves connected to growing nexuses involved in gradient destruction. The ability to perceive new gradients must have conferred huge evolutionary advantages, selecting for intelligence. A blackjack player counting cards recognizes statistically unlikely preponderances of high cards and aces, and puts his money down in larger bets in the hope that the numerical playing card gradient will reduce itself, as is its statistical wont (Griffin, 1999). Arbitrageurs buy cheap in one place and sell dear in another, selecting for global communications and means of commodity transfer.

Indeed, commodification itself, the transformation of a desired and expensive luxury into a cheap and available product (and sometimes necessity), may be understood as a reduction in supply-demand gradients. The belief systems of societies, whose members feel kinship on the basis of interpretations and signs, and which battle each other, sometimes to the death, for access to resources, are perhaps also open to fruitful analyses in terms of gradient reduction theory. Gradient-based thermodynamics shows much promise for a variety of fields, including economics, evolutionary theory, ecology, and, of course, further Gaia studies, which began in James Lovelock's recognition of chemical atmospheric disequilibria.

Acknowledgments

The authors would like to thank Eric D. Schneider for inspiration and discussions, and acknowledge support from an NSF Graduate Research Fellowship to JHW. This is Lamont-Dohesty Earth observatory contribution number 6562.

Notes

1. A more complete discussion of some of the topics discussed here will be available in Eric D. Schneider and Dorion Sagan, *Energy Flow*, forthcoming, University of Chicago Press.

2. Copernican heliocentrism and vitalism's demise are arrayed with Darwinian evolution in a preliminary series of major scientific deconstructions in Margulis et al. (2002).

3. For example, Rod Swenson (1997), recognizing the inanimate basis, in thermodynamic behavior, of what has too readily been linked anthropocentrically with life and cognition, contrasts "Descartes' dualistic world [which] provided the metaphysical foundation for the subsequent success of Newtonian mechanics and the rise of modern science in the seventeenth century [but which defined] psychology and physics ... by their mutual exclusivity" with the energetically based "active, end-directed, or intentional dynamics of living things," errs in promulgating a "Law of Maximum Entropy Production." (pp. 217, 221–225). Entropy is neither maximized nor easy to measure in many complex open systems.

4. The natural computing functions of cycling thermodynamic systems unconsciously "seeking" equilibrium (but instead forming highly complex processes when their foundational gradients are maintained) may be the most interesting (if not elsewhere mentioned) example of complexity theorist Stephen Wolfram's (2002) claim of a universal equivalence of computing abilities among natural systems which are not "obviously simple." We would argue, however, that computer algorithms, far from generating real-world structures, let alone the second law of thermodynamics, represent a subset of natural energetic equilibrium-seeking processes (Sagan and Whiteside, 2002).

5. Computer technology consultant Peter Bennet's (1998) story "Jamie the Prospector" is about a financial wizard who uses a computerized trading system that instantaneously identifies disparities among stocks, commodities, bonds, and currencies. Jamie's computer system translates the price disparities into a three-dimensional cyber landscape over which Jamie flies in virtual reality; by using a joystick, he levels the hilly regions, which represent gradients. On the eve of the new millennium, the Far East shuts down its financial exchanges to avoid mishaps due to the projected year 2000 computer glitch. A great hilly region appears, which Jamie quickly levels, pocketing hundreds of millions of dollars in a matter of minutes.

6. Derrida's (seen but not heard) use of the term *différance*, as well as the importance of similar differences in Heidegger, Wittgenstein, Bohm, and others, can be found in the web article *Tracing the Notion of Difference* at http://tyrone.differnet.com/experience/append.htm.

7. Gradients are necessary but not sufficient for complex teleological, teleonomic, or teleomatic behaviors (Lenton and Lovelock, 2000).

8. The "just rightness" of the steepness of gradients, leading to the sensitive appearance of complex behaviors only under certain conditions, can be (and has been) mistakenly assumed to mean that human-style conscious awareness must be in the vicinity (Sagan, 2000).

9. The comment was made at a colloquium organized by Richard Lewontin at Harvard University in the early 1980s.

10. Weinberg's "simple but impersonal principle" statement is from his lecture at the nineteenth annual Key West Literary Seminar, Science & Literature: Narratives of Discovery, January 11–14, 2001.

References

Barlow, C. C., ed. 1992. *From Gaia to Selfish Genes: Selected Writings in the Life Sciences*. MIT Press, Cambridge, MA.

Baross, J. A., and S. E. Hoffman. 1985. Submarine hydrothermal vents and associated gradient environment as sites for the origin and evolution of life. *Origins of Life*, 15:327–345.

Bataille, G. 1988. *The Accursed Share: An Essay on General Economy*. vol. 1, *Consumption*, Robert Hurley, trans. Zone Books, New York.

Bateson, G. 1979. *Mind and Nature: A Necessary Unity*. Dutton, New York.

Bennet, P. 1998. Imprint: random thoughts and working papers (Global Business Network), 2:1–11.

Blum, H. F. 1968. *Time's Arrow and Evolution*. Princeton University Press, Princeton, NJ.

Chaisson, E. 2001. *Cosmic Evolution: The Rise of Complexity in Nature*. Harvard University Press, Cambridge, MA.

Clarke, A. C. 1972. *Report on Planet Three and Other Speculations.* http://public.logica.com/~stepneys/cyc/l/law.htm.

Corliss, J. B. 2001. *Life Is a Strange Attractor: The Emergence of Life in Archaean Submarine Hot Springs.* http://www.syslab.ceu.hu/corliss/Origin_of_Life_Intro.html.

Dawkins, R. 1982. *The Extended Phenotype.* W. H. Freeman, San Francisco.

Derrida, J. 1980 [1967]. *L'Écriture et la Différance: Writing and Difference,* Alan Bass, trans. University of Chicago Press, Chicago.

Doolittle, W. F. 1981. Is nature really motherly? *CoEvolution Quarterly,* 29:58–63.

Gribbin, J., with M. Gribbin. 2000. *Stardust: Supernovae and Life—the Cosmic Connection.* Yale University Press, New Haven, CT.

Griffin, P. 1999. *The Theory of Blackjack,* 6th ed. Huntington Press, Las Vegas.

Guillen, M. 1995. *Five Equations That Changed the World: The Power and Poetry of Mathematics.* Hyperion, New York.

Harder, B. 2002. Water for the rock: Did Earth's oceans come from the heavens? *Science News,* 161:184.

Harding, S. P., and J. E. Lovelock. 1996. Exploiter mediated coexistence and frequency-dependent selection in a geophysiological model. *Journal of Theoretical Biology,* 182:109–116.

Kauffman, S. A. 2000. *Investigations.* Oxford University Press, Oxford.

Kelly, K. 1995. *Out of Control: The New Biology of Machines, Social Systems and the Economic World.* Perseus Press.

Kirchner, J. W. 1991. The Gaia hypotheses: Are they testable? Are they useful? In *Scientists on Gaia,* S. H. Schneider and P. J. Boston, eds. MIT Press, Cambridge, MA.

Kirchner, W. H. 1993. Acoustical communication in honey bees. *Apidologie,* 24(3): 297–307.

Koschmieder, E. L. 1993. *Bénard Cells and Taylor Vortices.* Cambridge University Press, Cambridge.

Lenton, T. M., and J. E. Lovelock. 2000. Daisyworld is Darwinian: Constraints on adaptation are important for planetary self-regulation. *Journal of Theoretical Biology,* 206:109–114.

Lenton, T. M. 1998. Gaia and natural selection. *Nature,* 394:439.

Lovelock, J. E. 1979. *Gaia: A New Look at Life on Earth.* Oxford University Press, Oxford.

Lovelock, J. E. 1991. *Healing Gaia.* Harmony Books, New York.

Margulis, L., and D. Sagan. 1997a. *Slanted Truths: Essays on Gaia, Evolution and Symbiosis.* Copernicus Books, New York.

Margulis, L., and D. Sagan. 1997b. *Microcosmos: Four Billion Years of Microbial Evolution.* Berkeley: University of California Press.

Margulis, L., D. Sagan, and J. H. Whiteside. 2002. From cells to cities: The evolution of composite individuality—early life leaves its record in the rocks. *Kosmos: Review of the Alexander von Humboldt Foundation,* July:23–24.

Mayr, E. 1982. *The Growth of Biological Thought.* Harvard University Press, Cambridge, MA.

Morowitz, H. 2002. *The Emergence of Everything: How the World Became Complex.* Oxford University Press, New York.

Morowitz, H. J. 1979. *Energy Flow in Biology: Biological Organization as a Problem in Thermal Physics.* Ox Bow Press, Woodbridge, CT.

Nørrestranders, T. 1991. *The User Illusion: Cutting Consciousness Down to Size,* Jonathan Sydenham, trans., pp. 21–22. Viking, New York, 1991.

Russel, M. J., A. J. Hall, A. G. Cairns-Smith, and P. S. Braterman. 1998. Submarine hot springs and the origins of life. *Nature* 336:117.

Sagan, D. 1990. *Biospheres: Reproducing Planet Earth.* Bantam, New York.

Sagan, D. 1992. Metametazoa: Biology and multiplicity. In *Incorporations (Zone 6; Fragments for a History of the Human Body),* Jonathan Crary and Sanford Kwinter, eds., pp. 362–385. New York, Zone Books.

Sagan, D. 2000. The Goldilocks paradox: Thermodynamics and Gaia. In *Abstracts Guide, 2nd Chapman Conference on the Gaia Hypothesis.* American Geophysical Union, Valencia, Spain, June 19–23. American Geophysical Union, Washington, DC.

Sagan, D., and L. Margulis. 1991. Gaia: A "good four-letter word." *Gaia* magazine, 3:4–6.

Sagan, D., and L. Margulis. 1993. Gaia hypothesis. In *McGraw-Hill Yearbook of Science & Technology 1993.* McGraw-Hill, New York.

Sagan, D., and E. D. Schneider. 2000. The pleasures of change. In *The Forces of Change: A New View of Nature,* pp. 115–126. National Geographic Society, Washington, DC.

Sagan, D., and J. Whiteside. 2002. A skeptical view of the "sciences of complexity." *Swift* (online newsletter of the James Randi Educational Foundation), http://www.randi.org/jr/011802.html, January 18, 2002; http://www.randi.org/jr/012502.html, January 25, 2002; and http://www.randi.org/jr/022202.html, February 22, 2002.

Schneider, E. D., and J. J. Kay. 1989. Nature abhors a gradient. In *Proceedings of the 33rd Annual Meeting of the International Society for the System Sciences,* P. W. J. Ledington, ed., vol. 3, pp. 19–23. International Society for the System Sciences, Edinburgh.

Schneider, E. D., and J. J. Kay. 1994a. Life as a manifestation of the second law of thermodynamics. *Mathematical Computer Modeling,* 19:25–48.

Schneider, E. D., and J. J. Kay. 1994b. Complexity and thermodynamics: Towards a new ecology. *Futures,* 26:626–647.

Schwartzman, D. 1999. *Life, Temperature, and the Earth.* Columbia University Press, New York.

Schopf, J. W., ed. 1992. *Major Events in the History of Life.* Jones and Bartlett, Boston.

Shock, E. L. 1990. Geochemical constraints on the origin of organic compounds in hydrothermal systems. *Origins of Life,* 20:331–367.

Skoyles, J., and D. Sagan. 2002. *Up from Dragons: The Evolution of Human Intelligence*. McGraw-Hill, New York.

Smolin, L. 1997. *The Life of the Cosmos*. Oxford University Press, New York.

Swenson, R. 1997. Thermodynamics, evolution, and behavior. In *The Encyclopedia of Comparative Psychology*, G. Greenberg and M. Haraway, eds. Garland, New York.

Von Frisch, K. 1967. *The Dance Language and Orientation of Bees*. Harvard University Press, Cambridge, MA.

Watson, A. J., and J. E. Lovelock. 1983. Biological homeostasis of the global environment: The parable of Daisyworld. *Tellus*, 35B, 284.

Wicken, J. 1987. *Evolution, Thermodynamics and Information: Extending the Darwinian Program*. Oxford University Press, New York.

Wiener, N. 1961 [1948]. *Cybernetics or Control and Communication in the Animal and the Machine*. MIT Press, Cambridge, MA.

Wolfram, S. 2002. *A New Kind of Science*. Wolfram Media, Champaign, IL.

16
Gaia and Complexity

Lee F. Klinger

Abstract

Complexity, the body of theory on self-organization and self-regulation, and Gaia, the view of the Earth as a living system, are considered together here in an examination of the Earth and its ecosystems for patterns of symmetry and self-similarity. A conceptual model of the duality in nature that elaborates on the symmetry, asymmetry, and fractality of systems is shown to be readily extendable to a wide range of phenomena, from molecules of water to planets in our solar system. Furthermore, this model shows how certain Eastern traditions of thought, specifically the Chinese principle of yin-yang, closely correspond to patterns in symmetry, entropy, free energy, and fractality expressed in nature.

Evidence for symmetry and fractality in the Earth and its ecosystems, focusing in particular on gross structural properties, is presented and discussed from a complex systems perspective. Gaia exhibits both symmetry and asymmetry in its surface structures, consistent with the dual action of symmetry-building (yin) and symmetry-breaking (yang) processes. Self-similarity is clearly seen in the spherical layering of Earth's crust, mantle, and core. Ecosystems, too, possess symmetry and fractality in the overall structure and composition of biomes and landscapes. Fractality in the canopy architecture of ecosystems across northern Asia is shown in ecological survey data collected from northern China and from the Hudson Bay lowland in Canada. While evidence of criticality of the Earth is suggested by the fractality of certain global phenomena such as earthquakes, atmospheric carbon dioxide concentration, and surface temperature, the degree of fractal symmetry, a parameter expected to vary more or less predictably according to whether a system is ordered, critical, or chaotic, is not yet characterized well enough to indicate in which of these states the Earth may reside. Overall, the lines of evidence presented here indicate that complexity theory provides fair guidance in understanding patterns and behaviors observed in and on the Earth.

All of science is, in some way, the study of dynamical systems, some simple, most complex. That, no doubt, is why scientists from so many fields are compelled to consider and study complexity. Complexity represents a collaboration across disciplines of related theories that link the phenomena of self-organization, self-replication, life, evolution, morphogenesis, chaos, and more into a unified set of principles to explain the emergence of order in systems. That these organizational principles appear not to violate the laws of thermodynamics is no less remarkable than their ability to account for the exquisite beauty seen in snowflakes and seashells. Inspired by such mavericks as James Lovelock, Lynn Margulis, Stuart Kauffman, Brian Goodwin, and Per Bak, I hope to extend, here, the theory of complexity to Gaia, the theory that the Earth is a living system. Furthermore, as the concepts of complexity reach new arenas of science and culture, they are bound to touch non-Western traditions of thought. Hence, while many of the fundamental notions of complexity have been traced to early Western science and philosophy (Capra, 1996), it is fair also to consider whether similar notions appear in the traditional Eastern views of nature (Capra, 2000; Klinger and Li, 2002).

The relevance of complexity to Gaia has been mentioned repeatedly in the literature on complexity theory since the 1990s. Indeed, several authors have remarked that the science of complexity makes Gaia theory more plausible (Goodwin, 1994; Coveney and Highfield, 1995; Bar-Yam, 1997; Kauffman, 2000), while others have even suggested that complexity is the underlying theory for Gaia (Bak, 1996; Levin, 1998). The relevance of complexity has also been noted in the Gaia literature (Lovelock, 1995; Klinger, 1996b; Klinger and Short, 1996; Klinger and Erickson, 1997; Von Bloh et al., 1997; Downing and Zvirinsky, 1999; Harding, 1999), which echo the opinion that complexity provides a theoretical basis for Gaia. Yet none of these works have closely examined the overlapping realms of Gaia and complexity. Considering that both theories share several common themes, including self-organization, self-regulation, emergence, and coevolution, it seems fitting to explore their shared domains. If under scrutiny the theories are found to be facets of the same general phenomena in nature, then further breakthroughs in our understanding of complex systems are all but inevitable.

A timely treatment of Gaia and complexity by Lenton and van Oijen (2002) provides an excellent review of the two theories and lays out a solid foundation of evidence supporting the view of Gaia as a complex adaptive system. Complex adaptive systems (CAS) are defined by Levin (1998) as complex systems with (1) sustained diversity and individuality of components; (2) localized interaction among those components; and (3) an autonomous process that selects from among these locally interacting components a subset for replication or enhancement. It is this last criterion that distinguishes CAS (including living systems) from other complex systems. Lenton and van Oijen present a conceptual and mathematical framework of complex systems (including CAS) and their various behaviors that this chapter will build upon. They classify complex systems according to three behavioral states: ordered, critical, or chaotic. Critical (or criticality) refers to the state of a system that lies in the ordered regime near the edge of chaos (Bak, 1996). They then assess whether Gaia possesses certain emergent properties and behaviors characteristic of complex adaptive systems, such as self-organization, criticality, continual adaptation, hierarchical organization, generation of perpetual novelty, and far from equilibrium dynamics (Kauffman, 1993; Bak, 1996; Holland, 1995). In addition, they present results of complexity-based models which demonstrate how these characteristics may emerge on an imaginary planet with life, such as Daisyworld (Watson and Lovelock, 1983). In conclusion, they argue that "Gaia is a complex system with self-organising and adaptive behavior, but ... it does not reside in a critical state"; that is, Gaia is not poised at the edge of chaos.

In this chapter I, too, argue that Gaia is a complex adaptive system, but that it *does* reside in a critical state. This argument stems from the insights gained by examining complex systems for predicted patterns of symmetry and fractality in a conceptual model of system transformation based on the main tenets of complexity, evolution, and chaos. This model attempts show how the forces that build symmetry and those that break symmetry interact in the expression of self-similar (fractal) structures and behaviors in complex systems. Following a description of the model and its potential relevance in other disciplines, I present recent findings on the fractal structure of the whole Earth and on the fractal structure in ecosystems of Asia to assess whether this model is valid for these larger systems.

A Conceptual Model

The model assumes that the overall symmetry and fractal properties of systems vary in prescribed ways according to the principles of complexity and the laws of thermodynamics. Fractals are widely seen throughout nature as shapes within shapes and as cycles within cycles. Self-similarity in the form and behavior of both living and nonliving things is a phenomenon recognized long before the discovery of fractals. However, the recent attention to fractals now provides us a clearer conceptual and mathematical framework for the description of nature's repeating patterns. Fractals are often characterized as scale-invariant patterns or probabilities occurring along contractions of the multidimensional state space of a given system. Mathematical proof of fractals is depicted in the wide array of recurring diorama that emerge from the Mandelbrot set, and in the self-similar patterns seen in countless other geometric solutions of certain simple equations that incorporate both real and imaginary numbers.

In physics, fractals emerge in David Bohm's treatments of wholeness and the implicate order, resulting in an ontological theory explaining the self-similar or holographic nature of the universe in a manner consistent with quantum physics (Bohm and Hiley, 1993). Fractality is linked to chaos theory in the appearance of strange attractors, probability distributions of the state space of a chaotic system that upon close inspection are found to be self-similar (Hénon, 1976). Fractals are also features of complex systems in ordered states, as can be seen in the self-similar patterning in mineral crystals, and of complex systems in critical states, as in the power-law relationship of the frequency and magnitude of catastrophic events (Bak, 1996). In the real world, such fractals typically span scales of one or two orders of magnitude (Avnir et al., 1998). Thus, whether in the ordered regime, in the chaotic regime, or near the edge of chaos, complex systems exhibit some degree of fractality.

Complex systems also must obey certain symmetry principles. In particle physics, gauge symmetry demonstrates that the known laws of physics are closely associated with symmetry principles (Greene, 1999).

In general, the symmetry of any system relates to the degree of correspondence among its parts. While conventionally used to describe the geometry of shapes in nature, the concept of symmetry can also be extended to describe the behavior of a system, that is, the degree of correspondence in the states of a system through time. Symmetry, then, may be better considered as a statistic, a measure of the likelihood that, within a system, a particular part will possess a form similar to its adjacent or corresponding part(s), or the probability of a future behavior determined from knowledge of past behavior. It is apparent that the many statistical tests and measures (e.g., probability functions, tests of central tendency, and time series analyses) are, in fact, tools that scientists use to characterize and quantify the degree of symmetry or asymmetry in nature (Vicsek, 1989). With respect to fractals, the basic premise of statistics—that a sample of a system can be used to describe the whole system—rests on the inherent assumption of self-similarity in nature.

The concept of symmetry in fractals relates simply to the degree to which self-similar (or self-affine) patterns are expressed along a given range of scales. Thus, if a fractal pattern of a system is seen to be regular, appearing at every scale across ten orders of magnitude, the fractal symmetry of that system would be greater than one where the fractal is expressed irregularly at, say, only two or three scales across the same ten orders of magnitude. With regard to fractality in complex systems, symmetry principles should still apply. Hence, in the transition from an ordered to a critical state or from a critical to a chaotic state, the symmetry of a system, including its fractal symmetry, is always broken (MacCormac, 1998).

For any given system let us consider two parameters that can best describe its overall symmetry: (1) the geometric symmetry (S_g), a combined function of the mirror symmetry, rotational symmetry, translational symmetry, fractal symmetry, and so on, and (2) the behavioral symmetry (S_b), a combined function of the temporal symmetries. While these two parameters characterize the different dimensions of a system, they are not independent. Spatial and temporal symmetries are often correlated. In architecture, for instance, the more symmetrical the design of a structure, the more likely it will remain standing over time.

How might the geometric and behavioral symmetries vary and interact throughout the lifetime of a complex system? Figure 16.1a shows a hypothetical plot of a complex system trajectory through time in the state space defined by these two parameters. The model suggests that, in transformation, system trajectories in this state space will spiral between more symmetric and more asymmetric states, each spiral being some characteristic periodicity or "life cycle" of a complex system. The plot can be adapted for living organisms by adding the elements of conception, death, and reproduction to the system trajectories (figure 16.1b).

This model points outs a simple dichotomous classification of systems that emerges from an interpretation of the Eastern concept of yin and yang that follows from symmetry principles (table 16.1). The duality in nature elucidated in this table extends from the observation that all the various dynamics and behaviors of complex systems can ultimately be expressed as one of two processes: symmetry-building or symmetry-breaking. Thus, self-organization, self-regulation, autocatalysis, and crystallization all relate to building and maintaining symmetry in complex systems, while bifurcation, turbulence, morphogenesis, and evolution are all symmetry-breaking processes. By recognizing the correspondence of symmetry-building with the principle of yin and of symmetry-breaking with the principle of yang, it would seem that the resulting patterns of symmetry/asymmetry can be used to form an image of nature that nicely complements both Eastern and Western traditions of thought (figure 16.2). Clearly, this ancient Chinese principle deserves further attention from scientists attempting to understand the dualistic nature of systems (Kleppner, 1999; Capra, 2000).

A key feature of this model is in the novel way it links entropy and order. Contrary to the conventional depiction of a universe tending toward increased entropy and disorder, an increase in entropy is seen here as leading toward greater symmetry and, therefore, more order in the universe. Thus, if the "heat death" of the universe is followed to its ultimate conclusion, then the result would be a universe in complete thermodynamic equilibrium, one where all matter is distributed evenly and frozen in space (i.e., perfect symmetry). That an increase in entropy can lead to greater order in systems has been pointed out by Prigogine (1996) in his reevaluation of the second law of thermodynamics.

This conceptual model lends itself nicely to the classification of complex systems referred to in Lenton and van Oijen (2002). Table 16.2 shows how the principles of symmetry and fractality can be applied to ordered, critical, and chaotic systems in a logical manner. For instance, the ordered, critical, and chaotic behaviors in molecular systems can reasonably

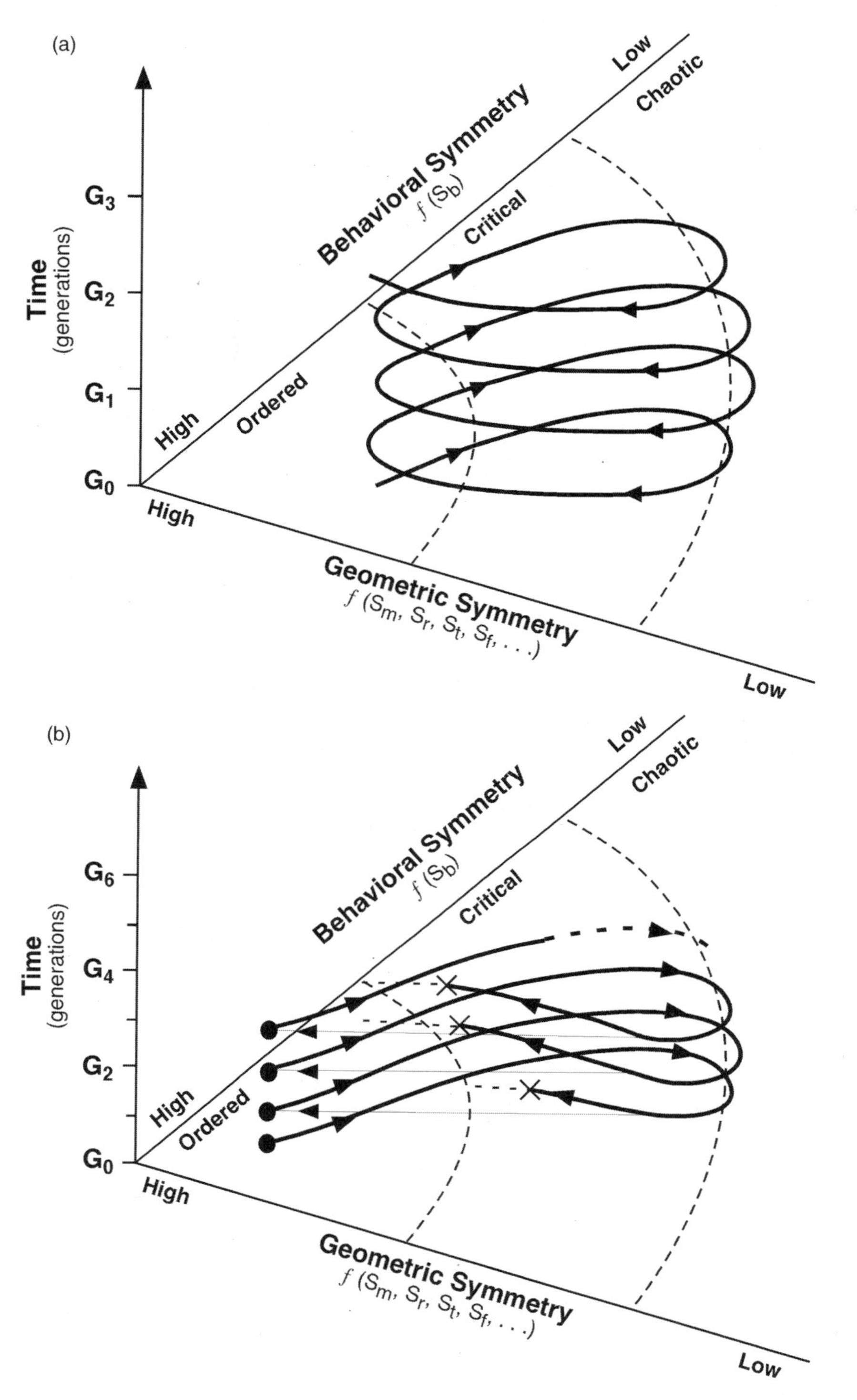

Figure 16.1
Hypothesized trajectories through time of complex systems as a function of changes in their geometric symmetry (S_g) and behavioral symmetry (S_b); (a) generic complex system, and (b) complex adaptive system, showing conception (●) and death (×).

Table 16.1
Dichotomy in nature with respect to both Eastern and Western principles

	Yin	Yang
Symmetry	↑	↓
Entropy	↑	↓
Free energy	↓	↑
Information*	↓	↑
Characteristic qualities	order	chaos
	dark	light
	cold	hot
	frozen	volatile
	winter	spring
	old	young
	persistent	ephemeral
	receptive	expressive
	feminine	masculine
Characteristic processes and behaviors	negative feedback	positive feedback
	stabilization	perturbation
	crystallization	vaporization
	ontogeny	phylogeny
	maturation	reproduction
	uniform growth rate	variable growth rate
	recycle	assimilate
	work	play
Numerical descriptors	Cantor set	Mandelbrot set
	periodic attractor	Lorenz attractor
Examples in nature	ice	fire
	Pluto	sun
	plain	mountain
	peatland	grassland
	microbe	mammal

* Algorithmic information (Gell-Mann, 1994).

Figure 16.2
Yin and yang, an ancient Chinese symbol of the fundamental, complementary, and opposing dual forces that, in dynamic tension, bring about transformation in nature (Cirlot, 1962; Liu, 1979; Capra, 2000). The symbol contains obvious symmetries and asymmetries in shape and in color. The sigmoidal curve delineating the yin and yang represents the dynamic way in which the forces interact. The smaller circles within the symbol are a reference to self-similarity of the parts with the whole.

be seen in the frozen, liquid, and gaseous states of matter, respectively. This table also suggests that the dynamics of the neighboring planets Mars, Earth, and Venus might be viewed in a similar way. Is this a fair representation of planetary behaviors, or is a wider examination of our solar system in order?

Life at the Edge of Chaos

"Life at the edge of chaos" has become a mantra for scientists investigating complexity in living systems, the idea being that living systems, and other complex adaptive systems, evolve toward the edge of chaos, searching for new stable existences in the unexplored and ever-changing terrain near chaos (Kauffman, 2000). The edge of chaos is where living systems are most likely to find new means and forms of existence in their pioneering efforts to domesticate chaotic landscapes. Evolution is about finding some kind of stable existence in places ever further from equilibrium by experimenting with new forms, new behaviors, new ideas. Evolution conducts these experiments through mechanisms which break the structural and behavioral symmetries of systems. However, symmetry-breaking processes often create instabilities in systems, which is why so many evolutionary experiments end in failure.

At the same time, living systems grow, develop, and mature. Early growth and development of lifeforms, however chaotic and unpredictable, eventually gives way later in life to more predictable, orderly traits and behaviors. These functionally more stable, ancestrally derived states become reinforced through repetition of the deterministic convergent pathways of maturation occurring over countless generations. Development, then, may be viewed as a symmetry-building process in which complex systems proceed toward deep basins of attraction originating from and reinforced by more stable, primordial forms and behaviors. From a complex systems perspective, maturation and crystallization are analogous processes in that they are both symmetry-building, self-organizing behaviors.

Table 16.2
Fractals and symmetry principles applied to complex systems

	Behavioral states of complex systems		
	Ordered	Critical	Chaotic
Degree of symmetry	high	moderate	low
Fractal symmetry	continuous, regular	discontinuous, ±regular	discontinuous, irregular
Fractal dimensions	few	many	singular
Fractal types	self-affine, deterministic	self-affine, statistical	self-similar, deterministic and statistical
Examples in nature*			
matter	frozen	liquid	gaseous
water	snowflake	whirlpool	steam
viscous liquid	stratified	convective	turbulent
sounds	monotonic	music	static
planets	Mars	Earth	Venus

* Examples are grouped according to their relative behavioral states.

The depictions here of evolution and development, or ontogeny, suggest that these processes act in opposing manners, with evolution driving systems toward chaotic regimes, and development driving systems toward ordered regimes. Thus, change or transformation in nature is proposed here to result from evolutionary and developmental forces acting together on a complex system. So how, then, do we explain periodicities in the behavior of complex systems? Do the periodicities seen in living organisms (e.g., heartbeats, life cycles) simply emerge from the dynamic tension of these forces working to maintain a complex system somewhere between absolute order and total chaos? And where, exactly, is life with respect to the edge of chaos? Let us ponder this last question while returning to our own solar system.

Orbiting the outer reaches of the solar system, Pluto represents one end of a spectrum of planetary bodies. Far from the radiative fires of the sun and relatively free of cataclysms from asteroids and comets, Pluto may well possess some of the oldest "landscapes" under the sun. I envision this tiny planet's surface covered in forests of methane and nitrogen ices, ancient crystals with fractal canopies extending into the rarefied air. As for the chance of finding biological life in such a frozen landscape, it is probably nil. Living things require more active material and energy flows in order to move, metabolize, and evolve. Other than the occasional collapse of an overgrown crystal, the dynamic forces of chaos and disturbance are in short supply. Pluto, for all its order, is far too static a place for life to arise or survive.

At the other end of the planetary spectrum is the sun, its tremendous nuclear fires burning so bright that it can be seen from millions of light-years away. The sun is a place where commotion and chaos rule. Hot gases stream and swirl across its surface with unimaginable heat and fury. Order emerges for only brief periods, apparent in the solar flares and sunspots associated with solar storms. As for the chance of finding biological life in such a hellacious place, again it is probably nil. The dynamo of the sun lacks the conditions of stability needed to sustain the processes that bring order to living systems.

Somewhere between the crystalline forests of Pluto and the fiery chaos of the sun is a planet composed of both fire and ice, and of intermediate things, like water. It is a planet with conditions stable enough to allow the fine expression of gemlike order, and dynamic enough to really stir up that order every now and again. Crystalline order, the result of symmetry-building processes in nature, and chaos, brought on by forces that break symmetry, conspire to transform the planet as it spins and spirals in a fractal free fall through space, skirting, at times, the edge of chaos. It is Earth.

Indeed, every planet in this solar system may be viewed as a complex system, self-organized and self-regulating in a state some distance from equilibrium. As near as we can tell, all the planets have dynamic surfaces and atmospheres exhibiting some degree of orderly structure. Besides exhibiting self-organization, the Earth, as a living planet, possesses an atmosphere that is particularly far from chemical equilibrium with its surface. But what about the Earth's thermodynamic equilibrium relative to the other planets? Is not the sun much farther from thermodynamic equilibrium than is the Earth? Can planets be distinguished as living or nonliving based on their states of equilibrium vs. nonequilibrium (or states of order vs. disorder)? If not, what other properties do living and nonliving systems share that may be useful in helping to understand the essence of all complex systems?

Symmetry and Fractality of Gaia

If degrees of symmetry and fractality, as quantifiable parameters, are to prove useful in the characterization of Gaia as a complex system, then let us first consider those features and behaviors of the Earth which are obviously symmetrical, asymmetrical, and/or fractal. Poised at the edge of chaos, a living system must be acted on by both strong evolutionary (symmetry-breaking) forces, which drive the system toward a critical state, and strong stabilizing (symmetry-building) forces which prevent the system from becoming chaotic. Regarding the evolutionary forces, the asymmetry seen in the Earth's wobbly orbit around the sun is the result of a powerful symmetry-breaking event about 4.5 billion years ago when a massive bolide impacted Earth, knocking loose the debris that eventually formed the moon. A surface comprised of both oceans and continents, and an atmosphere characterized by variable cloudiness, are earthly asymmetries not evident on other planets. Gaia, too, exhibits powerful symmetry-building forces in plate tectonics. These forces brought about the breakup of Pangea about 200 million years ago, resulting in the present, more symmetric, distribution of continental crust. Regulation of atmospheric oxygen and ocean salinity through the eons indicate a high degree of symmetry in the behavior of Gaia. The tectonic activity and chemical regulation mentioned here seem to be uniquely Gaian within our solar system.

Let us now consider whether the Earth is fractal. A better understanding of the fractality of Gaia may allow one to predict where, when, and at what scales fractals may appear. As a complex system, Gaia should exhibit fractality, for instance, in its gross structure.

An initial search for fractals in the structure of the Earth suggests they are fairly common (Korvin, 1992). Coastlines are known to be fractal, as are the drainage networks of landforms (Turcotte and Newman, 1996). Data from superdeep boreholes reveal fractal features in the depth profiles of a suite of crustal properties, from magnetic susceptibility to bulk density (Leonardi and Kumpel, 1999). A germane example of fractality of Earth's morphology is

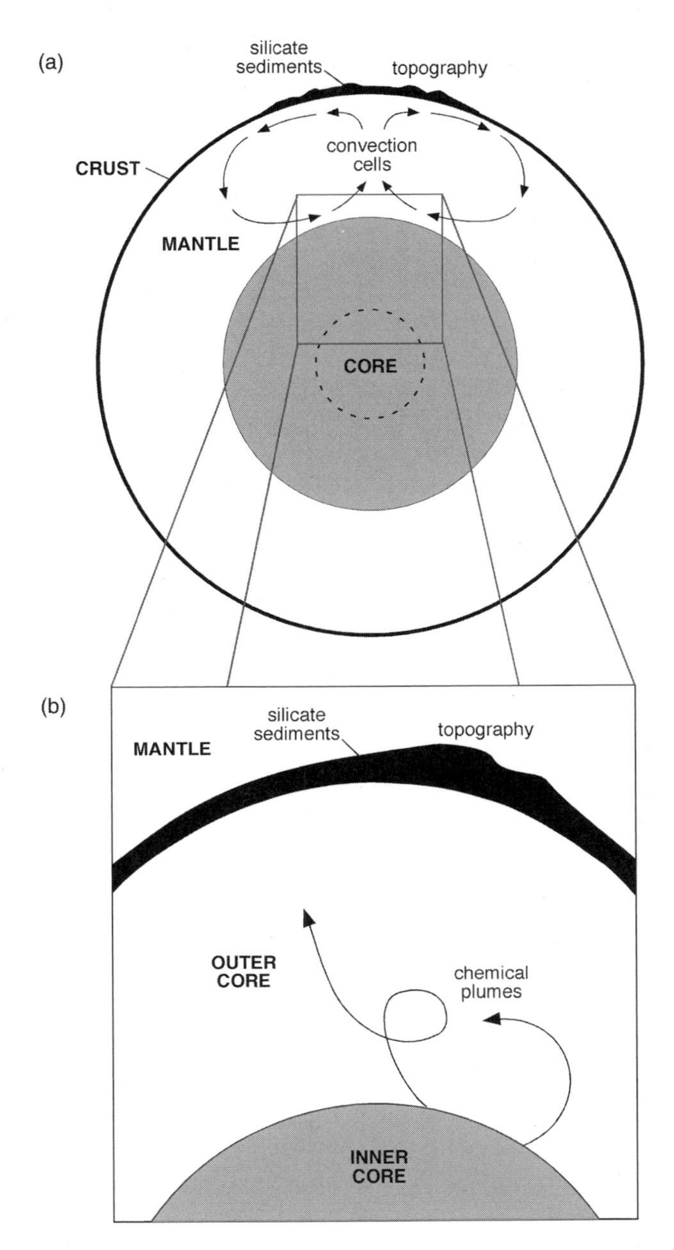

Figure 16.3
Self-similar structure of the earth as shown in cross-sectional views of (a) the entire lithosphere, and (b) the core region of the lithosphere, redrawn from Buffett et al. (2000).

presented in figure 16.3a and 16.3b. Figure 16.3a is a standard textbook rendering of the Earth, depicting its primary structural features in a meridional cross-section. The Earth is divided into three main zones: the core, the mantle, and the crust. The crust is composed of silica-rich sediments with a range of topographies. It floats atop the fluid mantle that exhibits convective movement. Figure 16.3b is a schematic of the Earth's substructure near its core, redrawn from Buffett et al. (2000). It depicts the internal structure as consisting of an inner core, an outer core, and a layer of silica-rich sediments, with topography, floating atop the fluid outer core that undergoes convective movement. Self-similarity appears in the siliceous crusts found at both the surface and the mantle-core boundary, and in the heat-driven transport mechanisms connecting these crusts to their respective core regions. This fractal, real or apparent, readily emerged from a quick search on Earth structure, a search guided by complexity theory. While it remains to be seen whether this finding is coincidental or is true evidence for fractality of Gaia, let me carry on the search for fractal forms and behaviors to include ecosystems, the first-order subsystems of Gaia.

Symmetry and Fractality of Ecosystems

Gaia is sometimes described as the theory of the Earth as a superorganism (Lewin, 1996; Rayner, 1997). However, to me, Gaia is more a kind of superecosystem emerging from the coupling of ecosystems, particularly the major biomes. This is in contrast to the view that global self-regulation is mainly the result of organism–environment coupling (Lovelock, 1995; Lenton and van Oijen, 2002). The superorganism concept, I believe, more aptly applies to ecosystems, as was recognized by Frederic Clements (1916) in his climax theory of succession, which forms the basis of my approach to ecology and geophysiology.

Ecosystems are coherent groups of organisms possessing dynamic boundaries and interacting among themselves and with their environment. Since they exhibit emergent properties of self-organization and self-regulation, they can clearly be treated as complex systems. While many ecologists are still cautious of both Gaia and complexity theories (Klinger, 1997), a growing number are being drawn to study ecosystems from the more holistic perspective of complexity. This school of systems ecology, which originated with Clements (1916) and has continued with the work of Margalef (1963) and Odum (1969), has long maintained that ecosystem organization, self-regulation, and development (succession) are emergent properties of interacting organisms and their environment. Further backed by the work on self-organized criticality in rain forests (Solé and Manrubia, 1995) and on the far from equilibrium thermodynamics of ecosystems (Jørgensen et al., 1992), the systems approach remains on firm conceptual ground in ecology. To the synecologists who would reject this holistic view of ecosystems, I ask simply whether they have reassessed climax theory in light of the new findings that peat bogs, not forests (as Clements had presumed), are climax ecosystems. Having done such an assessment (see discussion below), I remain confident in the validity of the systems approach employed here.

Complexity theory breathes new life into the classical notion that ecosystems, like organisms, develop in predictable fashion toward a mature, "climax" state. The successive stages of ecosystem establishment and growth may now be viewed as the complex deterministic behavior of groups of organisms with local feedback connections, getting larger, more ordered, and more robust with age, and, when undisturbed, converging onto a mature, highly ordered, and tightly self-regulated state, a climax ecosystem. Here, I wish to bring to bear on the subject of succession an assessment of my findings from ecosystems in Alaska, Colorado, California, Wisconsin, Tennessee, Georgia, North Carolina, New York, Missouri, Texas, Puerto Rico, Ontario, Québec, Spitzbergen, Brazil, Congo, and China. These and a bevy of findings by other authors indicate to me that, from the arctic to the tropics, long-term ecosystem development converges toward climax peatland ecosystems, and that characteristic and predictable changes occur in vegetation structure along the way. Peat bogs are very old ecosystems, several thousand years at least, and are typically preceded by old-growth evergreen forest. They are extremely large, containing many times the biomass of an old-growth forest, and support a rich abundance of mosses, lichens, fungi, and algae. These cryptogams function together with the various aerobic and anaerobic microbes to regulate the flows of energy, water, and nutrients through peatlands (Klinger, 1991, 1996a, 1996b; Klinger et al., 1994; Klinger and Erickson, 1997).

As noted above, other ecologists also recognize that ecosystems are, like other living systems, complex adaptive systems (Levin, 1998; Jørgensen and Müller, 2000; Bradbury et al., 2000). Indeed, Levin (1998) states that the emergent order and self-regulation seen in ecosystems make them "prototypical examples" of complex adaptive systems.

Regarding symmetry in ecosystems, the latitudinal zonation of the major terrestrial biomes exhibits a gross mirror symmetry around the equator. In savannas it is observed that successional processes at the prairie–shrubland boundary act to build symmetry in the soils and vegetation (Allen and Emerman, 2003). Conversely, modeling studies of the grassland–

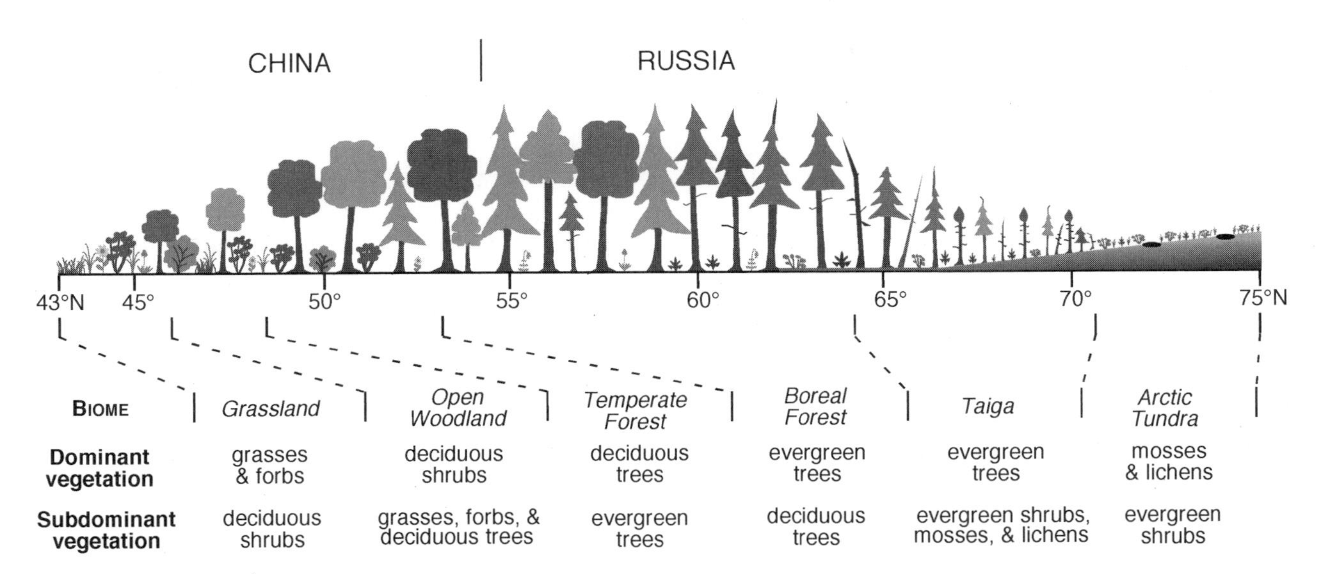

Figure 16.4
The major biomes of northern Asia and their dominant vegetation types within the 124–126°E longitude band (based on Olsen et al., 1982).

shrubland transition zone in northwestern Africa suggest that symmetry-breaking processes involving disturbance can account for much of the observed distribution of vegetation cover (Lejeune et al., 1999; Lejeune and Tlidi, 1999).

As complex systems, ecosystems must also be fractal. Studies examining ecosystem structure for fractality are few, but again the results are promising. Gap patterns in tropical rain forests have been shown to exhibit fractal properties which are thought to be related to self-similar dynamics of forest biomass (Solé and Manrubia, 1995). Fractality is also reported in the spatial distribution of remotely sensed net primary productivity values of forest and grassland ecosystems in the western United States (Ricotta and Avena, 1998). In my own research, using gradient analysis techniques to sample a wide range of landscapes, I have discovered a recurring pattern in ecosystem structure that, in all respects, appears fractal. To illustrate this, I present findings on the measured abundance of the major plant forms along zonally distributed ecosystems in Asia.

From space one can observe distinct zones of terrestrial biomes that extend, often latitudinally, across continents. This banded arrangement is particularly evident across Asia north of 40°N, where at least six different biomes extend from east to west across 90° of longitude (from 40°E to 130°E). The biomes are, from south to north: grassland, shrubland, temperate forest, boreal forest, taiga, and arctic tundra. These biomes, which appear along gradients of up to 10^4 km in length in Asia, are evident as well in Europe and North America. Figure 16.4 is a graphic depiction of these Asian biomes along the 124°E to 126°E longitude band, listing the vegetation types and major growth forms according to Olsen et al. (1982).

Ecologists are keen to point out that a similar pattern is often found on mountainsides where these same vegetation zones, from grassland to tundra, are arranged sequentially along gradients of increasing elevation. This pattern is consistent with the conventional view in ecology that, at larger spatial and temporal scales, climate controls the distribution of the world's major ecosystems. However, from both a Gaian and a complex systems perspective, plants and climate are closely coupled through a suite of feedback interactions. Thus, this sequence of biomes is viewed as an ordered set of stable or quasi-stable vegetation types that emerge repeatedly from this coupling. Complexity theory suggests that the similarity in the sequences of vegetation along both latitudinal and elevational gradients is simply an expression of fractality in ecosystems.

To investigate further this question of fractality, let us home in on the ecosystems of northern Asia. Straddling the China–North Korea border is the Changbai Mountain massif, a huge composite shield volcano that rises to over 2900 m in elevation. The mountain has a sacred status in the traditions of local peoples and, thus, has been afforded some protection from human impacts. The main ecological zones of this 8,000-km^2 region (centered on 42°24′N latitude and 128°15′E longitude) were surveyed in 1998 as part of an ongoing study to characterize the composition, abundance, and volatile organic compound emissions of vegetation for the major ecosystems of China (Klinger et al., 2002). Here I consider the leafy (green) biomass determinations for sixty 100-m^2 plots at nine sites along an elevational gradient from 700 to 1250 m asl on the northern flank of the Changbai massif. Figure 16.5a depicts the relative biomass of each of the major plant growth forms in plots stratified by elevation. The observed order of vegetation with increasing elevation is meadow, willow thicket, poplar/oak forest, pine/spruce forest, larch/heath peat forest, and *Sphagnum* bog. The spatial arrangement of dominant growth forms in the Changbai region (figure 16.5a) closely matches the sequence of biomes across northern Asia (figure 16.4). This self-similar character of Asian vegetation is presented as limited evidence for fractality of ecosystems. In this case, however, fractality cannot be established quantitatively because comparable abundance data for growth forms of the Asian biomes are not available.

Comparing these results from China with similar vegetation analyses from other regions of the world, a consistent pattern begins to emerge. Along a 200-km landscape gradient in the Hudson Bay lowland of northern Ontario, Canada, the focal point of the 1990 Northern Wetlands Study, vegetation was characterized using ground surveys and Landsat TM satellite data (Klinger et al., 1994; Klinger and Short, 1996). The following vegetation types were observed on former beach ridges (westward from the shore of James Bay): coastal meadow, willow thicket, poplar forest, spruce forest, spruce/larch bog forest, and *Sphagnum* bog. Again, the dominant growth forms observed along this sequence (figure 16.5b) closely match the fractal sequences in northern Asia (figures 16.4 and 16.5a). What is interesting about the Hudson Bay lowland is that key environmental parameters such as elevation, temperature, precipitation, and substrate type vary negligibly along the gradient. Ecosystem age, however, does vary appreciably and systematically along the gradient. The vegetation pattern above lies along an east-to-west gradient of surface ages from modern to over 5000 years old, derived through gradual development of lands continually arising from James Bay, as the lowland still rebounds from a depressed glacial state beneath the Laurentide ice sheet. The result is a chronosequence, an orderly sequence of characteristic plant forms appearing in both space and time (Klinger and Short, 1996).

Chronosequences are natural experiments that are commonly used in gradient analysis by ecologists studying succession. They are obvious places to look for evidence of a more widespread occurrence of this ecosystem fractal. An examination of this literature indicates that chronosequences and other gradient analyses in Alaska (Van Cleve and Viereck, 1981; Klinger, 1988), Colorado (Marr, 1961), Arizona (Whittaker and Niering, 1965), the Yukon (Birks, 1980), British Columbia (Tisdale et al., 1966), New Zealand (Wardle, 1991), and central Africa (Klinger et al., 1998) reveal more or less the same fractal pattern. To be clear, this small sample of studies is not offered as evidence that the fractal described here is common or widespread. Such an assessment will require a detailed examination of ecological gradients in many more regions. That strong hints of this fractal also appear in the stratigraphy of both plant and insect remains found in peat cores (Klinger and Short, 1996; Klinger et al., 1990) suggests that paleoecological data sets may also provide useful and robust means of evaluating the complex behaviors of ecosystems.

Discussion

The investigation of fractal patterns in Gaia and its first-order components, ecosystems, reveals structural self-similarities which can be readily observed and quantified in these systems. Reports that the frequency and magnitude of a wide range of phenomena, from earthquakes to mass extinction events, follow certain power laws are further indication of the importance and utility of self-similarity in the characterization of complexity in the Earth system (Bak, 1996). Both sedimentation and erosion processes are found to give rise to features with distributions that follow power law functions (Richards et al., 2000; Pelletier, 1999). Analyses of ice cores and sediment cores reveal that temperature and CO_2 time series also can be described by power laws (Fluegeman and Snow, 1989; Cronise et al., 1996; Pelletier, 1997). Such behavioral self-similarity is thought to be characteristic of complex systems in a critical state, near the edge of chaos. If so, then this evidence is pointing to the existence of criticality throughout the Earth system.

But uncertainty arises if one assumes that fractality is an inherent property of all complex systems, in a critical state or not. Do supercritical, chaotic behaviors like those of strange attractors appear any more or less fractal than the subcritical, orderly behaviors

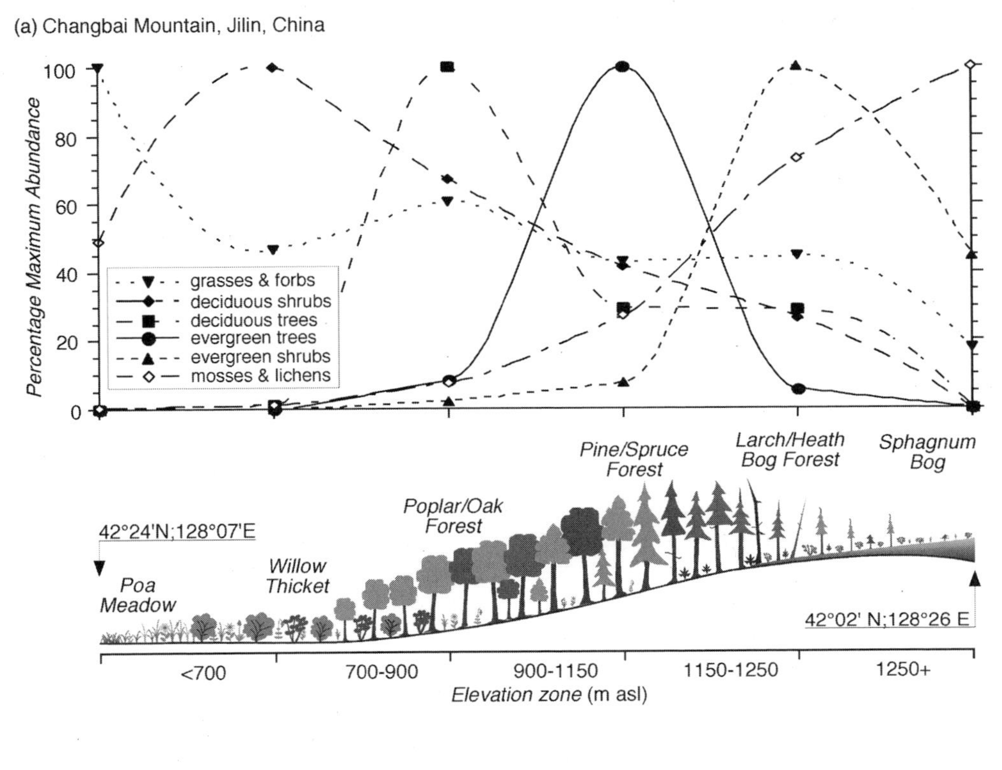

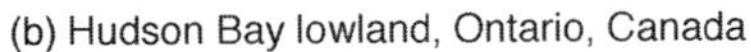

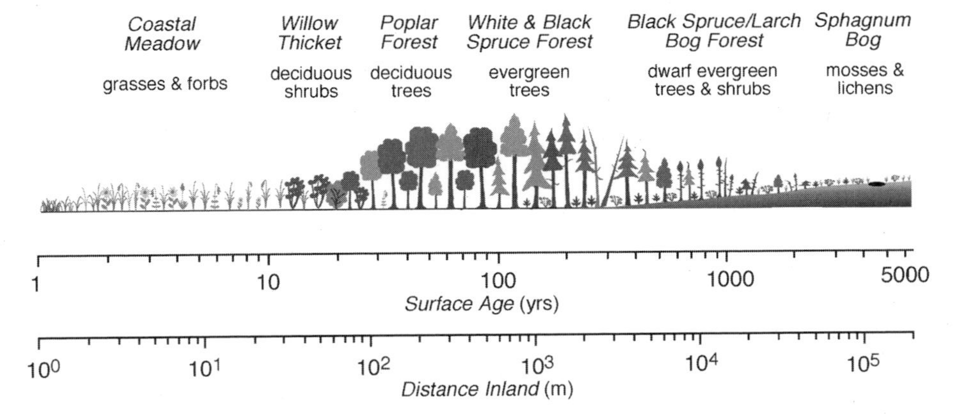

Figure 16.5
Self-similarity in ecosystems as indicated by patterns in vegetation structure and composition along environmental gradients in (a) the Changbai Mountain region of northern China and (b) the Hudson Bay lowland of northern Canada.

exhibited by crystals? What is the real significance of fractals? For whatever they signify, fractallike patterns are clearly evident in the structure and behavior of the Earth and its components.

Considering Gaia as a complex adaptive system, how can our understanding of fractals help to distinguish whether or not the planet is adaptive, or even living? To answer this, we might wish to consider how evolutionary forces, particularly natural selection, act on the fractal symmetry of organisms. Fractal symmetry refers to the scale-invariant properties describable by power laws (Turcotte and Newman, 1996). Systems with a high degree of fractal symmetry possess features expressing self-similarity continuously along extended spatial or temporal gradient. Systems with a lesser degree of fractal symmetry (or possessing greater fractal asymmetry) show self-similar features that, while still describable by power laws, tend to be expressed discretely along a given scale. Fractal symmetry is only one of several types of symmetry, including mirror, rotational, and translational symmetries, all of which may apply in the following discussion (Turcotte and Newman, 1996).

Primitive taxa of organisms are observed typically to be symmetric in form and behavior. Globular clusters of the primitive *Nostoc*, algae that grow along margins of ponds in bogs, are the most symmetric in shape and most fractal of all multicellular growth forms I've encountered. They are also behaviorally symmetric, exhibiting little change from day to day. Higher taxa, on the other hand, are considerably more asymmetric in form and behavior due to the symmetry-breaking forces of evolution. Humans can clearly see this of themselves and other recently evolved creatures. Fractal symmetry in primitive organisms tends to be expressed in a rather continuous fashion, whereas in higher taxa symmetry-breaking forces result in the fractals becoming highly differentiated and occurring at fairly discrete scales. While self-similarity remains a fundamental feature of the Earth's components, evolutionary forces have brought about a diversification of these components, a breaking of their various symmetries. Thus, Gaia as a living, evolving system might be expected to show somewhat greater asymmetry, including fractal asymmetry, than a nonliving planet. However, while this work does reveal some degree of fractal symmetry of the Earth, comparable information for other planets is lacking, so little can yet be said of Gaia's behavioral state based on its overall fractal symmetry.

All systems, living or not, must evolve in order to persist in an ever-changing universe. But evolution as a symmetry-breaking force must not be allowed to run rampant, or else chaos would prevail. A unique aspect of living systems is that they have achieved a means to facilitate and moderate evolution through the process of natural selection. Thus, as living systems search their state space, whether near or far from the edge of chaos, for new basins of attraction, new stable existences, natural selection provides a means to regularly assess the fruitfulness of this search, and to respond quickly and accordingly if the search is unsuccessful. Life will take a pile of sand and play with it, breaking its symmetry as it finds new stability in the far from equilibrium shape of a medieval castle. Life has acquired this unique ability to control and enhance evolutionary exploration through the cyclic processes of recombination, reproduction, and reorganization as well as through play behavior. The cycles of life are characterized by alternating shifts between more chaotic and more stable states. Whether we consider the life cycle of an organism, the successional cycle of an ecosystem, or the ice age cycle of a living planet, these oscillating states are profoundly expressed in a deterministic and predictable manner.

Conclusion

Gaia and its component ecosystems exhibit self-organization, self-regulation, symmetry, and fractality consistent with complexity theory. Yet other planets, too, possess the same defining properties of complex systems. Self-organization is clearly evident in the rings of Saturn. Self-regulation is seen in the persistent red spot on Jupiter. Mars, Venus, and the moon all have surface features with symmetrical and fractal properties (Bruno et al., 1992). Hence, all planets might be considered complex systems. Earth appears different only in the manner and degree in which symmetry and fractality are expressed. Fractal symmetry of the Earth, though not yet well understood, may prove to be a useful parameter in distinguishing this and other Earth-like planets from the ones without life. In all, complexity does offer us fair guidance in recognizing patterns and behaviors in the Earth system that might otherwise go unnoticed.

Acknowledgments

I am grateful to S. Canney, M. Cornish, C. Wiedinmyer, T. Karl, S. Emerman, and two anonymous reviewers for their valuable comments on this manuscript. Thanks to the Chinese Academy of Sciences and the National Center for Atmospheric Research for support of this work. The National Center for Atmospheric Research is sponsored by the U.S. National Science Foundation.

References

Allen, N. S., and S. H. Emerman. 2003. The control of the nitrogen cycle as a mechanism for the self-organization of prairie ecosystems. In *Proceedings of the 18th North American Prairie Conference*. Kirksville, MO: Truman State University Press.

Avnir, D., O. Biham, D. Lidar, and O. Malcai. 1998. Is the geometry of nature fractal? *Science*, 279, 39–40.

Bak, P. 1996. *How Nature Works*. New York: Springer-Verlag.

Bar-Yam, Y. 1997. *Dynamics of Complex Systems*. Reading, MA: Addison-Wesley.

Birks, H. J. B. 1980. The present flora and vegetation of the moraines of the Klutlan Glacier, Yukon Territory, Canada: A study in plant succession. *Quaternary Research*, 14, 60–86.

Bohm, D., and B. J. Hiley. 1993. *The Undivided Universe*. London: Routledge.

Bradbury, R. H., D. G. Green, and N. Snoad. 2000. Are ecosystems complex systems? In *Complex Systems*, edited by T. R. J. Bossomaier and D. G. Green, pp. 339–365. Cambridge: Cambridge University Press.

Bruno, B. C., G. J. Taylor, S. K. Rowland, P. G. Lucey, and S. Self. 1992. Lava flows are fractals. *Geophysical Research Letters*, 19, 305–308.

Buffett, B. A., E. J. Garnero, and R. Jeanloz. 2000. Sediments at the top of the earth's core. *Science*, 290, 1338–1342.

Capra, F. 1996. *The Web of Life*. New York: Anchor Books.

Capra, F. 2000. *The Tao of Physics*, 4th ed. Boston: Shambhala.

Cirlot, J. E. 1962. *A Dictionary of Symbols*. New York: Philosophical Library.

Clements, F. E. 1916. *Plant Succession: An Analysis of the Development of Vegetation*. Publication 242. Washington, DC: Carnegie Institution of Washington.

Coveney, P., and R. Highfield. 1995. *Frontiers of Complexity*. London: Faber and Faber.

Cronise, R. J., D. A. Noever, and A. Brittain. 1996. Self-organized criticality in closed ecosystems: carbon dioxide fluctuations in Biosphere 2. *International Journal of Climatology*, 16, 597–602.

Downing, K., and P. Zvirinsky. 1999. The simulated evolution of biochemical guilds: reconciling Gaia theory and natural selection. *Artificial Life*, 5, 291–319.

Elgin, D. 1993. *Awakening Earth*. New York: William Morrow and Company.

Fluegeman, R. H., Jr. and R. S. Snow. 1989. Fractal analysis of long-range paleoclimatic data: oxygen isotope record of Pacific core V28-239. *Pure and Applied Geophysics*, 131, 307–313.

Gell-Mann, M. 1994. *The Quark and the Jaguar*. New York: W. H. Freeman.

Goodwin, B. 1994. *How the Leopard Changed Its Spots*. New York: Charles Scribner's.

Greene, B. 1999. *The Elegant Universe*. New York: W. W. Norton.

Harding, S. P. 1999. Food web complexity enhances community stability and climate regulation in a geophysiological model. *Tellus*, 51B, 815–829.

Hénon, M. 1976. A two-dimensional mapping with a strange attractor. *Communications in Mathematical Physics*, 50, 69–77.

Holland, J. 1995. *Hidden Order*. Reading, MA: Addison Wesley.

Jørgensen, S. E., and F. Müller. 2000. Ecosystems as complex systems. In: Handbook of Ecosystem Theories and Management, edited by S. E. Jørgensen and F. Müller, pp. 5–20, New York: Lewis Publishers.

Jørgensen, S. E., B. C. Patten, and M. Straskraba. 1992. Ecosystems emerging: toward an ecology of complex systems in a complex future. *Ecological Modelling*, 62, 1–27.

Kauffman, S. A. 1993. *The Origins of Order*. New York: Oxford University Press.

Kauffman, S. A. 2000. *Investigations*. New York: Oxford University Press.

Kleppner, D. 1999. The yin and yang of hydrogen. *Physics Today*, 52, 11–13.

Klinger, L. F. 1988. *Successional change in vegetation and soils of southeast Alaska*. Ph.D. dissertation, University of Colorado, Boulder.

Klinger, L. F. 1991. Peatland formation and ice ages: a possible Gaian mechanism related to community succession. In: *Scientists on Gaia*, edited by S. H. Schneider and P. J. Boston, pp. 246–255, Cambridge, MA: The MIT Press.

Klinger, L. F. 1996a. The myth of the classic hydrosere model of bog succession. *Arctic and Alpine Research*, 28, 1–9.

Klinger, L. F. 1996b. Coupling of soils and vegetation during peatland succession. *Arctic and Alpine Research*, 28, 380–387.

Klinger, L. F. 1997. The silence of the ecologists. *Bulletin of the Ecological Society of America*, 78, 172–173.

Klinger, L. F., S. A. Elias, V. M. Behan-Pelletier, and N. E. Williams. 1990. The bog climax hypothesis: fossil arthropod and stratigraphic evidence in peat sections from southeast Alaska. *Holarctic Ecology*, 13, 72–80.

Klinger, L. F., and D. J. Erickson, III. 1997. Geophysiological coupling of marine and terrestrial ecosystems. *Journal of Geophysical Research/Atmospheres*, 102, 25359–25370.

Klinger, L. F., J. Greenberg, A. Guenther, G. Tyndall, P. Zimmerman, M. M'Bangui, J.-M. Moutsambotè, and D. Kenfack. 1998. Patterns in volatile organic compound emissions along a savanna-rainforest gradient in central Africa. *Journal of Geophysical Research/Atmospheres*, 103, 1443–1454.

Klinger, L. F., and Q.-J. Li. 2001/2002. Gaia in China. *Gaia Circular*, Winter, 13.

Klinger, L. F., Q.-J. Li, A. B. Guenther, J. P. Greenberg, B. Baker, and J.-H. Bai. 2002. Assessment of volatile organic compound emissions from ecosystems of China. *Journal of Geophysical Research*, 107 (D21), 4603, doi:10.1029/2001JD001076.

Klinger, L. F., and S. K. Short. 1996. Succession in the Hudson Bay lowland, northern Ontario, Canada. *Arctic and Alpine Research*, 28, 172–183.

Klinger, L. F., P. R. Zimmerman, J. P. Greenberg, L. E. Heidt, and A. B. Guenther. 1994. Carbon trace gas fluxes along a successional gradient in the Hudson Bay lowland. *Journal of Geophysical Research/Atmospheres*, 99, 1469–1494.

Korvin, G. 1992. *Fractal Models in the Earth Sciences*. Amsterdam: Elsevier.

Lejeune, O., P. Couteron, and R. Lefever. 1999. Short range cooperativity competing with long range inhibition explains vegetation patterns. *Acta Oecologica*, 20, 171–183.

Lejeune, O., and M. Tlidi. 1999. A model for the explanation of vegetation stripes (tiger bush). *Journal of Vegetation Science*, 10, 201–208.

Lenton, T. M., and M. van Oijen. 2002. Gaia as a complex adaptive system. *Philosophical Transactions of the Royal Society*, B357, 683–695.

Leonardi, S., and H. J. Kumpel. 1999. Variability of geophysical log data and the signature of crustal heterogeneities at the KTB. *Geophysical Journal International*, 135, 964–974.

Levin, S. A. 1998. Ecosystems and the biosphere as complex adaptive systems. *Ecosystems*, 1, 431–436.

Lewin, R. 1996. All for one, one for all. *New Scientist*, December 14, 28–33.

Liu, D. 1979. *The Tao and Chinese Culture*. New York: Schocken Books.

Lovelock, J. 1995. *The Ages of Gaia*. New York: W. W. Norton.

MacCormac, E. R. 1998. Symmetry and asymmetry in science and technology. *Journal of the Society for Philosophy and Technology*, 4, 53–61.

Margalef, R. 1963. On certain unifying principles of ecology. *American Naturalist*, 97, 357–375.

Marr, J. W. 1961. *Ecosystems of the East Slope of the Front Range in Colorado*. University of Colorado Study Series in Biology, 8. Boulder, CO: University of Colorado Press.

Odum, E. P. 1969. The strategy of ecosystem development. *Science*, 164, 262–270.

Olsen, J. S., J. A. Watts, and L. J. Allison. 1982. *Carbon in Live Vegetation of Major World Ecosystems*. Publication ORNL-5862. Oak Ridge, TN: Oak Ridge National Laboratory.

Pelletier, J. D. 1997. Analysis and modeling of the natural variability of climate. *Journal of Climate*, 10, 1331–1342.

Pelletier, J. D. 1999. Self-organization and scaling relationships of evolving river networks. *Journal of Geophysical Research*, B104, 7359–7375.

Prigogine, I. 1996. *The End of Certainty*. New York: The Free Press.

Rayner, A. D. M. 1997. *Degrees of Freedom*. London: Imperial College Press.

Richards, A., P. Phipps, and N. Lucas. 2000. Possible evidence for underlying non-linear dynamics in steep-faced glaciodeltaic progradational successions. *Earth Surface Processes and Landforms*, 25, 1181–1200.

Ricotta, C., and G. C. Avena. 1998. Fractal modelling of the remotely sensed two-dimensional net primary productivity pattern with annual cumulative AVHRR NDVI data. *International Journal of Remote Sensing*, 19, 2413–2418.

Solé, R. V., and S. C. Manrubia. 1995. Are rainforests self-organized in a critical state? *Journal of Theoretical Biology*, 173, 31–40.

Tisdale, E. W., M. A. Fosberg, and C. E. Poulton. 1966. Vegetation and soil development on a recently glaciated area near Mount Robson, British Columbia. *Ecology*, 47, 517–523.

Turcotte, D. L., and W. I. Newman. 1996. Symmetries in geology and geophysics. *Proceedings of the National Academy of Sciences of the USA*, 93, 14295–14300.

Van Cleve, K., and L. A. Viereck. 1981. Forest succession in relation to nutrient cycling in the boreal forest of Alaska. In *Forest Succession: Concepts and Application*, edited by D. C. West, H. H. Shugart, and D. B. Botkin, pp. 185–211. New York: Springer-Verlag.

Vicsek, T. 1989. *Fractal Growth Phenomena*. Singapore: World Scientific.

Von Bloh, W., A. Block, and H. J. Schellnhuber. 1997. Self-stabilization of the biosphere under global change: A tutorial geophysiological approach. *Tellus*, 49B, 249–262.

Wardle, P. 1991. *Vegetation of New Zealand*. Cambridge: Cambridge University Press.

Watson, A. J., and J. E. Lovelock. 1983. Biological homeostasis of the global environment: The parable of Daisyworld. *Tellus*, 35B, 284–289.

Whittaker, R. H., and W. A. Niering. 1965. Vegetation of the Santa Catalina Mountains, Arizona: a gradient analysis of the south slope. *Ecology*, 46, 429–452.

17
Gaia and Observer Self-selection

Andrew J. Watson

Abstract

Many of our perceptions about the properties of life at the planetary scale are biased by "observer self-selection"—the fact that we are able to observe the Earth only because life has persisted on it for billions of years and complex observers have evolved here. To account for this bias, it is necessary to be aware of the implicit assumptions we make. In this chapter I follow the consequences of assuming a variant of the "anthropic principle," that the Earth is typical of planets that evolve observers. (Such planets may, however, be very unusual in comparison with most planets, or even most planets that evolve life.) I suggest the following defensible speculations based on this assumption:

• Observer species such as humans may evolve only on "Gaian" planets—that is, ones where the biosphere tends to regulate planetary conditions. This follows because observers are unlikely to evolve on any planet in a very much shorter period than we ourselves have evolved. During such a period ($\sim 5 \times 10^9$ years) stars evolve and substantially increase their output of energy, while planets are subject to major disruptive impacts. Nonregulating biospheres may exist (Mars may have had one), but their life spans may be too short for observers to evolve. In this sense, the planetary regulation that seems to have happened on Earth may be a consequence of observer self-selection.

• Over long periods Earth history is "Gaia-like" and life has tended to stabilize the environment. There is no need, however, to argue that this is an innate property of life at the planetary scale ("innate Gaia")—a simpler explanation is that it is chance that a few critical feedbacks are operating this way ("lucky Gaia"). What we know of Earth history is consistent with these negative feedback mechanisms being assembled by chance. In support of this, there appear to have been several transition periods where newly evolved styles of organisms disrupted negative feedbacks and temporarily destabilized the environment. So far the system has always settled to a new stable state amenable to life after such disruptions, but this need not be the inevitable outcome. The Neoproterozoic "snowball Earth" hypothesis is a potential test case here. In its extreme form this hypothesis is incompatible with "innate Gaia" because it invokes an entirely nonbiological mechanism to save the Earth from being permanently frozen over.

• The timing of the major stages that paced the evolution of life on Earth suggests that the origin of prokaryotic life may not itself be a critical limiting step. The most difficult stages in evolution on Earth took place after the establishment of bacteria. Prokaryote biospheres may therefore be rather easily established and relatively common, but complex life may be rare. In particular, scenarios to explain the early appearance of prokaryotes suggest that the Mars probably also had a prokaryotic biosphere in its past.

Introduction

Lovelock and Margulis (1972) originally proposed that the environment of the Earth's surface was "homeostated, by and for the benefit of, the biota." This implied that life on Earth had a major influence on the environment, an idea that can be traced back at least to Hutton (1795). Furthermore this influence was almost always in the direction of tending toward greater stability. Since the 1970s, the first of these implications has come to be accepted as mainstream science. We all agree that life does influence the Earth environment, through the composition of the atmosphere, for example. However, the second implication, that life tends to stabilize the environment, remains controversial. It clearly has not *always* been the case, because we can point to specific events (such as the rise of oxygen in the atmosphere) where biological processes have brought about major change. We may ask nevertheless, whether there is a tendency for life to stabilize the environment, and if so, whether there is any more fundamental reason for this than pure chance, perhaps a very fortunate coincidence.

Does Gaia exist, and if it does, is it "just lucky," or is the Gaia tendency "innate" in life on Earth? By "just lucky," I mean that regulation is not a property that biospheres in general have, and that it is simply good luck that some key feedbacks are negative and have lengthened the life span of the biosphere on Earth. By contrast, if Gaia is "innate," this would mean that there is some intrinsic tendency of a biosphere toward regulation (for example, that regulation is an "emergent property" of biospheres), and we would expect biospheres that evolve on other planets to have this tendency also.

When asking such general questions, which we hope will be applicable to life wherever in the universe it evolved, we are hampered by the fact that we have knowledge of the history of life on only one planet. Furthermore, our observation is subject to an inescapable anthropic bias: our own existence is dependent on the fact that life on Earth evolved to produce organisms as complex as ourselves. We therefore cannot consider the Earth to be typical of planets that have life on them (by "typical" I mean specifically that the Earth would be an example chosen in a random, unbiased way from the population of all planets on which life evolved). At best, we can only consider it to be typical of the smaller subset of planets on which complex observer species have evolved. Furthermore, we do not know the size of this subset of planets, except that it contains at least one member—it may, for all we know contain *only* one member, and we are then unique in the universe.

Is there anything we can say that might have general applicability to life elsewhere in the universe, given that our only example is biased in this fundamental way? We can make some statements that will have general applicability, providing we correctly account for the bias in our observations, which we can do by applying Bayes's theorem of conditional probabilities. In the present context, lacking further a priori constraints, we may reasonably assume our observations of Earth are representative of the subset of planets on which observers evolve, but not of some wider set of planets. For short, let us call this the ETPO assumption—standing for "Earth is a typical planet with observers."

In the remainder of this chapter I explore a few consequences of this assumption relevant to the debate over Gaia. The ETPO assumption is obviously a close kin to the various "anthropic principles" discussed by many authors since the term was introduced by Carter (1974). It is, for example, stronger than the "weak anthropic principle," which is uncontroversial in that it states simply that the universe must be compatible with our existence as observers (Barrow and Tipler, 1986). It is related to the "self-sampling assumption" Bostrom (2000, 2002), according to which each individual observer is, or should reason as if, he or she is a typical observer selected at random from all observers in the same reference class. In the context of discussion of the history of life on Earth, the ETPO assumption, which applies similar reasoning to the Earth as a whole, seems appropriate. While ETPO is not inevitably correct—Earth could be a very *atypical* type of planet with observers—in the absence of other a priori information it is a reasonable assumption, one that much of the relevant discussion in the literature, including Carter's seminal work, seems implicitly to make.

How Long Would We Expect Observers to Take to Evolve?

An important argument on which we can make progress is the probable time it would take for complex organisms to evolve on any planet, assuming that the basic processes occur by mechanisms similar to those that have occurred here on Earth. Adapting an argument originally made by Carter (1983), we can conclude, if we adopt the ETPO assumption, that the expectation time for observers to arise, is of the order of, or greater than (probably much greater than), 10^{10} years.

Carter considered three cases, denoting the expectation time for observers to evolve on an earthlike planet as τ_o and the lifetime of stars on the main sequence as τ_{ms} (of order 10^{10} years):

$$\tau_o \ll \tau_{ms} \tag{1}$$

$$\tau_o \sim \tau_{ms} \tag{2}$$

$$\tau_o \gg \tau_{ms} \tag{3}$$

Case (1) is incompatible with the ETPO assumption because, if true, the average time for observers to arise would be much less than the 4.5×10^9 years that it has taken on Earth, so Earth could not be a typical planet with observers. Case (2) Carter argued to be a priori unlikely because there is no physical relation between the biological processes giving rise to observers and the processes determining the lifetime of stars. This leaves case (3) as the most likely. Case (3) implies that intelligent observers are very rare, with most planets failing to evolve to that stage before being obliterated by their evolving stars. However, among those few planets on which observers do arise, the period required would normally be $\sim\tau_{ms}$, with observers normally arising near the end of the life of the planet. The observation that we have evolved on the Earth in a time $\tau_e \sim \tau_{ms}$ is then consistent with ETPO in case (3). The ETPO assumption is also compatible with case (2), and we do not, for the purposes of this chapter, have to accept the reasoning that led Carter to reject this case. Either case (2) or case (3) implies that observers take several billion years to evolve on their planets, arising toward the end of their star's main sequence life. The difference would be that in case (2), most biospheres that survive several billion years will produce observers, whereas in case (3) they mostly will not.

If we are correct in assuming that observers require several billion years to evolve, it follows that a planet with liquid water at its surface over such a period will be an essential requirement to reach observer status. Given that the output of stars changes substantially over this time, planets on which negative feedback mechanisms operate to stabilize the environment are more likely to sustain long-lived biospheres. Thus most planets on which observers arise are likely to have Gaia-like properties. It seems probable, therefore, that most observers will see such a "self-regulatory" history of life on their planet. In this sense, the Gaia hypothesis could be regarded as a consequence of observer self-selection: biospheres that are not self-regulatory may simply tend to expire before they evolve observers.

Innate Gaia or Lucky Gaia?

The surface temperature of the Earth has been kept within habitable bounds, apparently by regulatory mechanisms, for billions of years. A full understanding of these mechanisms, especially in the early history of the Earth, is still lacking, but it seems nearly certain that they will involve life at a fundamental level. For example, an important part of the temperature regulation is through the adjustment of the CO_2 greenhouse effect via the carbonate-silicate weathering cycle. This negative feedback mechanism was originally conceived by Walker et al. (1981) to be purely inorganic in its operation. It was soon pointed out, however (Lovelock and Watson, 1982), that the biota substantially enhance the rate of carbonate-silicate weathering, so that today this is a biologically mediated process. In the Archean, this mechanism may have been supplemented by the presence of other greenhouse gases, since evidence from paleosols indicates that CO_2 may not have been at high enough concentrations to maintain temperatures against the weaker solar flux of that time (Rye et al., 1995). Possibly, increased concentrations of (biogenic) methane may have compensated for the lack of CO_2 (Pavlov et al., 2000). It is therefore difficult to escape the conclusion that the biota has played an important part in maintaining the temperature of the planet through most of its history.

Is long-term temperature regulation an inevitable property of biospheres, or is it just chance that the Earth system has this property? If biospheres are in general nonregulatory, or too weakly regulatory to greatly extend their life spans, then the probability that observers could arise on a given planet, already small if we follow Carter's reasoning, drops still farther. But if biospheres intrinsically regulate their planetary environments, this would substantially improve the odds that some might eventually produce observers.

An examination of the history of Earth's temperature regulation offers some clues to this question. Earth's temperature does not behave as if regulated by a well-adjusted central heating unit, but is punctuated by a number of major glacial periods. The most important of these are the Huronian, the Neoproterozoic, Late Ordovician, Permo-Carboniferous, and the modern cooling which began in the late Cenozoic. Figure 17.1 sketches the Earth's temperature through time, and compares it with the timing of major events in evolution. It is intriguing that at least three of the cold periods appear to follow major evolutionary innovations. The Huronian glaciation occurred at about the time of the appearance of free oxygen in the atmosphere, possibly because of the passing of a methane-dominated greenhouse following the development of oxygenic photosynthesis (Pavlov et al., 2000). The Neoproterozoic glaciations occurred at around the time of the radiation of metazoans and immediately before the "Cambrian explosion," and the Permo-Carboniferous has been modeled as a response to the increased weathering due to the radiation of vascular plants (Berner, 1990, 1991).

This history does not give much support to the view that temperature regulation is an innate property of the biosphere. Pursuing the analogy of a domestic heating unit, we would have to conclude that Earth's system was not installed by a competent heating engineer and there is no overriding principle that ensures that it works. It looks more like the invention of a barely competent householder who periodically indulges in do-it-yourself home improvements.

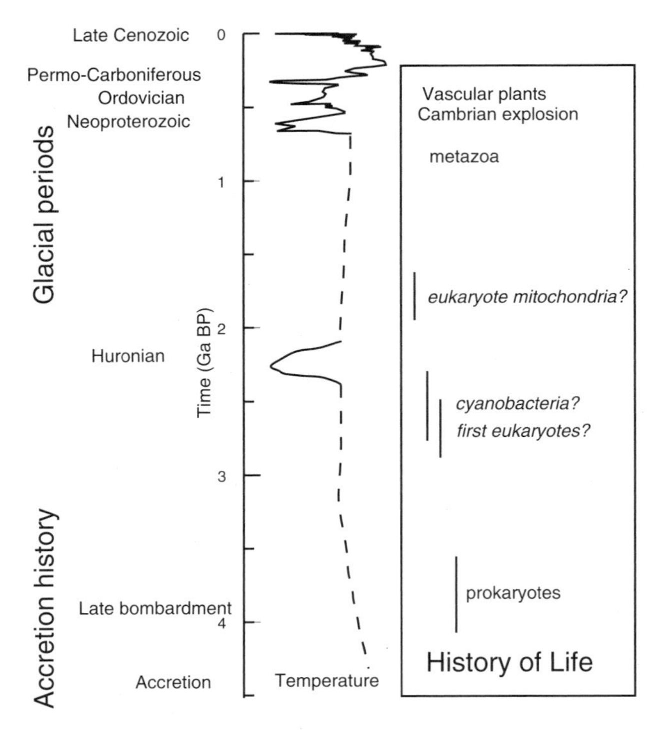

Figure 17.1
Time line of the Earth's temperature and accretion history compared with major evolutionary steps. Timing of evolutionary steps in italics are taken from the genomic analyses of Hedges et al. (2001). Nonitalicized evolutionary steps are from first appearances in the fossil record. Dashed lines in the temperature record indicate periods of poorly known temperatures where no glacial deposits are evident.

Feedback mechanisms seem to have regulated the temperature in "normal" periods, but at certain transition periods, biological innovation may have actively destabilized the system.

Could such a biological destabilization be sufficient to kill all life on Earth? Lovelock (1979) observed that this is probably unlikely, because it would require that a change brought about by life should kill all life, but in this case the organisms causing it would also kill themselves and thus would be self-limiting. However, in the case of runaway glaciation, this objection is not valid. It is perfectly possible to imagine that a biological innovation (such as enhanced weathering due to emerging land plants) would result in an extreme glaciation, with the biota causing the change mostly residing in the unfrozen tropics. At some point, runaway ice-albedo feedback might induce a "snowball Earth," exterminating all, or almost all, life. The current "snowball Earth" hypothesis of the Neoproterozoic glaciations proposes that such a runaway did occur (though not necessarily initiated by a biological innovation). In this theory, the planet was brought back from the icebound state only by the near-complete cessation of silicate weathering under the global ice sheets. An inorganic mechanism is thus postulated as the ultimate safety net for an Earth on which the normal temperature regulation mechanisms have broken down. "Snowball Earth" is a viable but far from proven hypothesis at present. As such it might be seen as an important test case for the Gaia hypothesis, because if it turns out to be correct, it would be a severe challenge to the notion of efficient temperature regulation by the biota.

The best-documented climate changes in Earth history are the late Quaternary glaciations. We know in detail how atmospheric composition has changed and how the temperature of the planet has varied over the last 420,000 years. We are also reasonably certain about the ultimate cause of these temperature variations: relatively minor changes in the pattern of insolation due to Milankovitch oscillations in the Earth's orbital parameters. Figure 17.2, redrawn from Petit et al. (1999), shows the variations of temperature, CO_2, and methane, and compares them with a measure of the orbital forcing. For the late Pleistocene, therefore, current understanding is that the net biogeochemical response is a positive feedback on global temperature, tending to increase the amplitude of the changes above that which would be expected for the purely physical climate system. Atmospheric CO_2 and methane increase in the atmosphere during the warm phases and decrease during the cold phases, reinforcing the changes due to ice–albedo response to the Milankovitch cycles. Here, too, therefore, there is plenty of evidence that the biota is active in altering global climate, but not that they tend to stabilize it.

To summarize, the temperature history of the Earth shows evidence for Gaian-like temperature regulation over long intervals, punctuated by periods where the biota apparently destabilize the planetary temperature. The detailed record of the late Quaternary also suggests that biological mechanisms provided positive, destabilizing feedback during this time. (These observations are not inconsistent with the properties of the Daisyworld model, as introduced by Watson and Lovelock [1983] and further explored by Lovelock [1987] and Lovelock and Kump [1994]). The biosphere thus appears to have a sometimes efficient, but frequently erratic, thermostat. Though the thermostat has extended the lifetime of the biosphere to several billion years, it is hard to make a case that it is sufficiently reliable to virtually guarantee that out-

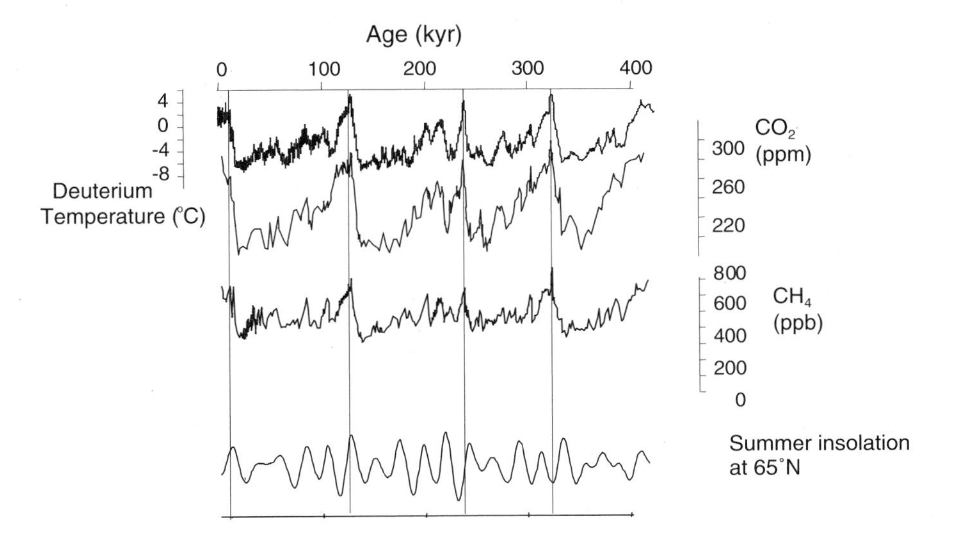

Figure 17.2
The Vostok ice core record of local temperature (as recorded by deuterium in ice, topmost curve), atmospheric CO_2 (second from top) and methane (third from top), over the last 420,000 years. The record is compared with sumertime insolation at 65°N (bottom curve). Redrawn from Petit et al. (1999). The coherence of the climate and atmospheric concentrations with Milankovitch cycles indicates that the ultimate cause of the glacial–interglacial cycles is probably orbital forcing. The CO_2 and methane response increases the magnitude of the temperature cycle, indicating that the biogeochemical system is providing positive feedback to temperature change during these cycles.

come. I suggest, therefore, that it is not inevitable that biospheres in general are efficiently self-regulating, though they probably have some tendency in that direction. "Lucky Gaia" seems a simpler and more defensible hypothesis than "innate Gaia."

Critical Steps in Evolution

Carter (1983) argued that sentient observers have taken several billion years to evolve on Earth because there were certain necessary, but inherently unlikely, critical steps in their evolution. These steps are defined as being unlikely in the sense that the expectation time to reach each one is much longer than the period of Earth history that has actually elapsed. In the case of the Earth (see figure 17.1), candidate critical steps might include the evolution of prokaryotes with a DNA-based genetic code, the origin of photosynthesis, eukaryotes, metazoa, and finally symbolic language in humans. In practice, while it is clear that some such set of critical steps exists, it is far from easy to enumerate them or rank them in order of difficulty. Evolution by natural selection is a powerful mechanism for producing organisms that could never be assembled by pure chance, so simply examining the structural complexity involved in achieving a critical step is not a good guide to how likely it is to occur. In practice, complex organisms can evolve rapidly by a sequence of incremental steps, each of which occurs in a (geologically) short time, provided each of those increments is itself favored in evolutionary terms (Dawkins, 1986). This difficulty is reflected in the literature: Carter (1983) was of the opinion that there were probably no more than two critical steps (defined as above) leading to human observers on Earth, but Barrow and Tipler (1986) attempted to enumerate ten and considered it possible that there were thousands.

A better guide to the critical steps than complexity alone might be the length of time between the steps that actually occurred on the Earth. We would expect that these difficult steps would be widely, and more or less evenly, spaced through Earth history. Figure 17.1 shows the timing of the possible critical steps listed above. Prokaryotic life appears to have been in place by 3.5 Ga bp (Knoll and Barghoorn, 1977; Schopf and Packer, 1987). This is only 0.4 Ga after the cessation of the late bombardment, the period of intense impacts that melted the surface of the moon and would have done the same for the Earth—it seems unlikely that life could survive through that period.

Currently, the time for the origin of oxygenic photosynthesis is unknown. If we accept that the organisms in the Woonawarra formation are cyanobacteria, as suggested by Schopf and Packer (1987) on morphological evidence, then photosynthesis followed very rapidly after the establishment of prokaryotes. This proposal has always been controversial, and one can be skeptical of it. Recently Brasier et al. (2002) have raised doubts as to whether these fossils were formed by living cells at all. However, there still seems to be little doubt that the nearly contemporaneous African microfossils (Knoll and Barghoorn, 1977) are evidence for life, though not for photosynthesis. Recent studies using genomic analysis to establish a timescale suggest a much later origin for cyanobacteria (Hedges et al., 2001). The first unequivocal evidence for oxygenic photosynthesis does not occur until the appearance of free oxygen in the atmosphere at the close of the Archean, ~2.2 Ga bp. Eukaryotes evolved sometime before about 1.8 Ga bp, possibly in two stages (Hedges et al., 2001); metazoa before 0.8 Ga bp; and language at ~0.001 Ga bp. This chronology suggests the true identity of some at least of the critical steps.

In particular, the early establishment of prokaryotes, only ~0.4 Ga after the late bombardment, is puzzling. One would think that there must be at least one, and probably several, critical steps before self-replicating DNA-based organisms became established. The rapid establishment of bacterial life could be a coincidence, but this is inconsistent with ETPO (since the Earth would not then be typical). Perhaps there is some reason why life could arise only in this first few hundred million years, but we do not at present have any reason to suspect this. Two further explanations for the early appearance of prokaryote life would be the following:

1. The sequence of events required to produce a living cell is sufficiently likely that it will probably occur, given a suitable environment and a few hundred million years (e.g., the origin of prokaryotes is not so unlikely as to be "critical" in the sense defined above).

2. Prokaryote life evolved elsewhere during some indeterminately long period before the Earth was formed, and was "seeded" onto the planet—panspermia.

In either case, prokaryote life would then be expected to have arisen on other planets as well, though in case (2) these may be localized to nearby solar systems, or perhaps just this solar system. In either case it is not unlikely that there has been prokaryote life and a primitive biosphere on Mars, since Mars clearly has periodically had liquid water on its surface.

The early evolution of prokaryote life on Earth is therefore a pointer suggesting that life is not spectacularly uncommon in the universe. However, the "Fermi paradox"—that we can see no evidence for extraterrestrial intelligences—can be interpreted as an indication that intelligent life is rare. The lack of success of the SETI programs in detecting any signals from extraterrestrial intelligences, also seems to indicate this. These observations would be naturally explained if the expectation time to pass the key stages between prokaryotes and observers is very much longer than the lifetime of main sequence stars. In this case prokaryote life may be common, but observers are rare because it is difficult to get from simple life to complex life. The Fermi paradox might also be explained by postulating that technological civilizations have very short lifetimes (before they destroy themselves, or lose interest in exploring the universe). However, this would not explain why the time we have taken to evolve is of the same order as the lifetime of the Earth as a habitable planet.

Since I first wrote this chapter, Ward and Brownlee (2000) have published their "rare Earth" hypothesis, that the evolution of complex life is an unlikely event, though simple (bacterial) life may be much more common. They use several lines of argument to support this case, and while none of these by itself is conclusive, together they are persuasive. The most telling of their arguments is essentially the one advanced here: that the timing of the emergence of simple and complex life is most compatible with this interpretation. Provided there are several steps that must occur before the evolution of complex life (as expected from the discussion above), Carter's critical step theory leads naturally to this "rare Earth" view.

The origin of photosynthesis is a candidate for an early critical step, and perhaps a better one than the origin of prokaryotes. This is particularly true if we assume that its origin is not constrained by the Woonawarra fossils and may therefore have occurred a billion or more years after life first appeared. Before photosynthesis evolved, the only sources of energy available to the protobiosphere would have been chemical or heat gradients created during the accretion of the planet, or by inorganic reactions in the atmosphere. Photosynthesis would have increased by many orders of magnitude the amount of energy flowing through the global ecosystem. As such, it would have increased the rate of genetic replication and therefore the pace of evolution (since evolution occurs by mutation during the transcription of DNA). A substantial energy source was also necessary before the biota could materially alter the composition of the atmosphere and oceans. For example, the maintenance of 1 ppm methane in the present atmosphere implies that ~0.3 percent of the carbon fixed at present returns to the atmosphere as methane (Lovelock, 1987; Watson et al., 1978). Production of this amount of methane requires far more energy than would have been available to a prephotosynthetic biosphere. In a more reduced atmosphere, the lifetime of methane would have been much longer, but the lifetime of oxidized species much shorter, than today. Thus the maintenance of an atmosphere far from thermodynamic equilibrium, characteristic of the modern Earth (Lovelock, 1987), would presumably not be a property of a planet on which the biota had not evolved photosynthesis. In fact, if the biology obtained its energy from chemical gradients, this would presumably tend to reduce the disequilibrium in the raw materials available to it, rather than enhancing it. Thus one might date the Gaian biosphere from the origin of photosynthesis rather than the origin of life.

Conclusion

Cosmologists are increasingly taking anthropic arguments seriously, and there is a growing literature on their application to the observed properties of the universe. Those of us involved in the Gaia debate, which is about the properties of life at the planetary scale, also need to be aware of the anthropic bias inherent in our particular place in the universe. The conclusions we may be led to are inevitably contingent and speculative, and a resolute reductionist might therefore dismiss them as a waste of time. This would be a mistake, however, for the many advances that will occur in our knowledge of Earth history and space exploration in the coming decades will mean that our speculations have a reasonable chance of being tested. This brings them firmly into the realm of Popperian "conjecture and refutation" and makes them a subject of scientific rather than purely philosophical inquiry.

Given our state of near-complete ignorance about the distribution and properties of life on other planets, the speculations in this chapter can only be regarded as provisional "best guesses." For example, a competent risk assessor asked to estimate the probability of life having existed on Mars, given present knowledge, would, I believe, use the reasoning and reach the conclusion that I have. However, a suitably planned mission to Mars that obtained reliable data on microfossils, or the lack of them, beneath the surface might radically change this conclusion. Similarly, good evidence for or against "snowball Earth," better evidence on the timing of evolutionary events in Precambrian history, and the results of proposed "Terrestrial Planet Finder" missions may all in time enable us to test and revise these conclusions.

Acknowledgments

I thank many people for stimulating discussions that have helped to shape the ideas in this chapter, particularly Tim Lenton, Noam Bergman, and my family—Jackie, Adam, and James Watson. I also thank Jim Kasting, and one anonymous reviewer, for helpful reviews of the manuscript.

References

Barrow, J. D., and F. J. Tipler. 1986. *The Anthropic Cosmological Principle*. Oxford University Press, Oxford.

Berner, R. A. 1990. Atmospheric carbon-dioxide levels over Phanerozoic time. *Science*, 249, 1382–1386.

Berner, R. A. 1991. A model for atmospheric CO_2 over Phanerozoic time. *American Journal of Science*, 291, 339–376.

Bostrom, N. 2000. Observational selection effects and probability. Ph.D. thesis, London School of Economics.

Bostrom, N. 2002. *Anthropic Bias: Observer Selection Effects in Science and Philosophy*. Routledge, New York.

Brasier, M. D., O. R. Green, A. P. Jephcoat, A. K. Kleppe, M. J. Van Kranendonk, J. F. Lindsay, A. Steele, and N. V. Grassineau. 2002. Questioning the evidence for Earth's oldest fossils. *Nature*, 416, 76–81.

Carter, B. 1974. Large number coincidences and the Anthropic Principle in cosmology. In M. S. Longair, ed., *Confrontation of Cosmological Theories with Observational Data*, pp. 291–298. Reidel.

Carter, B. 1983. The anthropic principle and its implications for biological evolution. *Philosophical Transactions of the Royal Society of London*, A310, 347–363.

Dawkins, R. 1986. *The Blind Watchmaker*. Harlow, UK: Longman Scientific.

Hedges, S., H. Chen, S. Kumar, D. Y.-C. Wang, A. S. Thompson, and H. Watanabe. 2001. A genomic timescale for the origin of eukaryotes. *Biomed Central Evolutionary Biology*, 1(4). http://www.biomedcentral.com/1471-2148/1/4.

Hutton, J. 1795. *A Theory of the Earth, with Proofs and Illustrations*, vol. 2, Creech, Edinburgh.

Knoll, A. H., and E. S. Barghoorn. 1977. Archean microfossils showing cell division from the Swaziland system of South Africa. *Science*, 198, 396–398.

Lovelock, J. E. 1979. *Gaia: A New Look at Life on Earth*. Oxford University Press, Oxford.

Lovelock, J. E. 1987. *The Ages of Gaia: A Biography of Our Living Earth*. Oxford University Press, Oxford.

Lovelock, J. E., and L. R. Kump. 1994. Failure of climate regulation in a geophysiological model. *Nature*, 369, 732–734.

Lovelock, J. E., and L. Margulis. 1974. Atmospheric homeostasis by and for the biosphere: The Gaia hypothesis. *Tellus*, 26, 2–10.

Lovelock, J. E., and A. J. Watson. 1982. The regulation of carbon-dioxide and climate—Gaia or geochemistry. *Planetary and Space Science*, 30, 795–802.

Pavlov, A. A., J. F. Kasting, L. L. Brown, K. A. Rages, and R. Freedman. 2000. Greenhouse warming by CH_4 in the atmosphere of early Earth. *Journal of Geophysical Research—Planets*, 105, 11981–11990.

Petit, J. R., J. Jouzel, D. Raynaud, N. I. Barkov, J. M. Barnola, I. Basile, M. Bender, J. Chappellaz, J. Davis, G. Delaygue, M. Delmotte, V. M. Kotlyakov, M. Legrand, V. Y. Lipenkov, C. Lorius, L. Pépin, C. Ritz, E. Saltzman, and M. Stievenard. 1999. Climate and atmospheric history of the past 420 000 years from the Vostok ice core, Antarctica. *Nature*, 399, 429–436.

Rye, R., P. H. Kuo, and H. D. Holland. 1995. Atmospheric carbon-dioxide concentrations before 2.2-billion years ago. *Nature*, 378, 603–605.

Schopf, J. W., and B. M. Packer. 1987. Early Archean (3.3-billion to 3.5-billion-year-old) microfossils from Warrawoona group, Australia. *Science*, 237, 70–73.

Walker, J. C. G., P. B. Hays, and J. F. Kasting. 1981. A negative feedback mechanism for the long-term stabilization of earth's surface-temperature. *Journal of Geophysical Research—Oceans and Atmospheres*, 86, 9776–9782.

Ward, P. D., and D. Brownlee. 2000. *Rare Earth: Why Complex Life Is Uncommon in the Universe*. Springer-Verlag, New York.

Watson, A. J., J. E. Lovelock, and L. Margulis. 1978. Methanogenesis, fires and the regulation of atmospheric oxygen. *Biosystems*, 10, 293–298.

Watson, A. J., and J. E. Lovelock. 1983. Biological homeostasis of the global environment—the parable of Daisyworld. *Tellus*, 35B, 284–289.

J. H. Weissenbruch, *Landscape at Noorden, near Nieuwkoop*, 1891. Reproduced with permission from Museum Jan Cunen, Oss.

18
Taming Gaia: The History of the Dutch Lowlands as an Analogy to Global Change

Peter Westbroek

Abstract

If we wish to predict the long-term future of human involvement with Gaia, solid scientific methodology is of little use. We are forced into the sloppy approach of prediction by analogy with interactions between nature and culture in the past. This chapter discusses the making of the Dutch landscape as an analogy to the global effects of humans in the future. It concentrates on a delightful natural reserve that is little more than an abandoned peat mining district—the result of 1,000 years of hard labor, miserable poverty, fighting, greed, and speculation. This country went through many cycles of environmental collapse and reclamation. By the end of the nineteenth century, painters were the first to recognize the beauty of the area. Their vision was essential in shaping our perception of beauty and for maintaining of the glory of this landscape. The moral of this story is that with good luck and wise management, something beautiful may emerge from the pludering of Gaia by humans. However, the vision of artists like Lovelock, who remind us of the unique beauty of this planet, cannot be missed.

Despite impressive advances in our understanding of Earth dynamics and the global effects of human activities, the challenge for the science of global change remains overwhelming. We are just beginning to see the global effects of human intervention, while the response of the system is not yet in. Furthermore, the model predictions become less reliable as we try to look farther ahead. Can we say anything meaningful about the future of this planet on the millennium scale? We now enter the realm of wild speculation, where the norm is usually set by ideological controversy and unspecified gut feelings, rather than by hard data. Is any holdfast remaining in this misty world of conjecture, so that we can put some tentative limits on the unbridled flow of fantasy? I see no other option than to use information on our historical heritage at the regional scale as an analogue to the long-term global relationship between nature and culture. This is a risky approach. We must keep in mind the warning by Kant (1998) that analogies tend to be misleading and should be avoided in core scientific practice. Thus, the message of this chapter is the result of informed speculation rather than solid research. But if we don't grasp the available holdfasts, however lean, we may better avoid thinking about the problem altogether.

I concentrate on a single case in point—the history of the landscape around Nieuwkoop, a township 20 miles from where I live. For this, I take the liberty to use a text I wrote in the past, which received only a limited distribution. I feel that my Nieuwkoop story has more than purely anecdotal significance.

Treading on the quicksand of analogies creates the obligation to specify the items to be compared. In this chapter, the natural conditions in the Nieuwkoop area before the colonization by humans are thought to represent Gaia before the advent of man; the local inhabitants stand for humanity; and the colonization process, for what I call the taming of Gaia. Although I admit that the analogy between Nieuwkoop and the global situation is deceptive in many ways, it may at least help to give some preliminary insights.

My use of "Gaia" rather than "Earth" in this context is intentional: the evocative overtones of the Gaia vision are more in line with the analogous approach of my narrative. To clarify this point, I have to give my personal credo on that subject. To me, Gaia has been, and continues to be, a new look at life and the Earth, to paraphrase the subtitle of Lovelock's 1979 book. It conveys the amazement of the Apollo 8 crew when, orbiting the moon, they took the now-classic pictures of the Earth over the barren lunar landscape. Those pictures have become clichés, but Gaia keeps the sensation alive and continues to be a source of inspiration. The Gaia view not only helps us scientists to develop a top-down systems approach to our planet and to discover global geobiological feedbacks, it also reminds us of our ignorance, the most familiar object in the solar system being the least understood. In addition, it keeps the general public aware that we live in a miraculous "oasis in space" (Cloud, 1988) which deserves the utmost respect.

From the start, Gaia has been presented as hardcore science—a hypothesis or a theory—though the idea could be expressed only in terms of analogies such as "the self-organizing planet," "the superorganism," or "Daisyworld." The tragic result has been a glorious vision bogging down in endless and pointless discussions on testability and definition. Gaia evades testing and definition: over billions of years its behavior has emerged from zillions of interactions and feedbacks, only a handful of which are presently under investigation. Gaia gives us no more than a glimpse of our planet as it is—enough, though, to stimulate exciting science. I propose that we adopt the biologists' attitude. They have learned to avoid defining their ultimate subject, "life," and to concentrate on particular biological phenomena. Likewise, we may leave our research to Earth system science, while always keeping Gaia in the back of our minds.

I like to see Gaia as a work of art with particular significance to science and society, one that emerged from the deep intuition of James Lovelock. This is no depreciation by any means. As I shall argue later in this chapter, the power of art is to reveal the unseen emotional significance of the world, and so to frame our behavior. Science needs to join forces with the arts in order to get to the heart of people.

An Artist's Perception

Jan Hendrik Weissenbruch (1824–1903) was one of the most distinguished Dutch painters in the nineteenth century. Skies, shores, and landscapes were his passion, in particular the wide, wet polders—stretches of land reclaimed from the sea. He never had to travel far, because all this beauty was abundantly available around The Hague, where he had lived all his life. From his home he could walk to the famous collection of Dutch paintings at the Mauritshuis Museum in five minutes, and as a young man he spent many hours there contemplating and even copying the works of his seventeenth-century idols, Johannes Vermeer and Jacob Ruisdael. Although he remained faithful to these great examples all his life, his unrestrained abandonment to nature forced him to develop his own view of the world. "At times, nature gives me a real blow," he used to say. In such moments drawing and painting was easy. He sketched his impressions in charcoal, so that later, at home, he could work them out in paint. Over the years, his style changed from meticulous renderings to a highly personal impressionism. What strikes the eye is the subtle balance between joyful and spontaneous virtuosity and compositional grandeur. In particular his monumental skies are unforgettable, with their infinite variety of blues and grays. He brought the polders to life, and taught us to feel at home in this flat, green land of mud and water.

Although public recognition came only toward the end of his life, Weissenbruch was one of the most prominent members of The Hague school, a rather loose association of painters whose heyday was from 1870 to 1900. Jozef Israëls, Jacob Maris and Anton Mauve were other well-known members. For many years, the seventeenth-century artistic blooming in the Low Countries had paralyzed rather than stimulated painting there, but The Hague school brought a revival. Inspired by the nationalism that kept all of Europe in its grip, these men rediscovered the beauty of the Dutch landscape and everyday life. With approaches ranging from realism to impressionism, they depicted a nostalgic view of Holland, not as it really was, but as they wanted to see it. It was a period of industrialization, but in their paintings, chimneys or trains at the horizon, or swimmers in the water are rarely seen. Neither were portraits a favorite subject. People were mostly part of the landscape, together with their villages, houses, cattle, and implements. They were farmers or fishermen, resting in the fields or in their humble dwellings. Cities and towns were shown from a distance, or as peaceful street corners and intimate gardens. These painters abhorred glamour and avoided the dynamics of modern life.

No wonder they soon found The Hague too busy a place for inspiration. To find a peaceful environment, they had to move into the surrounding countryside. One such place was the township of Noorden, at Lake Nieuwkoop. Toward the end of his life, Weissenbruch lived there during the spring and autumn, and he produced his finest works there. Much of that landscape is still intact. Over the years, Weissenbruch's paintings have contributed to the preservation of this land of pastures and waters, although their original purpose was solely aesthetic.

Was it really nature that Weissenbruch looked for in Noorden? The paintings reveal something different. By themselves, the skies, waters, and fields look natural enough, but at the same time one perceives the strong hand of mankind. The pastures, waters, cattle, and reeds were rigorously planned and maintained by hard labor. It was a productive landscape, a utility, created to provide the inhabitants with a living. Never was it meant to be beautiful. The aesthetic qualities of this land, as they emerged from the marriage of nature and man, were not immediately apparent, but had to be discovered by artists of Weissenbruch's stature.

Now, a century after Weissenbruch, the human involvement with nature is perceived as a vicious problem. Are we destroying the natural balance that keeps the Earth alive? Are we evil and greedy; is nature good and pristine? Weissenbruch's paintings show us how something delightful and new may emerge from human interference. This insight is deepened when we place his favorite landscape in a historical context. With all our longing for nature, it is good to realize what we really ask for.

On a Rowboat

A good way to approach the area is from the southwest. A little stream, the Meije, winds through the meadows. One can ride a bicycle atop the bordering dike. The Meije is on the left, and on the right, at the foot of the dike, is a long string of farmhouses, many of which are beautifully preserved specimens from the seventeenth century. Several of them have thatched roofs and are surrounded by pleasant orchards, shrubs, and vegetable gardens. Beyond the farmhouses, narrow canals separate long, green meadows, forming parallel rows that stretch away from the road. The meadows are often about 100 meters wide and 1 kilometer long. At their far end, away from the road, rows of trees interrupt the otherwise monotonous landscape. Then follows another row of meadows, trees, and so on. In the distance are the towers of Woerden and Bodegraven.

To the left, on the other side of the Meije, is a similar pattern of meadows and canals (figure 18.1). The landscape here is more loosely organized. The bushes at the end of the meadows are more haphazard than the trees on the other side, and the canals are not so straight. Reeds grow along the watercourse, together with patches of willows and a wealth of water flowers. Cows peaceably stare at infinity.

Rowboats can be rented everywhere, so one can continue over water, across the Meije and along one of the canals toward the bushes and the lake beyond. The meadows are soaked and nearly level with the water in the canal. They give way to fields of reeds, moss, and flowering plants: real boglands. A rectangular network of canals, ditches, and rows of alder, birch, and willow divide the area into a system of regular patches. The flowers bloom and the birds sing—this is nature at its best.

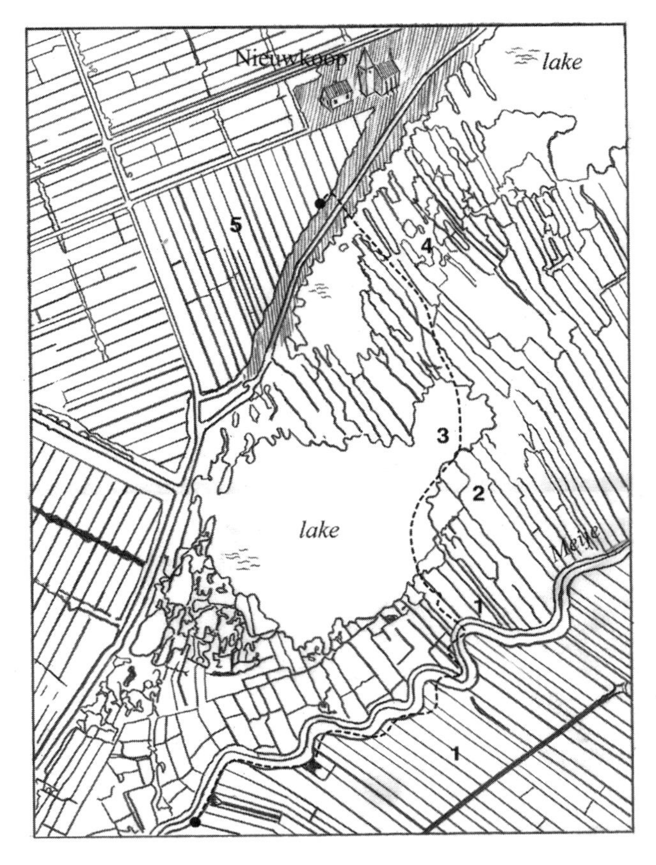

Figure 18.1
Map of the surroundings of Meije and Nieuwkoop, with five characteristic landscapes: (1) farmlands exploited since the Middle Ages; (2) reed lands exploited since the nineteenth century; (3) a lake; (4) remains of peat exploitations (mainly eighteenth century to the beginning of the nineteenth century); (5) the nineteenth-century polders at Nieuwkoop.

The canal now widens and ends in open water, one of the many lakes in the region. This one is two kilometers or so across, and about four meters deep. Reeds and clusters of trees line the shore. Across the lake we float into another new world, a complex labyrinth of narrow ridges of land alternating with waters up to 100 meters wide. Again, the land ridges are covered with reeds and trees. Then comes the dike that cuts off the bog area. And Nieuwkoop, 700 years old: dignified buildings, small houses, old and new, sailboats, restaurants, and hamburger palaces. The village forms a long ribbon along the dike and merges, several kilometers farther, into Noorden, where Weissenbruch used to stay.

Look for a spot on the dike from which you can view the polder beyond. The difference in height makes you dizzy, however flat this land may be. The polder is at least four meters below the surface of the waters we have just crossed. It was a lake until two centuries ago. Blocks of rectangular meadows are separated by meticulously arranged sets of perpendicular roads, dikes, rows of trees, and large farmhouses. The scale is larger, and the planning more efficient and modern, than the historic panorama where we started our trip.

This little excursion allows one to cross five types of landscape characteristic of the western region of the Netherlands: farmlands exploited since the Middle Ages; reed lands that originated in the nineteenth century; a lake; peat exploitations (mainly seventeenth century); and the nineteenth-century polder at Nieuwkoop. It is a carefully designed system of multilevel waterways, polders, and dikes—the result of a struggle of centuries between humanity and the elements. This is Weissenbruch's nature, yet there is nothing purely natural here. If left unattended, the whole area would soon be under water.

An interesting paradox underlies these terrains. A thousand years ago this was the "wilderness," a virtually impenetrable and uninhabitable region between the sandy hills to the east and the low sand dunes along the coast. In the seventeenth century, however, this area provided the economic basis for a mighty empire: Holland in the Golden Age.

Figure 18.2
Map of the Netherlands, including the deltas of the Rhine, Meuse, and Scheldt. The shaded area is lower than mean sea level plus one meter, and floods if left unprotected. Note the position of the dunes along the coast, the mudflats between the string of islands in the north and the mainland, and Nieuwkoop.

Natural Causes

To understand the development of this region, one must go back more than 10,000 years, when the last ice age ended and the present warmer period, the Holocene, began. Figure 18.2 shows the area in the larger geographical context. A major part of the Dutch territory can be viewed as a river delta merging into the North Sea. The area has been subsiding for a long time, and during the past million years, the rivers that flow down into this delta, particularly the Rhine and the Meuse, have filled the space that became available with sands and clays, debris from the Alps and other high regions upstream.

During the last glacial period, ice covered large continental areas around the Arctic; the southern boundary of that ice sheet ran across northern Germany and southern Denmark, and into the North Sea. So much water was tied up on land as ice that the level of the sea was some 100 meters lower than today. Much of the North Sea was dry, and in the Netherlands a polar desert or a tundralike regime prevailed. Most of the land was covered with sand brought down by the rivers and tossed around by the wind.

When temperatures moderated at the beginning of the Holocene period, the ice caps started to melt. The sea level rose, and reached the present Dutch coastline about 7,000 years ago. The sea went even farther inland and then was pushed back by the steadily accumulating sediment. The rivers brought down huge masses of clay that were swept into the sea and accumulated there in a thick blanket along the coast—a huge mudflat that widened over time, edging toward the land.

About 3,000 years ago, low sand dunes started to develop along the western and northern coast of the Netherlands, protecting the original mudflats from marine incursions. At this stage a zone behind the dunes, 20 to 40 kilometers wide, was transformed from mudflats into a huge marshland where large masses of peat could accumulate. The "wilderness"

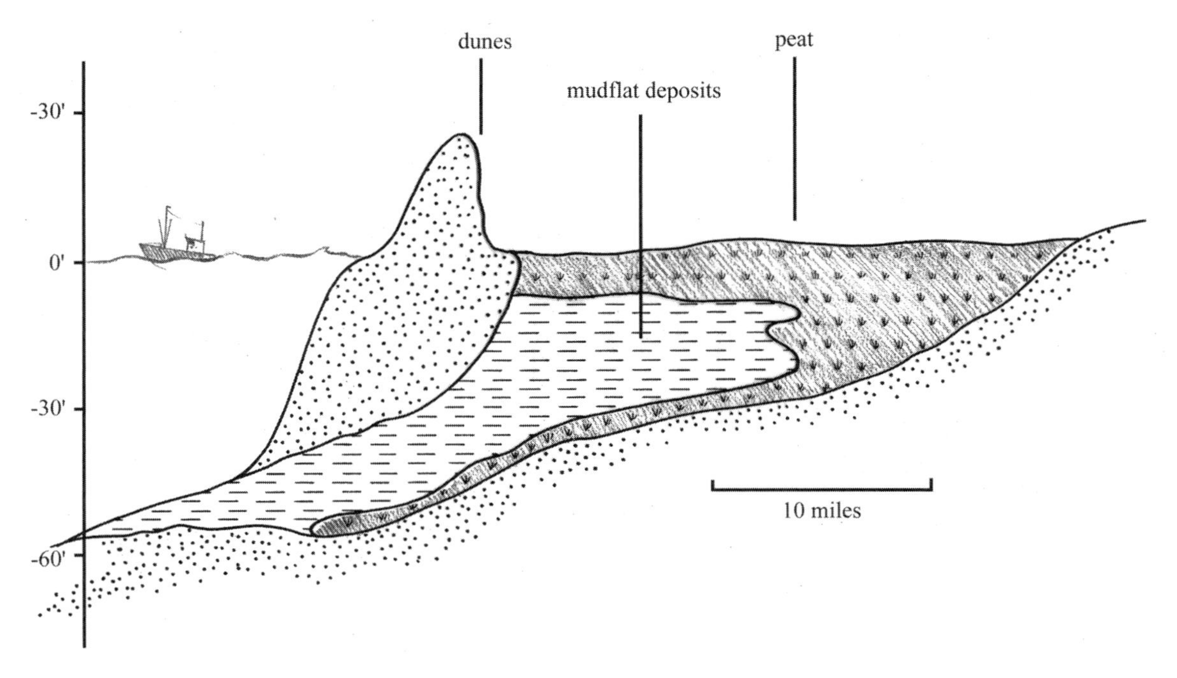

Figure 18.3
East–west cross section through young sediments in the western part of the Netherlands, showing extensive peat development behind coastal dunes.

was born. All that is left of the original mudflats is now in the north of the country, between the string of islands and the mainland. Figure 18.3 is an east-west cross section through the Holocene sediments in the middle of the country; from it you can deduce this sequence of events. First, the marine clay expands over the sandy underground, separated from the coast by a narrow strip of peat. The thin peat deposit at the base of the Holocene sediments is the result. Then come the dunes and the peat that covers the clay.

Peat is essentially a water-soaked mass of plant remains. When its production exceeds its decomposition, large deposits may form. Just underneath the plant cover, proliferating bacteria and fungi attack the black sludge and convert it into carbon dioxide and water. The oxygen required is poorly soluble in water, so the pore water becomes anaerobic, oxygen depleted. The consumers suffocate, so that their work must be taken over by bacteria that thrive in this anaerobic environment. Because anaerobic breakdown yields less energy, it is less efficient. Organic acids and carbon dioxide accumulate, and then the same thing happens as in the production of yogurt—the breakdown comes to a standstill. Throughout the Holocene period, the conditions in large parts of Holland were highly favorable for peat accumulation. In this deltaic area, rivers and rainfall provided abundant water. Temperatures were moderate under the prevailing marine climate and drainage was poor.

In this region a lake will fill up by itself with peat and become land. The kinds of life change in a well-defined sequence. In general, floating water plants and algae form the first debris to be accumulated in the lake. When the water is less than two meters deep, reeds can take over, and at a depth of half a meter, sedges dominate. Finally, the ground is high enough for trees to develop, leading to a type of peat full of roots and stems.

These successive plant communities all depend on groundwater for their development, and cannot grow much above the surface. With enough rain, however, peat moss (sphagnum) may dominate the scene. It has a very peculiar structure which allows it to hold water. In hot, dry periods it uses up its water reservoir and appears brown and dead. But during a shower it sucks up large quantities of water, and appears green and healthy again. Sphagnum also recycles its food very efficiently because it can grow high above the groundwater level, fed only by the nutrient-poor rainwater. It stores large quantities of extra nutrients in its cell walls, depriving its competitors of essential nourishment. Peat moss may rapidly suffocate the trees and shrubs that were forming the wood peat. As it does, it can produce mossy cushions up to seven

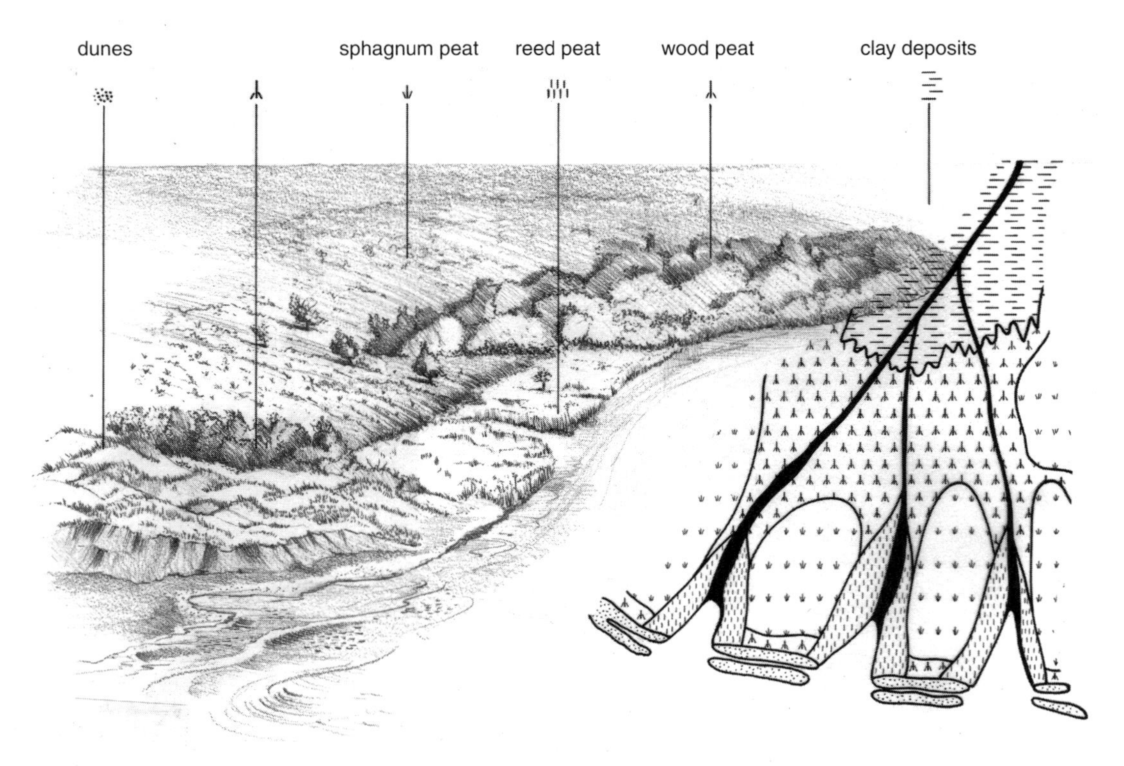

Figure 18.4
Distribution of various types of peat in a deltaic area such as Holland.

meters high, covering hundreds of square kilometers. These are curious constructions: gigantic water mattresses, pervaded and kept in place by a fine network of organic remains, with a thin veneer of living tissue on the surface.

This is what happened on a huge scale in Holland. The distribution of the different types of peat is shown in figure 18.4. The zones bordering the rivers and streams were regularly flooded and received a good share of nutrients and clay. Under these conditions, wood peat dominated. Close to the river mouths, where the sea turned the water brackish, reed peat was laid down; sphagnum cushions developed in the large areas in between. This development of peat clearly illustrates the role that life has played in this area during the past few thousand years. There are few other geological forces besides the growth of peat cushions that rival life in raising, by several meters, a stretch of land of this size in such a relatively short period of time.

Some 1,000 years ago, this terrain was wilderness. The Meije was just one of many streams in the region that removed superfluous water from the peatlands. Away from its banks, the surface gradually rose and was covered with swamp woods. Then, at about the present location of the open lakes, a huge sphagnum cushion with very few trees started to form, as would be expected. The contrast with the present is dramatic (see figure 18.5). The only thing that seems the same is the course of the Meije. How did the present situation evolve from the earlier, natural one?

The Impact of Culture

The wilderness used to be a forbidding place; even the Romans avoided it. Their settlements were only along the sandy levees of the main rivers. At the beginning of the thirteenth century, however, increasing population made exploitation inevitable. The area was brought under feudal control of the counts of Holland and the bishop of Utrecht, and a methodical cultivation system was initiated. Colonists were recruited from among the serfs, and in exchange for the heavy life, they were exempted from feudal obligations. Thus, a spirit of liberty and enterprise was born in the Dutch swamps while everywhere else in Europe serfdom was still the rule.

From the start, drainage was the major problem for cultivation. But at Nieuwkoop, at the edge of the sphagnum cushion, this was easily overcome; it is here

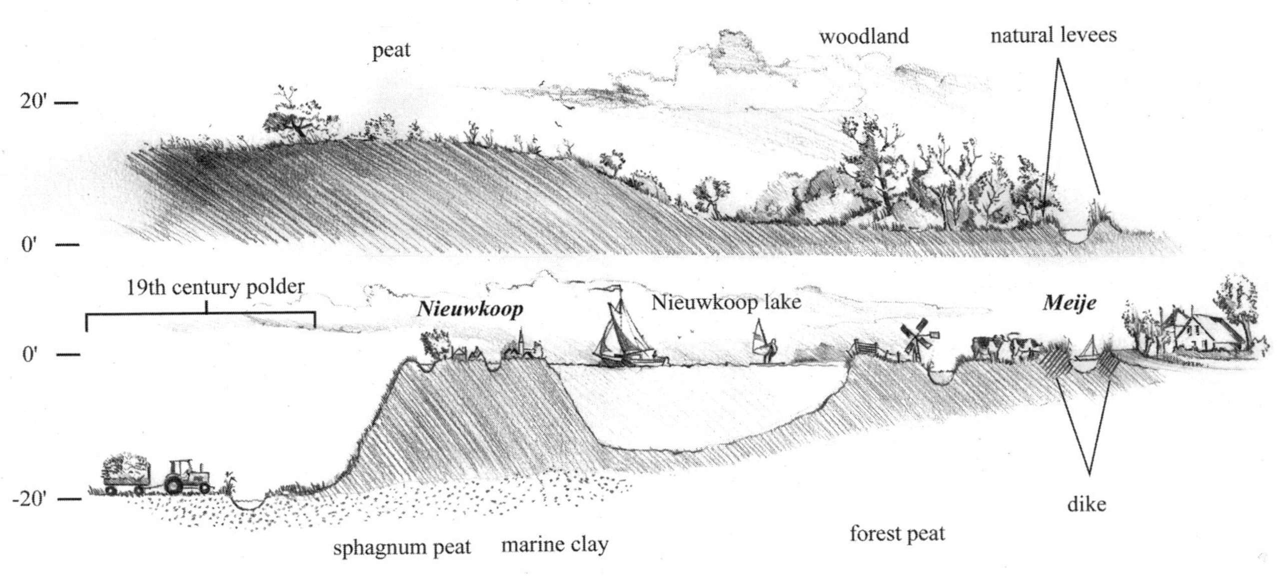

Figure 18.5
Cross sections through the Nieuwkoop and Meije areas 1000 years ago and at present. Redrawn from J. Teeuwisse, *De Ontwikkeling van het Landschap van 1000 tot 2000 in Nieuwkoop. Beelden en Fragmenten* (Noorden: Post, 1982).

that local exploitation began. Farmhouses were built in a row along the edge of the cushion, and narrow plots, 100 to 200 meters wide and separated by ditches, were extended into the bog for a stipulated distance, generally 1250 to 2500 meters. Reclamation started from the farmhouses as the first settlers dug drainage ditches to lower the water level. The peat excavated from the ditches was mixed with manure and spread over the land, and the vegetation was burned for fertilization. The settlers cultivated grains and kept sheep, not only for their own consumption but also to sell to the growing wool industry in such nearby towns as Leiden. More than a generation elapsed before a single row of fields was brought into cultivation. When the terminal line was reached, it became the starting point for a second generation of exploitation. As a result, the countryside was divided into a remarkably uniform sequence of parcels. Even now, this is a very characteristic feature of the Dutch landscape—we saw it at the beginning of our trip, on either side of the Meije.

A large part of peat—as much as 80 percent in sphagnum peat—is water. Drainage causes a swamp to shrink, a process that is accelerated by the increased exposure to oxygen, which stimulates the breakdown of plant debris. The settlers soon discovered that their activities caused their lands to subside and be drowned to an ever greater extent. They deepened the ditches several times, and removed water from the surface, using manual labor. A vicious cycle began: the improved drainage caused further collapse, forcing the peasants to take more severe measures. The existing swamp streams, such as the Meije, could not remove all of the excess water. Drainage works in one area brought flooding elsewhere, which caused many skirmishes between the local landlords. But the flooding also imposed the construction of an intricate system of land drainage and reclamation and, concomitantly, it led to the creation of novel technologies in water management. A complex network of new waterways, canals, ditches, and dikes appeared in the landscape, and as a result, new exploitation became feasible. By the end of the Middle Ages, the whole region around Nieuwkoop was in cultivation. In the course of this development, the original conflicts of interest among the inhabitants gave rise, by pure necessity, to a spirit of collaboration in matters of water management.

The invention that revolutionized the development of the Dutch landscape was the wind-driven water-pumping mill. The earliest version appeared at Alkmaar in 1408. It was small, driving a paddle wheel or a scoop wheel, and could carry water some two meters upward. A series of two to four such mills was needed to drain deeper water. More sophisticated and much more efficient windmills using Archimedes' screws superseded the older types, and were a common feature by the seventeenth century. They made it possible to drain large terrains that otherwise would have fallen victim to occasional incursions of the sea. New types of polders came into being. They were surrounded by ring dikes and, outside these, ring canals into which the windmills discharged their surplus water and from which the water could be transferred to the main rivers. The improved drainage allowed the polders to be used as grasslands, which made cattle farming profitable. The development of Dutch cheese went hand in hand with the introduction of the windmills.

One important element in the development of this landscape has not yet been mentioned. The peasants lived atop a thick blanket of peat, an excellent fuel. Originally, they dug away some of it to meet their own needs, and it was readily replenished by renewed accumulation. But the demand for fuel increased as the population grew. More fuel was needed, in particular, to support the growing towns in a wide variety of industrial activities: beer brewing; pottery, metal, and cement industries; brickworks; and so on.

In the early days, only a superficial layer of peat was dug away, but in 1530 an important innovation was introduced that would leave deep scars in the Dutch landscape. It was the *baggerbeugel*, a long-handled metal net that allowed peat to be dredged from several meters below the water level. Figure 18.6 shows the peat mining operation. Large, deep rectangular pools were excavated, and the peat was spread out as a slurry on narrow strips of land in between, dried, and carved into blocks. The complex labyrinth of land ridges and waters near Nieuwkoop are the remains of such exploitation. One reason for the resounding success of peat mining was that the fuel could be transported very cheaply by boat over the elaborate system of canals. Peat ecology supplied the energy that stoked the industrial centers of the seventeenth century, known as the Golden Age in Holland. If horsepower had been needed for transport, the enormous feeding expenses would have precluded the development of this remarkably prosperous economy.

Unfortunately, the hunger for fuel in the cities of Leiden, Gouda, and Amsterdam led to widespread destruction of the land. The skeleton of narrow strips of land left in the wake of the underwater excavations was an easy target for wave erosion in stormy weather. Poverty encouraged the rural population to sacrifice the meager long-term benefits of the land for attractive short-term profits. For the majority of people, the consequences were detrimental. Poverty increased, and depopulation of the area followed. Regulations issued by the local authorities to curb the destruction were circumvented. Large territories laboriously brought into cultivation in earlier days gave way to steadily growing lakes that could barely be kept in check. The advent of the more powerful windmills in the seventeenth century meant that some of the land could be reclaimed as polders; the last remains of turf were sold, and the exposed marine clays were used for agriculture or grassland for cattle. It was not until the nineteenth century, with the introduction of the steam pump, that the deterioration of the landscape could be brought under control.

Nowadays, the country is drained by an elaborate system of electrical pumps. Large polders have been reclaimed in recent years, and there are only a few major works that remain to be carried out. Some lakes, such as those between Nieuwkoop, Noorden, and the Meije, were preserved—their exploitation would not have been profitable. In the 1930s, the local government decided to use Lake Nieuwkoop as a dump for the rubbish of the towns and cities in the area. *Guide for Nieuwkoop and Noorden* (1935) states that no harm was done: "Only little rubbish is dumped, but we cover it up with peat so that nobody can see it. And best of all, it gives work to quite some people." Although this activity met with growing opposition from the nascent movement for environmental protection, dumping continued until the 1960s. Cleansing is under way now—at an estimated total cost of 10 million euros.

Delta Metropolis

Today, the integrity of Nieuwkoop's surroundings is under renewed attack. Since Weissenbruch's time, the population of the Netherlands has increased from 4 million to 16 million. People have not only become more numerous; in addition, their wealth and mobility have vastly increased. Airplanes draw their chalk stripes on the sky, and you can hardly avoid the sound of automobiles. In the western Netherlands, fragmented urban centers are combining to form a new, embracing construct, the Delta Metropolis. It is a ring-shaped region with 6 million inhabitants, comprising the "main ports" of Amsterdam and Rotterdam, as well as The Hague, Utrecht, Leiden, and Gouda. This urbanized ring surrounds the Green Heart of Holland, with Nieuwkoop right in the middle. This constellation is thought to favor living conditions in the emerging metropolis. To reach the countryside, you don't have to travel through endless stretches of suburbia, as inhabitants of London or Paris do. You can just take your bicycle.

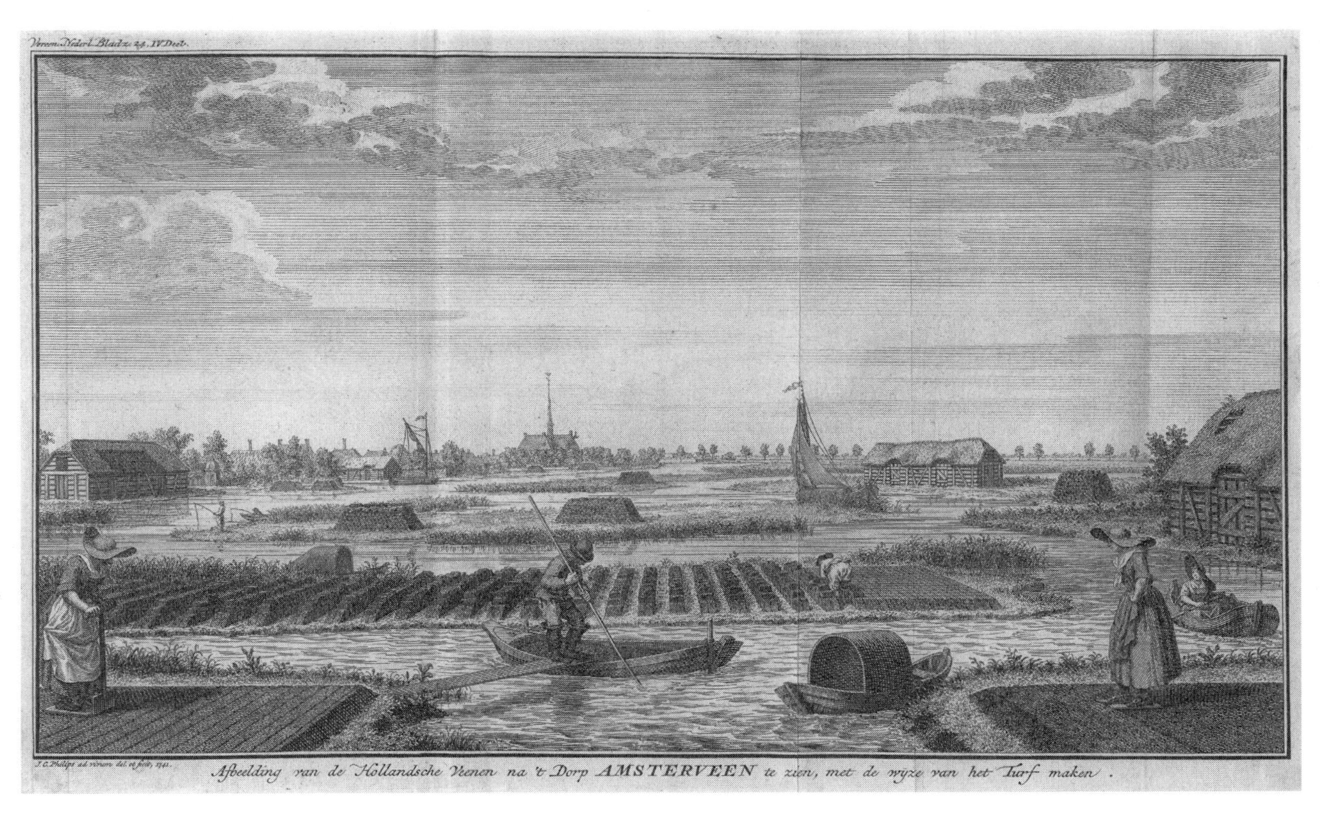

Figure 18.6
The peat mining operation in Holland. The man in the rowboat is digging for peat with a net on a long pole, the *baggerbeugel*. The peat is spread out over elongated strips of land, dried, and prepared for transport by boat. Engraving by J. C. Philips (1741).

Clearly, further development of the Delta Metropolis requires coherent planning by the government and leaves little space for local initiatives. The metropolis is conceived as a major center within northwestern European economic networks. Large-scale amplification of the existing facilities for transport and communication is foreseen, both between the urban centers within the ring and with other agglomerations. Intellectual centers of excellence must create a favorable climate for international investment. To accommodate the expected growth in population, innovative architecture will bring new life to expanding agglomerations. The execution of these plans is well under way, and has to be completed within a few decades.

What, in all this turmoil, is to become of Nieuwkoop, Noorden, and the Meije? The rural character of Weissenbruch's days is already evaporating. Pollution, sewage, and tourism are constant threats to the diversity of plants, birds, and fish, and encroaching urbanization undermines the comfortable peace of the olden days.

The present population is older and richer than it was in the past. While the young people escape to the cities, retired citizens move in. They build comfortable imitation farmhouses, or transform the old ones into cozy homes and weekend cottages. Farmers, once central to this community, now see their position marginalized. A large part of the countryside has received the status of a natural monument. In the past, hundreds of reed cutters used to work around the lake. Now, only ten of them are left to supply the roof building trade and keep the landscape open. The old villages are becoming city districts, and the area is changing from a productive landscape into a consumption commodity. It is unlikely that this land can sustain these changes for long, without losing its intimate dignity.

The Green Heart of Holland is literally central to the scheme of the Delta Metropolis. Judicious exploitation of the area's natural amenities will boost the quality of life for the entire population. In contrast to the urban ring, the Green Heart must bring home the delightful peace of the Dutch landscape. A new balance is sought between human demands and the natural environment.

A 250-million-euro project has been adopted to protect the remaining wetlands from deterioration and to reinforce its infrastructure. Twenty years are thought to be required for its implementation. Extensive agricultural domains are converted into natural parks and recreational areas. Elsewhere, modern combinations of agricultural and ecological management are implemented. The quality of environment and water are improved, and existing natural centers combined into larger reserves capable of supporting the unique vegetation and fauna.

The Gaia Analogy

I broadly distinguish between two contradictory visions about the relationship between culture and nature, no matter the scale—local, regional, or global. The culture-centered, and probably most widespread, view places us humans at the focal point and makes nature subordinate to our interests. Nature is there for us to use. Environmental problems are an invention of weak-hearted dreamers. The alternative nature-centered view emphasizes our dependence on the natural world. We must live in harmony with nature in order to survive. In its extreme form, we are seen as a greedy breed that is digging its own grave by destroying the subtle natural balance. The environmental movement thrives on the widespread awareness that only the pristine natural world was beneficial, and that we should try and find a way back to it. The idea of global management in particular is seen as a dangerous and misguided illusion.

The Nieuwkoop narrative illustrates how misleading these extreme viewpoints are. In this area, culture and nature are not perceived as opposite, but as intertwined. The present state of "nature" around Nieuwkoop is not the original wilderness, but something more humane—the result of almost a millennium of meddling by the inhabitants with nature. The story shows that unbridled nature was not tailored to human demands and could support very few people. Unavoidably, the wilderness had to be tamed if the inhabitants wished to survive. The end result was delightful.

If our ancestors could forge new alliances with nature so well, why couldn't we do it again? If they managed to create delightful surroundings locally, why could we not do so on a global scale? Taming Gaia certainly is a risky enterprise, but so was the taming of the Nieuwkoop wilderness. The idea of a cultivated Gaia and a sustainable development at the global scale is not illusory per se, but is a matter of common survival. We cannot go back to nature or brutally impose our will on it; we have to proceed by cautious interaction.

It is helpful to perceive the finer structure of culture–nature interactions as they are revealed by Nieuwkoop's past. We discern a regular pattern in the area's history. Human interference was destructive as soon as it began. The earliest attempts at cultivation brought about a dangerous imbalance in the natural equilibrium. Later, the ruinous effects were curbed by protective countermeasures and, in due time, a new human-maintained balance was achieved. The Netherlands has gone through a long succession of such cycles of destruction and reclamation. Its history is punctuated with alternating times of change and stability. If we extrapolate this pattern to the global scale, we may perceive the taming of Gaia not as the realization of a master plan, but as an open-ended process with many failures and minor steps ahead. Time and again new threats will emerge, which one by one will have to be overcome. The struggle will last as long as there are humans around.

In the Dutch lowlands, flooding forced the inhabitants into collaboration. Even today the authorities for water management are the best organized in the country. Similarly, global threats require a coordinated global response. The biggest problem standing in the way of the taming of Gaia may be in the coordination of the global human community. Humanity will not stand a chance as long as it remains divided by differences of culture, wealth, and resources. Since the three richest people in the world have more income than the 48 poorest nations together, the social contradictions have never been as acute as today. We now see globalization emerging in the interest of states and industries. This movement can become viable only if supported by the world population at large. As the threats of global change become more acute, many of our social problems may turn out to be soluble.

Abrams and Primack (2001) pointed to a cosmology for the twenty-first century emerging from science. I see the significance of the Gaia metaphor in this context. Gaia, rather than Earth, appeals to millions of people, and reminds us that our planet is our only home in the universe. As this work of art fills us with respect for the billions of years in which life could emerge and proliferate, it helps us to overcome our differences. Too much has Gaia been depicted as the splendidly pristine heavenly body now being destroyed by mindless human activities. The taming of Gaia should be part of the picture. We are like the first settlers in Nieuwkoop, who cultivated the wilderness, unaware of the long story of disasters and successes they set into motion. Of course we don't know whether Gaia will ever be tamed. But with luck our children will have a delightful planet to live on.

Acknowledgments

I am indebted to Dr. Siebrand Tjallingii for introducing me to the story of Nieuwkoop and to Dr. G. O. van Veldhuizen, the present mayor of this municipality, for explaining the newest developments. Bram van der Vlugt made me aware of inspiring initiatives by farmers and citizens to keep the Nieuwkoop area a wonderful place. I would like to thank Claartje de Loor and Mariette Jitta (Gemeentemuseum, The Hague) for introducing me to Weissenbruch. Judith de Jong and Kees Pleij made highly valued suggestions. Edwin Jacobs, of the Museum Jan Cunen in Oss, made the reproduction of Weissenbruch's painting available. The pencil drawings are by Cees van Nieuwburg. This chapter is an update of the opening chapter of my book *Life as a Geological Force* (1991). I can safely assume that this text is unknown to the large majority of the readers of this volume, since my book has been out of print for years and sold poorly; the only versions still available are in French, Japanese, and Catalan (a Czech translation is under way). The Commonwealth Fund made available the copyrights. The critical comments of three anonymous reviewers were invaluable.

References

Abrams, H. C., and J. R. Primack. 2001. Cosmology and 21st Century Culture. *Science*, 293, 1769–1770.

Cloud, P. 1988. *Oasis in Space: Earth History from the Beginning*. Norton, New York and London.

Gruyter, J. de. 1968. *De Haagse School*, vol. 1. Lemniscaat, Rotterdam.

Jacobs, E., H. Hanssen, and M. van Heteren, eds. 1999. *J. H. Weissenbruch, 1824–1903*. Waanders Uitgevers, Zwolle.

Kant, I. 1998 [1787]. *Critique of Pure Reason*. Translated and edited by P. Guyer and A. W. Wood. Cambridge University Press, Cambridge.

Lambert, A. M. 1971. *The Making of the Dutch Landscape: An Historical Geography of the Netherlands*. Seminar Press, London.

"Natuur in De Venen". Rapport van de Werkgroep Natuur in het Kader van het Stimuleringsproject Kerngebied De Venen. 1994. H. van Arkel, ed. Prov. Utrecht Dienst Ruimte en Groen, Utrecht.

Plan van Aanpak en Convenant De Venen. Nota voor Statencommissie, SCGW 98-137, verg. 13-11-1998, agendapunt 10.

"Ruimte maken, ruimte delen": Vijfde nota over de Ruimtelijke Ordening 2000/2020. Ministerie van Volkshuisvesting, Ruimtelijke Ordening en Milieubeheer, The Hague.

19
Gaia and the Human Species

Crispin Tickell

Abstract

Gaia theory has become an integrative factor in modern science. It brings the physical and biological environments together into a single self-regulating system, allowing for both temporal and spatial variability. The impact of our animal species on the Earth is enormous and increasing. In trying to cope with the problems thus created, we should remember that Gaia has no special tenderness for humans. We need to treat its current configuration with respect and understanding.

Gaia theory arises from the relationship between life and its physical environment on the surface of the Earth. It is at once a very old idea and a very new one. What was first a kind of analogy has now become an integrative factor in modern science.

Looking back, it is strange how uncongenial the observation was to the practitioners of the conventional wisdom when it was put forward in its present form over a quarter-century ago. Unfamiliar ways of looking at the familiar, or any rearrangement of the intellectual furniture, tend to arouse emotional opposition far beyond rational argument: thus opposition to evolution by natural selection, continental drift and tectonic plate movement, and, more recently, cometary or asteroid impacts from space. Gaia theory challenges current habits of reductionism and the tendency of some academics to put their subjects into boxes, shut the lid, and ignore what is going on in other boxes. Most of us are better at looking at the constituent elements of problems than at seeing the connections between them and understanding how the resulting system works.

Gaia theory is a supreme example of interdisciplinarity. It comes at a time when science as such is under challenge from a society somewhat disenchanted with it. Whether it be over civil nuclear power, asbestos, thalidomide, or "mad cow disease," public skepticism over scientific claims has greatly increased. This has made many scientists defensive and reluctant to embrace new interpretive ideas of which Gaia is a good example.

There have been many definitions of Gaia theory, and I will not venture a new one. For good working purposes, I suggest that put forward by James Lovelock and Lynn Margulis in a joint paper in 1984: "... the evolution of a tightly coupled system whose constituents are the biota of their natural environment, which comprises the atmosphere, the oceans and the surface rocks." Or, more recently, by Greg Hinkle: "symbiosis seen from space."

The mysteries of the relationship between the living and the nonliving environments were recognized from the earliest days, and in many respects are the stuff of religion. Gods and goddesses were seen to embody specific elements, ranging from the vastness of the sky to the most local spring, and the notion that the Earth itself was alive came up regularly in Greek philosophy. Leonardo da Vinci saw the human body as the microcosm of the Earth, and the Earth as the macrocosm of the human body. He did not know as well as we do now that the human body is itself a macrocosm of tiny elements of life—bacteria, parasites, viruses—often at war with each other and together constituting around half our body cells. Giordano Bruno was burned at the stake just over 400 years ago for maintaining that the Earth was indeed alive, and that other planets could be too. The geologist James Hutton saw the Earth as a self-regulating system in 1785, and T. H. Huxley saw it likewise in 1877. Others, notably in Russia, have done so in a dispersed way ever since. But it was James Lovelock who in the 1960s brought forward Gaia theory in its present form and, with Lynn Margulis, has worked on it ever since.

What is in a name? I remember a conversation with a distinguished scientist keen to trash "all that Gaia nonsense." When I protested and offered to rename it "geophysiology," "Earth systems science," or something similar, he brightened up and eventually confessed that "most of it must be right." The choice of the Greek goddess Gaia rather than of some Greek-derived scientific polysyllable, or worse some acronym, was a risk. On the one hand, it was just too attractive for those in search of a new religion at a time when traditional religions were breaking down; on the other, it was just too repulsive for those who liked to hide their science in coded vocabulary. The result was that some New Age travelers jumped aboard, and some otherwise sensible scientists jumped off. This is probably still the case. But as a theory, Gaia is now winning.

The scientific communities of the four great international global change research programs—the International Geosphere-Biosphere Programme, the International Human Dimensions Programme on Global Environmental Change, the World Climate Research Programme, and the International Biodiversity Programme—met at Amsterdam on July 13, 2001. They then adopted a Declaration on Global Change, signed by almost a thousand people, which stated squarely that "the Earth System behaves as a single, self-regulating system, comprised of physical, chemical, biological and human components. The interactions and feedbacks between the component parts are complex and exhibit multi-scale temporal and spatial variability."

Here indeed is Gaia theory. The same goes for the Earth systems science which is now the concern of the Geological Society of London (with which the Gaia Society recently merged). Whatever the label, Earth systems science, or Gaia, has now become a major subject of inquiry and research, and no longer has to justify itself.

It was, I think, Lynn Margulis who described Gaia as "a tough bitch." So she is. Over 3.8 billion years, her robustness is impressive and reassuring. She has survived the great extinctions from outside the Earth, and the great catastrophes from within it. This has required a remarkable resilience whereby physical and biological mechanisms have adapted to new circumstances. Gaia is a lady who has remained broadly the same underneath, but can wear many clothes for many weathers and many fashions. She has no particular tenderness for humans. We are no more than a small, albeit immodest, part of her. Only in the last tick of the clock of geological time did humans make their appearance, and only in the last fraction of it did they make any impact on the Earth system as a whole.

Only now do we know how vulnerable our little planet is to human depredations. A periodic visitor from outer space would find more change in the last 200 years than in the preceding 2000, and more change in the last 20 years than in the preceding 200. The association between humans and their environment, including the micro world in and around them, has changed at every change of human evolution: from vegetarians to meat eaters, from hunters and gatherers to farmers, and from country to city dwellers. But the most radical divide was the beginning of the industrial revolution in Britain some 250 years ago. Before then, the effects of human activity were local, or at worse regional, rather than global. All the civilizations of the past cleared land for cultivation, introduced plants and animals from elsewhere, and caused a variety of changes. The southern and eastern coasts of the Mediterranean are a case in point. The soils have now become sand, the trees are often camel grass, animals of all kinds have disappeared, and the clouds sail overhead to drop their rain somewhere else.

With the industrial revolution, all began to change. I suspect that our generation is the first in which the magnitude of the effects on the Earth as a whole has become manifest. This clearly emerged in the Amsterdam Declaration. There it was stated that changes brought about by human activities

> . . . to Earth's land surface, oceans, coasts and atmosphere and to biological diversity, the water cycle and biogeochemical cycles are clearly identifiable beyond natural variability. They are equal to some of the great forces of nature in their extent and impact. Many are accelerating? . . . [They] have the potential to switch the Earth's system to alternative modes of operation that may prove irreversible and less hospitable to humans and other life. The probability of a human-driven abrupt change in Earth's environment has yet to be quantified but is not negligible.

What are these changes? They fall into five main categories, all interlinked. First, there has been a giddy-making increase in human numbers, rising from around 1 billion at the time of Thomas Malthus (who first drew attention to the relationship between population and resources) at the end of the eighteeth century, to 2 billion in 1930 and now over 6 billion. The world population is increasing by over 80 million people each year. There are over 450 million new people on the Earth since the Rio Conference on Environment and Development in 1992. More than half our species now lives in cities, which are themselves like organisms drawing in resources and emitting wastes. In short, we are spreading like dandelions, or any other species on a reproductive rampage. Indeed, it has been suggested that human multiplication is a case of malignant maladaption in which a species, like infected tissue in an organism, multiplies out of control, affecting everything else. In terms of factors of increase in the last century, air pollution rose by around 5, water use by 9, sulfur emissions by 13, energy use by 16, carbon dioxide emissions by 17, marine fish catch by 35, and industrial output by 40.

All this has profoundly affected the condition of the land surface. More people need more space and more resources. Soil degradation is widespread, and deserts are advancing. Such degradation is currently estimated to affect some 10 percent of the world's current agriculture area. Although more and more land, whatever its quality, is used for human purposes, increase in food supplies has not kept pace with increase in population. Today many of the problems are of distribution. But even countries generating food surpluses can see limits ahead. Application of biotechnology, itself with some dubious aspects, can never hope to meet likely shortfalls. In the meantime, industrial contamination of various kinds has greatly increased. To run our complex societies, we need copious amounts of energy, at present overwhelmingly derived from dwindling fossil fuel resources laid down hundreds of millions of years ago. We also have to deal with the mounting problems of waste disposal, including the toxic products of industry.

Next there has been increasing pollution of water, both salt and fresh. No resource is in greater demand than freshwater. At present such demand doubles every twenty-one years and seems to be accelerating. Yet supply in a world of over 6 billion people is the same as at the time of the Roman Empire in a world of little more than 300 million people. We are at present using some 160 billion metric tons more water every year than is replenished.

Then there have been changes in the chemistry of the atmosphere. Acidification from industry has affected wide areas downwind. Depletion of the protective atmospheric ozone layer permits more ultraviolet radiation to reach the surface of the Earth, with so far unmeasured effects on organisms unadapted to it. Greenhouse gases are increasing at a rate which could change average world temperature, with big resulting variations in climate and local weather as well as sea levels. According to the estimates of the Intergovernmental Panel on Climate Change, we could be altering the global climate at rates far greater than would have occurred naturally in the last 10,000 years, with unforeseeable consequences. Carbon levels in the atmosphere are now the highest in the last 160,000 years, and rising fast.

Last, humans are causing extinction of other organisms at many times the normal rate. Indeed, the rate of extinction is reminiscent of the dinosaur extinction of 65 million years ago. The rising damage to the natural services on which we, like all species depend, is immeasurable. There is no conceivable substitute for such services. At present there is a creeping impoverishment of the biosphere. According to the Living Planet Index put forward by the World Wide Fund for Nature in 2000, the state of the Earth's natural ecosystems has declined by about a third since the 1970s, while the ecological pressure of humanity has increased by about a half during the same period.

Environmental change rarely proceeds in curves. It goes in steps and thresholds. Due, perhaps, to the shortness of our individual lives and our lack of imagination, we tend to believe that what we know—the current diversity of life and the climate around us—will change only within narrow limits, and that if nature is allowed to take its course, things will revert to where they were. Unfortunately, history gives no foundation for this belief. As was well said in the Amsterdam Declaration, "The nature of changes now occurring simultaneously in the Earth System, their magnitudes and rates of change are unprecedented. The Earth is currently operating in a no-analogue state." Again Gaia has no special tenderness for our species. How things will eventually settle down, with or without us, is a matter for speculation.

A question often asked is the measure of human responsibility for what is happening, and whether we have some God-given role in the process. There has, for example, been some talk, notably among the religiously inclined, about an alleged human obligation of "stewardship" of the Earth. If so, the Earth has had to wait a long time for the arrival of the stewards. Certainly the trilobites managed for over 250 million years without them. Looking at the human record of predation, exploitation and extinction of other forms of life since the current version of hominids appeared less than 150,000 years ago, I am reminded of James Lovelock's remark that "humans are about as qualified to be stewards of the Earth as goats are to be gardeners."

Certainly humans carry heavy responsibilities, but stewardship is not the best way to describe them. For most people they relate primarily to the interest we all have in looking after ourselves. On this reckoning we have two excellent reasons for trying to treat the current configuration of Gaia with more respect and understanding. In particular, we need to maintain our own good health as well as that of the plants and animals, big and small, on which we depend for food. We pride ourselves on our medicine as if it were somehow detached from the natural world, but it is an obvious product of it. As well as conserving biodiversity at the level of species and ecosystems, we need to cherish the genetic diversity that occurs within them. Modern agricultural techniques have led to an excessive dependence on a few miracle strains of even fewer plants and animals. Without a large natural genetic reservoir, we make our food supplies vulnerable to disease, as the Irish potato growers learned to their cost in the nineteenth century.

Just as important are the ecological issues. At present we take as cost-free a broadly regular climatic system with ecosystems, terrestrial and marine, to match. We rely on forests and vegetation to produce soil, to hold it together, and to regulate water supplies by preserving catchment basins, recharging groundwater, and buffering extreme conditions. We rely upon soils to be fertile, and to absorb and break down pollutants. We rely on coral reefs and mangrove forests as spawning grounds for fish, and on deltas as shock absorbers for floods. There is no conceivable substitute for these natural services, but often we hardly notice them. In many cases we do not know the threshold which, once passed, leads to their collapse or radical change. Yet we cannot continue to assume that they will continue to come for free forever.

All this raises obvious issues of human values. Such values constitute each person's worldview. We tend to believe that greater material prosperity and welfare are overriding human priorities, that resources can be indefinitely exploited, and that economic growth on the traditional definition is good in itself: in short, ever upward and onward with freer markets, freer trade, and continuously rising consumption. With this goes an almost religious belief in technology as the universal fix: an extension of human capacity to adapt to and cope with whatever may arise. We live in what has been described as the technosphere, in which the technomarket blindly rules with little regard for human needs or aspirations. We tend to forget that information does not replace knowledge, and that knowledge is not the same as wisdom.

There is an accompanying spread of a culture of rising expectations, nourished by worldwide use of information technology through radio, television, E-mail, the Internet, and the press. One consequence is a drive toward industrialization as a synonym for "development," and the catchall answer to the world's manifest ills. With it has come globalization and an increasing homogenization of human culture, as well as a widening gap between rich and sophisticated, on the one hand, and poor and unsophisticated, on the other. As has been well said, globalization represents a kind of mutation in human civilization.

Another consequence is change in evolution itself. Human activity is changing the processes of natural selection, mutation, and symbiosis, not just through genetic engineering and modification of organisms, but also through large-scale extinction of species and the ecosystems in which they have a place. We have yet to see whether there is any realistic prospect of developing a subspecies of superhumans with genes tailored to specific requirements, but it is certainly not impossible in the long future. In his fantasy *The Time Machine* (1898), H. G. Wells foresaw a genetic division of humanity into Eloi (or Upper Worlders) and Morlocks (or Lower Worlders) in perpetual struggle against each other. Of course Gaia was still there, but so far as humans were concerned, it was not a Gaia we would happily recognize.

In fact, human damage to the current life system of the planet is not incurable. Most of the solutions to the problems we have created are already well known. Take human population increase. Overall, population is still rising, but in several parts of the world it is leveling off. The main factors are the improvement in the status of women, better provision for old age, wider availability of contraceptive devices, and lower child mortality and better education, especially for girls and young women. Take degradation of land and water. We know how to cope if we try. We do not have to exhaust topsoils, watch them erode into the sea, rely upon artificial aids to nature, destroy the forests with their natural wealth of species, or poison the waters, fresh and salt. Take the atmosphere. We do not have to punch holes in the protective ozone layer. We do not have to rely on systems of energy generation which will affect climate and weather in such a fashion that change, even for the better, might put an overcrowded world at risk. Take human relationships. We do not have to widen the gaps between rich and poor, or even to think of creating a genetically favored master class. Take the way in which we conduct most scientific inquiry. We do not have to break down issues into water tight compartments, and thus miss the internal dynamics of the life system as a whole.

Moreover, understanding of the Gaian approach is already spreading fast, whether it be labeled Gaia or not. An example of the need for it is in the field of economics, where fashionable delusions about the supremacy of market forces are deeply entrenched, and the responsibility of government to set the framework for economic activity and protect the public interest is often ignored. At present there is an astonishing failure to recognize true costs. Markets are marvelous at determining prices but incapable of recognizing costs. Definition of costs requires a Gaian approach to economics and measuring values, and this has to be brought back into pricing. In addition to the traditional costs of research, process, production, and so on, prices should reflect the costs involved in replacing a resource or substituting for it—and of course the costs created by the environmental problems associated with it. Governments also have to get rid of perverse subsidies—for example, those to the fossil fuel industry, the price of whose products is artificially low in relation to the damage their use does to the environment. In short, we need a paradigm shift in which politicians and ecologists alike recognize that humans are more than mere producers or mere consumers.

One of the key points in the Amsterdam Declaration was that a new ethical framework was urgently needed:

> The accelerating human transformation of the Earth's environment is not sustainable. Therefore the business-as-usual way of dealing with the Earth System is *not* an option. It has to be replaced—as soon as possible—by deliberate strategies of management that sustain the Earth's environment while meeting social and economic development objectives.

If we are eventually to achieve a human society in harmony with nature, we must be guided by respect for it. No wonder that some have wanted to make a religion of Gaia or life as such. At least we need an ethical system in which the natural world has value not only to human welfare but also for and in itself. The British poet D. H. Lawrence wrote, "I am part of the sun as my eye is part of me. That I am part of the Earth, my feet know perfectly; and my blood is part of the sea."

Let the same exhilaration take us all.

IV
QUANTIFYING DATA

20
Daisyworld Homeostasis and the Earth System

S. L. Weber and J. M. Robinson

Abstract

This chapter examines the Daisyworld model's steady-state solution with emphasis on the mechanisms that stabilize the system. This shows that to a large extent the model's formulations decouple the equilibrium temperature from incoming radiation. Such formulations obviously stabilize the planetary temperature subject to variations in incoming radiation. The heat transport formulation used in the Daisyworld model is found to be essential to the model's homeostatic behavior around a set point. Finally, the chapter reviews work that has altered the assumption of a fixed set point (the biological optimal temperature), showing that adapting daisies are no longer homeostatic. We hope that this analysis stimulates the search for a model based on more realistic concepts.

Introduction

The Daisyworld model (Watson and Lovelock, 1983; henceforth WL or the Daisymodel) was constructed to demonstrate that a biota could stabilize a planetary environment subject to large variations in external variables. This property is called homeostasis. As such, the Daisymodel is a very simple model representing a very complex system. It has attracted wide attention, and since its introduction in 1983, many authors have proposed modifications and extensions of it.

Simplified models can be useful tools, provided they are based on realistic concepts and that they give new insight into the implications of the underlying concepts (Petersen, chapter 3 in this volume). The concepts underlying Daisyworld are phrased in its core equations. They include the equilibrium planetary heat budget or the temperature-dependent function for plant growth. As shown by WL and many others, the combination of concepts implies homeostasis. Although Daisyworld's equations appear simple, it is not immediately apparent how the system of equations achieves homeostasis. This chapter first makes the Daisymodel more transparent by solving its equations and relating its behavior to its mathematical structure. Thereafter, it probes some of the assumptions implicit in Daisyworld's mathematics, reviews work that has altered these assumptions, and arrives at some conclusions about the requirements for stability.

The second section of this chapter reviews Daisyworld's core equations. For a full presentation of the equations, we refer to WL. The third section solves the model under the condition of population equilibrium to reveal the structure that makes it homeostatic. The fourth section defines homeostasis and distinguishes it from the mathematical property of attractivity. The fifth section compares Daisyworld's energy transfer mechanisms with those of a standard box model and concludes that the formulation used in Daisyworld departs from basic physics. The sixth section describes what happens when one of Daisyworld's key reference points, the biological optimal temperature, is allowed to vary freely and in a constrained fashion. It specifically shows that adapting daisies are no longer homeostatic. The seventh section provides a summary. It concludes that models based on more realistic concepts than the Daisymodel are required to study the homeostatic properties of an evolving biosphere.

The Daisymodel

Daisyworld is a hypothetical planet whose surface is partly covered by black daisies, partly by white daisies, and partly by bare ground. The daisies grow into (recede from) the bare ground in proportion to the areas they occupy and their rates of expansion (dieback). That is,

$$\frac{dx_b}{dt} = x_b[x_g\beta(T_b) - \gamma] \quad (1)$$

$$\frac{dx_w}{dt} = x_w[x_g\beta(T_w) - \gamma] \quad (2)$$

$$x_b + x_w + x_g = 1 \quad (3)$$

where x_b is the fraction of the planet covered by black daisies, x_w is the fraction of the planet covered by white daisies, and x_g is the fraction of bare ground. The expansion rate $x_g\beta - \gamma$ is given by the available bare land x_g and the fraction of bare land claimed per time step β, and by the rate γ at which daisies are replaced by bare ground.

The fraction of each unit of bare ground that is colonized by each color of daisies in a given time period is a parabolic function of the local temperature:

$$\beta(T_{b,w}) = 1 - \frac{(2(T_{opt} - T_{b,w}))^2}{(T_{max} - T_{min})^2} \tag{4}$$

where T_b is the local temperature of the black daisies and T_w is the local temperature of the white daisies. As noted below, local temperatures differ depending on surface albedo.

Three temperature constants affect growth rates: the optimal temperature for daisy growth T_{opt} and the limiting temperatures T_{min} and T_{max}. Outside the range $T_{min} < T_{b,w} < T_{max}$ it is assumed that daisy populations cease to grow ($\beta(T_{b,w}) = 0$), and die back in proportion to a constant value, γ.

Each surface type has a different albedo, and thus absorbs a different amount of shortwave radiation. This results in different local temperatures and different amounts of emitted longwave radiation. Thus, daisies affect the radiative budget and the climate of Daisyworld. The total amount of absorbed shortwave radiation and emitted longwave radiation is the sum of the contributions of the three surface types. The planetary heat budget is expressed in terms of the planetary area-weighted average temperature ($T_{planet} = x_b T_b + x_w T_w + x_g T_g$), and the planetary area-weighed average albedo ($A_{planet} = x_b A_b + x_w A_w + x_g A_g$), where T_g is the temperature of bare ground and $A_{b,w,g}$ are the albedos of the black and white daisies and of bare ground.

WL specify a Stefan Boltzman formulation for longwave emissions in the energy balance equation (i.e., they make emissions proportional to the fourth power of the absolute temperature). However, for analytical purposes, they linearize this to the form

$$(1 - A_{planet})S = E + \lambda T_{planet} \tag{5}$$

where S is the incoming shortwave radiation and E and λ are the coefficients of the (linearized) longwave radiation emission. As noted by WL and others (North et al., 1981; Saunders, 1994), the use of a linear emissions formulation does not significantly change model behavior.

Table 20.1
Parameter values of Daisyworld

Longwave radiation constant	E (W/m^2)	130
Longwave emission parameter	λ (W/m^2 °C)	2.0
Albedo of white daisies	A_w	0.75
Albedo of uncovered ground	A_g	0.5
Albedo of black daisies	A_b	0.25
Minimum temperature	T_{min} (°C)	2.5
Optimal temperature	T_{opt} (°C)	20.0
Maximum temperature	T_{max} (°C)	37.5
Death rate	γ	0.3
Redistribution parameter	q (°C)	20

To complete the system of equations, WL use a simple formulation to relate planetary temperature and albedo to their local counterparts:

$$T_b - T_{planet} = q(A_{planet} - A_b) \tag{6}$$

$$T_w - T_{planet} = q(A_{planet} - A_w) \tag{7}$$

For later use we rewrite this, subtracting (6) and (7):

$$T_b - T_w = q(A_w - A_b) \tag{8}$$

The underlying assumption is that the system redistributes energy from warmer, black daisy regions to colder, white daisy regions in such a manner that the local temperature differences are proportional to the local albedo differences. The proportionality constant, q, is a measure of the degree to which heat resists redistribution among the different surface types. For values and definitions of the constants used in the model, see table 20.1.

Before solving the system of equations, it is useful to consider the timescales inherent in the coupled biota–climate system. There are two timescales: one dictated by the response time of the climate and the other determined by the response time of the biota. It is assumed that the climate responds much more quickly than the biota. This is represented by the absence of a time derivative in (5)—that is, absorbed shortwave radiation is always exactly balanced by outgoing longwave radiation (the planetary heat budget is always in equilibrium). Consequently, the biota set the rate of change of the coupled system.

In addition to the two internal timescales, there is the external timescale given by the rate of change of the solar luminosity. Formulations used by Caldeira and Kasting (1992) indicate an increase in solar luminosity by 5–15 percent of its present-day value per Gyr. WL assume that this timescale is long compared to the adaptation timescale of the biota. Therefore it is justified to focus on the equilibrium solutions to (1) and (2), the system of equations describing daisy population dynamics. That is, the time derivatives can be set to 0 and we can assume that daisy growth $x_g\beta$ equals death γ for each daisy type. In the following we will solve the equilibrium equations for a range of values of the solar luminosity.

The Analytical Solution When Black and White Daisy Populations Are in Equilibrium with Insolation

This section derives a relatively simple equation for the planetary temperature which makes explicit how the system achieves homeostasis (Weber, 2001). This is done by rewriting and analytically solving the model equations (see also Saunders, 1994). As the first step, we follow the previous work of WL, who showed that the local temperatures of the black and white daisies are independent of the level of incoming solar radiation. For the sake of completeness, the argument is repeated here.

When daisy populations are in equilibrium, by definition their respective net expansion rates (i.e., growth minus death) are 0. This gives

$$x_g^{eq}\beta(T_b^{eq}) - \gamma = 0 \tag{9}$$

$$x_g^{eq}\beta(T_w^{eq}) - \gamma = 0 \tag{10}$$

Therefore, $\beta(T_b^{eq}) = \beta(T_w^{eq})$. According to (4), the equilibrium local temperature of the white daisies is lower than the optimal temperature, whereas the equilibrium local temperature of the black daisies is higher by the same amount:

$$T_b^{eq} + T_w^{eq} = 2T_{opt} \tag{11}$$

From (11) and (8) the equilibrium local temperatures T_b^{eq}, T_w^{eq} follow as

$$T_b^{eq} = T_{opt} + q\left(\frac{A_w - A_b}{2}\right) \tag{12}$$

$$T_w^{eq} = T_{opt} - q\left(\frac{A_w - A_b}{2}\right) \tag{13}$$

Using the present parameter values given in table 20.1, $T_b^{eq} = 25°C$ and $T_w^{eq} = 15°C$. It follows that the equilibrium daisy temperatures depend only on the biological parameters T_{opt}, q, and the local albedo contrast $A_w - A_b$. They are independent of the solar radiation S, and could be said to exhibit perfect homeostasis.

As the second step, we derive a direct relation between the planetary albedo and temperature. Substituting the expressions for the local temperature (12) and (13) back into the heat transport relations (6) and (7), it follows that

$$T_{planet}^{eq} - T_{opt} = q\left(\frac{A_b + A_w}{2} - A_{planet}^{eq}\right) \tag{14}$$

Thus, at equilibrium, departures of planetary albedo from the fixed value $(A_b + A_w)/2$ drive planetary temperature away from the biological optimal temperature T_{opt}. It is clear that the coupled biota–climate system contains a reference temperature, T_{opt}, and a reference albedo, $A_{ref} = (A_b + A_w)/2$, which are entirely predetermined.

At this point it is useful to introduce the concept of a reference value for the solar radiation, S_{ref}, defined as that value of incoming solar radiation which would result in a planetary temperature equal to the optimum temperature, T_{opt}, in an abiotic system (i.e., a planet entirely covered by bare ground) with a fixed planetary albedo A_{ref}. This is a specific case of the planetary energy budget (5). Thus,

$$(1 - A_{ref})S_{ref} = E + \lambda T_{opt} \tag{15}$$

Given (from table 20.1) that $A_{ref} = (A_b + A_w)/2 = 0.5$, it follows that $S_{ref} = 340\ W/\mathrm{m}^2$, a value similar to the average energy flux at the top of Earth's atmosphere ($342\ W/\mathrm{m}^2$).

The third step is to combine (14), (15), and the heat budget (5) to derive an expression for the equilibrium planetary temperature T_{planet}^{eq} as a function of the insolation:

$$T_{planet}^{eq} = T_{opt} - \left(\frac{S}{q} - \lambda\right)^{-1}(1 - A_{ref})(S - S_{ref}) \tag{16}$$

This equation shows how T_{planet}^{eq} follows from the parameters of the system, and provides an expression for how far the planetary equilibrium temperature is from the optimal temperature for daisy growth. The coupled climate–biota system responds to a change in solar radiation with respect to S_{ref} by a temperature adjustment with respect to T_{opt}. The amplitude of the temperature response is inversely proportional to $S/q - \lambda$. This latter term reflects the two mechanisms by which the system adjusts to changes in incoming radiation: a redistribution of heat (S/q) from areas of lower albedo to areas of higher albedo, and an increase in outgoing longwave radiation (λ) with increasing temperature. Of these, the temperature redistribution mechanism is dominant, since S/q is much larger than λ. A system which can optimally redistribute a surplus of solar radiation to the most

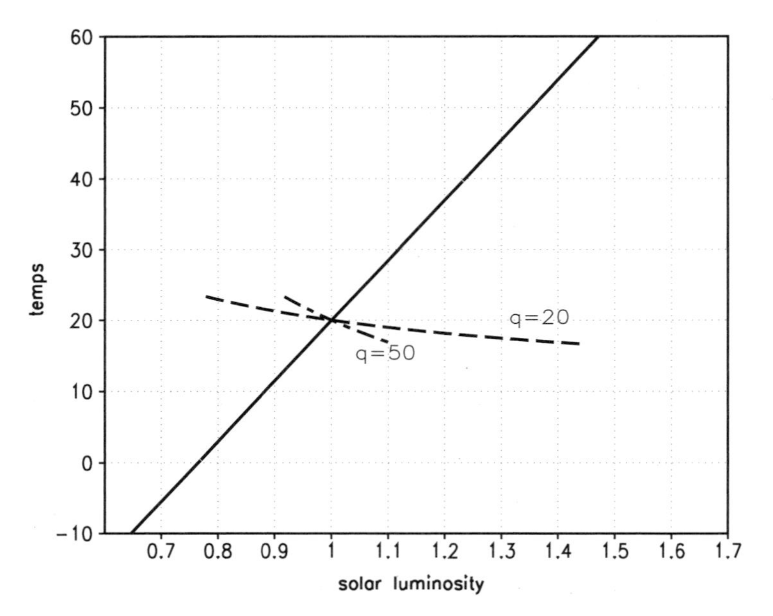

Figure 20.1
The equilibrium planetary temperature T^{eq}_{planet} (in °C) as a function of the solar luminosity: the dashed line labeled $q = 20$ gives the two-daisy solution for the standard parameter values in table 20.1, and the line labeled $q = 50$, the two-daisy solution for modified parameter values representing a slower heat diffusion than in the standard case. For comparison, the no-daisy solution (solid line) is shown as well.

reflective region (q small) is best able to stabilize its temperature.

The dashed lines in figure 20.1 relate the planetary equilibrium temperature (16) to solar radiation, which is expressed as a fraction of the reference value S_{ref}, for two sets of parameter values. One is the standard set as given in table 20.1. For the other the parameter q has been changed from 20 to 50°C and the local albedo contrast $A_w - A_b$ has been changed from 0.5 to 0.2. The local temperature difference $T_b - T_w$ is 10°C in both cases. As apparent by comparing the two dashed lines in figure 20.1, or by study of equation (16), increasing q weakens the system's ability to regulate temperature. The range of solar luminosities over which the two-daisy solution exists decreases for larger q and, to a much lesser extent, for smaller $A_w - A_b$ (Weber, 2001). Thus we find that when heat diffusion is slow, the system can achieve moderate thermal homeostasis only for a small range of solar luminosities.

In sum, three features make Daisyworld homeostatic: First, the property that local equilibrium temperatures depend on the prescribed optimal temperature for plant growth T_{opt} according to (12) and (13), but are independent of solar radiation. Second, the assumed direct link between local and planetary characteristics given by (6) and (7) and the underlying mechanism of heat exchange between the different surface types. Third, the energy budget established by equation (5). The third feature is fairly standard. The second and first features will be examined further below. Finally, we note that homeostasis is a prominent feature of Daisyworld because the heat exchange mechanism is assumed to be very efficient (q small).

Why Does a Numerical Solution Method Work?

The preceding section looked at the states of Daisyworld when daisy populations are in equilibrium. Here "equilibrium" means the static condition (or stasis) in which population growth $x_g\beta$ equals population death γ and daisy populations are unchanging over time. Mathematically, these states are found by setting the time derivative of the left-hand sides of (1) and (2) to 0 and solving to find the steady or stationary states. "Homeostasis" refers to a property of the equilibrium state: the planetary temperature is insensitive to variations in the solar luminosity over a large range of values. This property is alternatively described in the literature in terms of thermal "self-regulation" or the "stabilizing" effect of the biota on the temperature.

In the preceding section the equilibrium state was determined by analytically solving the stationary equations. An alternative way to find the equilibrium state is by numerical integration of the time-

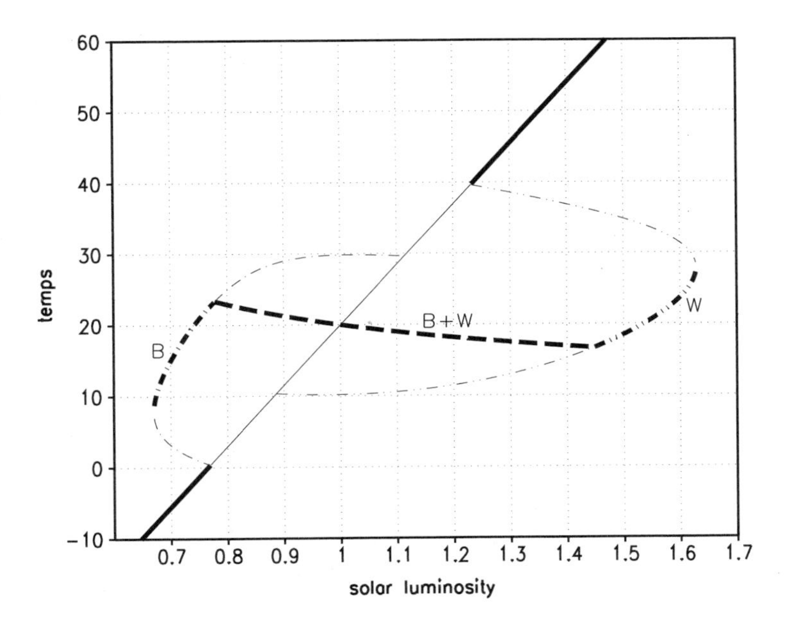

Figure 20.2
The equilibrium planetary temperature T_{planet}^{eq} (in °C) as a function of the solar luminosity for the standard parameter values given in table 20.1. Attracting states are indicated by thick lines and repelling states by thin lines. No daisies (solid line), white and black daisies (dashed line labeled B + W), black daisies only (dot-dashed line labeled B) and white daisies only (dot-dot-dashed line labeled W). Figure reproduced from S. L. Weber, "On Homeostasis in Daisyworld," *Climatic Change*, 48 (2001): 469, with the kind permission of Kluwer Academic Publishers.

dependent equations forward in time, until the time derivative becomes 0 and the stationary solution is reached. This approach works because the equilibrium state is a strong attractor. That is, the system is "attracted" to a certain range of equilibrium values, and has a mathematical tendency to damp out small deviations with respect to such an attracting equilibrium value. "Repelling" is the opposite of "attracting." ("Attracting" and "repelling" are also called "stable" and "unstable" in mathematical textbooks; however, these terms tend to cause confusion with the concept of "stabilizing" introduced above in association with homeostasis.) Under repulsion, deviations grow. Numerical integration cannot find repelling states because the system moves away from them.

Figure 20.2 shows Daisyworld's attracting and repelling states for bare soil, two-daisy populations, black daisies only, and white daisies only. The two-daisy state is always attracting. The bare-soil state and the one-daisy branches are partially attracting and partially repelling. When more than one attracting solution exists for a given value of solar luminosity, the system jumps from one branch to another in the numerical integration at the point where an attracting solution becomes repelling. This figure has been widely shown—for example, WL or Saunders (1994)—to demonstrate that the two-daisy system constrains the planetary temperature within narrow bounds.

The numerical method is illustrated in figure 20.3. To find the planetary temperature at equilibrium for a given value of the insolation $S + \delta S$, the calculation is begun using the equilibrium state for a slightly lower value of insolation S. At a certain point δS is added. The system initially responds to the increase in insolation by increasing T_{planet}. At once $T_{b,w}$ increase by a similar amount as T_{planet} (upper panel), because local temperatures are connected to the planetary temperature by (6) and (7). Higher local temperatures result in a higher growth rate of the white daisies and a lower growth rate of the black daisies (middle panel), according to (4). Small adjustments in the areas occupied by black and white daisies follow, causing an increase in planetary albedo (lower panel), a decrease in absorbed solar radiation, and a reduction of planetary temperature. WL call this mechanism the daisy feedback. Analysis shows that it is the key mechanism underlying the attractivity of the two-daisy equilibrium state (Weber, 2001). This feedback dampens the difference between the initial state (i.e., the equilibrium corresponding to S) and the stationary solution which is reached at the end of the integration for T_{planet} and A_{planet}. This is the equilibrium state corresponding to $S + \delta S$. The daisy feedback

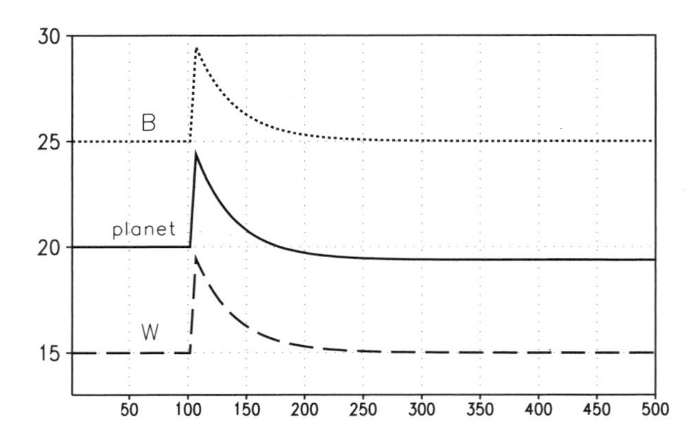

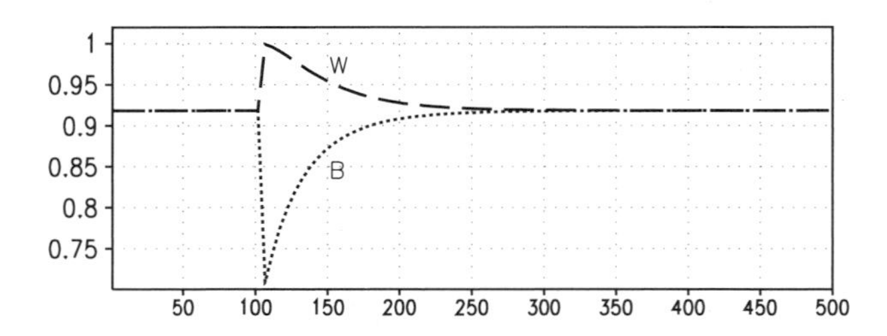

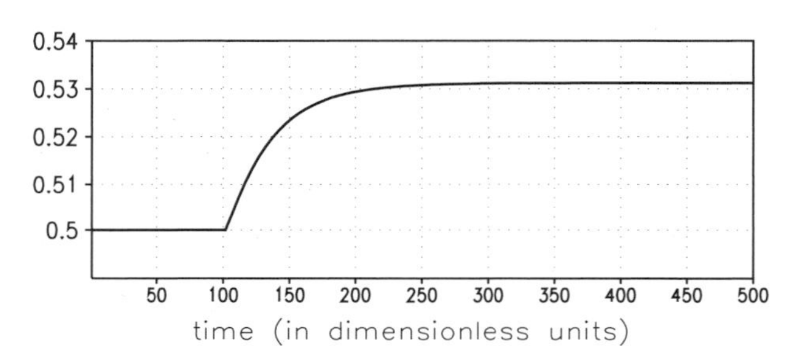

Figure 20.3
The evolution of the Daisymodel when solar radiation is increased by $\delta S = 20\ W/m^2$ at $t = 100$: (upper panel) the planetary temperature T_{planet} (solid line) and the local temperatures of the black (short-dashed line) and white (long-dashed line) daisies (all in °C); (middle panel) the corresponding growth rates of the black (short-dashed line) and white (long-dashed line) daisies; (lower panel) the corresponding planetary albedo A_{planet}.

drives the system to the new equilibrium state and makes the numerical solution method work.

Now we ask which factors determine the amplitude of the difference between the initial state and the end state of the integration. For the daisy temperatures (and growth rates) the difference is 0 according to (12) and (13). For the planetary temperature let us call the amplitude, or the size of distance in phase space, δT_{planet}. In the example depicted in figure 20.3, we find that $\delta T_{planet} = -0.6°\mathrm{C}$, with $\delta S = 20\ W/\mathrm{m}^2$ and $S = S_{ref}$ as the starting point. A general expression for δT follows easily from the equation that specifies the planetary equilibrium temperature (16) as

$$\delta T_{planet} \simeq \frac{q}{S}(1 - A_{ref})\delta S \tag{17}$$

As discussed in the previous section, the amplitude of the temperature response to a change in insolation is primarily determined by the efficiency of heat redistribution from relatively warm, black daisy regions to relatively cold, white daisy regions. Because the heat redistribution is very efficient, δT_{planet} is small and the planetary temperature remains close to its reference point, the optimal temperature for plant growth.

The system is also attracting with no daisies. In this case blind physics provides negative feedback and sets up an attractor. The higher the temperature, the greater the outgoing longwave radiation; the greater the outgoing radiation, the lower the temperature. Unlike the two-daisy state, however, the abiotic state is not homeostatic. Its equilibrium position is highly sensitive to incoming solar radiation S.

The fact that solutions to Daisyworld's equations are commonly obtained by numerical integration has led to a confusion between attractivity and homeostasis. It must be emphasized that the attractivity of an equilibrium state does not explain its location in phase space. Numerical integration finds an equilibrium because the system is attracting. The equilibrium states corresponding to two slightly different values of insolation S are close to each other according to (17) because the system is homeostatic.

Diffusion Among the Daisies

Daisyworld's heat transport relations (6) and (7) reduce the atmospheric general circulation to a few equations of a form reminiscent of the diffusion term used in box models of the planetary energy budget. We now compare WL's formulation against that of a classical two-box model. Such a model might, for example, be used as a first approximation of the coupled heat budgets of low and high latitudes, distinguished by different amounts of received solar radiation, or the coupled heat budget of land and ocean, distinguished by different heat capacities. In such models, differences in net radiation or in response time to the seasonal cycle create temperature gradients and promote down-gradient energy flows.

Here we use equations from North et al. (1981), covering a case with one white and one black box. As in the above treatment of the Daisymodel, we consider the equilibrium state:

$$(1 - A_b)S - (E + \lambda T_b) - \frac{D}{x_b}(T_b - T_w) = 0 \tag{18}$$

$$(1 - A_w)S - (E + \lambda T_w) + \frac{D}{x_w}(T_b - T_w) = 0 \tag{19}$$

The two boxes are distinguished by their different surface albedos. The first two terms in each equation represent the local absorbed shortwave radiation and emitted longwave radiation, respectively. The third term denotes the heat exchange between the two boxes. Consistent with Newton's law of cooling, heat flow is proportional to the temperature difference. Here D is the constant of proportionality. Although this formulation ignores the complexities that rotational forces, vertical fluxes, and latent heat transport introduce into planetary climates, it may be the best that can be done in two simple equations. Note that the area-weighted average of the separate heat budgets (18) and (19) gives a heat budget equation for the planetary temperature and albedo similar to WL's formulation (5).

Standard heat budget models specify boxes of fixed size. Daisyworld, on the other hand, has dynamic geography. Daisies expand into or recede from bare ground: daisy fractions $x_{b,w}$ are model variables, not constants. Two additional equations (such as the equations describing daisy population dynamics used in the section "The Daisymodel") are needed to solve (18) and (19) for variable box sizes. At this point we are primarily interested in the mechanisms that determine the temperature difference between the black box and the white box, since this can be compared with the formulation specified by WL. The temperature difference can be calculated by subtracting (18) and (19):

$$T_b - T_w = \frac{S(A_w - A_b)}{\lambda + \frac{D}{x_b} + \frac{D}{x_w}} \tag{20}$$

This equation can be understood as follows. The difference in absorbed solar radiation between the two boxes, $S(A_w - A_b)$, causes a temperature difference $T_b - T_w$. The system counteracts the temperature difference by (a) outgoing longwave radiation—the warmer, black box emits more radiation to space than the cooler, white box; and (b) by horizontal transport—energy flows from the warmer, black box to the cooler, white box. The efficiency of these two mechanisms is specified by λ and $D/x_b + D/x_w$, respectively. A higher total efficiency results in lower temperature differences, given the difference in absorbed solar radiation.

The spatial distribution of daisies is important to heat transport. If daisies grew in a mosaic of small patches, the distance between patches would be small, heat exchange between them would be rapid, and temperature differences between black and white daisies would be relatively small. Conversely, if each daisy population occupied one large contiguous block, the average distance between surface types would be large, heat exchange would be relatively slow, and temperature differences would be relatively large. In the box model, the spatial distribution of daisies can be shown to be implicit in the value chosen for D. The diffusion terms in (18) and (19) are based on the consideration that the net heat transport into a box is given by the divergence of the heat flux, where the heat flux is proportional to the spatial derivative of temperature (North et al., 1981). Explicitly writing this down and then scaling this term gives

$$C\kappa \frac{\partial^2 T}{\partial x^2} \sim C\kappa \frac{(T_b - T_w)}{L^2} \tag{21}$$

Here $C = \rho Hc$ is the heat capacity; with $\rho = 1.25$ kg/m^3 the air density; $H = 8000$ m a scale height; and $c = 10^3$ $J/(\text{kg}^\circ\text{C})$ the specific heat of air. The heat capacity is assumed to be the same for black and white daisies. The heat diffusion coefficient, κ, commonly takes on values between 0.5 and 3 10^6 m^2/s. It is clear now that $D = C\kappa/L^2$ incorporates a measure of length L.

If the boxes were daisy boxes, the length scale L would depend on the spatial distribution of the daisies. At one extreme, the planet is divided into two separate regions of black and white daisies, and L has hemispheric scale. At the other extreme the planet is covered with a fine-scaled mosaic of black and white daisies. For purposes of discussion, we consider L in the range from 500 to 5000 km. This results in values of D from 100 $W/\text{m}^{2\circ}\text{C}$ (smaller lengths) to 1 $W/\text{m}^{2\circ}\text{C}$ (larger lengths).

Finally, we compare temperature differences in the box model (20) with those of the Daisymodel (8). In both models, the local temperature difference is proportional to the local albedo difference. In the box model (20) the proportionality factor $F_{boxmodel}$ depends on S, λ, D and the daisy fractions $x_{b,w}$, whereas in the Daisymodel (8) the proportionality factor $F_{Daisymodel}$ is given by the constant q:

$$F_{boxmodel} = \frac{S}{\lambda + \frac{D}{x_b} + \frac{D}{x_w}} \quad \text{and} \quad F_{Daisymodel} = q \tag{22}$$

Comparison of the two formulations points to some deficiencies in Daisyworld's formulation of heat transfer. First, in basic physics temperature gradients scale with the heating gradient $S(A_w - A_b)$, and not with the albedo gradient $A_w - A_b$. Therefore, the proportionality factor should contain the term for incoming radiation, S. Second, as discussed above, heat transfer depends on the areas of daisies and on the sizes of daisy patches. The proportionality factor F should thus be dependent on the model variables $x_{b,w}$, and the parameter D. D, in turn, implicitly depends on how daisy regions are distributed over the planet and the associated length scale.

The value of $F_{boxmodel}$ varies considerably with the length scale in the system. Taking $S = S_{ref}$ and disregarding the effect of variable daisy fractions (assuming constant and equal daisy fractions), we find that $F_{boxmodel}$ can vary from values close to 1 for a fine-scale mosaic to values of around 50 for hemispheric-scale daisy regions. It was shown in the section "The Analytical Solution ..." that the temperature regulation in the system strongly depends on q, and thus on the proportionality factor F. Therefore, we conclude that the implicitly assumed spatial scale strongly affects the stability of the system. Parameterizing F with the constant, q, thus disregards spatial distribution and hides a physically-important feature of the system.

Rewriting Daisyworld's heat transport equations (6) and (7) to make them physically realistic—for instance, along the lines of equations (18) and (19)—requires that q be converted to a variable, as in (20). In addition, core equations should be reformulated so that the mean daisy temperature depends on solar radiation S consistent with (18) and (19), instead of on T_{opt}, as in (11). It does not seem promising to pursue this issue further in the context of the WL box model. Von Bloh et al. (1997) proposed an alternative approach to make spatial distributions explicit by using a standard energy balance equation on a high-resolution spatial grid, coupled to probabilistic growth and decay equations for the biota. Their model is too complex to be transparant, and the mechanisms responsible for numerically obtained results are difficult to disentangle.

Adapting Daisies

Free Adaptation

Robertson and Robinson (1998, henceforth RR) propose that if the fittest survive, a population's optimum temperature will evolve toward the temperature of the local environment. For example, the work of Weis and Berry (1988) on the photosynthetic enzyme rubisco shows that rubiscos of plants from hot climates have higher temperature optima than those of plants from cool climates. To represent adaptive evolution, RR changed Daisyworld to include continuous adaptation of the optimum temperature for daisy growth toward the local daisy temperature:

$$T_{opt;b}(t) = T_{opt;b}(t-1) + a[T_b(t-1) - T_{opt;b}(t-1)] \tag{23}$$

$$T_{opt;w}(t) = T_{opt;w}(t-1) + a[T_w(t-1) - T_{opt;w}(t-1)] \tag{24}$$

This modifies the growth rate (4), making it depend on how well the daisies have adapted (i.e., how closely their optimum temperatures have adjusted toward the environmental temperature). Thereby RR introduce a third internal timescale into the system, the evolutionary timescale of the biota. We assume that this timescale is much shorter than the rate of change of the solar luminosity. If that is so, we can restrict the problem to the equilibrium case where the daisies are perfectly adapted:

$$T^{eq}_{opt,b} = T^{eq}_b \quad \text{and} \quad T^{eq}_{opt,w} = T^{eq}_w \tag{25}$$

With free adaptation the optimal temperature for daisy growth $T^{eq}_{opt;b,w}$ is converted from a constant to a model variable. It is identical to the local temperature $T^{eq}_{b,w}$ for each daisy type. The relations (12) and (13), which hold in the original Daisymodel, are no longer valid for the free-adaptation case, and $T^{eq}_{b,w}$ now depend on the insolation S. This implies that there is no prescribed reference temperature in the free-adaptation system. As a consequence, it is no longer homeostatic.

Constrained Adaptation

Lenton and Lovelock (2000, henceforth LL) object to RR's formulation on the grounds that carbon-based life is inherently constrained to temperatures well below the boiling point of water, and functions poorly in freezing conditions. To represent this argument, they bound adaptation by introducing a parabolic multiplier function $M_{b,w}$ for the growth rate function, β:

$$\beta(T_{b,w}) = M_{b,w}\left(1 - \frac{(2(T_{opt;b,w} - T_{b,w}))^2}{(T_{max} - T_{min})^2}\right) \tag{26}$$

where

$$M_{b,w} = 1 - \frac{(T_{opt} - T_{opt;b,w})^2}{(77.5°\mathrm{C})^2} \tag{27}$$

The multiplier $M_{b,w}$ reaches a maximum value of 1 when the adapted optimum temperature $T_{opt;b,w}$ equals the original, prescribed optimum, T_{opt}, and it declines to 0 when the adapted optimum temperature reaches values of 77.5°C above or below the original optimum, T_{opt}.

In equilibrium LL's constrained daisies, like RR's daisies, are perfectly adapted. This implies that both daisy populations have adapted their optimum growth temperature so that it equals their local temperature—that is, equation (25) holds. Substituting (25) in (26), it follows that $\beta^{eq} = M^{eq}$. Now the system can be solved easily, following the approach used in the section "The Analytical Solution ..." for the original Daisymodel. From the requirement of equal growth rates $\beta(T_b) = \beta(T_w)$ at equilibrium, it follows that both types of daisies are equally well constrained:

$$M^{eq}(T_b) = M^{eq}(T_w) \tag{28}$$

Combining (25), (27), and (28), the mean equilibrium daisy temperature is:

$$T_b^{eq} + T_w^{eq} = 2T_{opt} \tag{29}$$

This is the equivalent of (11), derived for the original Daisyworld model. Equations (12)–(16) hold for the constrained-adaptation variant as well. It is therefore not surprising that the LL variant of Daisyworld is homeostatic.

Limits to Evolutionary Adaptation?

Whether, given 10^9-plus years of extremely gradual temperature increase, life would remain "as we know it" or whether some branch of life could evolve to survive boiling temperatures is difficult to test, and is outside the scope of this chapter. Questions relating to model structure, however, are germaine. Here we note that in the LL model, $M_{b,w}$ restores homeostasis by imposing a central value toward which adaptation gravitates because departures from this value reduce reproductive success. The introduction of lower and upper temperature limits on adaptation are largely irrelevant to homeostasis because the model seeks the central value. In LL, as in the original Daisymodel, homeostasis is due to the fixed reference temperature (T_{opt}) introduced in the planetary equilibrium heat budget.

The biological implications of $M_{b,w}$ are also questionable. A formulation in which rates of population expansion plummet as adaptation drives the population's attributes away from their original optimal state implies that adaptation to environmental change inherently carries a penalty. In effect, this says that the further life goes from the conditions it found optimal when it first evolved, the lower will be its reproductive success. Evolutionary history, however, shows many examples of necessity being the mother of invention. Oxygen-using metabolisms are more efficient than anoxic metabolisms; and taxa that have evolved to cope with the stresses of life on land are no less fit than those which have remained in aqueous environments.

It may be noted that the limits on biological adaptation need to be narrow if they are to protect the Earth system from physical catastrophe. As argued by Lovelock and Whitfield (1982) and later refined by Caldeira and Kasting (1992), a global average temperature above 40–50°C would invigorate the carbonate-silicate cycle, scrub CO_2 from the atmosphere, and push the silicate–carbon cycle into regions that make it difficult to sustain photosynthesis.

Summary and Discussion

The Daisymodel was developed to illustrate that a self-regulating planetary system can exist. The model is set forward as a parable. WL (1983, p. 284) warn that they "... are not trying to model the Earth, but rather a fictional world which displays clearly a property which we believe is important for Earth." That property is (1983, p. 286) "the ability to respond to a perturbation by restoring their local temperatures to prefixed values, despite the fact that no physically real reference temperature exists within the [Daisyworld] system."

We have examined the Daisymodel's steady-state solution with emphasis on the mechanisms that stabilize the system. This showed that the model's formulations implicitly decouple local temperatures from incoming radiation in the equilibrium state. Both local temperatures and planetary temperatures are tied to a fixed reference point. This does exactly what WL say Daisyworld doesn't do: it ties model behavior to the optimal growth temperature T_{opt}, which is a reference temperature. We also show that the heat transport formulation used in Daisyworld is essential to the model's homeostatic behavior, and that the formulation used departs from basic physics.

Life is adaptive. When RR replaced Daisyworld's prescribed optimal growth temperature with a variable that adjusts toward the local temperature, the system's reference point disappeared. As a result, local temperatures were linked to insolation and the system lost its homeostatic properties. LL reintroduced homeostasis into RR's formulation by constraining adaptation. We show that the central value imposed by LL is crucial for the resulting model behavior. The constrained-adaptation formulation thus achieves homeostasis by reintroducing a fixed reference point in the Daisyworld thermostat.

You may ask what this has to do with the science of Gaia. We prefer to turn that question around and ask, "What does Daisyworld have to do with the science of Gaia?" Suppose we have falsified claims made about a "fictitious world"; does this tell you anything about the real world?

For those who see Daisyworld as a cornerstone of Gaia's foundation, we hope the foregoing analysis raises doubts, or at least stimulates the search for a formulation based on more realistic concepts. The question of whether such a formulation would result in homeostasis should be an open one, not an a priori set goal. On the other hand, for those who regard Daisyworld as a cute (or annoying) distraction from the scientific questions raised by Gaia, we hope that it provides a basis for calling attention away from fictitious worlds and back to the study of real planets and realistic mechanisms.

Acknowledgments

We would like to thank Bill Scott for stimulating thoughts and help with LaTeX, and Tim Lenton and one anonymous referee for constructive and thoughtful reviews.

References

Caldeira, K., and J. F. Kasting. 1992. The life span of the biosphere revisited. *Nature*, 360, 721–723.

Lenton, T., and J. E. Lovelock. 2000. Daisyworld is Darwinian: Constraints on adaptation are important for planetary self-regulation. *Journal of Theoretical Biology*, 206, 109–114.

Lovelock, J. E., and M. Whitfield. 1982. Life span of the biosphere. *Nature*, 296, 561–563.

North, G. R., R. F. Cahalan, and J. A. Coakley. 1981. Energy balance climate models. *Reviews of Geophysics and Space Physics*, 19, 91–121.

Petersen, A. 2003. Models and geophysiological hypotheses. Chapter 3 in this volume.

Robertson, D., and J. Robinson. 1998. Darwinian Daisyworld. *Journal of Theoretical Biology*, 195, 129–134.

Saunders, P. T. 1994. Evolution without natural selection: Further implications of the Daisyworld parable. *Journal of Theoretical Biology*, 166, 365–373.

Von Bloh, W., A. Block, and H. J. Schellnhuber. 1997. Self-stabilization of the biosphere under global change: A tutorial geophysiological approach. *Tellus*, 49B, 249–262.

Watson, A. J., and J. E. Lovelock. 1983. Biological homeostasis of the global environment: The parable of Daisyworld. *Tellus*, 35B, 284–289.

Weber, S. L. 2001. On homeostasis in Daisyworld. *Climatic Change*, 48, 465–485.

Weis, E., and J. A. Berry. 1988. Plants and high temperature stress. In S. P. Long and I. Woodward, eds., *Plants and Temperature*, Society for Experimental Biology, Symposium XXXII, pp. 329–346. Cambridge: Cambridge University Press.

Salvaging the Daisyworld Parable under the Dynamic Area Fraction Framework

K. M. Nordstrom, V. K. Gupta, and T. N. Chase

Abstract

The Daisyworld model of Watson and Lovelock (1983) has received a great deal of attention over the years since it was first published. Because it is the primary parable for the Gaia hypothesis, this attention has been well deserved. However, the heat transport parameterization in Daisyworld contains assumptions that restrict its temperature-regulating properties to adjustments of the biota alone, and therefore applicability of these results to the Earth climate system is questionable, as pointed out by Weber (2001). The regulation of local temperatures, or "homeostasis," is proscribed by the biota independently of solar forcing. In addition, the strength of the model's homeostatic behavior, its main result, is forced to depend on an arbitrary parameter associated with the heat transport.

Here we construct a new version of the Daisyworld model based on local heat balances, using the simple heat transport parameterization of Budyko (1969). This Dynamic Area Fraction model (DAFM) formulation removes the assumption of perfect local homeostasis through the albedo-dependent local heat transfer equation. From this DAFM formulation we interpret the Daisyworld heat transport parameter physically, thus removing the artificiality in its parameter set. This representation results in globally similar temperature regulation, despite the removal of the assumption of perfect local homeostasis. These results are surprising, and lend weight to the Daisyworld parable.

Introduction: Modeling Hierarchy

The Daisyworld (DW) model of Watson and Lovelock (1983) has recevied a great deal of attention since it was published (e.g., von Bloh et al., 1997; Saunders, 1994; Meszaros and Palvolgyi, 1990; Zeng et al., 1990; Jascourt and Raymond, 1992; Nevison et al., 1999; Weber, 2001). Originally intended as a parable for the "Gaia hypothesis" set forth by Lovelock (1972) and Lovelock and Margulis (1974), it was a model of an extremely simple world in which competition between two species of daisy was capable of completely regulating the planetary mean temperature. DW's then-surprising results have played a major role in its visibility over the years as a mathematical illustration of how negative feedbacks from the biosphere could conceivably apply to the Earth climate system (ECS).

However, assumptions regarding an artificial heat transport parameterization in the model led it to proscribe perfect homeostasis in its local temperatures at steady state (Watson and Lovelock, 1983, eq. 14; Weber, 2001, eq. 11). Moreover, they forced the strength of the homeostasis, its main result, to rely upon a uninterpretable heat transport parameter. These assumptions fix one of only two local climate variables. Global homeostasis in the model is thus a result of the relative growth and decay of daisy areas alone, since regional temperatures are predetermined independently of solar forcing. Therefore, the applicability of the parable of DW to the ECS is jeopardized, since biota on earth do not have complete local control.

In this chapter, we take a new look at the DW model to remove this assumption by replacing the heat transport parameterization of Watson and Lovelock with the more Earth-like Newtonian cooling used in the energy balance model of Budyko (1969). We do this to investigate homeostatic behavior in DW without proscribing perfect local homeostasis, and to determine how much of the homeostatic behavior of the original DW can be attributed to this condition. In order to make this change, it is necessary to update local energy balances rather than to determine local temperatures from global energy balance. Although this represents a small additional complication to the original model, the resulting framework is still simple enough to be amenable to analytical treatment and rigorous error analysis. We will refer to this model as a Dynamic Area Fraction model (DAFM) (Nordstrom, 2002).

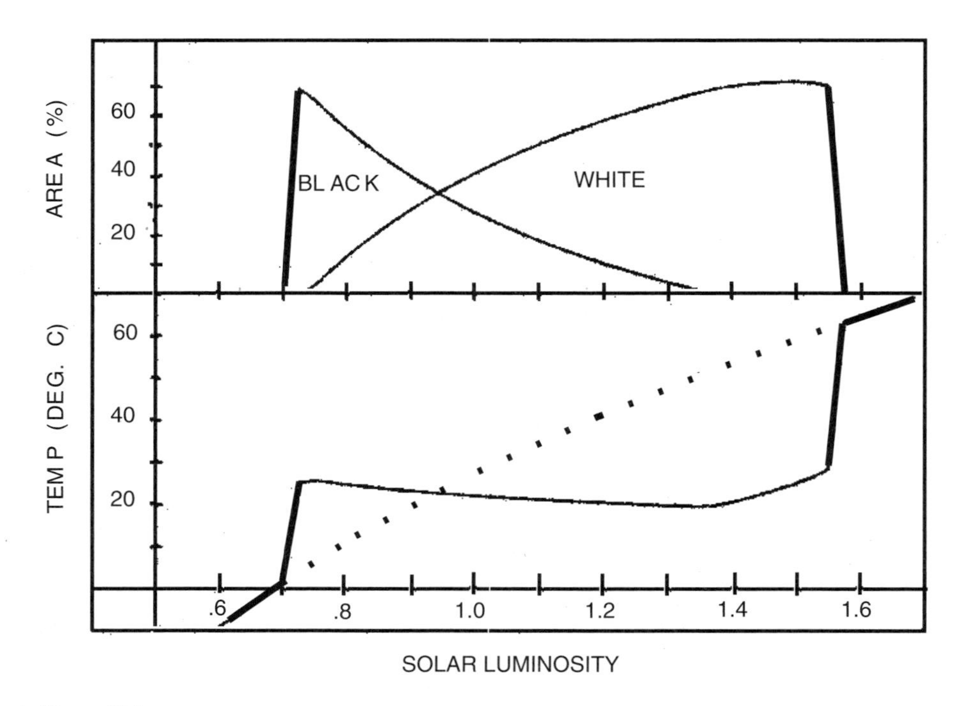

Figure 21.1
Homeostasis in DW. Daisies reduce the mean temperature of their planet (bottom) over increasing solar luminosity due to a negative feedback of the daisies (top), which maintain complete control of their local environments. Reprinted from Watson and Lovelock (1983).

This chapter examines homeostatic behavior in the DAFM and compares it with DW. First, we briefly discuss DW. Then we describe the development of the DAFM, and follow that with the results of our experiments and analytical work comparing it with DW.

Background

In 1983, Watson and Lovelock mathematically demonstrated the concept of homeostasis in the physical sciences with an extremely simple model they called Daisyworld. Homeostasis in biology refers to the tendency of an organism to maintain a relatively constant internal temperature despite external influences, and Watson and Lovelock's adaptation of the term to climate developed analoguously to its biological counterpart. Their concept of homeostasis in climate was based on the idea that the biosphere was capable of playing a first-order role, through the action of negative feedbacks with the system as a whole, in regulating and adapting its environment. (See figure 21.1.)

The DW model, developed for a planet covered with black and white daisies, consisted of the following six equations:

$$\dot{a}_i = a_i(\beta(T_i)a_d - \gamma) \quad i = b, w \tag{1}$$

$$\beta(T_i) = 1 - k(T_i - T_{opt})^2 \tag{2}$$

$$a_d = 1 - a_b - a_w \tag{3}$$

$$\sigma T^4 = SL(1 - A) \tag{4}$$

$$T_i = T + \delta(A - A_i) \tag{5}$$

$$A = a_b A_b + a_w A_w + a_d A_d \tag{6}$$

where β was the birthrate of a daisy species per available area; γ was the death rate; k was the sensitivity of the birthrate to the local temperature; T_{opt} was the optimal temperature for daisy birth; L was the solar luminosity, the ratio of solar shortwave radiation, S, to earth current, S_o; σ was the Stefan–Boltzmann blackbody constant; and δ was an inverse measure of heat transport. DW was divided into three dynamic regions representing black (subscript b), white (w), and bare-ground (d) populations, with areas normalized to be some fraction of the whole. These area fractions a_i varied between 0 and 1; and local albedos A_i were .75 for white daisies, .25 for black daisies, and .5 for bare ground. T was the global mean value for the temperatures T_i such that $T = \sum_i a_i T_i$, and A was the global mean value for albedos.

Watson and Lovelock showed that on this imaginary world, the biota were able to reduce the mean

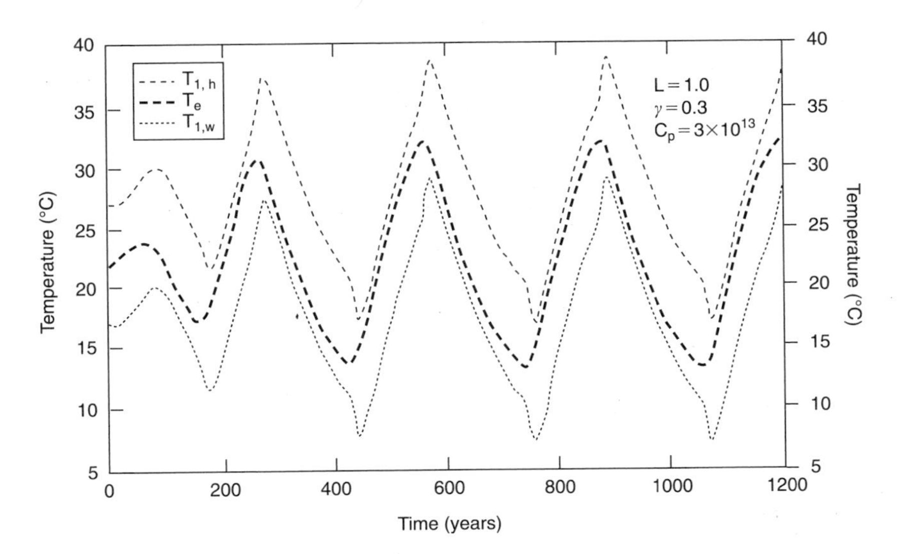

Figure 21.2
Oscillations in a DW with dynamic global radiative balance. Reprinted from Nevison et al. (1999).

temperature of their environment over a large range of solar luminosities (figure 21.1). Subsequent work with this model included extending it to 2-D (von Bloh et al., 1997), adding mutations of the biota with subsequent evolution of albedos (Saunders, 1994), and coupling a carbon greenhouse feedback (Meszaros and Palvolgyi, 1990). Zeng et al. (1990) pointed out chaos in a DW with discrete dynamics; and Jascourt and Raymond (1992) discussed this model and pointed out the existence of homeostasis in a mean sense under this chaos.

DW assumes an instantaneous temperature update through its energy balance equation (4). Such an assumption, equivalent to setting the heat capacity of the system to 0, is often deemed acceptable if the timescale of system adjustment is much smaller than that of its forcing. However, since we are interested in short-term behavior of the model as well (Nordstrom, 2002), we cannot make this approximation, even though the ECS adjusts on much faster time scales than those associated with the warming of the sun. (See figure 21.2.)

The heat capacity of the Earth, generally assumed to be dominated by that of the oceans at the surface, is commonly taken to be on the order of 10^{10} J/m^2K. Since the expansion of the Earth's atmosphere with solar heating is negligible compared with the earth's radius, the first law of thermodynamics requires that

$$\dot{Q} = c_p\dot{T} = R_{in} - R_{out}. \tag{7}$$

This equation is the standard energy balance relation in which T is the planetary mean temperature; R_{in} and R_{out} are the longwave and shortwave radiation terms, respectively; and $1/c_p$, the inverse of the heat capacity, plays the role of the timescale for changes in the planetary radiative balance. Since we take $c_p = 3 \times 10^{10}$ J/m^2K, after Harvey and Schneider (1987), $1/c_p$ corresponds to approximately 500 years. The behavior of DW with a dynamic temperature update was explored by Nevison et al. (1999), with $R_{in} = SL(1 - A)$ and $R_{out} = \sigma T^4$, as in the original model. Nevison et al. found oscillatory behavior of their model over a range of values for the daisy death rate γ (figure 21.2), which serves as an inverse timescale for daisy update. Nevison et al. also found that the mean value of the temperature oscillation over a cycle exhibited homeostasis in the sense of Jascourt and Raymond (1992). (See figure 21.3.)

The focus of this chapter, however, is a damaging criticism of DW made by Weber (2001), based on analytical work in the original paper by Watson and Lovelock. The original authors themselves pointed out that the method of parameterization of the heat transport in their model maintains steady-state local temperatures independent of solar radiation when both species of daisy are present. The argument itself is very simple and follows from the form of the birth rate $\beta(\cdot)$ (2) and the area update equation (1) in the steady state (steady-state conditions are denoted henceforth by an asterisk superscript), such that

$$T_b^* + T_w^* = 2T_{opt}, \tag{8}$$

since

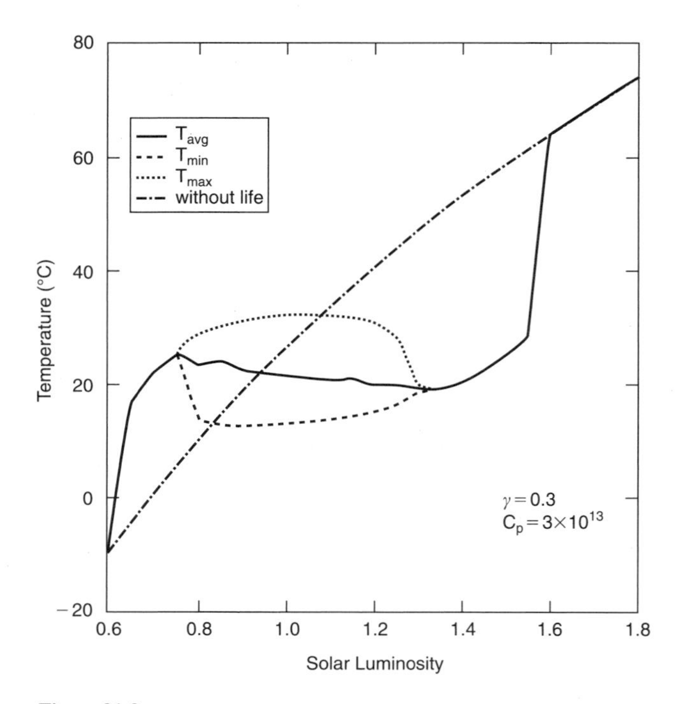

Figure 21.3
Homeostasis in cycle mean for a DW with dynamic global radiative balance. Reprinted from Nevison et al. (1999).

$$\beta(T_w^*) = \beta(T_b^*) = \gamma/a_d^*. \tag{9}$$

Consequently, (5) implies

$$T_b^* - T_w^* = \delta(A_w - A_b) \tag{10}$$

and therefore that local temperatures can be written (Watson and Lovelock, 1983, eq. 14) as

$$T_b^* = T_{opt} + \delta(A_w - A_b)/2 \tag{11}$$

$$T_w^* = T_{opt} - \delta(A_w - A_b)/2.$$

Here T_{opt}, δ, A_b, and A_w are all constants which do not depend upon the solar constant. Weber noted that this is an unsatisfactory situation if one wishes to apply DW's results to the ECS, since local temperatures on our planet clearly should be affected by radiative balances. This situation is equivalent to giving daisies perfect local homeostatic control over their environment, and is a direct result of the heat transport parameterization (5) and of the specification of a single optimal temperature for growth of both species of daisy in (2).

Of these two constraints, the more important for our purposes is the heat transport. To show this, let us examine the consequences of using different optimal temperatures. Instead of using T_{opt} as the optimal temperature for both species, suppose instead we use, say, T_{ob} and T_{ow}, for black and white daisies, respectively. Then, because (2) is a quadratic, any possible solution of (9) at the steady state must take either the form $T_b^* + T_w^* = K(T_{ow}, T_{ob})$ or $T_b^* - T_w^* = K(T_{ow}, T_{ob})$, in which $K(\cdot, \cdot)$ is some linear function only of the optimal temperatures. Any such function K will itself be a constant of the model.

From (10), it is immediately clear that there can exist no steady-state solution to (9) of the form $T_b^* - T_w^* = K(T_{ow}, T_{ob})$, except in the unlikely case that $K(T_{ow}, T_{ob}) = \delta(A_w - A_b)$. This leaves us with a steady-state solution that must take the form $T_b^* + T_w^* = K(T_{ow}, T_{ob})$, in turn implying that local temperatures in the steady state can be written

$$T_b^* = K(T_{ow}, T_{ob}) + \delta(A_w - A_b)/2 \tag{12}$$

$$T_w^* = K(T_{ow}, T_{ob}) - \delta(A_w - A_b)/2. \tag{13}$$

Since $K(T_{ow}, T_{ob})$ still does not depend upon the incident solar radiation, changing the second constraint is of no fundamental significance with regard to the removal of proscribed local homeostatic temperatures, and for modeling purposes the simplest assumption is therefore to make them equal. It would appear, therefore, that DW's locally proscribed homeostatic behavior is a direct result of the assumptions implicit in the heat transport parameterization (5). This assumption is not present in our DAFM.

Methods: Construction of a DAFM

A Dynamic Area Fraction model is constructed on a dynamic grid in the sense that the areas of the grid are not fixed but are allowed to grow and decay according to some prognostic equation. In a DAFM, the expansion and contraction of boundaries must be accounted for with a term in a local balance equation that not only reflects the noninertial nature of the reference frame but also averages in the intrinsic quantities specified at neighboring grid sites (e.g., temperature) for purposes of conservation (figure 21.4). In DW, three population types are represented, associated only with biota: black daisies, white daisies, and bare (arable) land. We therefore take these same three population types.

The practice of determining the local temperatures from the planetary mean, as is done in DW, is actually the extrapolation of a second moment—the difference in local temperatures—from a first moment. It thus requires additional information, in this case the a priori specification of the physics of heat transport (as in, for example, (5)). In a DAFM, we elect

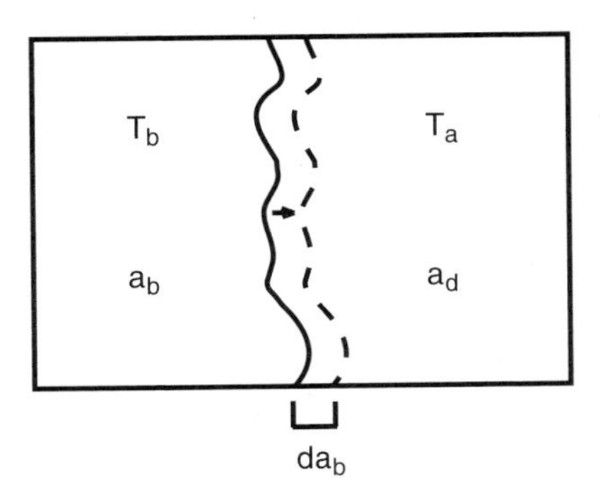

Figure 21.4
Schematic of boundary movement in a DAFM. As the boundary of the black daisy population (left) expands by da_b into the bare land region (right), it absorbs a quantity of heat *cpdabTd*, which must then be redistributed over the black region, thus decreasing *Tb* (which will be warmer than *Td*). The bare land region, however, has not expanded, and therefore absorbs no heat. Its heat per unit area remains constant, such that *Td* is unchanged.

instead to use radiative balance to determine local temperatures, then average over the model's boxes to determine the planetary mean. In this way, we not only eliminate a major physical inconsistency from the original model but also provide a richer framework for modeling, since with a DAFM it is possible to model a larger set of components of the ECS (Nordstrom, 2002).

Let the total heat per unit area of an region i in a DAFM be $q_i = c_p T_i$, with the total heat over an area $Q_i = a_i q_i$, such that

$$\dot{Q}_i = a_i[SL(1 - A_i) - \sigma T_i^4 + F_{T_i}], \tag{14}$$

where F_{T_i} is a diffusive term representing the transfer of heat between regions. If we sum Q_i over regions i, we can write the global energy conservation equation

$$\dot{Q} = \sum_i \dot{Q}_i = SL(1 - A) - \sigma \sum_i a_i T_i^4.$$

This can be written

$$\dot{Q} \approx SL(1 - A) - \sigma T^4, \tag{15}$$

since the variation of local temperatures T_i from the global mean T is small with respect to T. This is, of course, the radiative balance equation from the DW of Nevison et al. (1999) (7). Here $A = \sum_i a_i A_i$ is again the global mean albedo, and $\sum_i a_i F_{T_i} = 0$ is the condition that the interregional fluxes not add heat. For simplicity in our DAFM, we represent F_{T_i} with a linear relaxation, or Newtonian cooling term, after Budyko (1969),

$$F_{T_i} = D_T(T_i - T). \tag{16}$$

Now, we are interested in obtaining the temperature update equation from (14). In DW, the flux of heat across boundaries was proportional to the difference in the albedos of the regions involved (5), in such a way that the global mean temperature was dependent upon the distribution of population types only through the global mean albedo (11). Due to the dynamic nature of the surface model and the use of local radiative balances in a DAFM, such a condition is not appropriate. The global temperature is specified through $T = \sum_i a_i T_i$, and T_i should be proportional to Q_i, so the inclusion of any equation like (11) would overdetermine the model.

The nontrivial issue involved in determining an update equation for T_i is that as populations expand, their boundaries necessarily move into regions of different temperatures. Thus, we must derive a term in the local temperature update to account for losses and gains in area, a term that reflects the fact that the boundaries between regions in a DAFM are completely permeable mathematical constructs. To find such an equation in the DAFM, we require the global heat balance to be consistent with the energy conservation equation $Q = \sum_i Q_i = \sum_i a_i q_i$ introduced in (14). We must then redistribute the heat absorbed when we calculate T_i, the mean temperature over the region. This calculation, performed in appendix I, gives us for daisy-covered regions

$$\dot{T}_i = \frac{1}{c_p a_i}[\dot{Q}_i - \theta(\dot{a}_i)\dot{a}_i c_p (T_i - T_d)] \tag{17}$$

in which $\theta(\cdot)$ is the Heaviside function. For a bare-land region, the update is given by

$$\dot{T}_d = \frac{1}{c_p a_d}\left[\dot{Q}_d - \sum_i (1 - \theta(\dot{a}_i))\dot{a}_i c_p (T_d - T_i)\right]. \tag{18}$$

Although these equations appear to be complicated, they are actually quite simple. Heat balance in the DAFM is the sum of the solar radiative balance and regional (linear) heat transport (first term), and a volume expansion term (second term) reflecting the permeability of the artificial membranes placed between regions. This volume expansion term is 0 unless the local boundary has moved into another region, in which case it reflects the absorption and instantaneous averaging of absorbed heat.

The full model equations for the DAFM are thus

$$\dot{a}_i = a_i(\beta(T_i)a_d - \gamma) \quad i = b, w \tag{19}$$

$$\beta(T_i) = 1 - k(T_i - T_{opt})^2 \tag{20}$$

$$a_d = 1 - a_b - a_w \tag{21}$$

$$c_p \dot{T}_i = SL(1 - A) - \sigma T^4 - D_T(T_i - T) - K_{T_i} \quad (22)$$

$$A = a_b A_b + a_w A_w + a_d A_d \quad (23)$$

$$K_{T_i} = \begin{cases} \theta(\dot{a}_i)\dot{a}_i c_p (T_i - T_d)/a_i, & i \neq d \\ \sum_j [(1 - \theta(\dot{a}_j))\dot{a}_j c_p (T_d - T_j)]/a_d, & i = d \end{cases} \quad (24)$$

We have derived a model in which both climate system components, a_i and T_i, are allowed to update freely. This situation is in contrast to DW, in which only the a_i can change.

Next we compare the homeostatic behavior of DW with our DAFM in the absence of perfect locally proscribed homeostasis.

Comparison of Homeostatic Behavior in DW with That in the DAFM

The homeostatic behavior of the original DW appears to be a surprising result until one examines the assumptions underlying its parameterization of heat transport (5). As discussed above, this behavior then appears to be merely a direct consequence of this assumption, equivalent to assuming perfect local homeostasis (11). In our DAFM, we have removed this assumption and allowed heat to be transported in a manner similar to that in the ECS, thereby effectively adding a degree of freedom to the DW model of Watson and Lovelock. The question is, since local homeostasis is no longer proscribed, does homeostatic behavior still occur in a DAFM, and, if so, to what extent?

In order to compare the behavior of the two models, it is first necessary to find equivalent ranges for their parameter sets. That is, the DAFM relies upon the physically interpretable parameter D_T for heat transport, while DW relies upon δ, which is completely arbitrary. All other parameters remain the same. Thus, given a transport parameter D_T in a DAFM, we should be able to find a $\delta(D_T)$ under which DW lies in a similar regime. Such an equivalence is especially desirable in view of the fact that the strength of homeostasis in DW relies on the condition $\delta < SL/\sigma$ (Watson and Lovelock, 1983).

We can find a DW heat transport flux parameterization equivalent to that of a DAFM by noting (from 10) that $T_i - T_j = \delta(A_j - A_i)$ in DW for any two regions i and j. From (14) and (16), we note that at steady state in the DAFM, $D_T(T_i^* - T^*) + \sigma T_i^{*4} = SL(1 - A_i)$. Evaluating this expression at some reference luminosity and temperature pair, (L_{ref}, T_{ref}) at which the two models produce the most similar results, we find

$$\begin{aligned} T_i^* - T_j^* &= SL_{ref}(A_j - A_i)/[D_T + \sigma(T_i^{*2} + T_j^{*2}) \\ &\qquad \times (T_i^* + T_j^*)] \\ &\approx SL_{ref}(A_j - A_i)/[D_T + 4\sigma T_i^{*3}] \\ &\approx SL(A_j - A_i)/[D_T + 4\sigma T_j^{*3}] \\ &\approx SL_{ref}(A_j - A_i)/[D_T + 4\sigma T_{ref}^3]. \end{aligned} \quad (25)$$

From this we can extract

$$\delta(D_T) \approx SL_{ref}/[D_T + 4\sigma T_{ref}^3] \quad (26)$$

Then $T_i^* - T_j^* \approx \delta(D_T)(A_j - A_i)$ for the DAFM, as in DW.

We tune $\delta(D_T)$ such that the temperature–luminosity profile of DW is as close as possible to that of our DAFM. That is, with (L_{ref}, T_{ref}) we determine the $\delta(D_T)$ that minimizes the difference between the temperature profile in DW and the temperature profile in our DAFM. We define this difference to be the root mean square distance between the two profiles; given $T_{DW}^*(L_{ref}, L)$ for DW and $T_{DAFM}^*(T)$ for the DAFM, we are minimizing

$$D(L_{ref}) = \int_0^\infty dL (T_{DW}^*(L_{ref}, L) - T_{DAFM}^*(L))^2 \quad (27)$$

with respect to L_{ref} (or, equivalently, with respect to δ). The bounds of the integration can be reduced as well, since in the absence of daisies the only population in either model is the bare-land fraction. The steady state for both models is given by

$$T_{DW}^{*4} = T_{DAFM}^{*4} = SL(1 - A_d)/\sigma. \quad (28)$$

Therefore, the integration in (27) can be restricted to the range of luminosities at which daisies exist.

Analytically, (27) is formidable. We elect instead to solve it numerically. We do this by integrating DW to a steady-state temperature $T_{DW}^*(\delta, L)$ for increasing values of δ, over a range of luminosities containing the two-daisy state. We then approximate the integral (27) with a sum over these values of the root mean square difference between $T_{DW}^*(\delta, L)$ and $T_{DAFM}^*(L)$ for one of our values of D_T. For each case tested, numerical results suggest $(L_{ref}, T_{ref}) = (L_{opt}, T_{opt})$. We show the results of this calculation in the case $D_T = 1.3 \times 10^8$ J/m^{2}Kyr in figure 21.5.

Note that the expression for $\delta(D_T)$ (26) has the form

$$\delta(D_T) = \frac{SL}{\frac{\partial}{\partial T_i}(dQ_i/dt)|_{T_i = T_{opt}}},$$

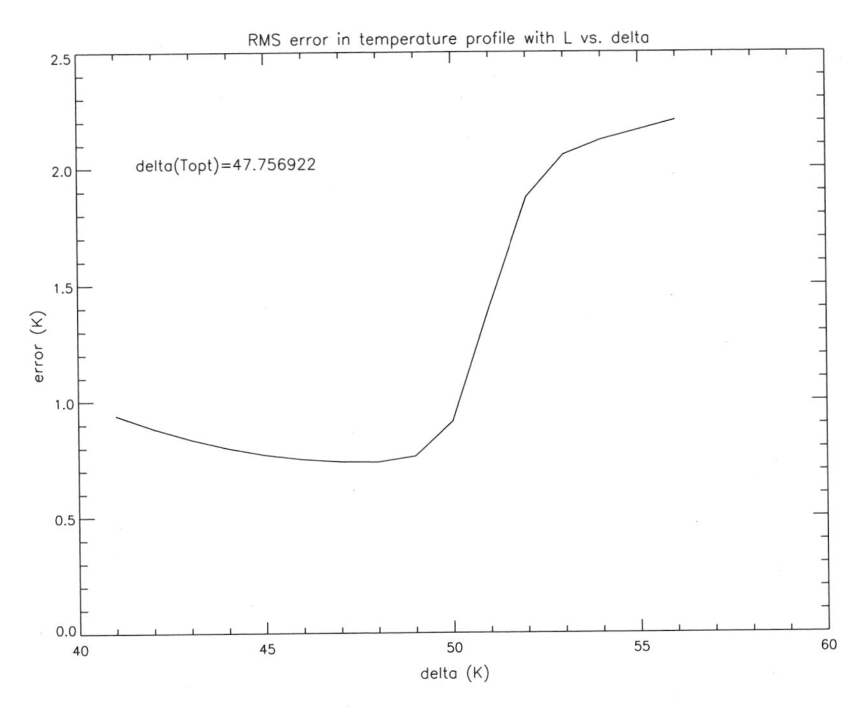

Figure 21.5
Results of minimization of RMS difference $D(L_{ref})$ ("error(K)"); the K refers to measurement in Kelvins between TDW and $TDAFM$ for $DT = 1.3 \times 10^8$ J/m^{2}K yr. Minimization is performed against $\delta(DT)$ ("delta(K)") rather than T_{ref}, since it is a more convenient measure. The function is clearly minimized for $\delta \approx 47$ K, implying $T_{ref} \approx T_{opt}$.

which suggests, using (22), a more general method for determining the value of δ for more realistic planetary atmospheres, and addresses more generally the issue of the interpretation of δ. The δ parameter appears to be the ratio of the solar radiation at the top of the atmosphere to the partial derivative of the heat balance with respect to temperature.

To investigate homeostasis in a more Earth-like situation, we made runs of the DAFM using smaller albedos of $A_b = .15$, $A_d = .2$, and $A_w = .25$, which on Earth would correspond to characteristic albedos for trees, soil, and grassland, respectively (McGuffie and Henderson-Sellers, 1997). At $L = 1$, with these albedos, all biota died off in the model, and so the steady-state global mean temperature was independent of D_T at $T \approx 263$ K. This is not surprising, since DW contains no greenhouse effect to trap solar radiation, and is consistent with the estimate made by, for example, Peixoto and Oort (1992) for Earth in the absence of a greenhouse effect. Larger values of L were therefore used.

We estimated the value for δ corresponding to four values for D_T in our DAFM. In table 21.1 we list the values studied, all of which were computed using $S = 1.07 \times 10^10$ J/m^{2}year, $L_{ref} = L_{opt} = 1.32$, $\sigma =$ 1.79 J/K^{4}year, and $T = 285$ K. With these values, we ran both DW and our DAFM, and compared the strength of the homeostasis in each by examining the behavior of the global mean temperatures. A semi-objective measure of the global homeostatic behavior of these models can be obtained by comparing the slope of the temperature anomalies in the two-daisy state, as found in Watson and Lovelock (1983) (see figure 21.6). This measure is only semi-objective, since the slope of both models in this range varies somewhat with L.

Table 21.1
Values of heat transport parameters δ and D_T used

Run	$\delta(K)$	D_T (J/m^{2}Kyr.)
1 (low)	3.41	3.98×10^9
2	25.05	3.98×10^8
3 (Earth)	47.76	1.3×10^8 (Budyko, 1969)
4 (high)	68.71	3.98×10^7

A more objective measure of the magnitude of homeostasis of both models can be taken as the total distance of the mean global temperature in the two-daisy state from the optimal temperature, T_{opt}. This

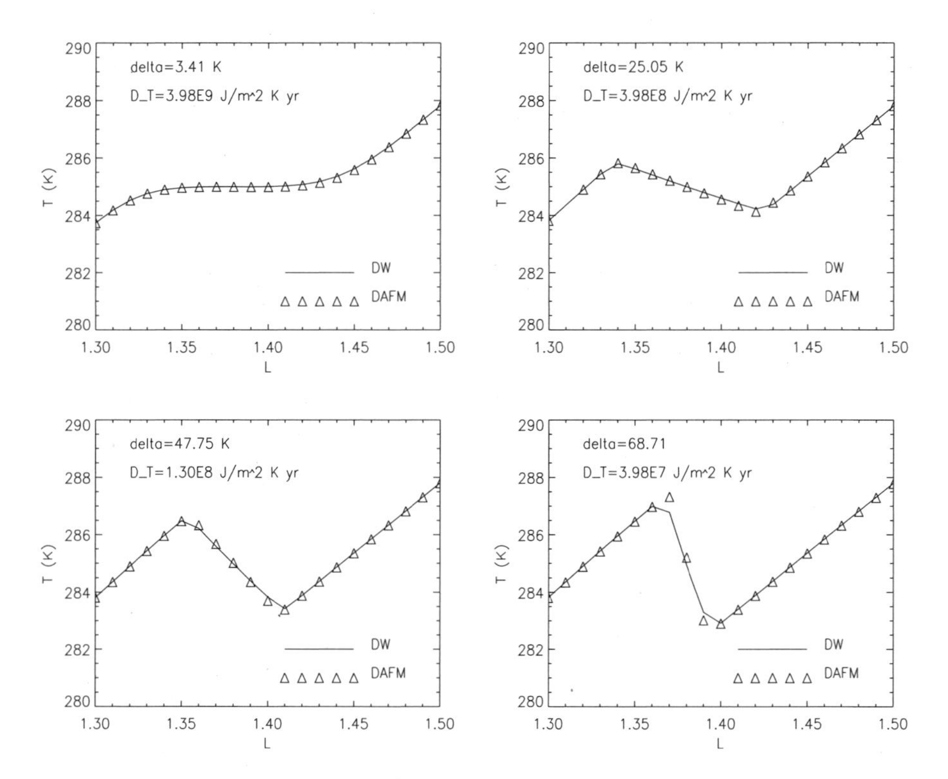

Figure 21.6
Temperature–luminosity profile magnified to show the two-daisy state in both DW (solid line) and DAFM (triangle). Note the consistent temperature anomaly (downward trend in temperature with increasing luminosity) in both.

measure we denote as

$$H^2(D_T) = \int_{\{L \mid a_b(L) \neq 0 \text{ and } a_w(L) \neq 0\}} (T^*(L) - T_{opt})^2 \, dL. \tag{29}$$

This measure also must be calculated numerically.

Comparative plots of local temperature contrast $T_b^* - T_w^*$ versus luminosity L for DW and our DAFM are shown in figure 21.7. The local temperature difference grows almost linearly with L in the DAFM, but remains constant in DW over the entire range of luminosities. As expected, the temperature differences in the DAFM are free to adjust to the influence of solar forcing, whereas those in DW are fixed.

Comparative plots of global mean temperature T^* versus luminosity L for all four values of D_T and δ are shown in figure 21.8. There is very good agreement between the curves in all four cases; differences can really be noticed only in the two-daisy state, which has been magnified in figure 21.6. The slope of the curve for the DAFM in these figures is close to the same as that in DW. Differences in the two profiles can be found at the edges of the two-daisy state, where the DAFM tends to reduce global mean temperatures more quickly toward the optimal temperature.

This similarity in behavior is interesting, since we have removed the heat transport condition (5) that led to proscribed homeostatic local temperatures (11), as pointed out by Weber (2001). Slopes for curves in the two-daisy state for both models are listed in table 21.2. In each case, these slopes are calculated with a second-order numerical derivative around the optimal point (L_{opt}, T_{opt}), though the increment of $\Delta L = .01$ is fairly crude. This is sufficient, since our intent is to show that slopes are on the same order. In appendix II, we analytically show that slopes for the two models generally will be approximately equal in this range.

In table 21.3 we give numerical results for the distance $H(D_T)$. The DW model is closer to the optimal temperature, T_{opt}, in all cases. Again, the measure of the DAFM is of (close to) the same order of magnitude.

Conclusion

We have introduced a Dynamic Area Fraction model (DAFM), a model developed to remove the inconsis-

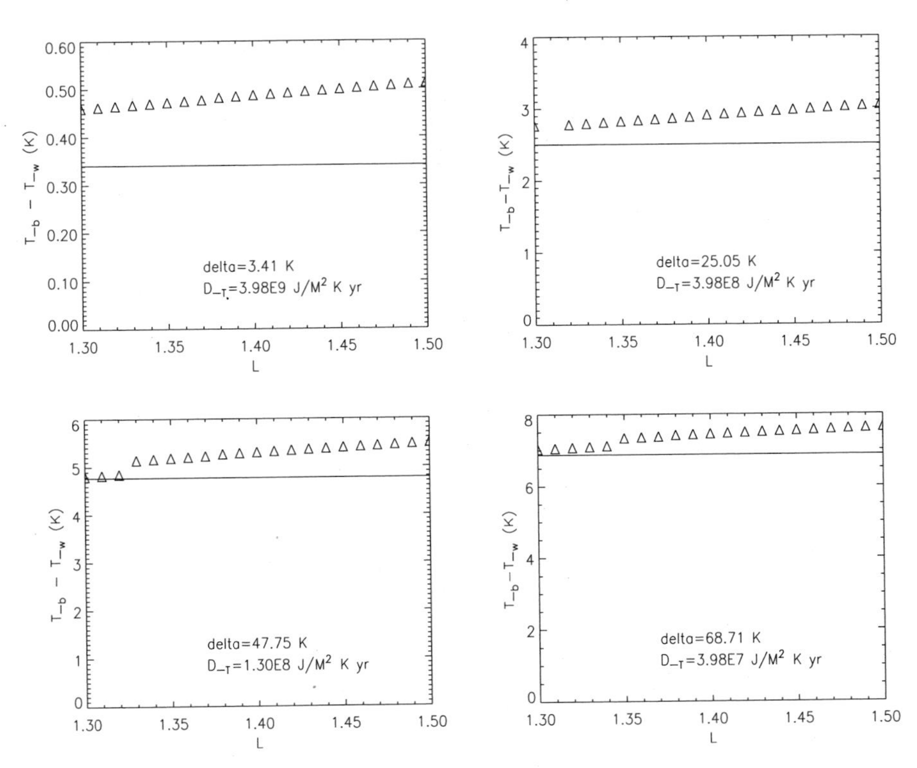

Figure 21.7
Black-white temperature contrast in DW (solid lines) and DAFM (triangles). The contrast in the DAFM is an increasing function of L throughout the two-daisy state in each case.

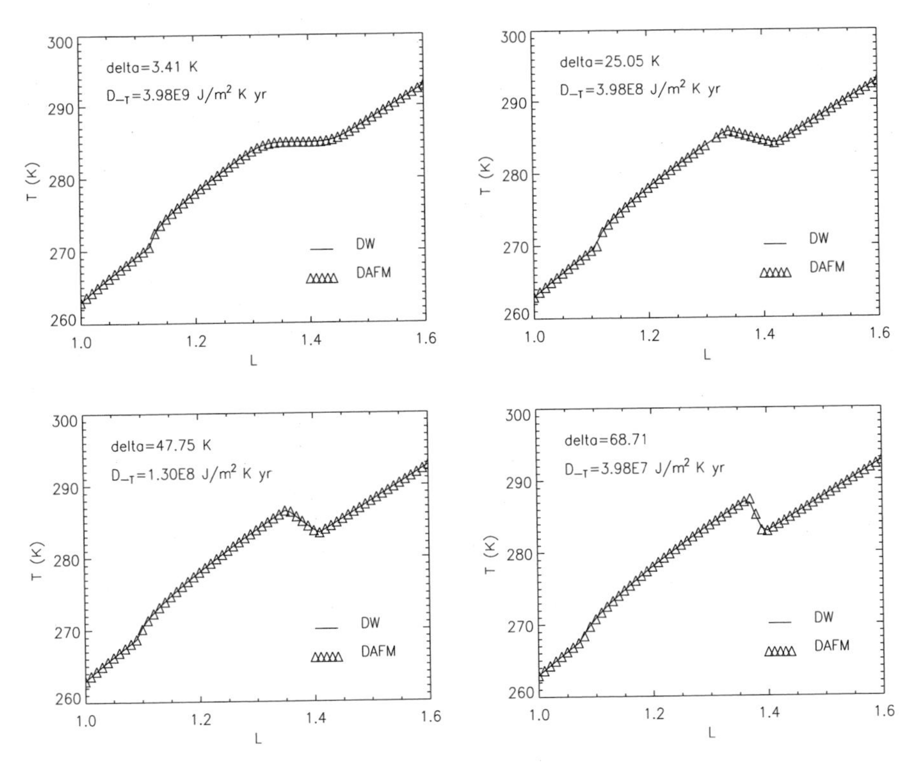

Figure 21.8
Temperature-luminosity profile for all cases studied. Agreement between the two models is quite good in global temperature, with differences occurring in the two-daisy state (see figure 21.6) and local temperatures (see figure 21.7).

Table 21.2
Values for two-daisy state slopes in DW and DAFM

Run	DW slope	DAFM slope
1 (low)	−0.06256	−0.3952
2	−20.20	−21.94
3 (Earth)	−59.63	−65.79
4 (high)	−174.5	−215.4

Table 21.3
Model distance from optimal temperatures as in $H(D_T)$ from (29)

Run	DW slope	DAFM slope
1 (low)	1.182	2.141
2	0.1001	0.1962
3 (Earth)	0.08641	0.3031
4 (high)	0.007233	0.1512

tent heat transport formulation that led to perfectly homeostatic local temperatures in the DW model of Watson and Lovelock. We have argued that the DAFM representation of DW's fundamental physics is more correct than that in the original DW. This is based on the fact that heat transfer between regions is allowed to occur through a Newtonian relaxation, a linear term that behaves like a diffusion. The one parameter required to specify this term, D_T, has been measured for the Earth climate system (ECS) by Budyko (1969); thus our model is better determined than the original DW, which relied upon a heretofore unmeasurable quantity, δ. From the formulation of the heat transport term in our model, we have provided a simple, approximate, analytical basis for relating DW's δ to D_T. This argument helps to interpret the original parameterization, and makes it possible to compare runs made with our DAFM with the original DW, since with it we can estimate equivalent regimes of the two fundamentally different models.

We have compared homeostatic behavior in the DAFM against that in the DW for similar parameter sets, and have shown that globally the homeostasis in the DAFM is of nearly equivalent strength. This is surprising, since homeostasis in DW evidently arose from allowing biota perfect control of their local environments, and the biota in a DAFM are more affected by their environments than that in DW as a result of the removal of this condition. These results imply that it might be possible for organisms in the ECS which exert a smaller degree of control over their local environment to affect the regulation of the planetary climate as effectively as an organism with more control.

The DAFM model developed in this chapter can also be viewed as a framework for the development of models with results directly applicable to the ECS (Nordstrom, 2002). Such a framework is ideally suited to energy balance approaches for investigating surface–atmospheric and hydrological feedbacks in climate.

APPENDIX I: DERIVATION OF DAFM BALANCE UPDATES

From $Q_i = a_i c_p T_i$ and $Q = \sum_i Q_i$, we can write, as a first guess,

$$\dot{Q} = \sum_i [\dot{a}_i q_i + a_i \dot{q}_i].$$

This predicts an update of

$$\begin{aligned} c_p \dot{T}_i &= SL(1 - A_i) - \sigma T_i^4 \\ &\quad - D_T(T_i - T) - c_p T_i \frac{d \ln(a_i)}{dt} \\ &= dq_i - c_p T_i \frac{d \ln(a_i)}{dt} \end{aligned} \tag{30}$$

from (4). However, this equation is incomplete because it has not taken into account the redistribution of heat due to boundary movement.

To see this, suppose that a cooler region j expands into a warmer one, k, without a change in the radiative balance. In such a case, the mean temperature over region j should have risen due to the absorption of heat from k. The mean temperature of region k should remain unchanged, having absorbed no heat per unit area from its neighbors ($dq_k = 0$), and the change in heat over the entire system should be $dQ = 0$. But, calculating the change in temperature in region j from (30) with $dq_i = 0$ for all i, we see that

$$\dot{T}_j = -T_j \frac{d \ln(a_j)}{dt}.$$

This quantity is negative because $\dot{a}_j > 0$, despite the fact that T_j should have risen. Therefore (30) is not the correct local temperature update.

Approaching the issue more carefully, we note that a permeable boundary is required to allow an expanding region to absorb a positive or negative quantity of heat from the region into which it expands, while the regional heat per unit area should remain constant as that region contracts. In an area interaction between more than two regions, this method of course requires the specification of the region into which another is expanding; but in DW, we have already made the assumption that one species of plant must die before its area can be considered for occupation by another ((1) and (3)). So necessarily regions must expand into the bare land fraction a_d. A nonbare region i—in DW, this refers to either the white or the black daisy fraction—that expands by da_i absorbs a quantity of heat equal to $da_i c_p T_d$, since it has expanded into region d by the above assumption. The temperature of region i was T_i before the expansion, so we replace a quantity $da_i c_p T_i$ in this process. Then the heat of the region changes by $c_p da_i(T_i - T_d)$ due to the expansion.

Now a nonbare region that contracts changes its heat by $c_p da_i(T_i - T_i) = 0$, since its boundary has not moved into a region of different temperature (see figure 21.4). Therefore, the local temperature update should be written, for a nonbare region, as

$$\dot{T}_i = \frac{1}{c_p a_i}[\dot{Q}_i - \theta(\dot{a}_i)\dot{a}_i c_p(T_i - T_d)]$$

$$= \frac{1}{c_p a_i}[\dot{Q}_i - \theta(\beta(T_i)a_d - \gamma)\dot{a}_i c_p(T_i - T_d)] \quad (31)$$

in which $\theta(\cdot)$ is the Heaviside function (included to enforce the condition that a contracting region exchanges no heat), and in which we have used the fact that $a_i > 0\ \forall i$. The bare region update can then be found from the global conservation requirement, which gives

$$\dot{T}_d = \frac{1}{c_p a_d}\left[\dot{Q}_d - \sum_i (1 - \theta(\beta(T_i)a_d - \gamma)) \times \dot{a}_i c_p(T_d - T_i)\right]. \quad (32)$$

Thus local heat balance in the DAFM is the sum of the solar radiative balance and regional (linear) heat transport (first term) and a volume expansion term (second term) reflecting the permeability of the artificial membranes placed between regions. These are the proper equations, taking into account an instantaneous averaging process for the heat transferred into each region. Note that regional temperatures are mean temperatures under this framework, and that including volume expansion terms of this form is equivalent to making the assumption that the timescale of local atmospheric temperature adjustments are small compared with other timescales in the model.

APPENDIX II: ANALYTICAL SLOPE COMPARISON

We will find closed form expressions for the slopes of the temperature–luminosity profiles in the two-daisy regime for both DW and the DAFM. From these profiles we will determine a ratio of slopes with respect to model parameter L to compare the extent of homeostasis in these models (Watson and Lovelock, 1983). In the two-daisy range, the slope should be negative for a homeostatic model and positive for a nonhomeostatic model. Additionally, the model with the smaller (more negative) slope clearly will be more homeostatic.

For the two-daisy state in DW, the global mean temperature is defined by (4) and the steady state for the local temperatures is defined from (2), (5), and (8) by

$$T_b + T_w = 2T_{DW} + \delta(2A - A_b - A_w) = 2T_{opt}, \quad (33)$$

in which we have formally dropped the asterisks for notational simplicity. We can eliminate the albedo A between these equations to obtain the steady-state temperature T_{DW} for Watson and Lovelock's model implicitly,

$$T_{DW} = T_{opt} + \frac{\delta\sigma T_{DW}^4}{SL} - \delta(1 - A_d). \quad (34)$$

We differentiate this equation with respect to L to find the slope of the DW temperature–luminosity profile,

$$T'_{DW} = \frac{dT_{DW}}{dL} = \frac{1}{L}\frac{\sigma T_{DW}^4}{(4\sigma T_{DW}^3 - \frac{SL}{\delta})} = \frac{S(1 - A)}{(4\sigma T_{DW}^3 - \frac{SL}{\delta})}. \quad (35)$$

We substitute $\delta = \delta(D_T) = SL_{opt}/(D_T + 4\sigma T_{opt}^3)$ to get

$$T'_{DW} = \frac{SL_{opt}(1 - A)}{(4\sigma[L_{opt}T_{DW}^3 - LT_{opt}^3] - D_T L)}. \quad (36)$$

In the DAFM, on the other hand, local temperatures for vegetated regions have a steady state defined by

$$\sigma T_b^4 + D_T(T_b - T_{DAFM}) = SL(1 - A_b) \quad (37)$$

$$\sigma T_w^4 + D_T(T_w - T_{DAFM}) = SL(1 - A_w). \quad (38)$$

Adding these two, we can use the steady-state condition for the area fractions $T_w + T_b = 2T_{opt}$ and the fact that $A_b + A_w = 2A_d$ to get

$$\sigma(T_b^4 + T_w^4) + 2D_T(T_{opt} - T_{DAFM}) = 2SL(1 - A_d) \quad (39)$$

Differentiating (39) gives

$$4\sigma(T_b^3 T_b' + T_w^3 T_w') - 2D_T T'_{DAFM} = 2S(1 - A_d), \quad (40)$$

but again $T_w + T_b = 2T_{opt}$ implies $T_b' = -T_w'$. Defining $\Delta = T_b - T_{opt} = T_{opt} - T_w$, (40) becomes

$$4\sigma(T_b^3 - T_w^3)\Delta' - 2D_T T'_{DAFM} = 2S(1 - A_d). \quad (41)$$

To eliminate Δ' in (41), we differentiate and rearrange (37) and (38) to get

$$(4\sigma T_b^3 + D_T)T_b' = S(1 - A_d) + D_T T'_{DAFM} + S(A_d - A_b) \quad (42)$$

$$(4\sigma T_w^4 + D_T)T_w' = S(1 - A_d) + D_T T'_{DAFM} + S(A_d - A_w). \quad (43)$$

Subtracting now, and again substituting $\Delta' = T_b' = -T_w'$, we find

$$\Delta' = \frac{S(A_w - A_b)}{(4\sigma[T_b^3 + T_w^4] + 2D_T)}. \quad (44)$$

Now using (41) we get, finally, the closed-form solution

$$T'_{DAFM} = \frac{dT_{DAFM}}{dL}$$

$$= \frac{S}{D_T}\left[\frac{2\sigma(A_w - A_b)[T_b^3 - T_w^3]}{(2D_T + 4\sigma[T_b^3 + T_w^4])} - (1 - A_d)\right]. \quad (45)$$

The first term in brackets in (45) is very small compared to 1 for all values of D_T we have considered, so

$$T'_{DAFM} \approx -\frac{S(1 - A_d)}{D_T}. \quad (46)$$

The ratio $r = \frac{T'_{DAFM}}{T'_{DW}}$ is thus given by

$$r = \frac{L}{L_{opt}} \frac{(1 - A_d)}{(1 - A)}\left(1 - \frac{4\sigma}{D_T}\left[\frac{L_{opt}}{L} T_{DW}^3 - T_{opt}^3\right]\right) \quad (47)$$

$$= \frac{L}{L_{opt}} \frac{(1 - A_d)}{(1 - A)}\left(1 - \frac{4SL_{opt}}{D_T}\left[\frac{(1 - A)}{T} - \frac{(1 - A_d)}{T_{opt}}\right]\right). \quad (48)$$

Now we expand in a small neighborhood of L_{opt} such that $\frac{L}{L_{opt}} = 1 + \lambda$, $\frac{T}{T_{opt}} = 1 - \tau$, and $\frac{(1-A)}{(1-A_d)} = 1 - \alpha$. Then r becomes, to first-order:

$$r = \frac{1 + \lambda}{1 + \alpha}\left[1 - \frac{4SL_{opt}(1 - A_d)}{D_T T_{opt}}\left(\frac{1 - \alpha}{1 - \tau} - 1\right)\right] \quad (49)$$

$$\approx (1 + \lambda)(1 + \alpha)\left[1 - \frac{4SL_{opt}(1 - A_d)}{D_T T_{opt}} \times ((1 - \alpha)(1 + \tau) - 1)\right] \quad (50)$$

$$\approx \left(1 + \lambda + \alpha - \frac{4SL_{opt}(1 - A_d)}{D_T T_{opt}}(\tau - \alpha)\right) \quad (51)$$

$$= \left(1 + \lambda + \alpha\left(1 + \frac{4\sigma T_{opt^3}}{D_T}\right) - \frac{4\sigma T_{opt^3}}{D_T}\tau\right). \quad (52)$$

Now, since $SL(1 - A)/SL_{opt}(1 - A_d) = \sigma T^4/\sigma T_{opt}^4$, we can relate parameters α, τ, λ with

$$\alpha = 4\tau - \lambda \quad (53)$$

Also, we have $T_i = T + \delta(A - A_i)$, so that

$$\tau = \frac{\delta(1 - A_d)}{T_o}\alpha. \quad (54)$$

Together, (53) and (54) reduce r to

$$r = 1 + \tau\left[1 + \frac{1 + \frac{4SL_{opt}}{D_T}x - \frac{4SL_{opt}\delta}{D_T}x^2}{1 - 4\delta x}\right], \quad (55)$$

where we have defined $x = (1 - A_d)/T_{opt}$. This ratio is close to 1, though the first-order term in λ is always positive. Thus, DW's homeostatic properties appear to have been reduced very slightly in the DAFM, by this measure, though they remain equivalent to 0-th order.

References

Budyko, M. 1969. The effect of solar radiation variations on the climate of the earth. *Tellus*, 5(9), 611–619.

Harvey, L., and S. Schneider. 1987. Sensitivity of internally generated climate oscillations to ocean model formulation. In A. Berger, J. Imbrie, J. Hays, S. Kukla, and B. Saltzman, eds., *Proceedings of the Symposium on Milankovitch and Climate*, pp. 653–667. Hingham, MA: ASI, and Dordrecht: D. Reidel.

Jascourt, S., and W. Raymond. 1992. Comments on chaos in Daisyworld. *Tellus*, 44B, 243–246.

Lovelock, J. 1972. Gaia as seen through the atmosphere. *Atmospheric Environment*, 6, 579–580.

Lovelock, J., and L. Margulis. 1974. Atmospheric homeostasis by and for the biosphere: The Gaia hypothesis. *Tellus*, 26, 2–10.

McGuffie, K., and A. Henderson-Sellers. 1997. *A Climate Modeling Primer*. New York: John Wiley and Sons.

Meszaros, E., and T. Palvolgyi. 1990. Daisyworld with an atmosphere. *Idojaras*, 94(6), 339–345.

Nevison, C., V. Gupta, and L. Klinger. 1999. Self-sustained temperature oscillations on Daisyworld. *Tellus*, 51B, 806–814.

Nordstrom, K. 2002. Simple models for use in hydroclimatology. Ph.D. thesis, University of Colorado at Boulder.

Peixoto, J., and A. Oort. 1992. *Physics of Climate*. New York: American Institue of Physics.

Saunders, P. 1994. Evolution without natural selection: Further implications of the Daisyworld parable. *Journal of Theoretical Biology*, 166, 365–373.

Von Bloh, W., A. Bloch, and H. Schellnhuber. 1997. Self-stabilization of the biosphere under global change. *Tellus*, 49B, 249–262.

Watson, A., and J. Lovelock. 1983. Biological homeostasis of the global environment: The parable of Daisyworld. *Tellus*, 35B, 284–289.

Weber, S. 2001. On homeostasis in Daisyworld, *Climatic Change*, 48, 465–485.

Zeng, X., R. Pielke, and R. Eykholt. 1990. Chaos in Daisyworld. *Tellus*, 42B, 309–318.

22
Food Web Complexity Enhances Ecological and Climatic Stability in a Gaian Ecosystem Model

Stephan P. Harding

Abstract

The relationship between complexity and stability in ecological communities has been a major area of speculation and research in ecology since the 1930s. Most mathematical models of communities have shown that stability declines as complexity increases, but so far modelers have not included the material environment in their calculations. Here, an otherwise conventional community ecology model is described which includes feedback between the biota and their climate. This "geophysiological" model is stable in that it resists perturbation. The more complex the community included in the model, the greater its stability in terms of resistance to perturbation and rate of response to perturbation. This is a realistic way to model the natural world because organisms cannot avoid feedback to and from their material environment.

Introduction

A central question in ecology concerns the relationship between the complexity of ecological communities and their stability. As pointed out by McCann (2000) in his comprehensive review of this topic, the pioneering ecologists of the 1950s suggested that complex communities should be more stable than simple ones with fewer species. Elton (1958) pointed out that crop monocultures were more prone to severe pest outbreaks than complex natural communities. MacArthur (1955) suggested that a species connected to many others through feeding relationships would retain a stable population size should one of these other species disappear, since in a diverse community ecologically similar species could readily fill the void.

The complexity-stability question was first given a rigorous mathematical treatment by May (1973), who overturned the perspective of Elton and MacArthur by showing that stability declines as complexity increases in randomly connected model communities. Most models since then have supported May's results, despite the inclusion of more realistic ecological relationships (May, 1981; Pimm, 1979a; Pimm, 1979b), but there are a few notable exceptions. Complexity has been found to lead to stability in models where the food supply is not affected by consumers (De Angelis, 1975), where perturbation involves the removal of species in lower trophic levels (Pimm, 1979a), or, most important, where weak trophic interactions are incorporated (McCann et al., 1998; Polis, 1998; McCann, 2000).

However, none of these models have taken the abiotic environment into account. It was Alfred Lotka (1925), one of the founders of population biology,who suggested that the dynamics of populations cannot be considered in isolation from the material environment in which they are embedded and with which they interact. He saw organisms and environment as a single evolving system in which each influences the other in a relationship of mutual feedback, and suggested that the mathematics of the evolution of this whole system would be more tractable than that of either population or environment taken separately.

A qualitatively different class of strongly stabilized community ecology models which build on Lotka's insight are those that include environmental feedback (Watson and Lovelock, 1983). In these models instability is the rare condition. In this chapter, experiments with a community ecology model are described which incorporate explicit mathematical feedbacks between biota and environment. In this class of model, greater complexity gave rise to greater stability. Thus both weak interactions (McCann et al., 1998) and biota-environment feedback (this study) have been found to increase stability.

Complexity is often defined in terms of the presence of more species, stronger interspecies interactions, and greater connection between species (Begon et al., 1996). Pimm's (1984) measures of stability, resilience, and resistance are used in the model described here. Resilience is taken to mean the speed with which a community returns to a former state after having been displaced from it by perturbation, while resistance is thought of as the ability to avoid such displacement.

Stable communities are those with high resilience and resistance. Resilience and resistance are not necessarily positively correlated. For example, a highly resistant system, when eventually displaced, may not be resilient because it requires a long time to return to steady state. In this chapter resilience is measured by a community's rate of return to steady state after a perturbation.

Daisyworld

The numerical model Daisyworld (Lovelock, 1983; Watson and Lovelock, 1983; Lovelock, 1992) was used as the basis of the experiments. The original Daisyworld is a mean field model of an abstract flat surface sown with seeds of two daisy types, one dark and the other light. A model sun warms the planet. Daisies germinate and grow between 5°C and 40°C with peak rates at 22.5°C, following a Gaussian curve. Once daisies begin to grow, they compete for space; as the available space decreases, so does their growth rate. The interaction between solar luminosity and the planetary albedo sets the temperature of the planet. Plants with low albedos (darker types) absorb more solar radiation than the bare surface would, and thus warm both themselves and the planet. Those with high albedos (i.e., whiter types) reflect solar radiation back to space and therefore cool both themselves and their environment.

This simple model system responds to changes in solar luminosity and keeps planetary temperature well within the limits tolerable by life even when solar luminosity varies over a wide range. The model is unusually stable to changes in initial conditions and to perturbation. Temperature regulation via the biota's albedo in Daisyworld is a particular example of the general biota-environment feedbacks of the real world. The mathematical foundations of Daisyworld are increasingly being used in other Earth system models, such as those of Lovelock and Kump (1994) and Lenton (1998). A frequent criticism of Daisyworld is that it is a toy model and does not apply to the real world. However, a similar vegetation albedo feedback on climate in the boreal forests has been modeled in a manner inspired by Daisyworld (Betts and Cox, 1997). Daisyworld is useful and practical because its stability and self-regulating properties do not depend on a particular mathematical formulation. Its growth equation was taken from a deterministic plant population dynamics model (Carter and Prince, 1981). Furthermore, Maddock (1991) has shown that self-regulation around a steady state arises in a variety of well-known population ecology models where there is tight coupling between biota and their environment.

Methods

The original multitrophic-level Daisyworld described by Lovelock (1992) was used as the basis for the experiments reported here. Twenty-three different daisy types were each given a unique and constant albedo ranging between 0.2 (dark) and 0.775 (light). Daisy-eating herbivores were also included, each type preying on daisies according to one of Holling's (1959) three classic functional responses, which describe the way in which predation rate varies with prey abundance. A type 1 response is linear, a type 2 response is inversely exponential, and a type 3 response is sigmoid. In the work described here, each functional response leveled off at a plateau of 0.5. As a result of each of these feeding strategies, rare daisy types experienced proportionally less predation from herbivores than did abundant ones. The herbivore functional responses used in this model are described mathematically in equations (6) to (8) below. Previous studies have looked at the impact on the system of a single herbivore operating under one of these three feeding strategies (Harding and Lovelock, 1996), but in this model three herbivores were introduced together and competed for food and space. A carnivore, which consumed the herbivores according to a type 3 functional response, could also be introduced. Neither herbivores nor carnivores could change their colors (i.e., their albedos), and so they had no direct feedback effects on climate. However, like the daisies, their growth rate was temperature-dependent.

The equations used in this model were based on those of the original model by Lovelock (1992), in which the equations for climate, competition among daisy types and daisy growth are given. Here, these basic Daisyworld equations were modified in various ways. First, the original quadratic growth equation was replaced with a more biologically realistic Gaussian temperature-dependent growth curve, as follows:

$$\beta_i = \exp(-(22.5 - T_i)^2 0.01) \quad (1)$$

where β_i is the growth rate of each of the k daisy types, and T_i is their local, albedo-dependent temperatures.

The equations for including higher trophic levels followed equations (8) to (10) in Lovelock (1992), but with two modifications. First, herbivore and carnivore growth rates (β_b and β_c, respectively) depended only on T_e, the overall temperature of the system, according to the following equations:

$$\beta_b = \exp(-(22.5 - T_e)^2 0.01) \tag{2}$$

$$\beta_c = \exp(-(22.5 - T_e)^2 0.01) \tag{3}$$

The second modification was to introduce the three classical functional responses (Holling, 1959) to model density-dependent impacts of herbivores on daisy populations and of the carnivore on herbivore populations, where A (plants), B (herbivores), and C (carnivores) refer to populations of each trophic level; (x, y, z) to the space unoccupied by these populations; a_i to the fraction of the planet covered by each of the k daisy types; and γ_i to the constant death rate for each daisy type, set at 0.15. The modified growth equation for daisy types was

$$da_i/dt = a_i[\beta_i x - \gamma_i - (B_1(p_{1i}\lambda_i) + B_2(p_{2i}\lambda_i) + B_3(p_{3i}\lambda_i)] \tag{4}$$

where λ_i is determined by one of the equations (6), (7), or (8), and p is a coupling matrix defined as:

$$p \equiv \left\{ \begin{matrix} p_{1,3} \cdots\cdots\cdots p_{1,23} \\ p_{2,3} \cdots\cdots\cdots p_{2,23} \\ p_{3,3} \cdots\cdots\cdots p_{3,23} \end{matrix} \right\} \tag{5}$$

where $p_{ij} \varepsilon (0, 1)$

If $p_{ij} = 0$, daisy j was not eaten by herbivore i, whereas if $p_{ij} = 1$, the daisy was eaten.

Here λ_i represents the functional response for a given herbivore, where λ_i modified the impact of B_i, the herbivore population, according to the following equations, in which F determined the slope of the functional response, up to a plateau of 0.5:

For a type 1 response:

$$\lambda_i = a_i F, \quad \text{up to a plateau of } \lambda_i = 0.5 \tag{6}$$

For a type 2 response:

$$\lambda_i = (1 - \exp(-a_i F))/2 \tag{7}$$

For a type 3 response:

$$\lambda_i = (1 - \exp(a_i^2 - F))/2 \tag{8}$$

The values of F used in the experiments are detailed in the legends to figures 22.2 and 22.3.

The equation used to model the growth of the three herbivores was a version of Lovelock's (1992) equation (9), modified to incorporate λ_i, the density-dependent predation by the carnivore, when present, and G_i, a growth increment, itself a function of E_i, the total amount of daisy biomass consumed by the herbivores. The new equation was as follows, where i varies from 1 to 3, corresponding to the three herbivores, and where the herbivores' death rate was held constant at $\gamma_i = 0.3$.

$$db_i/dt = b_i(A\beta_i y + G_i - \gamma_i - C\lambda_i) \tag{9}$$

where

$$G_i = 1 - \exp(-1.02E_i) \tag{10}$$

and

$$\lambda_i = 1 - \exp(-20B_i^2) \tag{11}$$

The total amount of daisy biomass consumed by each herbivore was calculated as follows:

$$dE_i/dt = B_i\left(\sum_{i=1}^{23} p_{in}\lambda_n\right) dt \tag{12}$$

where i refers to herbivores and n refers to daisy types, and

$$E_i = k + B_i \int \left(\sum p_{in}\lambda_n\right) dt \tag{13}$$

where k was a constant set at $k = 0.0001$, used to initiate the herbivore population.

Similarly, Lovelock's (1992) equation (10) for describing the growth of the carnivore population was modified to incorporate G, the carnivore's growth increment, a function of E, the total herbivore biomass eaten by the carnivore, with its death rate held constant at $\gamma = 0.05$:

$$dc/dt = c(B\beta z + G - \gamma) \tag{14}$$

where

$$G = 1 - \exp(-1.02D) \tag{15}$$

where D is the total amount of herbivore biomass eaten by the carnivore, calculated as follows:

$$dD/dt = C\left(\sum_{i=1}^{3} \lambda_i\right) dt \tag{16}$$

where

$$\lambda_i = (1 - \exp(-3B_i))/2 \tag{17}$$

Finally,

$$D = k + C\int \left(\sum \lambda_i\right) dt \tag{18}$$

where k was a constant, set at $k = 0.001$, used to initialise the carnivore population.

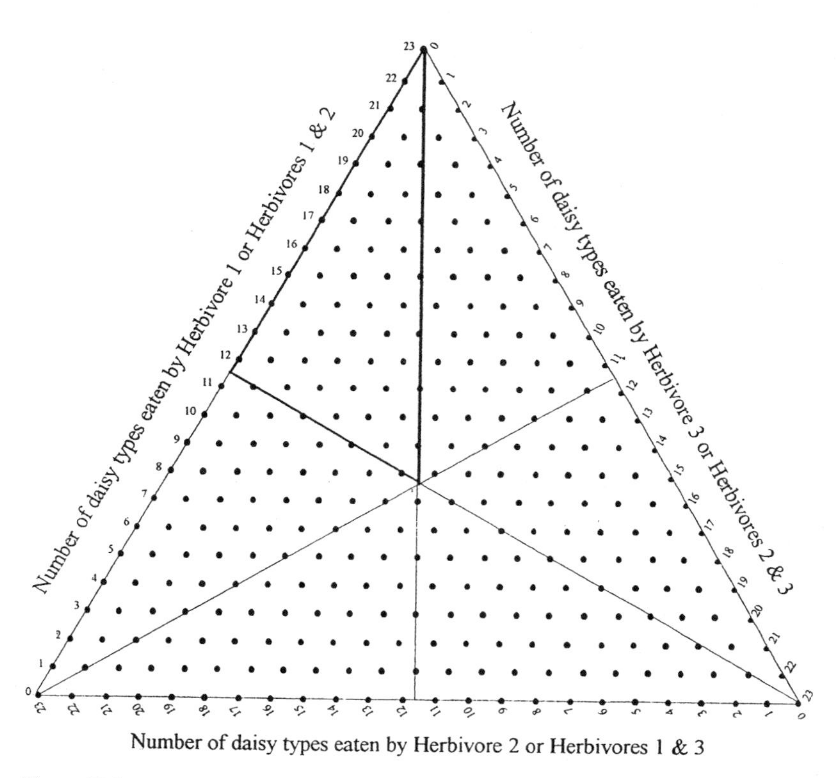

Figure 22.1
The parameter space of all possible food web configurations for $c = 1$ and $c = 2$ food webs, in the absence of carnivores, and with each herbivore allocated the same type 2 functional response, with $F = 2.2$. To obtain data for the entire parameter space, it was sufficient to sample the top left subtriangle (in heavy line), and then to extrapolate the results to the remaining space. This is because the food web configurations within each subtriangle are self-similar in terms of herbivore impacts on daisy types.

A critical parameter in the experiments reported here was food web connectance, c, which was varied from the maximum of $c = 3$, where each daisy type was eaten by each of the three herbivores, to the minimum of $c = 1$, where each daisy type was eaten by only one randomly selected herbivore. The connectance parameter, c, specified the values (0 or 1) in the coupling matrix, p. Intermediate values of c—$c = 1$ or 2 and $c = 2$ denoted that each daisy was eaten by one or two randomly selected herbivores, or by two randomly selected herbivores, respectively. These low ($c < 3$) food web connectances can be thought of as having arisen due to natural selection among daisies for anti-herbivore responses such as distasteful secondary compounds which are known to influence plant–herbivore dynamics in the field (Huntly, 1991).

The stabilizing effect of environmental feedback makes possible large models involving many evolving nonlinear differential equations. The price for this advantage is the need to explore the extended space occupied by the model variables, and it was decided to explore the effects of varying connectance patterns. There were 300 possible ways to link the 23 daisy types with the 3 herbivores, depending on the number of daisy types eaten by each herbivore. This was in the absence of carnivores and for all food webs, where each herbivore was allocated the same type 2 functional response, with slope $F = 2.2$ and a plateau of 0.5. The connectance c was set as either $c = 1$ or $c = 2$. Each of these 300 food web configurations specified the numbers of daisy types eaten by each herbivore in a given food web. The triangular diagram (figure 22.1) illustrates the part of the model space occupied by each food web configuration. This space was explored systematically by a computer program that stepped through a self-similar portion of the total space shown in figure 22.1. Linking the three herbivores with randomly chosen daisy types made connectance patterns. The program produced three of these patterns for each food web configuration. For all three patterns all variables were held constant while solar luminosity, J, was increased in 20 equal increments, from $J = 0.61$ to $J = 0.93$. The time taken for the diversity index to return to its steady state was used to measure the food web's resilience. This was observed for each luminosity step after the introduction of the three herbivores and, where appropriate, after the subsequent introduction of the carnivore. Resistance, the ability of the system to resist change, was measured by the amplitude of the temperature deviation from the optimum growth temperature of 22.5°C after the experimental perturbations. To calculate the stability of the model, the means of the 20 resilience and resistance measurements of each pattern were taken, and then the grand means of the three replicates were calculated as the final measures of resilience and resistance.

The effects of variations in herbivore functional response type and slope on the behavior of the model were explored by assigning the three herbivores in a given food web either identical or different functional responses, of high or low slope, and then measuring resilience. As before, three replicate connectance patterns were generated for $c = 1$ and $c = 1$ or 2 webs, with food web configuration held constant (see Harding, 1999 for details).

In addition, the effects of perturbing the model with a 5 percent step increase in solar luminosity were explored. As before, the functional response curves of figure 22.2 were used to generate these experiments. Five $c = 1$ webs and five $c = 2$ webs, each with a different randomized connectance pattern, were constructed, each with constant food web configuration (see legend to figure 22.4 for details). Each web was allowed to reach steady state under a low initial solar luminosity and was then perturbed. The time taken to return to a steady-state daisy diversity index was then measured; a rapid return meant high resilience. These resilience measurements were repeated for each of 20 equal increments in solar luminosity. For each web, the mean of these 20 measurements was taken as an index of its resilience.

Results

Figure 22.2a shows a typical example of population dynamics and temperature regulation for a fully connected ($c = 3$) food web, with solar luminosity held constant at a level comfortable for the biota. The figure illustrates the stability of models incorporating environmental feedback. Less stable behavior was found with models where the food webs were more loosely connected over a range of solar luminosities from 0.7 to 1.2 times the present solar output. Model stability was not much altered by changes in herbivore functional responses but was sensitive to different food web configurations and connectance patterns, although temperature regulation was still well within the bounds tolerable by life. At the extremes of solar luminosity, just before the system failed, it became unstable. There were rapid oscillations in daisy and herbivore populations and in temperature. This was particularly noticeable with steep types 1 and 2 functional responses, but no attempt was made to analyze these odd effects.

As the system in figure 22.2a evolved, it was dominated by two daisy types whose albedos best matched the given solar luminosity. The population dynamics were stable and temperature regulation was at the optimum. Upon addition of the three herbivores, their grazing reduced the abundance of the few dominant daisies, and allowed the less well-matched rarer types to flourish. Daisy diversity increased with a smooth gradation of daisy type abundances; no single type was overly dominant. Competition between the herbivores led to the dominance of a single type. Upon introduction of the carnivore, the dominant herbivore became its prey, so the other two herbivores increased until all three reached equal abundance. These changes in herbivore abundance had only a small effect on the daisies, although a few types became extinct. The model system kept its temperature close to optimal for the biota throughout these perturbations. Herbivores and the carnivore had little effect on Daisyworld's capacity to regulate climate.

When the food web connectance was decreased from $c = 3$ to $c = 1$ or 2, the system was less stable (figure 22.2b). Although the herbivore dynamics were similar to those in the fully connected web, the daisy dynamics were markedly different. After the introduction of the herbivores, four daisy types reached high abundance, with a roughly even gradation in the abundance of less common types. The appearance of the herbivores led to a temperature increase to a steady state about 1.0°C higher than the model optimum for plant growth of 22.5°C.

In this experiment, one of the herbivores again became dominant. This in turn decreased the grazing of daisy types eaten by its competitors, and these daisy types increased. Among the now more abundant daisy types were those less well-matched to the solar luminosity, and therefore temperature regulation was impaired. Adding the carnivore improved the model performance. It reduced the abundance of the dominant herbivore, allowing the others to increase until

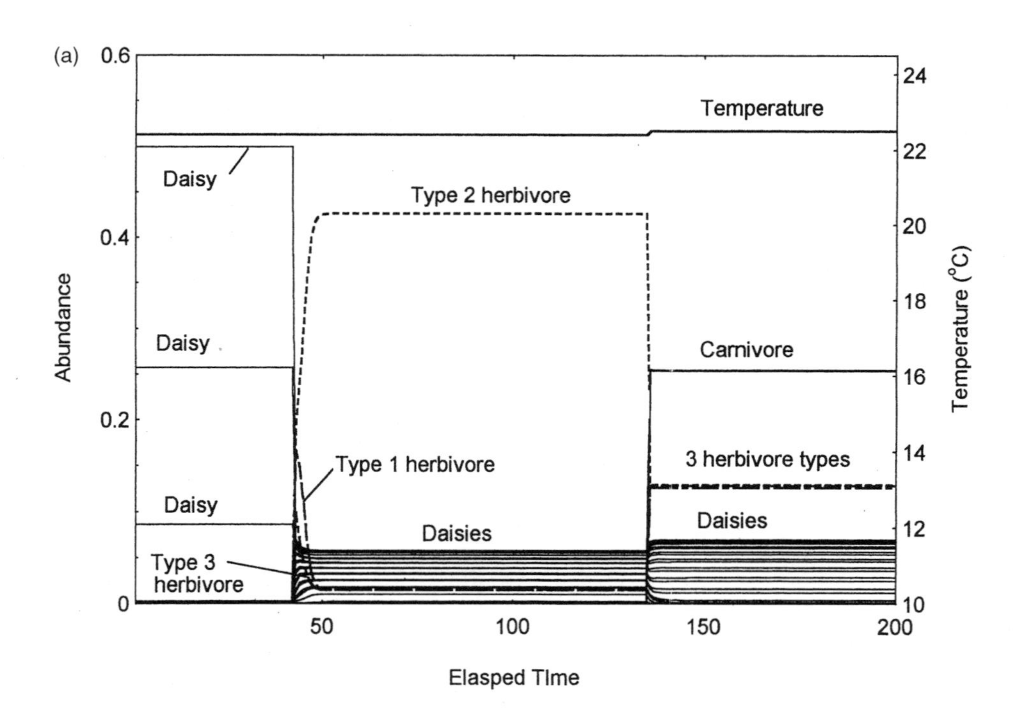

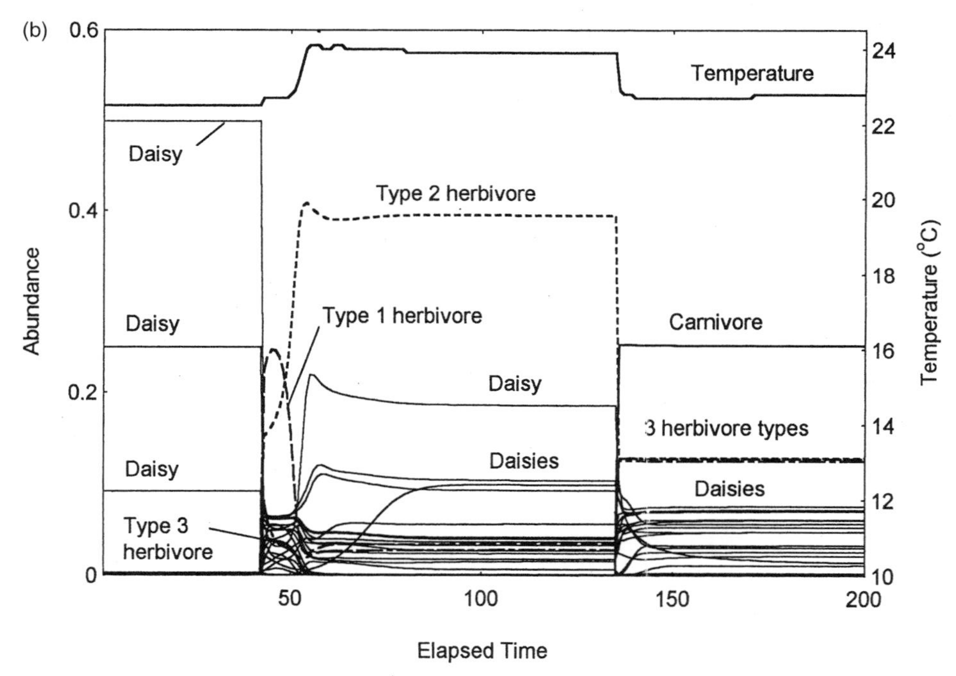

Figure 22.2
Population dynamics and temperature regulation in food webs with different values of c, the connectance parameter, among 23 daisy types, 3 herbivores, and 1 carnivore. Herbivore 1 was allocated the type 1 response with slope $F = 1$; herbivore 2, the type 2 response with slope $F = 3$; and herbivore 3, the type 3 response with $F = -3$. (a) Dynamics of a fully connected food web ($c = 3$). Solar luminosity was held constant at $J = 0.744$. (b) Dynamics of a more loosely connected food web ($c = 1$ or 2), with the following food web configuration: herbivores 1, 2, and 3 ate $n = 14$, $n = 12$, and $n = 14$ randomly assigned daisy types, respectively. Solar luminosity was held constant at $J = 0.744$. (c) Dynamics of a food web with the lowest connectance ($c = 1$); herbivores 1, 2, and 3 ate $n = 8$, $n = 8$, and $n = 7$ randomly assigned daisy types, respectively. Solar luminosity was held constant at $J = 0.826$. Similar results were obtained when each of the three herbivores was given the same type 2 functional response with slope $F = 2.2$.

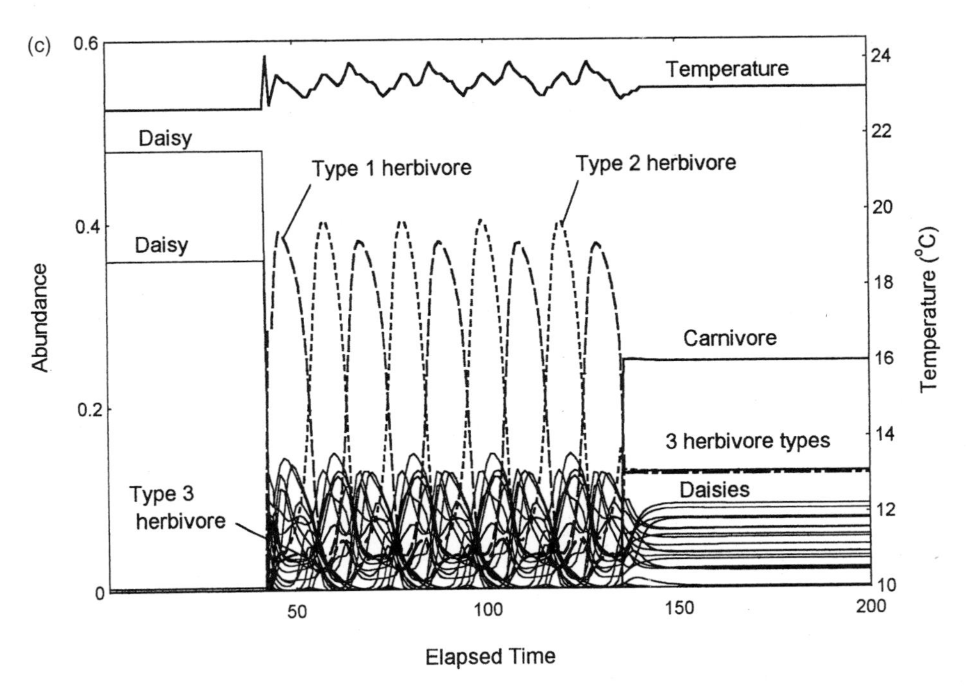

Figure 22.2 (continued)

the three types were equally abundant. The abundant daisy types that had previously flourished due to competitive suppression of their herbivores now declined. There was again an even gradation of daisy type abundances, and those with best matched albedos predominated, thereby improving temperature regulation.

Models with the most loosely connected ($c = 1$) webs were markedly less stable. At certain solar luminosities and in the absence of the carnivore, they sustained high-amplitude periodic dynamics. When this happened (figure 22.2c), the populations of two herbivores regularly oscillated out of phase with each other, while the third herbivore's numbers gently oscillated in step with one of the other herbivores. The oscillations of the model system also affected temperature, which was lightly modulated at the same frequency and over a range of 0.8°C. The mean temperature remained constant. These periodic fluctuations emerged unexpectedly in $c = 1$ webs. There appeared to be no relationship between the emergence of periodicity and solar luminosity. This form of instability seemed to reflect the randomness in the underlying connectance patterns. Sustained periodic dynamics were not seen in the more fully connected food webs, where $c > 1$. Adding the carnivore to $c = 1$ food webs obliterated periodic dynamics where they had been present (figure 22.2c). The presence of the carnivore brought the system to a steady state. The three herbivore types then became equally abundant, there was an even gradation in the abundances of daisy types, and smooth temperature regulation. The model displayed stability once more, as it did with the more connected food webs. Even with the least stable, loosely connected Daisyworlds the departure of the temperature from optimum was seldom more than 2.0°C over a wide range of solar luminosities.

Figure 22.3a demonstrates that despite some scatter in $c = 1$ webs, resilience and resistance showed a statistically significant, positive correlation throughout the parameter space of $c = 1$ and $c = 2$ food webs. Thus, in what follows, changes in resilience can be understood to imply similar changes in resistance, and hence in the overall stability of the model.

Despite wide variations in critical initial conditions of the model (i.e., in food web configuration, connectance pattern, and slope and type of the herbivore functional responses), low connectance generally gave rise to low resilience (figure 22.3b). However, there was a minor exception in that food webs varied in their resilience depending on their "herbivore connectance diversity." This quantity was highest in food webs in which each herbivore was linked to the same number of daisy types. For example, high herbivore connectance diversity would occur in a $c = 1$ food

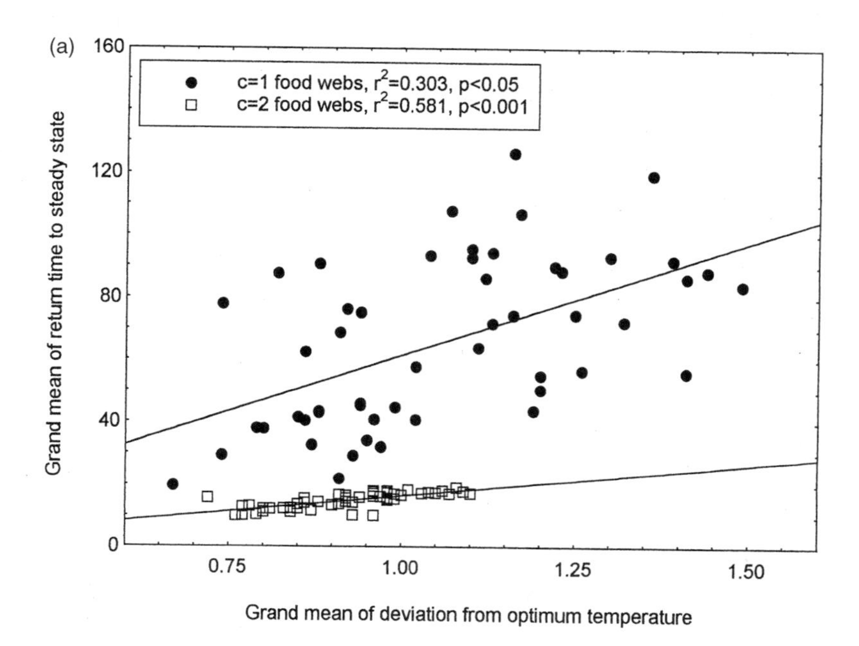

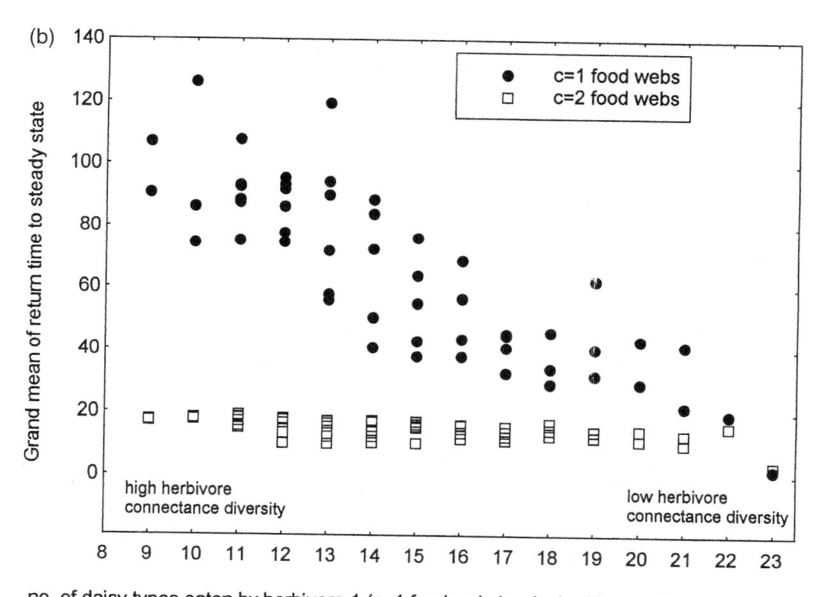

Figure 22.3
(a) The relationship between the grand means of the maximum deviation from optimum temperature (the inverse of resistance) and of the return time to steady state (the inverse of resilience) in the sets of sampled $c = 1$ and $c = 2$ food webs. The two variables show a significant positive correlation in both cases, although the relationship for the $c = 1$ food webs shows a relatively high degree of scatter. (b) The return time to steady state of the daisy diversity index in $c = 1$ and $c = 2$ food webs as a function of herbivore connectance diversity. The portion of the parameter space sampled was that enclosed by the top left-hand triangle in figure 22.1. Each herbivore was given the same type 2 functional response, in which the slope factor, F, was 2.2 (see equation 7). Each data point represents the grand mean of the three replicate random connectance patterns, generated for each food web configuration, as described in the text. The return time to steady state (the inverse of resilience) was markedly higher in $c = 1$ webs than in $c = 2$ webs, particularly when the herbivore connectance diversity was high, toward the left of the graph.

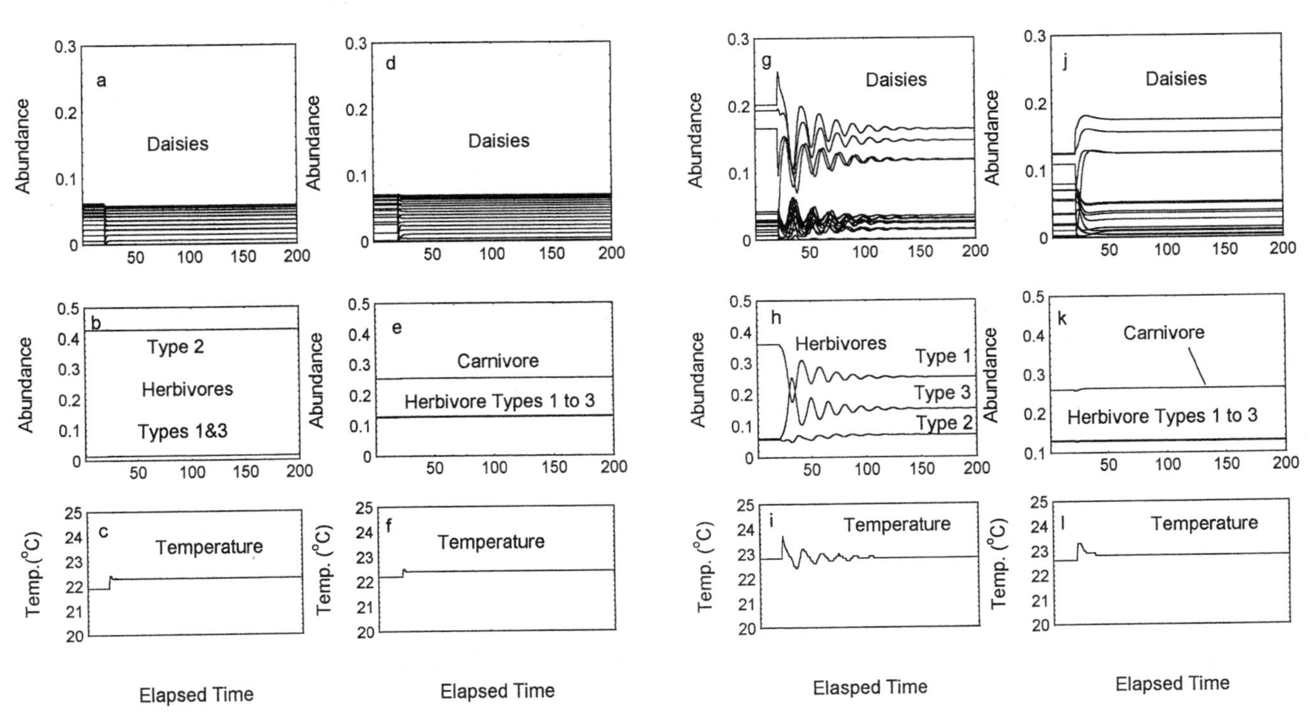

Figure 22.4
Examples of the resilience to perturbation from a 5 percent step increase in solar luminosity of food webs with different degrees of connectance, using the functional response values of figure 22.2. (a–c) For the fully connected food web ($c = 3$) without carnivores, and (d–f) with the carnivore. Note the very high resilience, with very rapid return to steady state. Initial solar luminosity was $J = 0.7$. (g–i) For a least connected ($c = 1$) food web without carnivores, using the food web configuration of figure 22.2c. Note the very long return time to steady state in both temperature and population dynamics, indicating low resilience. Initial solar luminosity was $J = 0.7$. (j–l) The same ($c = 1$) food web with the carnivore present. Note the much higher resilience. Initial solar luminosity was $J = 0.7$.

web if two of the herbivores ate eight daisy types each and the third herbivore ate the remaining seven daisy types. Low herbivore connectance diversity would occur in $c = 1$ webs if herbivore 1 ate all 23 daisy types, with the other two herbivores eating no daisies. Figure 22.3b reveals that $c = 1$ webs were markedly less resilient than $c = 2$ webs over a wide range of herbivore connectance diversities. Furthermore, in $c = 1$ webs, despite some scatter in the data, it was clear that resilience continually increased as herbivore connectance diversity declined. Thus, for $c = 1$ food webs, return time to steady state decreased as herbivore 1 became more prominent in a food web. By contrast, $c = 2$ webs displayed an extensive region of roughly constant high resilience, covering almost all of the parameter space, with far less variability in the data.

In the real world, climate change is a frequent perturbation. Increasing the solar luminosity by 5 percent simulated a change of magnitude roughly equivalent to that between glacial and interglacial climates. Fully connected food webs showed high resilience, reaching steady state almost instantly after the perturbation (figure 22.4a–22.4c). The presence of carnivores made no difference to resilience, but at some solar luminosities did slightly improve temperature regulation (figure 22.4d–22.4f).

In the absence of the carnivore and in food webs with the lowest connectance ($c = 1$), resilience fell to low levels. In these webs, at some luminosities, the climate perturbation set off a long-term periodic oscillation in population and in temperature. The oscillations did decay, but only after many time steps (figure 22.4g–22.4i). Adding a carnivore improved resilience, although the return time to equilibrium was longer than in more highly connected food webs with a carnivore present (figure 22.4j–22.4l). At other luminosities the perturbation did not give rise to periodic behavior in $c = 1$ webs, although both population and temperature changed and then took a relatively long time to return to steady state. Adding the carnivore again improved resilience. Resilience was least for $c = 1$ food webs with carnivores absent, and increased when the carnivore was present or when connectance was increased (figure 22.5).

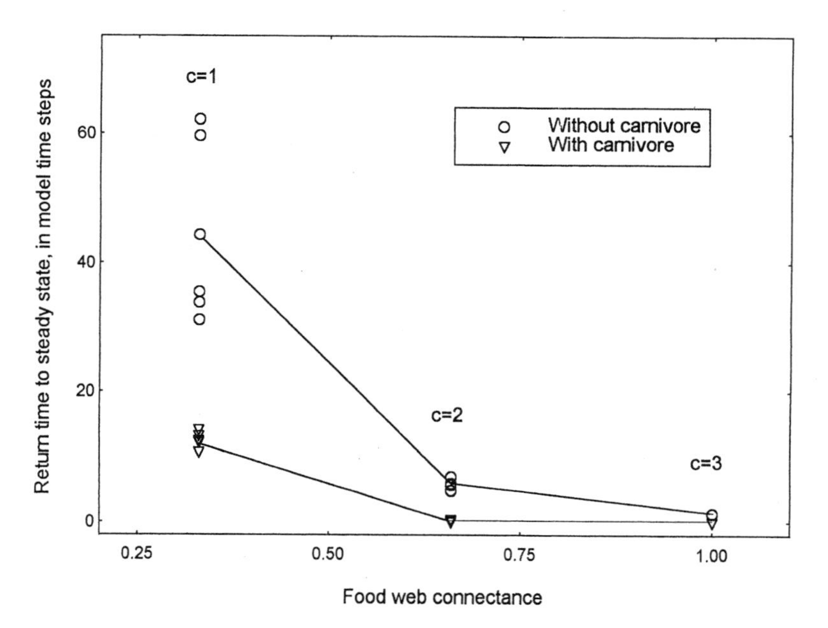

Figure 22.5
The relationship between food web connectance and resilience after perturbation by a 5 percent step increase in solar luminosity, using the functional response configuration of figure 22.2, with and without the carnivore. In food webs where $c < 3$, the herbivore connectance diversity was made as high as possible. The five resilience indices are plotted for each web where $c < 3$, with lines joining mean values. Similar results were obtained with a wide range of herbivore functional response configurations.

Discussion

So far almost all model experiments in community ecology have been made with communities where the only selection pressures were internal, that is, where the environment was defined by the organisms themselves. In the majority of these experiments, complexity and instability appeared to go hand in hand. However, a notable exception is that weak consumer-resource interactions increase stability when coupled to strong consumer-resource interactions (McCann, 2000).

In the model presented here, the environment includes the physical world. In addition, the organisms have the capacity to change their material environment, as well as to adapt to it. Daisy types (the system's primary producers) experience selection pressure mainly from environmental feedback, which acts "externally" as a strong stabilizing force (Lenton, 1998). Internal selection is present at two levels, from herbivore grazing and from competition for space among the daisy types themselves.

In the first level of internal selection, herbivores selected daisies according to their abundance, and as a consequence daisies "evolved" immunity to certain herbivores as a result of past grazing pressure. This internal selection pressure was independent of the material environment. Consequently, some effects on system stability and the quality of temperature regulation were expected and noticed. For example, the presence of daisies equipped with anti-herbivore defenses did sometimes lead to instabilities in both population dynamics and climate due to the loosening of food web connectance. Sometimes internal selection pressure from herbivores improved system stability. It did so by opening niche space for daisy types otherwise denied an influence on the system's temperature by previously abundant, well-matched types. Low internal selection pressure among daisy types also was potentially destabilizing because some daisy types which had less than optimal albedos could flourish, and hence upset temperature regulation. In most of the model experiments, environmental feedback was strong enough to drive the selection of daisy types best able to regulate climate, and this led to reasonably good climatic regulation and population stability.

Furthermore, in all the above experiments, increasing complexity by adding a third trophic level, the carnivore, improved stability. Pimm and Lawton (1980) described models that differed from those just described only by the absence of biota-climate feedback. Adding higher trophic levels to their models made them unstable.

In rare cases, freedom from low internal selection pressures among the daisy types can overturn the stabilizing influence of external selection, thereby plunging Daisyworld into severe instability. Von Bloh et al. (1997) modeled Daisyworld using a two-dimensional cellular automaton in which they were able to create varying degrees of habitat fragmentation. As the habitat fragmented, there was a reduction in competition between the daisies in the separated fragments. Ecological niches then arose in which daisy types with destabilizing climatic effects could flourish. As fragmentation proceeded, a sharply defined threshold appeared which marked a destabilized climate. Thus, the most diverse systems were the least stable climatically. As Von Bloh et al. point out, the price to be paid for very high biodiversity in their two-dimensional Daisyworld was a loss of self-regulatory capacity.

Unlike the traditional community ecology models with no environmental feedback, where instability is almost always a consequence of diversity, Daisyworlds are only rarely unstable. The presence of a diversity of daisy albedos is beneficial for effective regulation, as long as there is some internal selection pressure among the daisy types. On subjecting the model to a gradually increasing solar luminosity, the high diversity of daisy types ensured that there were always a few types whose albedos matched the current solar luminosity well enough to provide effective temperature regulation. This is akin to the observation that plant species diversity can stabilize ecosystems in the field because previously rare species replace dominants which have declined due to disturbances such as drought (Tilman, 1996), and to the original suggestion by MacArthur (1955) that connectance increases stability. Thus, in Daisyworld as in the field, species are individually vulnerable, but high species diversity ensures that the community as a whole is more likely to persist.

Community ecologists have debated the importance of the coupling between biota and the material environment. Although biota–environment feedbacks at individual and even ecosystem levels are recognized (Jones et al., 1994), large-scale physical aspects of the environment, such as temperature and atmospheric composition, continue to be viewed as givens to which organisms must adapt, but which they cannot influence to any significant extent. Geophysiologists recognize that feedback between the biota and their material environment is not always obvious, but assert that in certain cases it can be significant.

Could it be that systems with loose biota–environment coupling persist only because tightly coupled, fully geophysiological systems are actively regulating the material environment in the "background"? There are several good candidates for such "keystone geophysiological feedbacks," such as the seeding of cooling clouds by DMS-emitting marine algae (Charlson et al., 1987), the regulation of atmospheric oxygen by the bacterial community in the seabed sediments (Van Cappellen and Ingall, 1996) and through the weathering of phosphorus from rocks by land plants (Lenton and Watson, 2000), the regulation of planetary temperature through the biotic enhancement of rock weathering (Lovelock and Whitfield, 1982; Schwartzman and Volk, 1989), and the proposed linkages between marine and peatland ecosystems (Klinger and Erickson, 1997). Other possibilities for tight coupling exist at local levels among soil microorganisms. Here the impacts of the biota on the material environment remain localized enough to exert selection pressures, which in turn may lead to the regulation of soil conditions at a state conducive to plant growth. Since plant growth is essential for ecosystem functioning, such local feedback processes may be fundamental to the health of all terrestrial ecosystems.

A prediction stemming from this work is that a positive relationship should exist between complexity and stability in real ecosystems when there are feedbacks to and from the growth of the biota and the state of the material environment. Furthermore, at the level of the whole Earth, these relationships may well contribute to the stability of our planet's biogeochemical cycles.

Acknowledgments

I thank the Gaia Charity and the Dartington Hall Trust for supporting this work, and Brian Goodwin for stimulating discussions. I also thank the journal *Tellus* for giving permission to print this version of the original paper.

References

Begon, M., J. L. Harper, and C. R. Townsend. 1996. *Ecology: Individuals, Populations and Communities*, 3rd ed. Blackwell Science, Oxford.

Betts, R. A., and P. M. Cox. 1997. Contrasting physiological and structural vegetation feedbacks in climate change simulations. *Nature*, 387, 796–799.

Carter, R. N., and S. D. Prince. 1981. Epidemic models used to explain biogeographical distribution limits. *Nature*, 213, 644–645.

Charlson, R. J., J. E. Lovelock, M. O. Andreae, and S. G. Warren. 1987. Oceanic phytoplankton, atmospheric sulphur, clouds, albedo and climate. *Nature*, 326, 655–661.

De Angelis, D. L. 1975. Stability and connectance in food web models. *Ecology*, 56, 238–243.

Elton, C. S. 1958. *Ecology of Invasions by Animals and Plants.* Chapman and Hall, London.

Harding, S. P. 1999. Food web complexity enhances community stability and climate regulation in a geophysiological model. *Tellus*, 51B, 815–829.

Harding, S. P., and J. E. Lovelock. 1996. Exploiter mediated co-existence and frequency dependent selection in a numerical model of biodiversity. *Journal of Theoretical Biology*, 182, 109–116.

Holling, C. S. 1959. Some characteristics of simple types of predation and parasitism. *Canadian Entomologist*, 91, 385–399.

Huntly, N. 1991. Herbivores and the dynamics of communities and ecosystems. *Annual Review of Ecology and Systematics*, 22, 477–503.

Jones, C. G., J. H. Lawton, and M. Shachak. 1994. Organisms as ecosystem engineers. *Oikos*, 69, 373–386.

Klinger, L. F., and D. J. Erickson. 1997. Geophysical coupling of marine and terrestrial ecosystems. *Journal of Geophysical Research*, 102, 25359–25370.

Lenton, T. M. 1998. Gaia and natural selection. *Nature*, 394, 439–441.

Lenton, T. M., and A. J. Watson. 2000. Redfield revisited: What regulates the oxygen content of the atmosphere? *Global Biogeochemical Cycles*, 14, 249–268.

Lotka, A. J. 1925. *Elements of Physical Biology*. Williams and Wilkins, New York.

Lovelock, J. E. 1983. Gaia as seen through the atmosphere. In *Biomineralization and Biological Metal Accumulation*, P. Westbroek and E. W. de Jong, eds. D. Reidel, Dordrecht.

Lovelock, J. E. 1992. A numerical model for biodiversity. *Philosophical Transactions of the Royal Society*, B338, 383–391.

Lovelock, J. E., and L. R. Kump. 1994. Failure of climate regulation in a geophysiological model. *Nature*, 369, 732–734.

Lovelock, J. E., and M. Whitfield. 1982. Life-span of the biosphere. *Nature*, 296, 561–563.

MacArthur, R. H. 1955. Fluctuations of animal populations and a measure of community stability. *Ecology*, 36, 533–536.

Maddock, L. 1991. Environmental feedback and populations. *Tellus*, 43B, 331–337.

May, R. M. 1973. *Stability and Complexity in Model Ecosystems.* Princeton University Press, Princeton, NJ.

May, R. M. 1981. *Theoretical Ecology: Principles and Applications*, 2nd ed. Blackwell Science, Oxford.

McCann, K. 2000. The diversity–stability debate. *Nature*, 405, 228–233.

McCann, K., A. Hastings, and G. R. Huxel. 1998. Weak trophic interactions and the balance of nature. *Nature*, 395, 794–798.

Pimm, S. L. 1979a. Complexity and stability: Another look at MacArthur's original hypothesis. *Oikos*, 33, 351–357.

Pimm, S. L. 1979b. The structure of food webs. *Theoretical Population Biology*, 16, 144–158.

Pimm, S. L. 1982. *Food Webs*. Chapman and Hall, London.

Pimm, S. L. 1984. The complexity and stability of ecosystems. *Nature*, 307, 321–326.

Pimm, S. L., and J. H. Lawton. 1980. Are food webs divided into compartments? *Journal of Animal Ecology*, 49, 879–898.

Polis, G. A. 1998. Stability is woven by complex webs. *Nature*, 395, 744–745.

Schwartzman, D., and T. Volk. 1989. Biotic enhancement of weathering and the habitability of the Earth. *Nature*, 340, 457–460.

Tilman, D. 1996. Biodiversity: Population versus ecosystem stability. *Ecology*, 72, 350–363.

Van Cappellen, P., and E. Ingall. 1996. Redox stabilization of the atmosphere and oceans by phosphorus-limited marine productivity. *Science*, 271, 493–496.

Von Bloh, W., A. Block, and H. J. Schellnhuber. 1997. Self-stabilization of the biosphere under global change: A tutorial geophysiogical approach. *Tellus*, 49B, 249–262.

Watson, A. J., and J. E. Lovelock. 1983. Biological homeostasis of the global environment: The parable of Daisyworld. *Tellus*, 35B, 284–289.

23

Gaia in the Machine: The Artificial Life Approach

Keith L. Downing

Abstract

The field of artificial life (ALife) specializes in the study of emergent phenomena, wherein simple local mechanisms interact to form complex global structures. Gaia theory is the perfect specimen for ALife dissection, since any viable explanation for Gaia must satisfy basic reductionist criteria: natural selection at the individual level must be reconciled with the emergence of large-scale Gaian patterns such as recycling and homeostasis. ALife provides a variety of standard tools for testing Gaian claims in virtual worlds, where evolving digital genomes control the phenotypes of individual organisms that must then fend for themselves in simulated environments. When, through this *in silico* struggle, the organisms collectively seize control over some aspect of that environment, additional support for both Gaia theory and its compatibility with natural selection arises.

This chapter presents several examples of the evolutionary emergence of Gaian patterns in two different ALife simulators. It also briefly analyzes these systems with respect to several common primitive and emergent properties of Gaian phenomena.

Introduction to Artificial Life

Artificial Life (ALife) researchers are united by the common goal of understanding the dynamics of living and lifelike systems, regardless of substrate (Langton, 1989). Thus, ALife searches for common mechanisms at all scales of life, from cells to organisms to populations to ecosystems, and across a broad range of systems: social, economic, business, and political (Holland, 1995). The quest also ventures into virtual worlds of the computer and Internet, focusing on substrates such as cellular automata (Langton, 1986), Lindenmayer trees (Lindenmayer and Prusinkiewicz, 1989), Boolean networks (Kauffman, 1993), and computer memories (Ray, 1992). The latter is the breeding ground for diverse computer viruses, nonbiological but extremely lifelike entities of great relevance to ALife.

The unifying concept in ALife is emergence (self-organization): the formation of global patterns from the interactions of local mechanisms. These self-organizing and self-regulating systems involve many subunits or subsystems, but no global controller. Essentially, the local interactions spawn global structures which then serve as constraints for future interactions but do not possess, and cannot enforce, a global plan.

The nests and food-gathering routes of social insects are classic examples. Through local mechanisms such as the placement of individual cells in a honeycomb or the chemical (e.g., pheromone) signals laid along a path to food, individual insects influence the behaviors of nest mates in ways that gradually lead to the production of completed nests or optimal foraging trails. In economics, the emergence of stable prices and the clearing of markets via the interactions of producers and consumers bring Adam Smith's invisible hand under the careful scrutiny of ALife scientists. And at the cellular level, the wonder of morphogenesis, steered only by multitudes of individual cells, all with the same DNA program, excites the interest of the ALife community, both in terms of unlocking the basic biological truths and of using a similar morphogenetic process in powerful computational tools.

The types of emergence can be roughly classified relative to the nature of the interacting local entities. In inorganic emergence, physical forces govern the behavior of inorganic entities to derive interesting formations such as Rayleigh–Benard convection cells, crystals, and clouds. Many inorganic reaction–diffusion systems also fit into this category.

Organic emergence has several different subclasses, depending upon the adaptive capabilities of the organic units (agents). Innate emergence involves agents with hardwired reactive behaviors, such as bacteria or social insects, none of which appear to have significant learning abilities. Adaptive emergence involves populations of flexible agents that can learn and/or evolve. Economic, social, and political systems typify learned emergence, since agents in these systems exhibit adaptive responses to their environment. In both innate emergence and learned emergence, the global patterns are built during the lifetime of a single population. Conversely, evolutionary emergence involves the formation of interaction webs, such as ecosystems, over generational timescales. Here, the local mechanism of natural selection combines with genetic duplication, mutation, and recombination to govern global pattern formation. As the ecosystems diversify, the organisms may acquire learning capabilities, thus facilitating dual adaptive emergence: one driven by both learning and evolution.

From the ALife perspective, Gaia appears to demand an evolutionary emergent explanation—for example, one that clearly justifies global regulatory loops in terms of the evolving behaviors of simple, nonlearning organisms, such as bacteria. This inspires the standard ALife approach in which populations of simple evolving agents are simulated for hundreds or thousands of generations. Whereas most ALife researchers would then be interested in any emerging global patterns, those with Gaian aspirations would hunt for typical Gaian fingerprints, such as environmental regulation and nutrient recycling, among the output data.

Bio-Logic and Gaia-Logic

The Holy Grail of ALife is a general set of principles for explaining a wide variety of living and lifelike systems: a *bio-logic* (Langton, 1989) that applies equally well to cells, organisms, ecosystems, economies, and societies. Concepts such as emergence and self-regulation are clearly kernel elements of bio-logic, but no formal consensus definitions for these terms exist, let alone general descriptions of the processes underlying them.

Similarly, Gaia theory needs a *Gaia-logic* to help unify the different conceptions of Gaia under a common framework. In the ALife spirit, this section reviews several basic mechanisms and emergent patterns that appear in many proposed Gaian phenomena, both natural and simulated.

Basic Mechanisms

The first basic mechanism is trivial: organisms must be affected by their physical environments. This is often modeled as relationships between ambient factors and growth rates. An apparent implicit assumption in Gaia theory is that these ambients should consist of something other than available food resources, such as temperature, pH, or salinity.

The second mechanism is the inverse of the first: organisms must have some causal means of altering their physical environment. Normally, this influence occurs only at the aggregate level. For example, one daisy in Daisyworld cannot exhibit a significant influence on global temperature, but a large population can. In nature, a single plankton cannot produce enough DMSP to form a cloud, but millions of plankton can.

Let us denote that substance or physical condition produced by an organism which has some causal effect upon the environment as the *Gaian root*. In the examples above, the Gaian roots are local albedo and DMSP, respectively.

The third primitive component is a set of causal links relating those general environmental factors (GEFs) which are directly influenced by the biota to those ambients which govern biological growth rates. In some cases, the causal chain may be short or trivial (e.g., the GEFs are the ambients), while others involve a complex environmental model. Daisyworld has a relatively short pathway from local albedo to aggregate global albedo to global temperature and back to local temperature (a function of the global temperature and local albedo), which then controls growth. In the plankton-DMS example, the chain involves DMSP, DMS, cloud formation, shading effects upon temperature and wind, and so on.

The fourth mechanism is a local selective advantage of Gaian-root production. Since the global environmental effects of individual organisms are insignificant, if standard individual-based natural selection is to guide the evolution of Gaian phenomena, then the Gaian root must incur some selective advantage over those "cheaters" which harvest the benefits of global changes but do not expend energy to effect them. The classic example is the combination of (a) the estimate that plankton's metabolic cost of DMSP production exceeds the global climatic benefit by nine orders of magnitude (Caldeira, 1989) and (b) the finding that internal DMSP concentrations are critical to preventing osmotic water loss in phytoplankton (Liss et al., 1997). Here, in contrast to Daisyworld, the selective advantage of the Gaian root is different from the global climatic effect.

Buffering of an organism's interior milieu from the external world is often critical to achieving local selective advantage. Otherwise, the Gaian root or its immediate by-products simply diffuse into the envi-

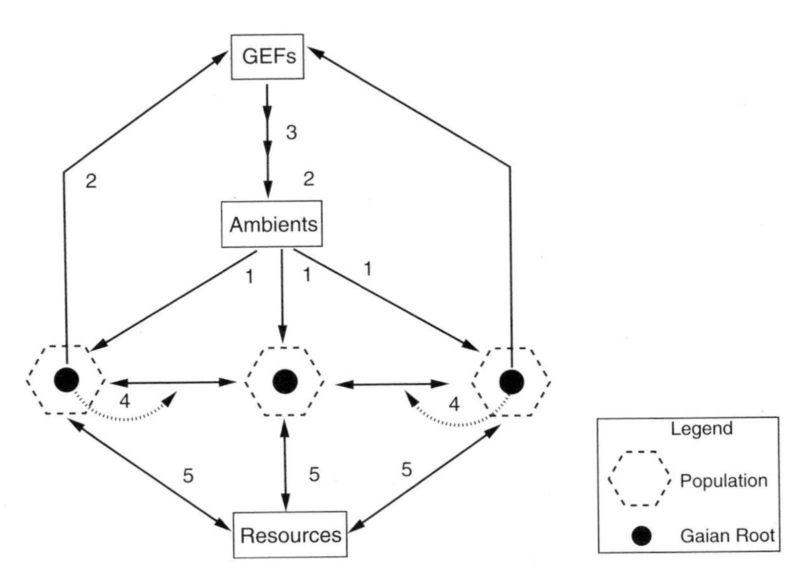

Figure 23.1
Summary of the basic causal mechanisms involved in many proposed Gaian phenomena: (1) ambient effects upon growth rate, (2) direct biotic effects of the Gaian root upon general environmental factors (GEFs), (3) GEF effects (possibly indirect) upon ambient factors, (4) local selective advantage of Gaian root production, and (5) biotic production and consumption of resources.

ronment. Daisyworld achieves this buffering by using local albedo to help compute local temperature, while the ALife organisms discussed below maintain internal chemical concentrations that affect other local growth conditions.

Finally, Gaian phenomena involve the consumption and production of resources by the biota. In Daisyworld, space is the resource, while chemical nutrients are key resources in the Guild and Metamic systems presented below. Similarly, a reasonable simulation of the plankton-cloud connection will include algal nutrient supplies and DMSP release. This not only supports the emergence of one Gaian pattern, resource recycling, but also adds realism to any biological simulation.

Figure 23.1 summarizes these five primitive characteristics, many of which appear in natural and artificial systems that exhibit Gaian behavior. Although some of these primitives could also be emergent in particular simulations or real-world phenomena, they are generally taken as givens in Gaian discussions. A Gaia-logic will presumably build on factors such as these and purported relationships between them such that emergent Gaian dynamics can be predicted from instantiations of these primitives in different biogeochemical contexts.

In general, Gaia research has focused on uncovering these primitive mechanisms and showing that they fit into fully plausible causal circuits. The key issue now becomes one of emergence: Given these primitive relationships, do the forces of nature actually drive systems toward biotic and environmental conditions that manifest powerful feedback and recycling states?

Emergent Patterns

The first emergent Gaian property is common to most biological systems: frequency-dependent population growth. The standard situation involves negative frequency-dependent selection in a typical Malthusian manner: higher population density entails more competition for resources and lower overall fitness. However, since Gaia embodies the notion of organisms making the planet more livable for each other, some positive relationships between population size and average fitness will often occur as well.

The second (and hallmark) emergent property of Gaian systems is biotic-abiotic feedback, with the emphasis being on stabilizing negative feedback. Essentially, the biota, Gaian root, GEFs, and ambients are connected in a homeostatic loop such that perturbations to population density or the environment are regulated back toward their normal values. Thus, the biota participate in the regulation of factors such as climate, the Redfield ratio, and marine salinity via population density and diversity. Furthermore, if a biota with plentiful resources has optimal growth under particular ambient conditions, then indirect biotic influences often combine with natural selection to push the ambients toward those preferred values.

The purported dominance of warming positive feedbacks at the end of glacial periods (along with the fact that the biota "permit" the large temperature swings of the glacial cycles) leads many to question the importance of homeostasis in Gaia theory. Furthermore, the central Gaian notion of the biota making the planet more favorable for life does not entail purely homeostatic connections to abiotic factors, since the explosive growth enabled by positive feedback is also important for the rapid colonization of environments following migrations or catastrophic environmental changes. Hence, biotic-abiotic feedback loops in general, whether positive or negative, appear to be key emergent Gaian patterns.

A third emergent property is Lenton's (1998) frequency-dependent selective feedback. He differentiates between the effects of population density upon growth and selective advantage. The former is the first emergent pattern above, while the latter is more specific to Gaia: if a population of organisms can incur environmental change, then these alterations may raise or lower its selective advantage over other populations.

For example, in Daisyworld, the proliferation of dark daisies warms the planet, which increases the selective advantage of the light over the dark varieties. In Guild, if one metabolic strategy dominates the population, it will often decrease its own selective advantage by altering global chemical concentrations so as to favor other strategies, as detailed below.

In general, a negative relationship between population density and selective advantage is important in Gaian systems such as Guild, where the emergence of diversity is a key prerequisite to Gaian control. In Daisyworld, the relationship is present but not as vital, since any single type of daisy can control global temperature over a given interval of solar irradiance.

A fourth emergent property, recycling, is gaining increasing popularity as a fingerprint of Gaian dynamics (Volk, 1998; Williams, 1996). The ability of diverse species (often microorganisms) to coordinate their metabolic inputs and outputs such that few resources go to waste can provide both a resource base that supports several orders of magnitude more biomass than the external nutrient fluxes alone, and considerable insurance against wild fluctuations in those exogenous flows.

Recycling can support Gaian control, since the populations become large enough that their aggregate environmental effects can be felt, and as the competition for food abates, natural selection can begin to operate on characteristics other than those directly related to resource gathering—for example, side effects of the Gaian root such as dehydration prevention.

Individually, the above primitive mechanisms and emergent patterns are not unique to Gaia. Collectively, the basic mechanisms are hardly sufficient conditions for Gaia, but certain combinations may turn out to be necessary, assuming that eventually Gaia is formally defined.

ALife Models of Gaia

The brilliant Daisyworld (Watson and Lovelock, 1983) model clearly proves the feasibility of distributed regulation. However, from an ALife perspective, the homeostasis in Daisyworld is hardly emergent, since all necessary phenotypes are hardwired into the differential equations. Hence, the model cannot convincingly reconcile Gaia theory with natural selection. Later work (Lenton, 1998; Robertson and Robinson, 1998) added a gene pool of possible albedos and retained Daisyworld's regulatory tendencies. But, interestingly enough, when the daisies' optimal growth temperature also became genetically determined, homeostasis vanished (Robertson and Robinson, 1998), as predicted by Saunders (1994).

Although these and other Daisyworld extensions add degrees of freedom to the original model, the genotypes/phenotypes that offer the best homeostatic possibilities for each solar input are easily computed, and the model is devised so that these genotypes are within easy reach of the original population. In short, the "emergent" results are fairly easily predicted. Conversely, in classic ALife simulations, genomes typically code for a wide variety of properties whose combinations offer only vague possibilities for speculation. Only by running the simulations, often for hundreds or thousands of generations, can one ascertain the emergent global patterns.

Below, we present two ALife models, Guild and Metamic. Both were designed to test the compatibility of Gaia theory and natural selection by showing the convincing emergence of Gaian patterns from the interactions of genetically diverse bacteria-like organisms.

The Guild System

The Guild system (Downing and Zvirinsky, 1999; Downing, 2000) simulates the evolutionary emergence of a diverse interacting web of species whose combined activity can control some aspect of the physical environment. It moves a few steps beyond Daisyworld by including a wide range of diverse genotypes and incorporating a popular new metric for Gaian activity, nutrient recycling (Volk, 1998).

Biogeochemical Motivations Another intriguing example of life's ability to create favorable conditions for more life involves the creation of efficient nutrient-recycling pathways. As detailed in Volk (1998), the external supplies of critical elements such as carbon, nitrogen, and phosphorus to terrestrial and aquatic ecosystems are far below the amounts actually required by the biota. The deficit is filled by recycling processes wherein C, N, and P atoms are shuttled among different compounds that are ingested and expelled by various organisms.

For example, carbon is taken up by photosynthesizing plants as CO_2 and used to build organic carbon compounds such as carbohydrates, which are then transferred to herbivores or detritus-consuming microorganisms, only to be returned to the atmosphere as carbon dioxide by respiration of plants, animals, bacteria, and fungi. A small percentage of the carbon sinks out of aquatic and terrestrial ecosystems as organic detritus and the calcium carbonate shells of buried microorganisms, returning millions (to hundreds of millions) of years later via geophysical processes such as volcanism, deep-sea thermal ventilation, and rock weathering (Schlesinger, 1997). In the case of weathering, the biota have been shown to significantly accelerate this key step of the carbon, nitrogen, and phosphorus cycles (Schwartzman and Volk, 1991), so many links of the circuit feel the biotic presence.

The net result of these recycling loops is that the biota annually consume 200 times more carbon, 500–1300 times more nitrogen, and 200 times more phosphorus than is supplied by external fluxes (Schlesinger, 1997; Volk, 1998). These numbers represent the cycling ratios for the three elements (computed as the intra-biota transfer rate divided by the external flux). Without this amplification, Earth's biota would be restricted to a fraction of their current total biomass; and without the biota, there would be no amplification.

In short, the coordination of biochemical processes across a diverse range of organisms enables life to thrive to a degree which dwarfs that of an uncoordinated, low-recycling environment. Furthermore, the abundance of critical nutrients adds stability to the environment, enabling the biota to endure periods of fluctuating external inputs. Once again, life begets life via its effects upon the environment.

If we classify organisms by the chemicals that they consume and produce (i.e., by their metabolisms), then each group constitutes a biochemical guild (Volk, 1998). The formation of recycling loops is therefore dependent upon the emergence of the proper complement of biochemical guilds such that the waste products of one guild become the resources of another.

Simulated Emergence of Recycling and Control Guild combines abstract models of chemistry, biological growth, and natural selection to simulate the emergence of both nutrient recycling networks and the regulation of global chemical ratios. We borrow one key mechanism from Daisyworld: the ability of organisms to create local buffers against the global environment, where the combined buffering effects of many organisms can then exert an influence on the global situation. However, we avoid much of Daisyworld's hardwiring by providing a large genotype space, defined by a genetic algorithm (GA) (Holland, 1992; Goldberg, 1989) chromosome.[1]

The simulations are seeded with a single species, so all additional genotypes must arise by mutation and crossover. Furthermore, the regulatory task is one involving the coordinated effort of a wide range of temporally coexistent genotypes; a single dominant species cannot do the job alone. Hence, it is the biotic community as a whole that regulates global conditions, and these heterogeneous communities emerge from a homogeneous seed population that is subjected to nothing more than competition for resources, reproduction (by splitting) of successful resource gatherers, and genetic operators. In addition, Guild illustrates the emergence of recycling networks. Thus both perspectives on Gaian activity, recycling and regulation, are commensurate with neo-Darwinism.

Guild employs a standard GA along with a simple model of chemical interactions. The environment consists of n nutrients/chemicals, $N_1 \cdots N_n$, with input and output fluxes I_k and O_k, respectively, and internal stores E_k for $k = 1 \cdots n$. An organism's genome determines both the chemicals that it feeds on and those it produces during metabolism; an organism cannot consume and produce the same chemical. Organisms reproduce by splitting; the genetic operators are mutation (during splitting) and crossover (via gene swaps between organisms). The growth, reproductive, and genetic dynamics are intended to

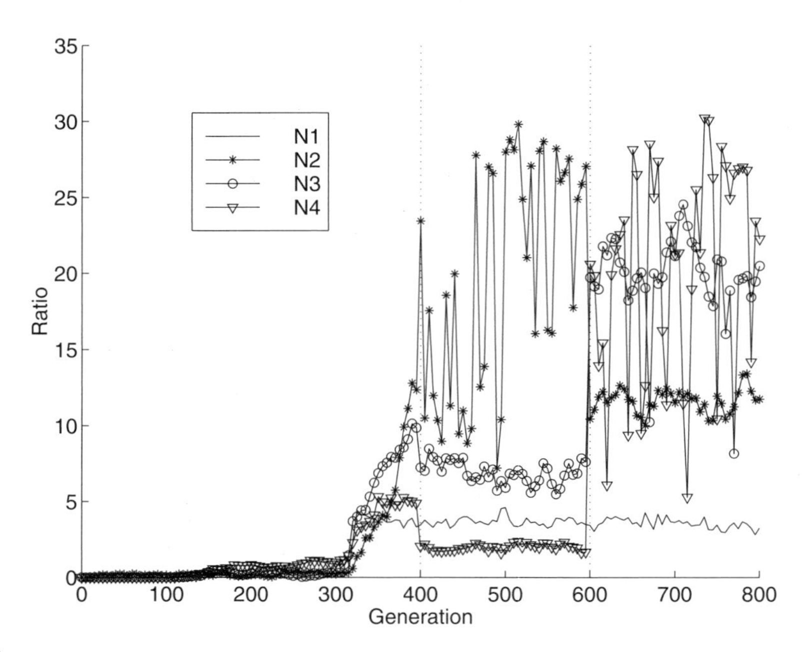

Figure 23.2
The evolution of cycling ratios (i.e., total interorganism transfer/external flux) for four nutrients in a run of Guild. The abrupt increase indicates the emergence of recycling in a community of biochemical guilds. Dashed vertical lines denote points of extreme change in external input fluxes.

mirror those of bacteria, which are the basis of Earth's primary biochemical guilds.

In addition, the organisms are assumed to be most active (i.e., have the highest feeding rates) when the relative fractions of the environmental chemicals, E_k, in the organism's immediate vicinity are near a particular user-defined optimal ratio. By producing and consuming chemicals, the organisms can create local ratios that differ from the global values, thus providing a semiprotective buffer against their surroundings. For example, organisms that consume N_1 and produce N_2 will have, respectively, lower and higher local amounts than the global values. Conceptually, the preferred ratio is analogous to an ambient factor such as pH, whose value is dependent upon many different chemical concentrations. Thus individual growth is governed by both the availability of food resources and the degree of satisfaction with the resource ratios within one's buffer.

As shown in Downing and Zvirinsky (1999), a variety of Guild runs exhibit the emergence of both nutrient recycling (i.e., high cycling ratios) and control of the global nutrient ratios at levels near the optima. Control is particularly evident when the environmental input fluxes are drastically perturbed, yet the biota maintain nearly optimal nutrient ratios.

Figures 23.2 and 23.3 show the results of a typical Guild run in which the initial population of size 100 consists of a single phenotype that produces N_1 and consumes N_2. The environment is initially devoid of nutrients, with input fluxes of (20, 20, 20, 20) units/time step for $N_1 \cdots N_4$, respectively, and output fluxes of 1 percent of the standing amounts, E_i, $i = 1 \cdots 4$. At generation 400, the input fluxes change to (5, 10, 25, 50), and then to (50, 25, 10, 5) at generation 600. The biota have optimal growth with ambient nutrient ratios (i.e., normalized E_i values) of .4, .3, .2, and .1.

Figure 23.2 illustrates the increase in cycling ratios as phenotypic diversity rises and the recycling loops form, while figure 23.3 shows the approach of nutrient ratios to their biota-preferred values (dotted horizontal lines) and their persistence in the face of the two large disturbances at generations 400 and 600.

As a brief causal explanation, competition drives the initially homogeneous biota toward greater trophic diversity (i.e., diversity of consumed nutrients), and since each organism must produce at least one nonconsumed chemical as waste, a diversity of outputs also emerges. This increasing biotic heterogeneity results in the fortuitous formation of recycling networks. When all of the pieces (i.e., guilds) of these networks fall into place, previously underconsumed (and thus accumulating) nutrients are taken into the food chain, fueling a population explosion and an in-

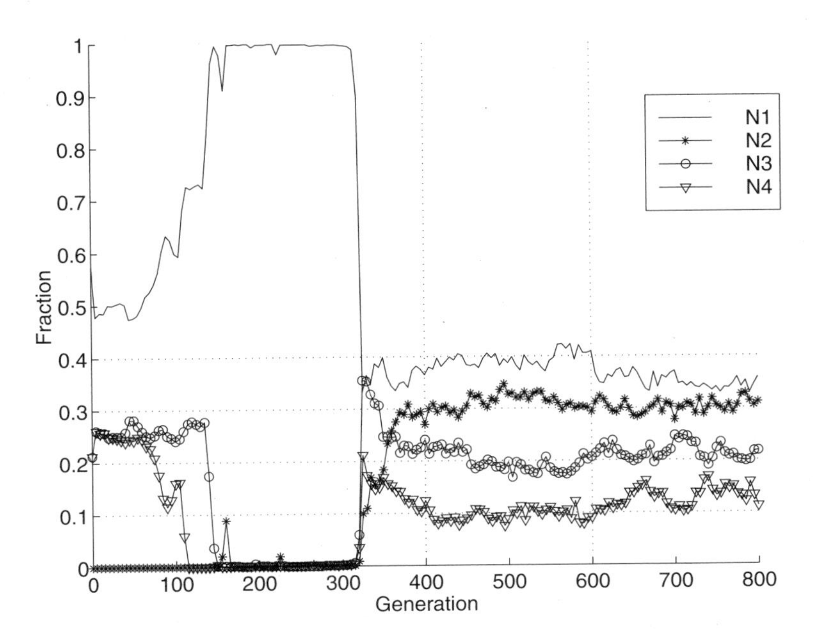

Figure 23.3
The evolution of the global fractions of four nutrients in a run of Guild, where all guilds have optimal growth when the proportions are (.4, .3, .2, .1) for nutrients $N_1 \cdots N_4$, respectively, as denoted by the dashed horizontal lines. Note the relative stability of optimal growth fractions despite major fluctuations in the external input fluxes in generations 400 and 600.

crease in cycling ratios. The elevated nutrient transfer within the recycling network then facilitates further population growth within each guild. The high transfer fluxes between these large interconnected guilds dwarf the environmental input and output fluxes, thus reducing the biota's sensitivity to external perturbations.

Competition within and between this diverse collection of well-populated guilds results in a frequency-dependent selection that enables the guilds to effectively control global chemical ratios via their cumulative production and consumption. For example, consider a guild G that consumes nutrient N_k. The local environment of G will have a lower proportion of N_k than the global environment, so changes in the total biomass of G will decrease E_k and, hence, the global proportion of N_k. Now if the global ratio is nearly optimal, then any major increase in G's biomass will push the global ratio away from optimal and give a selective advantage to the members of other guilds, those that produce N_k. This will then push E_k back toward the optimal proportion. In short, any deviations from optimal (i.e., error) of the global ratios will create an environment that favors guilds which (fortuitously) decrease the error. Importantly, the selective advantage of these guilds stems not from their global influence but from their ability to create pleasant local environments for their own growth. Thus, the interplay between the guilds, orchestrated by standard natural selection, achieves and maintains a stable optimal nutrient ratio.

In summary, natural selection operating on a collection of diverse competing biochemical guilds leads to the emergence and stability of both a self-sustaining nutrient recycling network and a distributed controller. (For an in-depth description of Guild, the parameter settings—such as mutation, crossover, metabolism, and feeding rates—for particular test cases, and full details of simulation results, see Downing and Zvirinsky, 1999).

Clearly, models of this simplicity cannot fully explain complex biogeochemical phenomena, but they can often illustrate the sufficiency of particular mechanisms for deriving similar patterns. This work shows that simple local interactions, under the scrutiny of natural selection, can lead to interesting cooperative arrangements. Since these particular cooperative results, efficient recycling networks and distributed global chemical regulation, are both viewed as fundamental examples of Gaia in action, our simulations lend support to the basic compatibility of Gaia and evolution. The Guild system's use of a legitimately large genotype space and more convincing evolutionary simulations makes a somewhat stronger argument for this compatability than Daisyworld's small set of hardwired genotypes.

Unfortunately, the Guild model, like the original Daisyworld, sidesteps the evolving-preferences issue (Saunders, 1994; Robertson and Robinson, 1998), since all guilds are assumed to have the same constant preferred chemical ratios. When included in the Guild genomes, these preferences inhibit regulation, because individuals simply evolve preferences to the current conditions. These "regulatory parasites" are clearly a problem.

Another weakness is Guild's reliance on an overly abstract chemistry that permits all intermolecular transformations, conserves mass only at the highest level of abstraction, and has no specific thermodynamic constraints upon reactions. For example, an organism is free to convert M units of compound A into M units of compound B, for all compounds A and B. The transformation does involve an energy cost, but one based solely on M, not on A or B. This permits a relative metabolic free-for-all in which diversity quickly arises, along with the concomitant recyling of compounds and environmental control.

Metamic

The Metabolically Abstract Microorganism system (Metamic) was designed to remedy Guild's chemistry shortcomings. It employs the MD-Chem (Modular Designer Chemistry) module (Downing, 2001) to generate random abstract chemistries that conform to user specifications and respect a few fundamental biochemical constraints: they differentiate between atoms and molecules, and the former are conserved in all reactions; and a simple thermodynamic constraint links the synthesis (breakdown) of larger molecules from (into) smaller ones with energy consumption (release). Hence, metabolism can be partitioned into anabolic (biomass-producing) and catabolic (biomass-burning) stages; and an organism's fitness reflects its ability to exploit the underlying chemistry and overall ecological situation (in terms of the other biochemical guilds present in its environment) to produce energy and build biomass.

Metamic Primitives Like Guild, Metamic is a simple box model with chemical inflows and outflows. The box constitutes an environment, E, for a population of individually modeled metabolizing agents (*metamics*) that interact in only two ways: indirectly, via chemical exchange with E, and directly, via gene-swapping during conjugation. Each metamic is modeled as a cell with a genetically determined metabolism, a local chemical buffer, and a semipermeable membrane that separates it from the environment.

Metamic employs MD-Chem chemistries as bases for all intra- and extracellular chemical activity. The chemical basis for a Metamic run is defined by the pair (C, R), where C denotes the set of legal compounds, and R, the legal reactions.

A parameterizable set of physical variables, Φ, provides a second pathway for bidirectional interactions between the organisms and their environment. The value of each $p_i \in \Phi$ is defined as a function of the normalized concentrations of some or all of the molecules in C. Values for all p_i are computed for the global environment, based on the global concentrations, and local values are computed for each cell, based solely on intracellular concentrations. As described below, the physical variables can influence metabolic rates and, hence, growth.

Each metamic's genetic algorithm (GA) genome encodes r_T, a subset of R, plus base reaction constants, k_r, for each inherited reaction. From the genome, the exothermic, r_x, and endothermic, r_n, reactions are separated, where $r_T = r_x \cup r_n$; together, they compose the organism's metabolism. As shown in figure 23.4, the abstract metabolic process consists of two phases: catabolism and anabolism. On each time step, metamics receive an energy request and begin catabolism, in which the exothermic reactions, r_x, run for the maximum of two durations: the official time step and the estimated time needed to generate the required energy (based on previous energy production rates of the r_x). If the former exceeds the latter, an energy surplus results, thus triggering the anabolic processes (i.e., the endothermic reactions, r_n), which run long enough to consume the available energy and build structure by reducing internal entropy.

Within a cell, the rate of each chemical reaction is determined by standard mass-action dynamics with an additional term, physical-factor satisfaction, which is an exponentially decaying function of the sum of the deviations of each physical factor $p_i \in \Phi$ from its optimal value (for metabolism), p_i^*. The p_i^* and the functions for computing the p_i from chemical concentrations are user-supplied and may differ considerably across experiments.

Those compounds which constitute biomass can vary between organisms and are essentially the largest few molecular types (with user-defined lower size bounds) that an organism can produce.

The cell membrane is semipermeable in that diffusion rates depend upon the molecule. A metamic's biomass cannot diffuse into or out of its cell. Also, any compounds that the organism is a net consumer

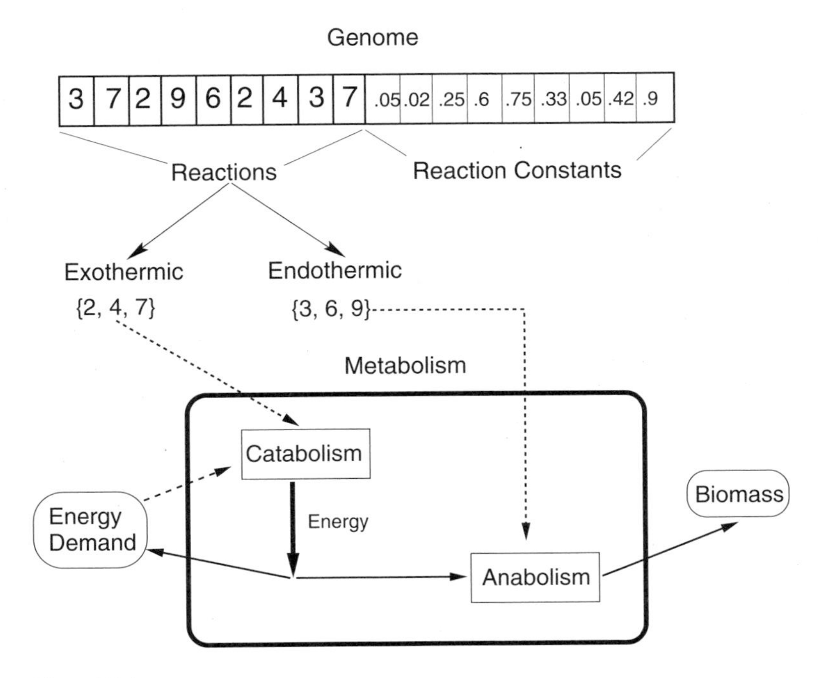

Figure 23.4
A metamic's genome determines the reactions and reaction constants for its metabolism. Exothermic reactions make up the catabolism, and endothermic reactions the anabolism. The energy produced by catabolism pays back the energy demand, and any remaining energy goes toward structure-building (i.e., anabolism).

of (in r_T) can diffuse in, but not out, of the cell. Conversely, compounds with net production in r_T can only diffuse out, not in. All other compounds can freely diffuse into and out of the cell.

Thus, the cell membrane provides a buffer zone for the cell. This zone is not perfectly protective, since diffusive chemical exchanges with the environment will move local and global chemical concentrations, and hence p_i values, closer, but the restrictions on diffusion do allow a cell to create an interior milieu that differs from its surroundings. This buffering is critical for cell growth, since the p_i can affect metabolic rates.

If an organism doubles its birth biomass, then reproduction occurs by asexual splitting, with possible mutation of both child genomes. Organisms also undergo a form of double bacterial conjugation by occasionally swapping GA chromosome segments with one another.

As summarized in figure 23.5, Metamic's primitive mechanisms support a feedback loop involving metabolism, chemical concentrations, and the physical factors. Mass–action dynamics govern the relationship between metabolic rates (and growth) and the intracellular chemical concentrations, while user-specified constraints relate concentrations to physical

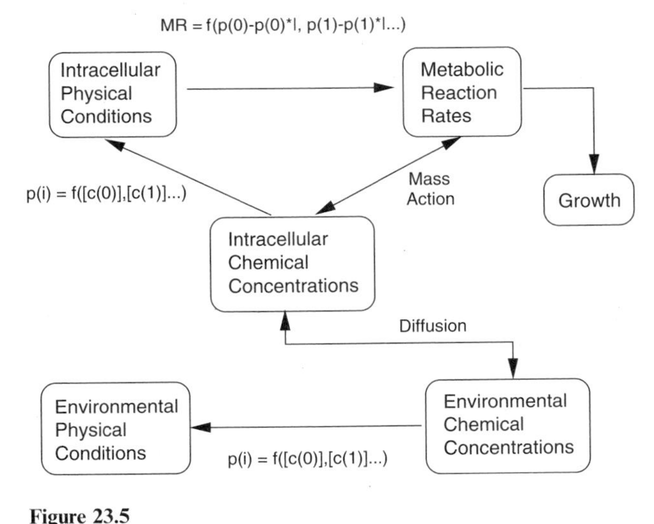

Figure 23.5
The primary interactions in Metamic that causally interlock the biota and biosphere. Arrows indicate influence of one factor upon another, with arrow labels indicating the underlying mechanism. The $p(i)$ are physical variables, the $p(i)^*$ are optimal values for those variables, the $[c(i)]$ are chemical concentrations, and MR is the metabolic rate.

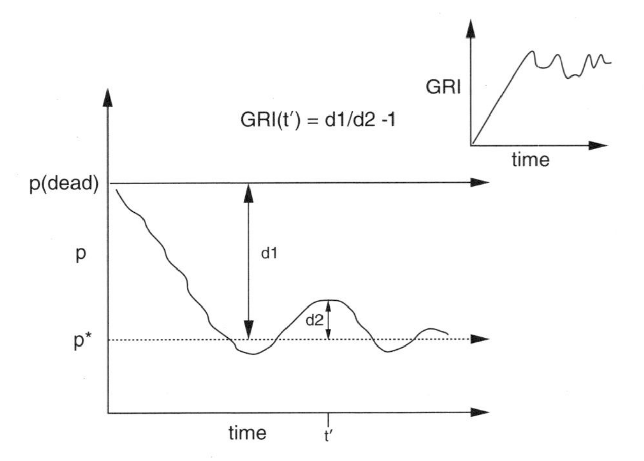

Figure 23.6
Illustration of the Gaian regulatory index (GRI). The large graph shows the trajectory of a physical variable p, along with the essential terms needed to compute the GRI at any time point t'. The smaller graph plots the time series of GRI values corresponding to the trajectory of p.

conditions. Deviations of physical variables from their growth optima then influence metabolic rates to close the loop.

GRI: The Gaian Regulatory Index

The Gaian Regulatory Index is a quantitative measure of the Metamic biota's ability to control the environment within a range that is most conducive to growth. In a nutshell, the GRI is the gain of the distributed regulator having p_{dead} as a base value and p^* as the target value, where (a) p_{dead} is the steady-state value of a focal physical variable, p, when the global environment is simulated using chemical inflows, outflows, and internal chemical reactions, but without organisms, and (b) p^* is the (arbitrarily chosen) metabolic optimum value for p.

As depicted in figure 23.6, the GRI at any time point t' is -1 plus the ratio of the distance from p^* to p_{dead} (i.e., the total distance that a perfect regulator would move p) to the actual distance from $p(t')$ to p^* (i.e., the error). Higher GRI values indicate better regulation.

Gaia Hunting

The Metamic research is motivated by a central question: How likely is the emergence of distributed environmental regulation in an evolving community of metabolizing organisms? We follow Kauffman's (1993) classic ALife approach of generating many possible sets of underlying interactions and then tallying the frequency of various emergent phenomena generated by these sets: MD-Chem generates a wide variety of abstract chemistries, and Metamic is run using each chemistry and tested for the emergence of distributed environmental control.

In all, 100 Metamic tests using 100 different life-supporting MD-Chem chemistries were performed. Figure 23.7 shows the results of four such tests. Each test consists of several runs:

1. G_{dead}—a short simulation of an organism-free environment to establish a *dead* value, p_{dead}, for the physical variable, p, in the environment, E.

2. G_{base}—a long simulation, with organisms but with the satisfaction term removed from all mass–action equations—making all metabolisms insensitive to p—to yield a baseline trajectory for p, with a trajectory-averaged p value, p_{avg}.

3. G_{out}—a long simulation with the satisfaction term back in the mass–action equations and with the metabolic optimum p value, p^*, set to $p_{avg} + d/2$, where $d = p_{avg} - p_{dead}$.

4. G_{in}—a similar long simulation with $p^* = p_{avg} - d/2$.

In all tests, the intracellular and environmental p values were calculated as

$$p = \tilde{c}_1 - \tilde{c}_2 + \tilde{c}_3 - \tilde{c}_4 \tag{1}$$

where $\tilde{c}_i$ $i = 1 \cdots 4$ are the normalized (over all ten compounds) concentrations of the four smallest compounds in the chemistry. This is an arbitrary choice for the functional relationship between the chemical and physical factors. The four smallest compounds were also the only ones with (equal) nonzero forcing fluxes into the global environment.

For each of the 100 life-supporting chemistries that were tested, the time-averaged GRI values, $\overline{GRI}$, were recorded. In all, 20 of the 100 chemistries gave $\overline{GRI}$ values above 5 (on either the G_{in} or the G_{out} run), indicating a reasonable degree of regulation. The average $\overline{GRI}$ of these 20 cases was 7.63, with a standard deviation of 2.71. The very best regulator had $\overline{GRI} = 14.97$, while 7 of the 20 had low values between 5 and 6.

In several scenarios, such as the upper row and leftmost column of figure 23.7, the regulatory run clearly distinguishes itself from the G_{base} case and maintains a stable value near the optimum, p^*. In other situations, such as the bottom right graph of figure 23.7, the base and regulatory cases are less clearly separated. Although this does not necessarily decrease $\overline{GRI}$, it weakens the homeostatic claim when

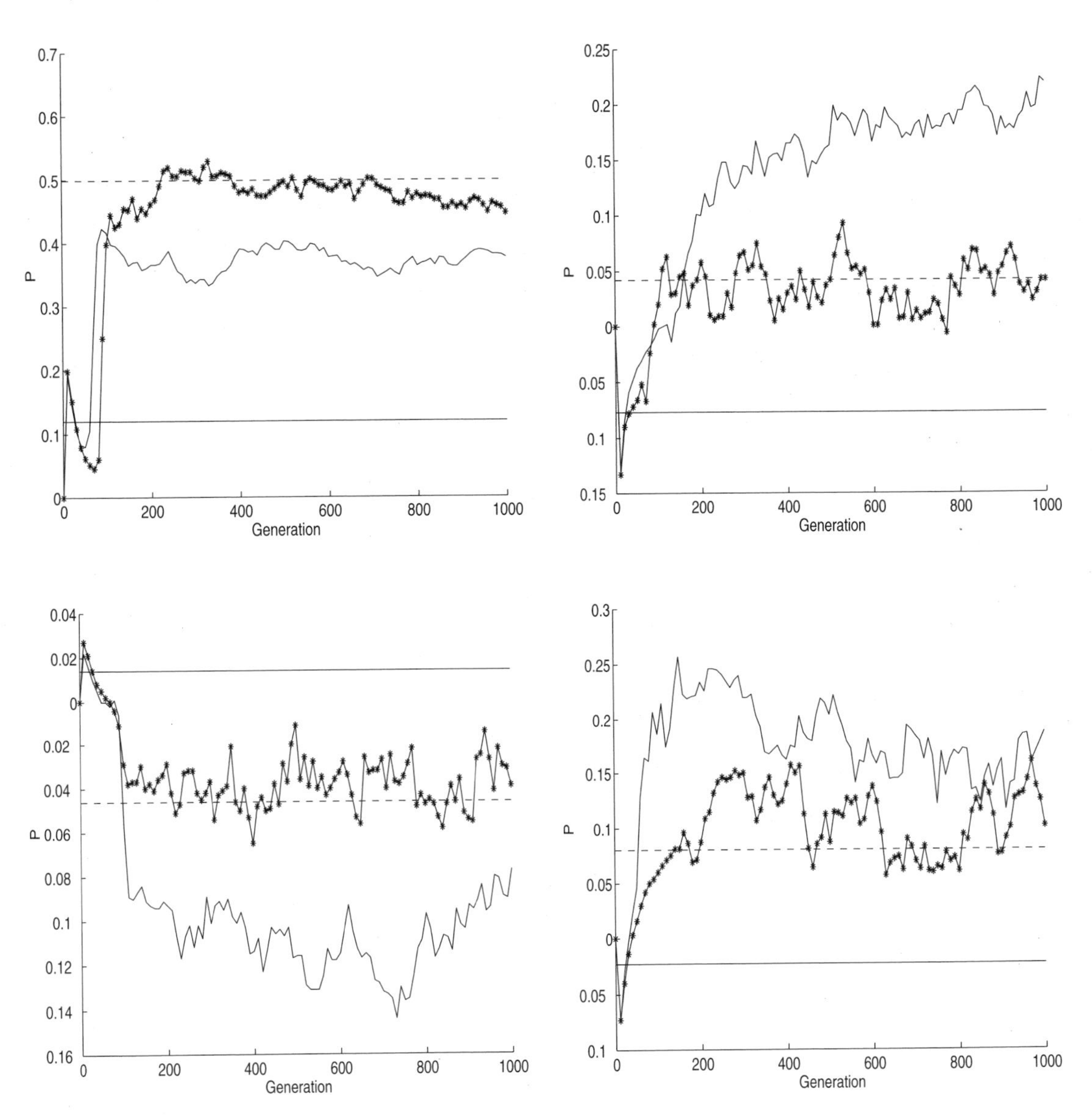

Figure 23.7
Emergent regulation ($\overline{\text{GRI}} > 5.0$) in 4 different 10-compound, 20-reaction MD-CHEM chemistries. A total of 100 life-suppoting chemistries were tested, and 20 percent gave $\overline{\text{GRI}}$ values above 5. In each graph, the straight solid line denotes p_{dead}, and the straight dashed line represents p^*. The solid curve is the trajectory of p in the G_{base} run, while the curve with asterisks shows p's trajectory in the G_{in} or G_{out} case.

the base case never strays too far from either p^* or the regulatory case. This was the impetus for choosing p^* at a distance from p_{avg}—but not too close to p_{dead}.

In general, these results indicate that emergent distributed control is hardly a rare occurrence in situations that include the basic Gaian primitive mechanisms discussed earlier. Still, it is certainly not a necessary consequence of metabolizing organisms in relatively closed environments. However, the basic phenomenon arises in 20 percent of the tested chemistries, with 7–10 percent of them giving rather convincing visual evidence of a deviation from normal behavior to an unconscious environment-controlling mode.

Remember that in the G_{base} runs, the metabolic rates are independent of the physical variable, p, so there is no selective advantage for local regulation of intracellular p values. In many G_{in} and G_{out} runs, the chemistry simply does not provide the flexibility: organisms cannot alter their metabolisms in order to attain near-optimal p values while simultaneously producing enough biomass to eventually reproduce. Hence, they must stick to the standard metabolisms but operate under suboptimal p regimes. Detailed analysis of individual chemistries and runs reveals many interesting trade-offs between metabolic efficiency and p-value control (Downing, 2002).

In Guild, the environmental control is robust to changes in external chemical fluxes. This stems from the lack of constraints on chemical transformations in that model. However, the reduced reaction possibilities in the above Metamic runs strongly restrict the metabolic options. Hence, regulation often breaks down with a change in exogenous flux.

The trade-offs observed between efficient metabolisms and those with Gaian side effects partially support the neo-Darwinian "cheaters will prosper" attack. Indeed, the p-value-controlling side effects can exact a heavy cost in Metamic, although they do provide an immediate local selective advantage. The empirical question is the standard one: Do the benefits exceed the costs? Clearly, no right-minded Gaian or anti-Gaian could give a definitive answer to this general question, and neither can Metamic. In some runs in some chemical worlds, the benefits dominated and control emerged as an apparent stable attractor; in others, it arose occasionally, as something akin to a saddle point; and in many cases, the chemistry simply did not support much metabolic choice, although sustained life was possible. So if Metamic has captured some essence of the real world, then one basic take-back-to-the-wild lesson from the computer runs is simply that Gaian homeostasis is neither tautological nor impossible, but clearly contingent on a host of physicochemical, and quite possibly historical, factors.

Model Summary

Table 23.1 summarizes Daisyworld, Guild, and Metamic with respect to the proposed Gaia-logic

Table 23.1
Summary of Gaia-logic characteristics in Daisyworld, Guild, and Metamic

	Daisyworld	Guild	Metamic
Primitive Mechanisms			
Ambients → growth	Temperature → growth	Chemical ratio → growth	local p → growth
Gaian root	albedo	chemical compounds	chemical compounds
GEFs → ambients	global albedo → global temperature → local temperature	global chemical ratios → local effective chemical ratios	global chemical concentrations + local metabolism → intracellular concentrations → local p values
Local selective advantage	daisy albedo buffers against global temperature	intracellular chemical ratio buffers against global ratio	growth enhanced when local $p \approx p^*$
Resource	space	all compounds	high-energy and biomass precursor compounds
Emergent Properties			
Frequency-dependent growth	Emerges from space competition	Emerges from nutrient competition	Emerges from nutrient competition
Biotic-abiotic feedback	Practically hardwired due to fixed, small phenotype space	Arises from emergent biodiversity and recycling	Arises from emergent p-controlling metabolisms
Frequency-dependent selective feedback	negative among the daisy populations	negative among the guilds	both negative and positive among guilds
Recycling (cycling ratio)	none	high (30–50)	moderate (5–15)

characteristics. Here, the intention is to illustrate the presence of most features in all three models, and not necessarily to compare the approaches. Space restrictions prohibit a classification of natural Gaian phenomena along these same dimensions, but Lenton's (1998) analysis reveals many possible matches, along with additional Gaia-logic primitives.

In general, the move from Daisyworld to ALife models adds so many degrees of freedom to the simulation that surprises are inevitable. The rampant nonlinearities incurred by hundreds or thousands of interacting metabolizing agents often makes analytic prediction impossible. And since positive feedbacks may turn chance fluctuations into dominant structures, the results of two identical (except for the random-number seed) simulations can quickly diverge. Thus, behind the beauty of emergent global patterns there often lurks a beast of contingencies and unpredictabilities, and long hours of parameter tuning. ALife researchers argue that these costs/risks are justified if the model can unlock the mysteries of a complex natural phenomenon, such as the evolutionary emergence of Gaia. So while Daisyworld serves as an existence proof of emergent distributed control, significant future advances in Gaia theory require a modeling paradigm that explicitly incorporates the evolutionary process. Guild and Metamic are first steps in that direction.

Conclusion

A few ardent Gaia believers envision a world in which biotic control over the biosphere is ubiquitous, optimal—they just have not quantified exactly *what* is being optimized—and the fundamental organizing principle of nature, above even natural selection. However, most Gaia sympathizers conservatively interpret the biogeochemical data and computer models as hinting at a phenomenon that is (a) prevalent, (b) far from optimal but generally satisfactory at maintaining livable conditions under an ever-increasing solar input on a planet where large-scale physical factors can easily erase much of the biota's best and worst work in a geological heartbeat; and (c) completely consistent with natural selection.

This research is grounded in the latter view. Our motivation is an understanding of some of the fundamental mechanisms by which evolving organisms could wrestle some degree of environmental control from the brute physical forces of nature. In short, how could Gaian phenomena emerge on an evolutionary scale? Hence, Guild and Metamic take a step beyond Daisyworld by assuming very little about the initial genotype/phenotype pool but still providing a virtual incubator for the emergence of distributed recycling and homeostasis, in full harmony with individual-based natural selection.

The ALife perspective calls for a detailed analysis of the frequency of emergent Gaian phenomena in natural and artificial systems, and the similarities between these occurrences. Only then can the prospect of Gaia as a general, possibly prevalent, phenomenon be thoroughly assessed. ALife facilitates this batch processing of Gaian alternatives, and popular ALife tools such as evolutionary algorithms and individual-based simulation (and the fast computers to run them) are now accessible to the Gaian community. As Daisyworld has shown, for understanding general principles, abstract computer simulations can be just as useful as expensive ice-core drillings. ALife offers further, richer inquiries of this silicon variety.

Note

1. The classic GA involves a population of chromosomes/genotypes represented as strings of binary digits. An often trivial developmental process decodes them into integers, real numbers, or any of a wide variety of meaningful phenotypes. The performance of each phenotype on a set of tests then determines its fitness, which governs the probability that its corresponding genotype will be explicitly reproduced to form the next generation. During reproduction, chromosomes are combined via crossover and altered via mutations of single bits or inversions of bit sequences. In ALife systems, the fitness is a more implicit measure of the agent's ability to gather resources. Reproduction then occurs as a natural part of the simulation when the agent grows to a mature size and either asexually clones itself or mates with another mature agent—again producing altered genotypes across the generations.

References

Caldeira, K. 1989. Evolutionary pressures on planktonic production of atmospheric sulphur. *Nature*, 337, 732–734.

Downing, K. 2000. Exploring Gaia theory: Artificial life on a planetary scale. In M. Bedau, J. McCaskill, N. Packard, and S. Rasmussen, eds., *Proceedings of the 7th International Conference on Artificial Life*, pp. 90–99. Cambridge, MA: MIT Press.

Downing, K. 2001. Modular designer chemistries for artificial life. In L. Spector, ed., *Proceedings of the 3rd Genetic Evolutionary Computation Conference*, pp. 845–852. San Francisco: Morgan Kaufmann.

Downing, K. 2002. The simulated emergence of distributed environmental control in evolving microcosms. *Artificial Life*, 8(2), 121–151.

Downing, K., and P. Zvirinsky. 1999. The simulated evolution of biochemical guilds: Reconciling Gaia theory with natural selection. *Artificial Life*, 5(4), 291–318.

Goldberg, D. 1989. *Genetic Algorithms in Search, Optimization and Machine Learning*. Reading, MA: Addison-Wesley Longman.

Holland, J. H. 1992. *Adaptation in Natural and Artificial Systems*. Cambridge, MA: MIT Press.

Holland, J. H. 1995. *Hidden Order: How Adaptation Builds Complexity*. Reading, MA: Addison-Wesley.

Kauffman, S. 1993. *The Origins of Order*. New York: Oxford University Press.

Langton, C. 1986. Studying artificial life with cellular automata. *Physica*, D22, 120–149.

Langton, C. 1989. Artificial life. In C. Langton, ed., *Artificial Life: Proceedings of an Interdisciplinary Workshop on the Synthesis and Simulation of Living Systems*, pp. 1–49. Reading, MA: Addison-Wesley.

Lenton, T. 1998. Gaia and natural selection. *Nature*, 394, 439–447.

Lindenmayer, A., and P. Prusinkiewicz. 1989. Developmental models of multi-cellular organisms: A computer graphics perspective. In C. Langton, ed., *Artificial Life: Proceedings of an Interdisciplinary Workshop on the Synthesis and Simulation of Living Systems*, pp. 221–249. Reading, MA: Addison-Wesley.

Liss, P., A. Hatton, G. Malin, P. Nightingale, and S. Turner. 1997. Marine sulphur emissions. *Philosophical Transactions of the Royal Society of London*, B352, 159–169.

Ray, T. S. 1992. An approach to the synthesis of life. In C. Langton, C. Tayler, J. Farmer, and S. Rasmussen, eds., *Artificial Life II*, pp. 371–408. Reading, MA: Addison-Wesley.

Robertson, D., and J. Robinson. 1998. Darwinian Daisyworld. *Journal of Theoretical Biology*, 195(1), 129–134.

Saunders, P. 1994. Evolution without natural selection: Further implications of the Daisyworld parable. *Journal of Theoretical Biology*, 166, 365–373.

Schlesinger, W. 1997. *Biogeochemistry: An Analysis of Global Change*. Boston: Academic Press.

Schwartzman, D., and T. Volk. 1991. Biotic enhancement of weathering and surface temperatures on earth since the origin of life. *Palaeogeography, Palaeoclimatology, Palaeoecology*, 90, 357–371.

Volk, T. 1998. *Gaia's Body: Toward a Physiology of Earth*. New York: Copernicus.

Watson, A., and J. Lovelock. 1983. Biological homeostasis of the global environment: the parable of daisyworld. *Tellus*, 35B, 284–289.

Williams, G. 1996. *The Molecular Biology of Gaia*. New York, Columbia University Press.

24

On Causality and Ice Age Deglaciations

Alexandre Casanovas and Vicent Gómez

Abstract

Of the several factors that influence ice ages, the most important is solar forcing; others, such as ocean circulation, are consequences of uneven solar radiation. The complexity of the orbital movement induces a complex distribution of the solar radiation over time over the globe. Complex insolation is related here with ODP site 659 sea core and Vostok ice core data.

To evidence causality, a Sugeno-type fuzzy inference system, which acts as an artificial neural network (ANN), relates the solar forcing time series to the time-delayed data of the geological record. ANN models help to elucidate very complex causal relations where several inputs could act synergistically. We consider that if the ANN is not capable of reproducing the data, there is no causal relation involved.

We found that Cape Verde ocean temperature can be predicted from solar forcing with less relative mean absolute deviation (MAD) than any other of the variables. Methane MAD reduces to half the previous value if we include Cape Verde seawater temperature in the prediction. The carbon dioxide error also reduces to almost half if we include the methane series and $\delta^{18}O$. Antarctica ice deuterium, which cannot be predicted accurately using only the insolation data, reduces to one-third the MAD if we use tropical ocean temperature and CH_4 and CO_2 mixing ratios.

We suggest the following Gaian scenario: clathrates accumulated during the long cold periods in tropical oceans become unstable as the ocean water warms, liberating greenhouse gases that fire the Antarctica deglaciation. This contribution is decisive at locations where the oceans cannot directly influence, such as over dry continents. The Vostok record helps us to confirm this issue for Antarctica.

Introduction

Lovelock's idea of Earth as a self-evolving and self-regulated living system was based on the fundamental change of the planet's atmosphere 3 billion years ago by bacteria and photosynthetic algae. This chapter points out that another Gaian process could have contributed to Quaternary deglaciations in Antarctica through release of methane from tropical sea methane hydrates (clathrates).

Deep-sea drilling had shown that the Earth's climate began to cool about 25 MYr BP, with the polar ice caps expanding steadily to the equator. Since ice growth has a positive feedback by rejecting more solar radiation from Earth, the process can culminate, as predicted by Budyko (1969), in a vast part of the planet, mostly the continents, being covered by ice, a situation known as a snowball or a slushball Earth. But the actual culmination of this process has been a series of sudden changes in temperature from cool to warm, with alternating glacials and interglacials, in a sawtooth manner. Something is correcting the trend.

It is difficult, as pointed out by climate modelers (Hyde et al., 2000; Crowley et al., 2001), for tropical oceans to freeze, and therefore the possibility exists of a continuous growing of clathrate deposits. High plankton productivity and high sedimentation rates yield large amounts of organic matter, which becomes the basis for the production of methane in the sediment. Clathrates are rigid enclosures of water molecules that include mostly methane, but also carbon dioxide or H_2S, and are stable only under high pressure and low temperature. Therefore they are stable on the seafloor if the bottom waters remain cool—less than, for instance, 4°C below 400 m deep ⟨www.gashydrate.de⟩. A cool ocean will conserve large amounts of clathrates. If liberated, methane is a severe greenhouse gas, and is converted to another greenhouse gas, carbon dioxide, in about 12 KYr. (Kennett et al., 2000) have found evidence of massive outgassing of unstable methane hydrates switched on by water temperature and thermohaline circulation.

It is not the first time that clathrates have been associated with methane release to the atmosphere. The variations on methane mixing ratio registered (e.g., on ice cores) are correlated with temperature, but causality is in doubt: Does the increase of temperature precede or succeed (as pointed out by Kennedy et al., 2001) methane release? To help to clarify this issue, the first part of the chapter suggests a new method to demonstrate causality for complex systems.

Of the several factors that influence ice ages, the most important is solar forcing. Others, such as ocean circulation (Paillard, 1998; Ganopolsky and Rahmstorf, 2001) are consequences of uneven solar radiation over the globe, and are related to solar forcing. By assuming that it was only necessary to account for the effect of solar forcing at 65°N latitude in summer, Milankovitch's theory reduced the influence of solar forcing owing to orbital changes of insolation. The idea behind this is that most of the Earth's land is in the northern hemisphere, and glaciation is easier over land. There is evidence against a pure Milankovitch theory, such as Devils Hole and other geological records (Karner and Muller, 2000). Henderson and Slowey (2000) had shown that the timing of the second-to-last ice age coincided with solar forcing at low latitudes in the southern hemisphere.

It is usual to reduce the checking of the solar forcing on geological records by searching for characteristic frequencies of the Earth's orbit. This is not a definite method. Solar forcing on Earth shows several periodicities induced by the orbital movement. An approximate 100 KYr periodicity is due to eccentricity, a 41 KYr to obliquity, and a 23 KYr to precession. The influence of the eccentricity, obliquity, and precession is not uniform over the globe. The effect of the obliquity is to amplify the seasonal cycle and is maximum at high latitudes, whereas the effect at the tropics is negligible. The 41 KYr period is perceptible in the Deep Sea Drilling Program site 607 sea core, at the latitude of the Azores Islands, whereas it is absent from Ocean Drilling Project site 659 sea core, at Cape Verde latitude. Only the 100 KYr periodicity is detectable on site 659 (Muller and MacDonald, 1997).

In fact, the complexity of the orbital movement induces a complex distribution of the solar radiation over time over the globe. The differences in radiation between latitudes feed circulations such as the Northern Atlantic deep water, and are also a forcing. Therefore, the effect of nonuniform solar radiation all over the globe should be considered, together with the differences of simultaneous values of solar radiation at high and low latitudes.

Some questions remain open, including three given below. Why do glacials and interglacials occur simultaneously on both hemispheres even though the orbital variations of solar radiation affect the northern and southern hemispheres differently? Why is the periodicity of ice ages almost 100 KYr even though the insolation changes due to eccentricity are low in relative terms? How does one explain the isotope stage 11 paradox (about 400 KYr BP) in which the summer insolation at 65°N is not high enough to switch on deglaciation? We hope to contribute to answering these questions in this chapter.

The first part of the chapter proposes a new method to evidence causality. In the second part we apply the method to determine the effect on Earth of solar forcing relating the insolation with ODP site 659 sea core (Tiedemann et al., 1994) and Vostok ice core (Petit et al., 1999) data. Finally, a scenario is suggested that includes a Gaian contribution to Quaternary deglaciations on the last 420 KYr BP.

Determining Causality

The Earth orbital movement, and therefore the received solar forcing, is chaotic (Laskar et al., 1999). For the Earth, that means the orbital movement is somewhat unpredictable at very long times, but is stable. For us, it means that solar forcing has a complex pattern that is different for each latitude, or that differences of solar forcing at different latitudes also have their own complex patterns. Differences of simultaneous values of solar radiation at high and low latitudes are also considered a forcing in this chapter.

Causality will be defined as follows. If two systems interact and are contemporary, a common pattern of evolution follows. If they are not contemporary and interact, the pattern of evolution of the first, the cause, can be detected in the pattern of evolution of the second after some time delay. This delay is obtained here from the correlation coefficient of the involved time series.

Causality can be better detected in complex time series, since it is highly improbable to find the complex causal pattern later on, in an accidental way. It is like accidentally finding the key for a lock.

We can expect that in the time series of a geological record there are influences of several forcings, with different delays. Therefore we cannot expect simply to find the same pattern anew, but several causal patterns added on in a complex way. Therefore the matching of the patterns cannot be detected visually, but only by using some ad hoc tool.

Artificial Neural Nets and Fuzzy Inference

The parallel processing of forcing at several latitudes (including differences in forcing) is an appropriate task for an artificial neural network (ANN), a network structure of simple elements (nodes) that operate in parallel and mimic the structure of neural cells. Each node is a processing unit that includes adaptive parameters, and the links between the nodes specify causal relationships between them. Unconnected nodes are not directly related.

ANNs can "learn" from the input data by adapting the parameters of the nodes and, after the learning process, can model the complex behavior of these data—including nonlinear processes. In the learning process, some links are enhanced and others weaken or even disappear.

We used a specific form of neural net, called adaptive neuro-fuzzy inference systems (ANFIS) by Jang et al. (1997), that has a dual nature: as well as modeling complex behavior it is also a fuzzy inference system. In ANFIS's ANNs there are only two layers, and only the first layer, defined by fuzzy categories, learns from the data. Learning consists of adapting the number of categories and the membership of the data to these categories. The learning process with an ANFIS net uses the information in the data to perform an optimized classification.

The second layer performs a linear regression of the data of the class to the output, weighted by membership. This is part of the Sugeno form of inference. Unlike conventional neural nets, these ANNs cannot "overlearn." The classes may be slightly modified by "learning" more, but the fit remains conventional. The first layer, by dividing into classes, introduces the nonlinear general behavior.

Common uses of ANNs and fuzzy inference systems are control and modeling. However, they were used here to establish whether cause-effect relationships exist between time series.

An ANN model was used to link insolation at several latitudes to temperature and other proxies. We used a Sugeno-type (i.e., adaptive) fuzzy inference system (Jang et al., 1997), which is included in a Matlab toolbox (Matlab, 1998), and since it can be self-configured, it is simple to use. This is an important point because it was necessary to adjust a large number of ANNs and we needed to assure equality of fits. Subjective decisions, when configuring networks, is known to reduce the quality of the results.

The use of ANNs was quite specific, and the error could not be reduced endlessly by training. This is because we did not try to predict a time series from past data of the same series, as is usually the case; instead, we predicted a time series from an unrelated, different time series (e.g., the ice volume in Antarctica from the methane mixing ratio). If we performed this process with two white noise (random) series, the relative error would be very high, because one is unpredictable from the other.

Rank of Results

Because we were mostly interested in the rank of the results, the full series could be used as test data. Also, the series are normally short; ice cores and sea cores contain some 500–1000 noise-corrupted data. The lesser error (calculated as RMS, root mean square, or MAD, mean absolute deviation) of calculated versus measured data was used as a figure of merit to compare results between scenarios. If the ANN was not capable of reproducing the data, we concluded there was no causal relation.

In order not to give excessive weight to the square of a big isolated error, we preferred the MAD to the RMS error. The figure of merit errors are relative (percent) errors, because the scales are widely different and we want to compare results. Therefore, the values were expressed in relative MAD.

Time Delay

The link is established between the solar forcing time series and the time-delayed data of the geological record or, in a second stage, between the solar forcing time series and delayed data of the geological record with other parameters of the cores. The reason for delays is that we need to consider the time necessary for, for example, heating the land or oceans. The time delay is dependent on each input, and its optimum value is determined from the correlation coefficient CC of two time series $x(i)$ and $y(i)$, i being successive steps in time,

$$CC(m) = E\{x(i)y(i+m)\}$$

where $E\{\ \}$ is the mean. Figure 24.1 shows the correlation coefficient for each of the six solar forcing values relative to seawater O_2 isotopic deviation. As can be seen, there are small differences in the time lag for the different inputs at Cape Verde, but most are about 3 KYr, a value related to ocean heat capacity. The choice of time delays is for the shorter one giving the maximum correlation coefficient.

We do not expect that every nonnull correlation means a causal process. The cause-effect relationship is evidenced by fruitful modeling of the effected time series. The link (if it exists) points to only one of the legs of the causal process; several others can act at the same time.

Besides the correlation coefficient, there are other ways to establish the relation between two time series,

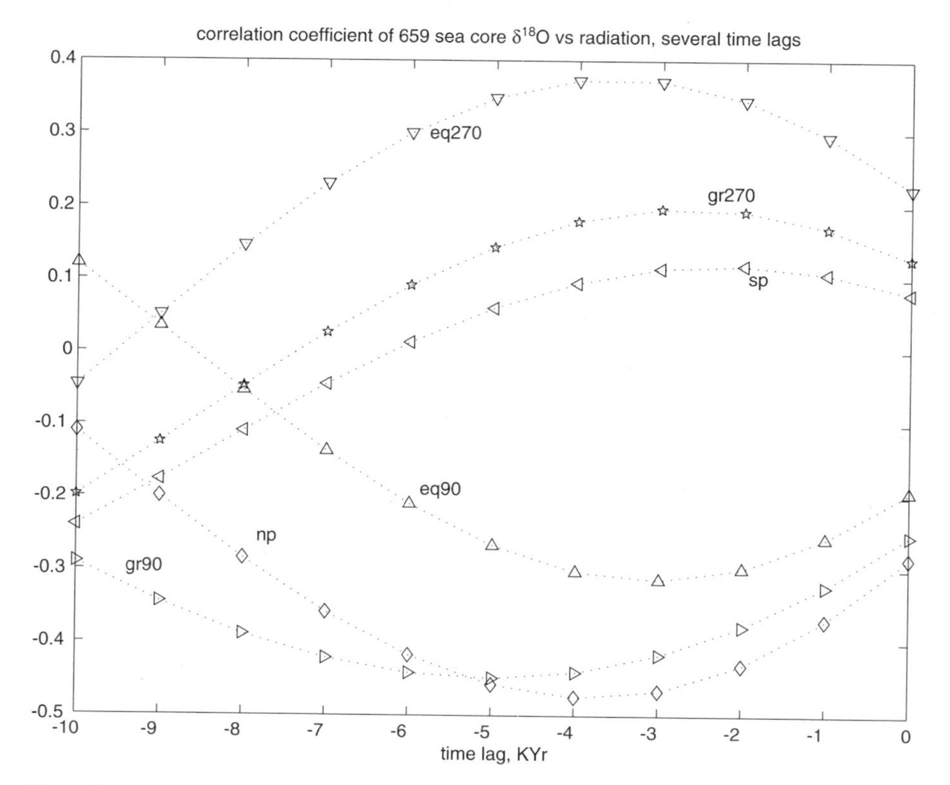

Figure 24.1
Variation of the correlation coefficient of the ocean δ^{18}O with radiation forcings and the time lag.

$x(i)$ and $y(i)$, such as the structure function (Monin and Yaglom, 1981) and mutual information (Kantz and Schreiber, 1999). The correlation coefficient yielded better results in our work than the other two, which nevertheless are theoretically better adapted to nonlineal series.

Hierarchy of Models

The hierarchy of models is an additional concept that we have applied. Simultaneous forcing at several latitudes is considered the primary input, and several delayed time series are sequentially added to input if, as a consequence, the predicted output reduces the MAD. In this way we can reach a reasonable error for some cases, but the process is very selective in rejecting nonexistent causal relations. Hierarchy of ANN models helps to establish very complex causal relations where several inputs could act synergistically.

As an example of the application of the hierarchy, we consider the case of δ^{18}O, CH_4 and CO_2. We intend to predict the experimental values of the cores with ANNs, and consider as input a group of six values representatives of the solar forcing. The minimum relative MAD is obtained for δ^{18}O, and therefore the bottom ocean temperature at Cape Verde is conditioned by global forcing.

If we calculate another variable, such as the CH_4 mixing ratio, we will find a better value if we use δ^{18}O and solar forcing (the six standard values) than if we use only forcing. Radiation is not excluded, in order to consider synergism related to the distribution of radiation over the globe. Finally, we can predict the CO_2 mixing ratio from the δ^{18}O, CH_4 mixing ratio, and solar forcing quite accurately, with better results than if we consider only solar forcing.

Common sense is also necessary to counterweight pure numbers. The CO_2 mixing ratio can be predicted with less MAD than seawater δ^{18}O. There are two reasons for this result. First, the series of the sea core cannot be of such high quality as the Vostok ice core. Interpolation or some filtering of the high frequencies by the sampling rate (Nyquist limit) can be noted. Second, solar radiation is heat, and cannot by itself fire the increase of the CO_2 mixing ratio. Solar radiations need to heat something or feed a heat flux (e.g., via an ocean circulation).

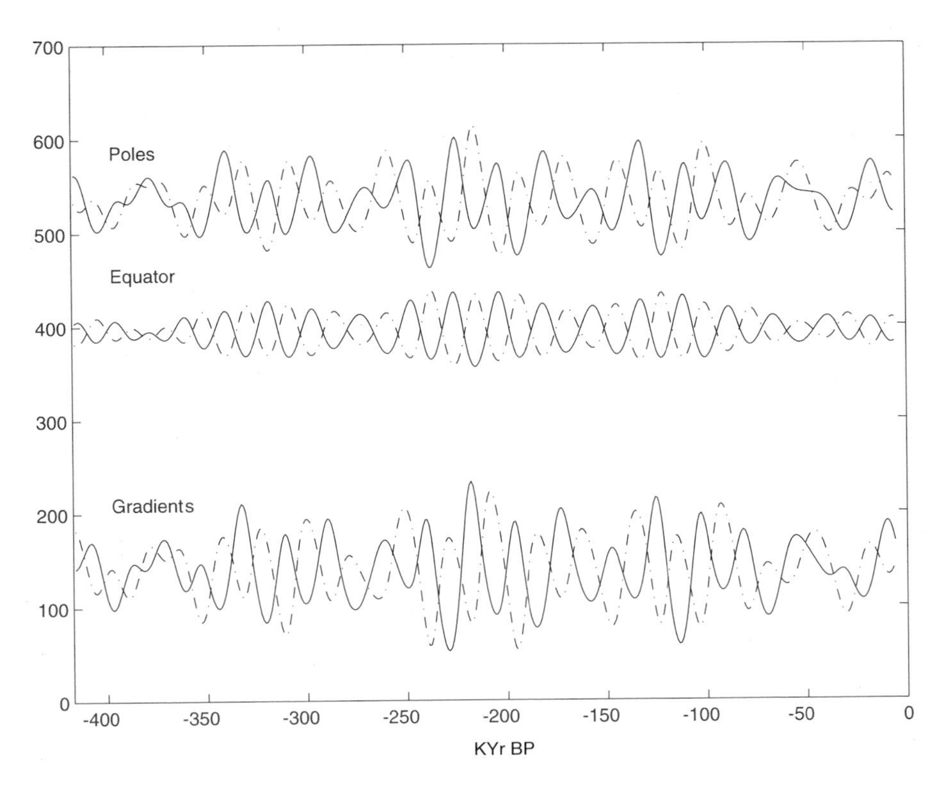

Figure 24.2
The variation of the daily irradiation for the period considered. The values are for summer in each hemisphere (June in the northern hemisphere [continuous line] and December in the southern hemisphere [dotted line]). The absence of night contributes to the high daily solar irradiation of the poles.

Results

We used data from the ODP site 659 (Cape Verde) sea core (about 18°N latitude, 21°W longitude; Tiedemannn et al., 1994) and from the Antarctica Vostok ice core (Petit et al., 1999). The forcing data are from Berger and Loutre (1991) and Laskar, Joutel, and Boudin (1993). There are no appreciable differences among the several calculations for the orbital parameters for the Quaternary, and Laskar, Joutel, and Boudin offer good latitudinal and time detail. The neural networks are used to establish causal relationships, if any, between solar forcing (and other suitable variables) and the Vostok and ODP site 659 time series.

We considered the influence of solar radiation all over the globe, but we needed to reduce the number of representative input values. It was necessary to consider the annual variability, and for the case we tried the summer values of each hemisphere. The latitude relative to poles determines orbital variability, and we used the extremes: 0 and 90 latitude. Therefore, the solar forcing inputs used were evaluated at 90° mean longitude of equinox (about June) for the northern hemisphere and 270° mean longitude of equinox (about December) for the southern hemisphere. The gradients that can feed circulations were evaluated through the differences between pole and equatorial insolation for each hemisphere. The representative inputs used are listed below.

The daily insolation at 90°N latitude and 90° mean longitude of equinox (abbreviated np [North Pole] in figure 24.1)

The daily insolation at 0° latitude and 90° mean longitude of equinox (abbreviated eq90 [equator at 90° mean longitude of equinox])

The daily solar insolation at 0° latitude and 270° mean longitude of equinox (abbreviated eq270)

The daily insolation at 90°S latitude and 270° mean longitude of equinox (abbreviated sp [South Pole])

The difference of daily insolation between 90°N and 0° latitude at 90° mean longitude of equinox (abbreviated gr90)

The difference of daily insolation between 90°S and 0° latitude at 270° mean longitude of equinox (abbreviated gr270).

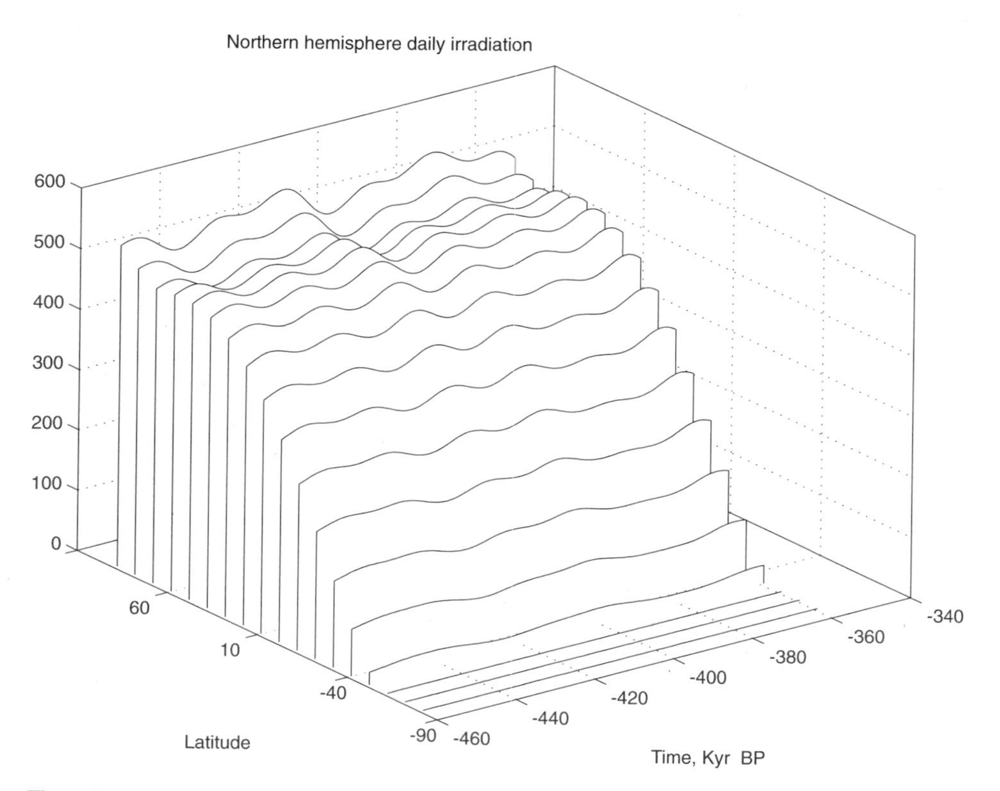

Figure 24.3
Detailed latitudinal variation of northern hemisphere insolation near isotope stage 11.3. The maxima are for 410 KYr BP for the North Pole and a similar (not shown) 418 KYr BP for the South Pole.

These values are represented in figure 24.2 and, with more latitudinal and time detail only for insolation, in figure 24.3 for stage 11.3 (solid lines for northern hemisphere, and dot-and-dash lines for the southern hemisphere). A counterintuitive issue is the high (summer) daily solar irradiation of the poles, due to the absence of night.

In table 24.1 these radiation inputs are abbreviated as "forcing." Representative results related to seawater $\delta^{18}O$ deviation, methane mixing ratio, carbon dioxide mixing ratio, and Antarctica ice deuterium deviation are presented here.

We found different time delays and cause–effect relations associated with these inputs, showing that pure Milankovitch theory based on daily solar insolation at 65°N latitude and 90° mean longitude of equinox is too simple an input to account for the Vostok and ODP site 659 data. Table 24.1 shows evidence of the complex relations involved.

The hierarchy of models was applied to successive stages. First, we intended to predict with minimum error one of the core variables, and we found that Cape Verde ocean temperature can be predicted with a 5 percent relative MAD. Sachs et al. (2001) report a tropical South Atlantic substantial warming before Termination I.

Table 24.1
The hierarchy of models

Input	Output	Relative mean absolute deviation
Forcing	ocean $\delta^{18}O$	5.4%
Forcing	CH_4	9.4%
Forcing + ocean $\delta^{18}O$	CH_4	5%
Forcing	CO_2	4.2%
Forcing + CH_4	CO_2	2.7%
Forcing + CH_4 + ocean $\delta^{18}O$	CO_2	2.3%
Forcing	δD	27%
Forcing + CH_4	δD	11.3%
Forcing + CH_4 + ocean $\delta^{18}O$ + CO_2	δD	8.6%

Except for the CO_2 mixing ratio, the other errors are much higher. The CO_2 experimental series can be more accurate than, for instance, the methane series. The methane error reduces to half the previous value if we include seawater temperature in the prediction.

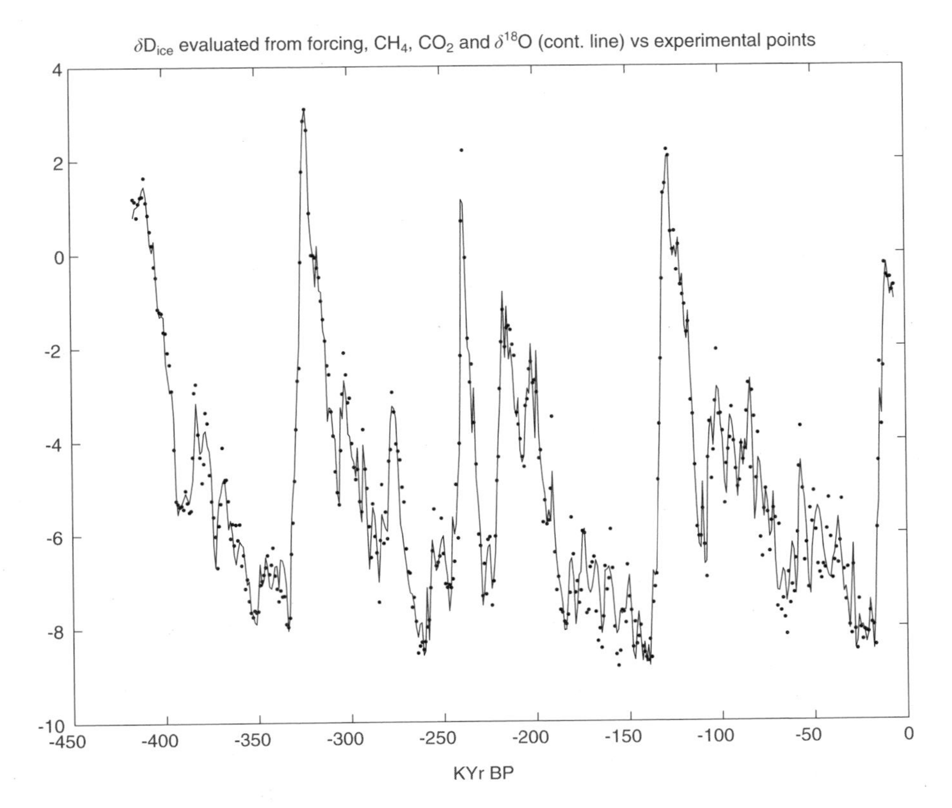

Figure 24.4
Anctartica deuterium isotope deviation δD calculated (continuous line) versus the Vostok data (dots). The abrupt changes are well reproduced, and there is more scatter on interglacials periods.

The carbon dioxide error reduces almost to half if we include the methane series and δ^{18}O.

Antarctica ice deuterium is a proxy of the Antarctica ice mass, and cannot be predicted accurately (27 percent relative MAD) using only the insolation distribution over the globe. But if we use CH_4 and CO_2 data, and tropical ocean temperature, the error reduces to a third. Noncausal relations are rejected by the test (e.g., "forcing" predicts atmospheric δ^{18}O with a 62 percent relative mean absolute deviation).

Therefore, the results suggest the following causal chain: forcing heats the tropical ocean with a delay of 3–4 KYr. The increase of the tropical ocean temperature precedes the methane release. Methane converts to CO_2 in about 11 KYr. All these factors together can explain the Vostok data for the proxy of ice mass in Antarctica.

Figure 24.4 shows predicted Antarctica ice deuterium deviation versus the Vostok data. As can be seen, most of the artifacts are found during interglacials, accounting for most of the error, whereas the abrupt changes are well reproduced. If we consider that the errors can accumulate along the chain, the results are very promising.

A part of the alternation of glacial and interglacial periods can be explained without resort to greenhouse gases: if solar forcing feeds northern and southern hemisphere circulations by heating the tropical ocean, the deglaciation is simultaneous in both hemispheres. If the ocean becomes warmer, the North Pole can melt: nuclear submarines have reported a sensible reduction of the North Pole ice thickness in a 20-year period. The melting can produce a very sudden and important change in the absorption of radiation from reflecting ice to nonreflecting water.

However, if the performed analysis is correct, greenhouse gases are needed to explain the evolution of the mass of ice over a continent, as in the case of Antarctica. The effect of greenhouse gases is not so noticeable if water vapor is present, as on a tropical ocean, because the water vapor is also a greenhouse gas. But over Antarctica the atmosphere is very dry, and high values of the carbon dioxide mixing ratio can be decisive to reduce the Anctartica ice.

We suggest that the Gaian scenario is, therefore, that clathrates accumulated during the long cold periods in tropical oceans can become unstable as the ocean water becomes warmer, and that the greenhouse gases thus liberated help the deglaciation. This contribution is decisive at locations that the oceans cannot directly influence, such as over dry continents. The Vostok record helps us to confirm this issue over Anctartica.

The spectrogram of the eccentricity shows peaks at 98 and 128 KYr, and the periodogram of the data also presents both frequencies (and some 41 KYr contribution), but the 98 KYr period is more noticeable, due to the shortness of the series for a 128 KYr period and to faults in the sampling that represent more important relative losses for the long periods. In any case, both frequencies are related to eccentricity (i.e., the prevailing orbital actuator on the equatorial zones).

Finally, the isotope stage 11 paradox (Blunier et al., 1998; Steig et al., 1998) perhaps can be explained only by solar forcing: figure 24.3 shows successive insolation maxima near the poles for about 12 KYr before stage 11 for the southern hemisphere and 4 KYr before stage 11 for the northern hemisphere. The latter coincides with the delay found for the ODP 659 seawater temperature proxy.

Conclusions

From the hierarchy of models, we conclude that ocean temperature and circulation play a primary role in triggering the other processes, such as release of methane (from seabed methane hydrates) that in a short period is converted to atmospheric carbon dioxide. Ocean temperature and ocean circulation are derived from astronomical inputs: latitudinal variable solar forcing, northern and southern hemisphere forcing (at mean longitude of equinox of 90° or 270°) and latitudinal gradients. The seawater temperature involved is the Cape Verde ocean temperature, a near-equatorial site.

The hierarchy chain—
forcing $> \delta^{18}O$ (ocean),
forcing $+ \delta^{18}O$ (ocean) $>$ methane,
forcing $+ \delta^{18}O$ (ocean) $+$ methane $> CO_2$,
forcing $+ \delta^{18}O$ (ocean) $+$ methane $+ CO_2 >$ deuterium deviation on Antarctica ice—
seems to confirm the influence of a growing methane mixing ratio in deglaciation. The magnitudes involved have different delayed responses, and their nonlinear contribution acts synergistically.

The hierarchy shows a direct influence of forcing and seawater proxy of temperature on methane, and of forcing, seawater proxy of temperature, and methane on carbon dioxide. CH_4 and CO_2 greenhouse gases can act as a thermal switch on Antarctica, where water vapor has a reduced contribution to downward longwave irradiance. Unstable methane hydrates can be a source of methane in the way of Kennett et al. (2000).

However, deglaciation is probably the result of the variable and nonlinear coupling of solar forcing, seawater temperature increase at the equator, and increased greenhouse gases acting in several ways. If seawater temperature is the source of North Pole ice melting, the North Pole albedo would therefore decrease, and the greenhouse effect would switch on through the increase in water vapor and higher surface temperatures.

Feedback is probably involved, but we have not tested possible feedback effects. Several variables seem to have mutual delayed influence between them, such as temperature and methane mixing ratio.

The release of methane (probably from methane hydrates) seems to play a fundamental role in temperature regulation of the globe. Concerning Gaia, clathrates of biological origin help to avoid a runaway cooling of the Earth. But the switch is the equatorial ocean.

References

Berger, A., and M. F. Loutre. 1991. Insolation values for the climate of the last 10000000 years. *Quaternary Sciences Review*, 10, 297–317.

Blunier, T., J. Chappellaz, J. Schwander, A. Dällenbach, B. Stauffer, T. F. Stocker, D. Raynaud, J. Jouzel, H. B. Clausen, C. U. Hammer, and S. J. Johnsen. 1998. Asynchrony of Antarctic and Greenland climate change during the last glacial period. *Nature*, 394, 739–743.

Budyko, M. I. 1969. Effect of solar radiation variations on climate of Earth. *Tellus*, 21, 611–619.

Crowley, T. J., W. T. Hyde, and W. R. Peltier. 2001. Earth CO_2 levels required for deglaciation of a "near-snowball" Earth. *Geophysical Research Letters*, 28, 283–286.

Ganopolsky, A., and S. Rahmstorf. 2001. Rapid changes of glacial climate simulated in a coupled climate model. *Nature*, 409, 153–158.

Henderson, G. M., and N. C. Slowey. 2000. Evidence from U-Th dating against northern hemisphere forcing of the penultimate deglaciation. *Nature*, 404, 61–66.

Hyde, W. T., T. J. Crowley, S. K. Baum, and W. R. Peltier. 2000. Neoproterozoic "snowball Earth" simulations with a coupled climate/ice-sheet model. *Nature*, 405, 425–429.

Jang, J. S. R., C. T. Sun, and E. Mizutaki. 1997. *Neuro-fuzzy and Soft Computing*. Englewood Cliffs, NJ: Prentice-Hall.

Kantz, H., and T. Schreiber. 1999. *Nonlinear Time Series Analysis*. Cambridge: Cambridge University Press.

Karner, D. B., and R. A. Muller. 2000. Paleoclimate—A causality problem for Milankovitch. *Science*, 288, 2143–2144.

Kennedy, M. J., N. Christie-Blick, and L. E. Sohl. 2001. Are Proterozoic cap carbonates and isotopic excursions a record of gas hydrate destabilization? *Geology*, 19, 443–446.

Kennett, J. P., K. G. Cannariato, I. L. Hendy, and R. J. Behl. 2000. Carbon isotopic evidence for methane hydrate instability during Quaternary interstadials. *Science*, 288, 128–133.

Laskar, J. 1999. The limits of Earth orbital calculations for geological time-scale use. *Philosophical Transactions of the Royal Society of London*, A357, 1735–1759.

Laskar, J., F. Joutel, and F. Boudin. 1993. Orbital, precessional, and insolation quantities for the Earth from −20 MYr to +10 MYr. *Astronomy and Astrophysics*, 270, 522–533.

Matlab Fuzzy Logic Toolbox. 1998. http://www.mathworks.com/, 2-70 to 2-98.

Monin, A. S., and A. M. Yaglom. 1981. *Statistical Fluid Mechanics*, vol. 2. Cambridge, MA: MIT Press.

Muller, R. A., and G. J. MacDonald. 1997. Glacial cycles and astronomical forcing. *Science*, 277, 215–218.

Paillard, D. 1998. The timing of Pleistocene glaciations from a simple multiple-state climate model. *Nature*, 391, 378–381.

Petit, J. R., J. Jouzel, D. Raynaud, N. I. Barkov, J. M. Barnola, I. Basile, M. Bender, J. Chappellaz, M. Davis, G. Delaygue, M. Delmotte, V. M. Kotlyakov, M. Legrand, V. Y. Lipenkov, C. Lorius, L. Pepin, C. Ritz, E. Saltzman, and M. Stievenard. 1999. Climate and atmospheric history of the past 420,000 years from the Vostok ice core, Antarctica. *Nature*, 399, 429–436.

Sachs, J. P., R. F. Anderson, and S. J. Lehman. 2001. Glacial surface temperatures of the southeast Atlantic Ocean. *Science*, 293, 2077–2079.

Steig, E. J., E. J. Brook, J. W. C. White, C. M. Sucher, M. L. Bender, S. J. Lehman, D. L. Morse, E. D. Waddington, and G. D. Clow. 1998. Synchronous climate changes in Antarctica and the North Atlantic. *Science*, 282, 92–95.

Tiedemann, R., M. Sarnthein, and N. J. Shakleton. 1994. Astronomic timescale for the Pliocene Atlantic delta-O-18 and dust flux records of Ocean Drilling Program Site-659. *Paleoceanography*, 9, 619–638.

25
Amazonian Biogeography as a Test for Gaia

Axel Kleidon

Abstract

The climatic conditions during the last ice age were considerably different from today. But by how much did the glacial changes affect the distribution of terrestrial vegetation types? Can these changes be used to test aspects of the Gaia hypothesis? Here I start by providing a quantifiable definition of "beneficial to life." I define it as the mean gross carbon uptake of the biota which measures biotic activity. This definition is then applied to a set of climate model simulations of the last ice age in which the extent to which vegetation is capable of recycling soil moisture is modified. I find that a larger capacity to recycle moisture leads to less continental aridity, suggesting higher carbon uptake and thus a more "beneficial" climate for life. This effect is particularly strong in the tropical parts of South America. At the same time that aridity is reduced, surfaces are cooler through enhanced evapotranspiration. This suggests that a definition of Gaia based on optimizing environmental conditions for biotic activity seems more appropriate than a definition based on homeostasis. Extensions of the definition to build a comprehensive framework for formulating and testing Gaian null hypotheses are discussed.

Ice Ages and Gaia

During the last ice age, the Earth's environment was drastically different from that of today. With large parts of the northern hemisphere covered by vast ice sheets, the northern vegetation zones were shifted considerably southward. Changes were not limited to the northern hemisphere; in fact, global oceans are believed to have been 2–5 K cooler than today, and atmospheric carbon dioxide concentrations to have been considerably lower (about 190 ppm), leading to global changes in climate and the biosphere (Gates, 1976; CLIMAP, 1976, 1981). In particular for the South American continent, it is generally believed that the Amazonian rain forest was widely replaced by savanna and grasslands due to enhanced continental aridity (Haffer, 1969; van der Hammen and Absy, 1994). This glacial aridity is derived from the assumption that cooler oceans transported less moisture to the continents, resulting in less precipitation over land and therefore leading to increased continental aridity.

Since water limitation is a primary determinant of shaping the geographical distribution of vegetation types in the tropics, it is concluded that the reduced moisture influx resulted in more arid vegetation types. This line of thought is substantiated by some limited reconstructions from pollen records taken at the margin of the present-day forest extent which indicate the presence of more drought-adapted vegetation (see, e.g., van der Hammen and Absy, 1994). Such a vast expansion of arid regions would certainly pose a challenge to the Gaia hypothesis, in that it would clearly suggest the lack of a negative, stabilizing feedback to counteract the effect of glacial conditions on the terrestrial biosphere.

Most of these studies have focused on a physical view of the climate system and do not explicitly deal with possible compensatory effects of vegetation, and it is therefore not adequate to use them for a test of Gaia. In particular, the buffering effect of rooting depth on water availability during dry seasons is only poorly considered. The rooting depth of the vegetative cover determines the depth to which water can be extracted from the soil and subsequently be transpired back into the atmosphere. During dry seasons, when water input to the soil by precipitation is limited, access to sufficiently large water storage in the soil is crucial to maintain transpiration. Deeply rooted vegetation of up to 68 meters depth is found in many ecosystems of the world (Stone and Kalicz, 1991; Canadell et al., 1996). In particular, it has been found that the Amazon rain forest is deeply rooted in various parts (Nepstad et al., 1994; Hodnett et al., 1996), and that the access to soil water below 2 meters depth substantially contribute to overall dry season transpiration.

The continuous transpiration during the dry season has direct effects on the local surface energy budget, leading to cooler surface temperatures, and far-reaching effects by enhancing continental moisture recycling. Climate model simulations which investigated the sensitivity of the climate system to soil water storage capacity in general (Milly and Dunne, 1994), and to deeply rooted vegetation in particular (Kleidon and Heimann, 2000), demonstrate that continuous transpiration during the dry season can lead to substantial enhancement of the continental hydrological cycle and the large-scale, atmospheric circulation. It has also been demonstrated that the enhanced summer dryness predictions for the temperate latitudes in an enhanced greenhouse climate are the result of small soil water storage capacities, and can be compensated for if sufficiently deep roots are considered (Milly, 1997).

Since changes in the water cycle are a central component in the argument for enhanced continental aridity during glacial periods, it would seem that rooting depth, the buffering capacity for the water availability on land, plays a particularly important role in counteracting glacial effects. The presence of deeply rooted vegetation would seem to provide a negative feedback mechanism which would potentially be strong enough to counteract the glacial changes. Because this deep root feedback would act to stabilize the biogeography during glacial times, it could be viewed as a Gaian mechanism. More important, it would act to substantially enhance the productivity of the vegetation by reducing the water limitation. This points out that the stabilizing aspect may not be as important an aspect of Gaia as the productivity enhancing tendency of life's feedbacks to Earth system processes (as proposed by Volk, 1998).

I will therefore start by providing a precise definition of "beneficial to life," based on recent work (Kleidon, 2002). I will then briefly describe the ice age model simulations of Kleidon and Lorenz (2001) and set them in context with the proposed definition of "beneficial to life." I will close with a discussion of how the proposed testing framework can be extended, and what the implications for a precise definition of the Gaia hypotheses will be.

Defining "Gaian" Effects

Before discussing whether a certain effect of vegetation is a "Gaian" phenomenon, it is important to define clearly what we mean by that. Obviously, a phenomenon can be seen as "Gaian" if it is beneficial to life. Life itself is carbon-based on Earth, so it makes sense to base a definition of what is beneficial to life on carbon fluxes. An appropriate flux to measure benefit would therefore be the gross carbon uptake by the biota, or gross primary productivity (GPP). This flux represents the amount of absorbed solar energy converted into organic carbon compounds which feed into the biota, and is therefore a measure of overall biotic activity. When integrated over the whole globe, this flux represents the total activity of the Earth's biota, which is appropriate for testing hypotheses regarding the overall effect of life on Earth system functioning. In the following I will refer to the globally integrated flux when referring to GPP. Note, however, that other variables, such as biomass or diversity, may also be valid measures (Schneider, 1986), but are not considered here.

With GPP as a measure of biotic activity, we can quantify whether a specific biotic effect is beneficial to life by comparing the values of GPP with and without this effect. Formally, I define a *Gaian effect* as one which leads to an overall increase in the GPP in the presence of the effect compared to the corresponding GPP value without the effect. The application of this definition seems most appropriate with the use of simulation models in which a particular effect can be taken out. In the next section I will apply this definition to the effect of deep-rooted vegetation during the ice age. Note that this definition extends the concept of "cycling ratios" of Volk (1998). Volk developed the concept of "cycling ratios" in order to show that the biota generally enhance the cycling of nutrients within the accessible part, thus leading to fewer nutrient constraints and a higher productivity. Using GPP extends Volk's approach by integrating different cycling ratios into a single variable, representing overall biotic activity. In the following, however, we will exclude nutrients from the analysis and focus on water limitation.

Deep Roots During the Ice Age: A "Gaian" Effect?

I investigated the effect of deep-rooted vegetation on the glacial climate by using an ice age version of the ECHAM 3 atmospheric general circulation model (Lorenz et al., 1996; Kleidon and Lorenz, 2001). The main changes incorporated into the model in order to simulate glacial conditions were the changes in orbital forcing, the reduction in atmospheric carbon dioxide concentration, the incorporation of massive ice sheets in the northern hemisphere, lowering the sea surface temperatures, and alterations of the land-sea mask to account for changes in land area due to the lower sea level. An additional model simulation was performed in order to determine the sensitivity of the glacial climate to rooting depth, which is used here to evaluate whether the effect of deep-rooted vegetation can be seen as a "Gaian" effect. In this simulation, the effect of deep-rooted vegetation was incorporated by increasing the soil water storage capacities to values which are sufficiently large to minimize seasonal water deficits. The details of the model setup can be found in Kleidon and Lorenz (2001). The effect of deep roots is then isolated by investigating the differences between the two simulated climates. These differences will also be compared with the ones between the glacial control climate to the present-day control simulation.

As mentioned in the introduction, deep roots allow plants to utilize soil water at greater depths for evapotranspiration during the dry season. This enhanced water flux from land regions leads to a more active continental water cycle, as reflected by the increase in annual mean precipitation shown in figure 25.1a. The enhancement is particularly pronounced on South America. The enhanced evapotranspiration during the dry season directly affects the surface energy balance through its coupling with the latent heat flux. Consequently, a substantial cooling can be found, primarily in the tropical regions (figure 25.2a). To put the magnitudes of change into proportion, the simulated changes in precipitation and temperature caused by the glacial boundary conditions compared to today are shown in figures 25.1b and 25.2b. In the tropical regions, the enhancement of the continental water cycle due to deep roots is of a magnitude similar to the reduction due to the glacial conditions. The magnitudes of temperature differences are also comparable, but of the same sign.

These changes do not tell us the potential implications for the productivity of the vegetative cover. Since most of the changes in climate occur in tropical regions, and water is the main factor affecting productivity in tropical regions, I used the simulated frequency of water stress as an index for productivity. The resulting annual means are shown in figure 25.3. Water stress is substantially reduced for large regions in the tropics in the deep-root simulation, suggesting that the effects of deep roots counteract the impact of glacial conditions on land aridity. There are some indications that the climate with deep roots is more consistent with reconstructions (for details, see Kleidon and Lorenz, 2001).

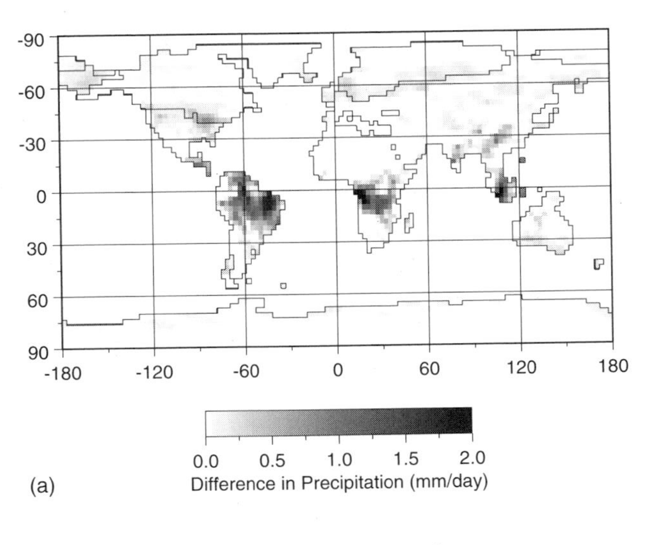

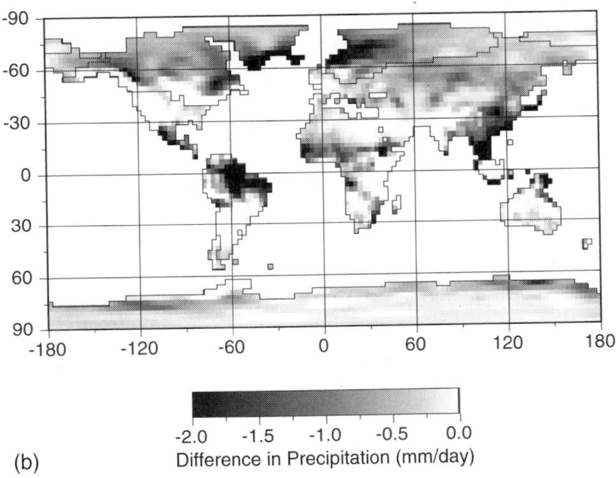

Figure 25.1
Simulated differences in annual mean precipitation during the last glacial maximum (a) due to the effect of deep rooted vegetation and (b) due to the imposed the glacial boundary conditions. Only the dominant direction of the difference is shown. The contour line denotes land regions as they are represented in the model resolution. The additional contour line in the high northern latitudes denotes glaciated regions.

Note, however, that the plant physiological effects of lower carbon dioxide concentrations during the glacial period on photosynthesis and water use efficiency are not considered here. Some modeling studies have suggested that the lower overall productivity may have reduced plant growth to such an extent that it affected the distribution of vegetation types (Jolly and Haxeltine, 1997). While the direct sensitivity of photosynthesis to atmospheric carbon dioxide concentrations in a laboratory environment (when other factors are not limiting) is well docu-

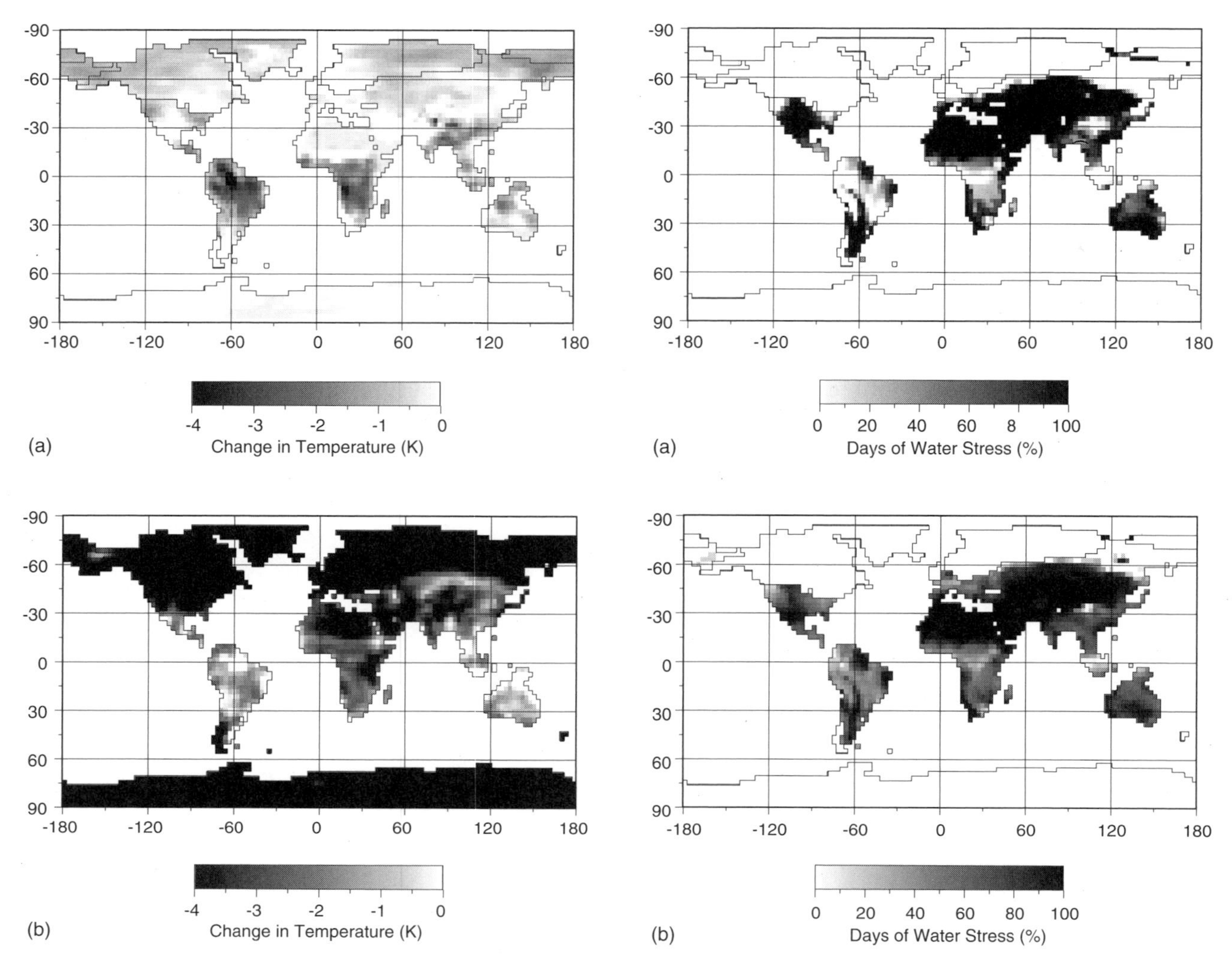

Figure 25.2
Simulated differences in annual mean temperature during the last glacial maximum (a) due to the effect of deep-rooted vegetation and (b) due to the imposed the glacial boundary conditions. Only the dominant direction of the difference is shown.

Figure 25.3
Simulated geographical distribution of water stress frequency for vegetation as an indicator of tropical productivity (a) with and (b) without the effects of deep roots.

mented, the net effect of glacial conditions, including substantially lower carbon dioxide concentrations, on photosynthesis in a natural environment is less certain. This aspect is therefore a major uncertainty of the results presented here.

The effect of deep roots essentially leads to a buffering effect which is summarized schematically in figure 25.4. At the extreme example of no continental evapotranspiration, precipitation P is determined solely by the influx IN of moisture from the ocean, which is balanced by the continental discharge of river basins R. With increasing values of ET, total precipitation is increasingly determined by ET, and the continental water cycle is increasingly decoupled from the oceanic source of moisture. This directly affects Budyko's (1974) radiative index of dryness (aridity index A), as shown in the figure, leading to more humid conditions with more continental evapotranspiration. Note that the continental discharge R remains unaffected by the intensity of continental moisture recycling and is affected only by the influx of oceanic moisture.

What this means is that a reduced moisture transport from the oceans as a consequence of glacial conditions directly reduces continental discharge R (which is consistent with many reconstructions; Farrera et al., 1999), but that we cannot infer the effect on precipitation P or aridity A as long as we do not know by how much evapotranspiration ET was

(a)
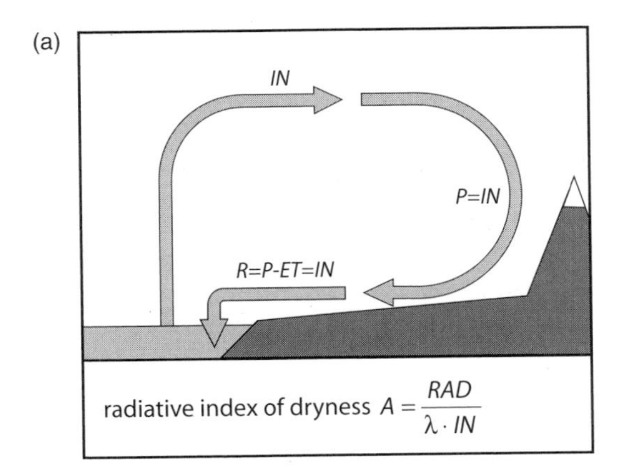

(b)
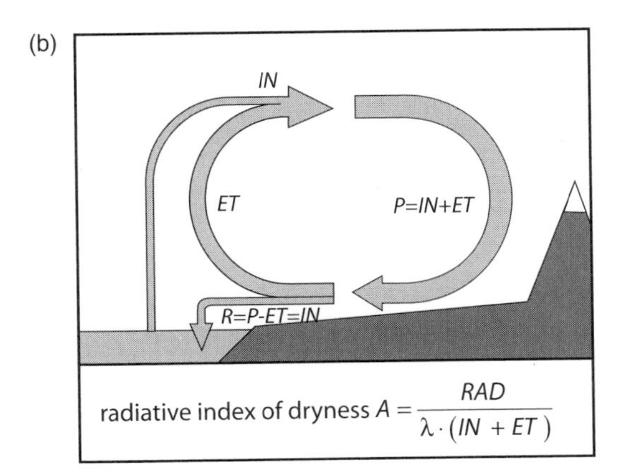

Figure 25.4
Schematic diagram of the continental branch of the hydrological cycle (a) without and (b) with evapotranspiration, and its implication for overall climatic aridity.

changed. In addition, atmospheric circulation feedbacks reduce the overall moisture convergence over land *IN* in the presence of deep roots—that is, the increase in evapotranspiration is not completely balanced by the increase in precipitation. Ultimately, more direct reconstructions of vegetation distributions are needed to clarify the extent of Amazonian rain forest. This, then, will also provide a direct means to test the importance of stabilizing biotic feedbacks, and therefore of Gaia.

Extending the Approach: Testing Gaian Null Hypotheses

The testing of one particular effect as a Gaian effect is clearly not sufficient to test the overall validity of the Gaia hypothesis. However, a clear and quantifiable definition of "beneficial to life" as it is done here points the way toward a more rigorous and quantitative formulation and testing of the Gaia hypothesis (Kleidon, 2002). In a complete assessment, we would need to test whether the sum of all biotic effects, given a particular fixed external forcing, would lead to a general enhancement of the globally integrated gross carbon uptake. Numerical model simulations seem well suited for such an assessment, because biotic effects can specifically be excluded from the simulation.

Note that the concept of regulation is no longer a part of this line of definition. Such a shift in focus from "regulation" toward "maximization" has already been suggested—for instance, by Volk (1998). This shift makes sense because vegetation can adapt to environmental change, and it has been demonstrated in the context of Daisyworld (Watson and Lovelock, 1983) that adaptation can counteract homeostatic behavior (Robertson and Robinson, 1998). Also, the suggested measure of carbon uptake for life's benefit makes more sense in that it is functionally related to life's "well-being." So why should, for instance, temperature, or any other variable, be regulated toward a target? The results from the climate model simulations here specifically point out that one effect, that of deep roots on climate, can counteract the reduction in water availability under ice age conditions while amplifying the cooling. In that sense, a "homeostatic" definition appears arbitrary and would seem to be easily falsifiable. A definition of Gaia which states that biotic effects enhance overall carbon uptake would seem to be more adequate.

Acknowledgments

This work was conducted during the author's stay at Stanford University, funded in part by a Feodor Lynen Fellowship of the Alexander-von-Humboldt Foundation. The author greatly appreciates the hospitality of Harold Mooney during the stay. The author acknowledges financial support from the American Geophysical Union (AGU) for attending the Second AGU Chapman Conference on the Gaia Hypothesis. The constructive comments of two anonymous reviewers are greatly appreciated.

References

Budyko, M. I. 1974. *Climate and Life*. New York: Academic Press.

Canadell, J., R. B. Jackson, J. R. Ehleringer, H. A. Mooney, O. E. Sala, and E.-D. Schulze. 1996. Maximum rooting depth of vegetation types at the global scale. *Oecologia*, 108, 583–595.

CLIMAP Project Members. 1976. The surface of the ice-age Earth. *Science*, 191, 1131–1137.

CLIMAP Project Members. 1981. GSA Map and Chart Series, MC-36. Boulder, CO: Geological Society of America.

Farrera, I., S. P. Harrison, I. C. Prentice, G. Ramstein, J. Guiot, P. J. Bartlein, R. Bonnefille, M. Bush, W. Cramer, U. von Grafenstein, K. Holmgren, H. Hooghiemstra, G. Hope, D. Jolly, S. E. Lauritzen, Y. Ono, S. Pinot, M. Stute, and G. Yu. 1999. Tropical palaeoclimates at the last glacial maximum: A new synthesis of terrestrial data. *Climate Dynamics*, 15, 823–856.

Gates, W. L. 1976. Modeling the ice-age climate. *Science*, 191, 1138–1144.

Haffer, J. 1969. Speciation in Amazonian forest birds. *Science*, 165, 131–137.

Hodnett, M. G., L. Pimentel da Silva, H. R. da Rocha, and R. Cruz Senna. 1996. Seasonal soil water storage changes beneath central Amazonian rain forest and pasture. *Journal of Hydrology*, 170, 233–254.

Jolly, D., and A. Haxeltine. 1997. Effect of low glacial atmospheric CO_2 on tropical African montane vegetation. *Science*, 276, 786–788.

Kleidon, A. 2002. Testing the effect of life on Earth's functioning: How Gaian is the Earth system? *Climatic Change*, 52, 383–389.

Kleidon, A., and M. Heimann. 2000. Assessing the role of deep rooted vegetation in the climate system with model simulations: Mechanism, comparison to observations and implications for Amazonian deforestation. *Climate Dynamics*, 16, 183–199.

Kleidon, A., and S. Lorenz. 2001. Deep roots sustain Amazonian rainforest in climate model simulations of the last ice age. *Geophysical Research Letters*, 28, 2425–2428.

Lorenz, S., B. Grieger, P. Helbig, and K. Herterich. 1996. Investigating the sensitivity of the Atmospheric General Circulation Model ECHAM 3 to paleoclimatic boundary conditions. *Geologische Rundschau*, 85, 513–524.

Lovelock, J. E., and L. Margulis. 1974. Atmospheric homeostasis by and for the biosphere: The Gaia hypothesis. *Tellus*, 26, 2–10.

Milly, P. C. D. 1997. Sensitivity of greenhouse summer dryness to changes in plant rooting characteristics. *Geophysical Research Letters*, 24, 269–273.

Milly, P. C. D., and K. A. Dunne. 1994. Sensitivity of the global water cycle to the water-holding capacity of land. *Journal of Climate*, 7, 506–526.

Nepstad, D. C., C. R. de Carvalho, E. A. Davidson, P. H. Jipp, P. A. Lefebvre, H. G. Negreiros, E. D. da Silva, T. A. Stone, S. E. Trumbore, and S. Vieira. 1994. The role of deep roots in the hydrological and carbon cycles of Amazonian forests and pastures. *Nature*, 372, 666–669.

Robertson, D., and J. Robinson. 1998. Darwinian Daisyworld. *Journal of Theoretical Biology*, 195, 129–134.

Schneider, S. H. 1986. A goddess of the Earth: The debate on the Gaia hypothesis. *Climatic Change*, 8, 1–4.

Stone, E. L., and P. J. Kalicz. 1991. On the maximum extent of tree roots. *Forest Ecology and Management*, 46, 59–102.

Van der Hammen, T. M., and L. Absy. 1994. Amazonia during the last glacial. *Palaeogeography, Palaeoclimatology, Palaeoecology*, 109, 247–261.

Volk, T. 1998. *Gaia's Body: Toward a Physiology of Earth*. New York: Copernicus and Springer-Verlag.

Watson, A. J., and J. E. Lovelock. 1983. Biological homeostasis of the global environment: The parable of Daisyworld. *Tellus*, 35B, 284–289.

26
Modeling Feedbacks Between Water and Vegetation in the North African Climate System

James R. Miller and Gary L. Russell

Abstract

Not only is water essential for life on earth, but life itself affects the global hydrologic cycle and, consequently, the climate of the planet. We use a global climate model to examine how the presence of vegetation can affect the hydrologic cycle in a particular region. A control for the present climate is compared with a model experiment in which the Sahara Desert is replaced by vegetation in the form of trees and shrubs common to the Sahel region. A second model experiment is designed to identify the separate roles of two different effects of vegetation: the modified albedo and the presence of roots that can extract moisture from deeper soil layers. The results show that the presence of vegetation leads to increases in precipitation and soil moisture in the western Sahara. In the eastern Sahara, the changes are less clear. The increase in soil moisture is greater when the desert albedo is replaced by the vegetation albedo than when both the vegetation albedo and roots are added. The effect of roots is to withdraw water from deeper layers during the dry season. From a Gaian perspective, one implication of this study is that the insertion of vegetation into North Africa modifies the hydrologic cycle by increasing the precipitation and soil moisture so that environmental conditions favorable for the vegetation are enhanced.

Introduction

The original form of the Gaia hypothesis states that the Earth's biota are part of a global self-regulating system in which environmental conditions remain favorable for the biota (Lovelock and Margulis, 1974). There is no question that the presence of biota has an impact on the climate system. Some of these impacts are chemical and arise through the cycling of elements by the biota ultimately leading to a biotic contribution to the composition of the atmosphere. There are also physical impacts. A vegetated land surface generally has a darker surface (lower reflectivity or albedo) than bare ground, and thus affects radiation and surface heating. Plants affect the hydrologic cycle by extending their roots into soil layers beneath the surface where water can be removed by transpiration. Plant structures affect the dynamics of the planetary boundary layer by altering the surface roughness. Whether the global cumulative impact of these vegetative effects leads toward regulation of the climate system or toward optimization of the climate system for the biota—and, furthermore, how we would know if it did—continues to be debated (Lenton, chapter 1 in this volume; Kirchner, 2002).

The relationship between water and vegetation within the context of the climate system is complex. Many modeling studies have addressed various aspects of this issue. At the global scale, Shukla and Mintz (1982) examined how the climate would be different, depending on whether the surface is saturated or dry. They found that there was a positive feedback when surface water was present because it leads to more precipitation and maintains the high water content in the soil. Kleidon et al. (2000) did a similar study, but instead of using wet and dry surfaces, imposed fully vegetated or fully desert conditions. They found that the hydrologic cycle intensified substantially for the fully vegetated case. Claussen (1998) combined a climate model and a biome model to investigate the stability of the atmosphere-vegetation system under present-day conditions of solar irradiation and sea surface temperature. He obtained two different equilibrium solutions when the model was initialized with different land surface conditions, and found that North Africa was the most sensitive region to the global-scale changes.

There have been many modeling studies of the regional relationships between vegetation and the hydrologic cycle. Among the first was that of Charney (1975), who modeled the impacts of human activity on desertification in the Sahel region of Africa. He suggested that increases in surface albedo would lead to increased atmospheric stability and less rainfall. Bonan et al. (1992) used simulations from a global climate model to show that high latitude boreal forests lead to warmer surface air temperatures in both summer and winter relative to simulations in which the forest is replaced with bare ground or tundra vegetation. Many of these earlier studies were restricted to examining only the impacts of changes in surface albedo.

Kleidon and Heimann (1998) used a global climate model to examine the role of rooting depths in net primary productivity. They found that greater rooting depths led to a substantial increase in net primary productivity, particularly in tropical regions where transpiration was enhanced during the dry season. Kleidon (chapter 25 in this volume) uses a global climate model to show that deeper rooting depths in South America during the last glacial maximum could have increased precipitation by about the same amount that other physical changes reduced it and also could have enhanced the cooling there. Sud et al. (2001) used a single column model to show that an increase in solar absorption and surface evaporation helps to increase local rainfall. A recent observational study by Bonan (2001) shows that the conversion of forest to grassland in the central United States has increased the albedo there and has led to cooling.

North Africa is a generally arid region, and there have been a number of studies of the relationships between vegetation and climate in the Sahel region (e.g., Charney, 1975) and, to a lesser extent, in the Sahara Desert. Zheng and Eltahir (1998), using a zonally symmetric atmospheric model, showed that the location of vegetation perturbations has a significant impact on the West African monsoon response. They found that changes in vegetation cover at the border between the Sahara Desert and West Africa had a smaller impact on monsoon circulation than does coastal deforestation in West Africa. Sud and Molod (1988) did model experiments in which albedo and soil moisture were modified in the Sahara Desert. They found that by increasing soil moisture or lowering the surface albedo, the local moist convection and rainfall would increase. Knorr et al. (2001) used a global model and obtained the same result in North Africa when they modified the albedo there. Bonfils et al. (2001) modified Saharan albedo to try to better represent climate conditions 6000 years ago, when North Africa was wetter and more vegetated. These studies support the conclusion that decreased albedo in the Sahara Desert would enhance convection and increase rainfall.

The purpose of this study is to examine the role of vegetation on the local water budget of the Sahara Desert, to determine whether the replacement of desert by vegetation changes the local water budget, and to determine the separate roles of albedo and rooting depth in these changes. Kleidon (2002) puts forth four null hypotheses ranging from "anti-Gaia," in which biotic effects worsen environmental conditions, to "optimizing Gaia," in which biotic effects enhance environmental conditions so that they are optimal for life. He introduces global gross primary productivity (GPP) as a measure to determine whether environmental conditions are favorable to life. Conditions that lead to higher GPP are assumed to be more favorable for life. We will discuss our results in the context of these null hypotheses.

The Climate Model

The model used in this study is the atmospheric component of the global coupled model described by Russell et al. (1995). This model includes nonlinear dynamics, advection, a full radiation scheme, parameterizations of moist convection and surface interaction, and treatments of subsurface reservoirs except for the ocean. The ocean surface temperature and sea-ice cover are specified from monthly climatology. The resolution of the model is four degrees in latitude, five degrees in longitude, and nine vertical layers. Heat and humidity have finer resolution because both means and prognostic directional gradients are defined and used during advection within each grid cell. Moist convection and condensation are performed on 2×2.5 degree horizontal resolution.

The ground hydrology scheme (Abramopoulos et al., 1988) handles snow, bare soil and vegetation, a canopy layer, six ground layers (depths are 0.1, 0.173, 0.298, 0.514, 0.886, and 1.529 m), and surface and underground runoff. The model's soil moisture is defined as the total water content of these six layers. In a grid cell, the bare soil and vegetated fractions are treated separately with separate variables. Over the bare soil fraction or dormant vegetation, water that infiltrates below the surface layer has difficulty returning to the surface, but continues to penetrate downward through the six layers until it becomes underground flow and is added to the grid cell's rivers and lakes. Over nondormant vegetation, roots can remove water from the lower soil layers and transpire directly from the canopy.

Formulation of Model Experiments

The control is a model simulation for the present climate. It includes ground and vegetation character-

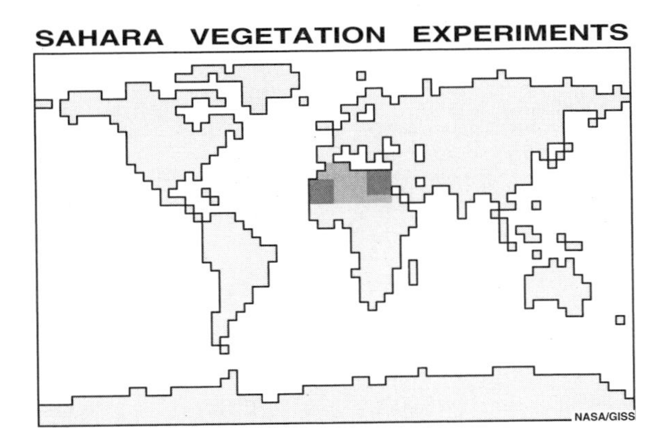

Figure 26.1
Shaded area represents region where desert is replaced with vegetation. Darker shaded areas are subregions of the study area referred to as western Sahara and eastern Sahara.

istics of the present-day Sahara Desert (85 percent desert, 15 percent shrubs and small trees). Where the land surface is categorized as desert, water can be removed from the top layer only by evaporation. Where it is categorized as shrubs and small trees, roots can extract water for transpiration from the top five layers. In most of the Sahara Desert, the model's bottom layer is bedrock so almost no water is extracted from that layer.

The first experiment replaces the surface characteristics in the region shown in figure 26.1 (Sahara Desert: 15W–35E, 16N–36N) with the same vegetation found in the Sahel region to the south of the Sahara. Because there are asymmetries in the model response to these changes between eastern Sahara (20E–35E, 20N–32N) and western Sahara (15W–0, 16N–28N), these two regions are examined in greater detail in the analysis. The replacement of desert by vegetation changes the seasonally varying surface albedo and adds roots that allow water to be extracted from the top five soil layers. Vegetation also affects the model's surface roughness, although in this study, it is not modified so that we can focus on the roles of roots and albedo.

Table 26.1 shows that the average albedo for the control for the entire region is about 32 percent, with little change between summer and winter. The surface albedo is specified from Hansen et al. (1983) and depends on vegetation type and season. For desert regions, the albedo also depends on whether the surface is wet or dry. Table 26.1 shows that the surface albedo is reduced to 15–16 percent when the desert albedo is replaced by vegetation having the same characteristics as vegetation native to the surrounding region. A second experiment replaces only the albedo in the control with that used in the vegetation experiment just described. This experiment allows us to sort out the changes to the hydrologic cycle that are caused by roots from those that are caused by albedo changes. The two experiments were run for 22 years, and are referred to as the vegetation and albedo experiments.

Impact of Vegetation on the Hydrologic Cycle in the Sahara Desert

In this section, the geographic patterns of the hydrologic impacts caused by replacing desert with vegetation are examined, both within and outside the study region. Figure 26.2 compares the control's seasonal variation of precipitation for western Sahara and eastern Sahara with the observed climatology of Legates and Willmott (1990). The model's annual cycle for western Sahara is similar to the observed, with little rainfall except in summer. The model's maximum precipitation occurs one month too late and is about twice the observed value. In eastern Sahara there is little rainfall in any season for both the model and observations.

In the remainder of this section we use the second decade of the model experiments to examine the spatial changes that occur in three components of the hydrologic cycle (precipitation, evaporation, and soil moisture) when desert is replaced by vegetation in the Sahara. Figure 26.2 and table 26.1 show that precipitation increases significantly in summer, with the largest increase (from 2 mm/day in September to 5.35 mm/day in August) occurring in western Sahara. In eastern Sahara, there is little change in precipitation except for the large increases between August and October. The spatial distribution of changes in precipitation is shown in figure 26.3. There is little change anywhere in the region in January, and there are increases in most of the region in July. There is an east–west gradient in the change with a generally increasing change moving from east to west.

Table 26.1 shows that there are significant changes in evaporation, too, with the largest changes again occurring in summer. The evaporation increases in February throughout much of the region and increases significantly in western Sahara in summer. In February, the increases in evaporation are larger than the increases in precipitation. This is possible because more water is stored from prior months and is available to be evaporated. The east-west gradient in the increased evaporation is consistent with that for precipitation.

Table 26.1
Model Variables Averaged over the Sahara Desert for Years 11 to 20

	Control		Vegetation		Albedo	
Sahara Desert	Feb	Aug	Feb	Aug	Feb	Aug
Cloud Cover (%)	13.2	13.9	9.8	43.0	11.2	48.3
Inc. Solar Rad. Surf. (W/m^2)	237.4	310.8	233.8	264.3	231.6	258.2
Abs. Solar Rad. Surf. (W/m^2)	160.8	209.2	198.9	221.1	197.1	216.0
Surface Albedo (%)	32.0	32.6	14.9	16.3	14.9	16.4
Surf. Air Temp. (°C)	20.2	32.6	22.4	29.7	22.5	28.8
Soil Moisture (cm)	49.4	47.6	56.9	58.9	78.8	81.5
Precipitation (mm/day)	0.16	0.41	0.10	3.00	0.11	3.64
Evaporation (mm/day)	0.20	0.39	0.43	1.61	0.30	2.46
Western Sahara						
Cloud Cover (%)	9.1	25.4	3.2	67.1	7.9	76.3
Inc. Solar Rad. Surf. (W/m^2)	250.1	305.4	247.8	246.1	242.5	237.6
Abs. Solar Rad. Surf. (W/m^2)	168.8	205.5	211.0	205.7	206.5	198.6
Surface Albedo (%)	32.5	32.7	14.8	16.4	14.8	16.4
Surf. Air Temp. (°C)	24.5	33.4	26.1	27.8	27.6	26.3
Soil Moisture (cm)	46.0	44.9	56.2	73.4	76.8	91.8
Precipitation (mm/day)	0.01	0.85	0.01	5.35	0.01	6.43
Evaporation (mm/day)	0.07	0.75	0.58	2.57	0.24	4.07
Eastern Sahara						
Cloud Cover (%)	15.3	2.5	12.9	13.3	12.2	14.1
Inc. Solar Rad. Surf. (W/m^2)	231.3	318.9	225.4	294.5	225.9	293.2
Abs. Solar Rad. Surf. (W/m^2)	153.5	210.7	192.0	247.2	192.4	246.0
Surface Albedo (%)	33.3	33.9	14.9	16.1	14.9	16.1
Surf. Air Temp. (°C)	18.1	32.6	21.1	34.1	20.2	34.1
Soil Moisture (cm)	40.9	40.2	42.5	39.2	70.1	67.4
Precipitation (mm/day)	0.19	0.01	0.09	0.45	0.10	0.54
Evaporation (mm/day)	0.22	0.05	0.20	0.36	0.22	0.44

For the vegetation experiment, the average soil moisture for the entire study region is higher in February and August than it is in the control. Table 26.1 shows that there are significant differences between eastern and western Sahara. The largest increases occur in western Sahara and are much larger in summer than in winter. In fact, the largest increases in the west occur just south of the study region. There is almost no change in soil moisture in eastern Sahara in summer or winter. The average increases in soil moisture for the entire region are dominated by the increases in western Sahara.

Role of Albedo and Roots in Vegetation Feedbacks on the Hydrologic Cycle

Changes in the hydrologic cycle induced by insertion of vegetation into the Sahara region are generally caused by the combined effects of changes in albedo, surface roughness, or extraction of water from deeper soil layers by roots. To sort out the effects of one of these, only the surface albedo is changed from the control. The albedo is specified according to the seasonally varying albedo of the vegetation in the Sahel region to the south of the Sahara Desert (table 26.1). In the vegetation experiment, both the albedo changes and the effects of roots are included. Both experiments use the same value for surface roughness as the control.

Figure 26.4 shows the second decade of the monthly precipitation for the control, vegetation, and albedo experiments. The February and August averages for the second decade of the experiments are shown in table 26.1. There is a significant change in the annual cycle of precipitation, with the greatest change occurring in summer, when the maximum precipitation occurs a month sooner and is much larger than for the control in figure 26.2. The sum-

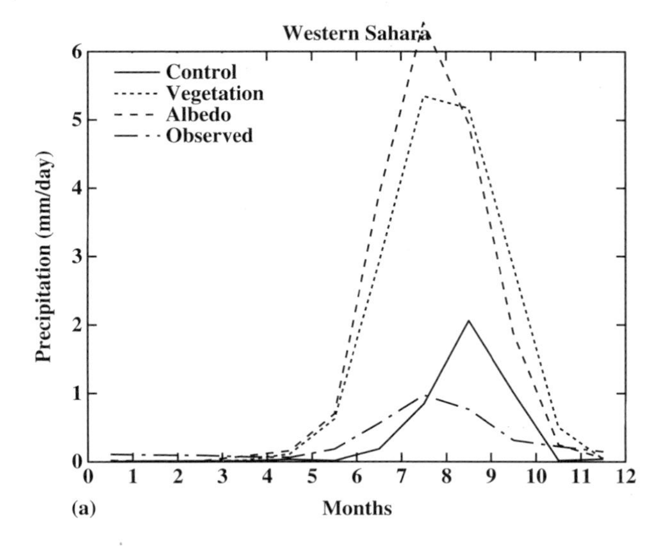

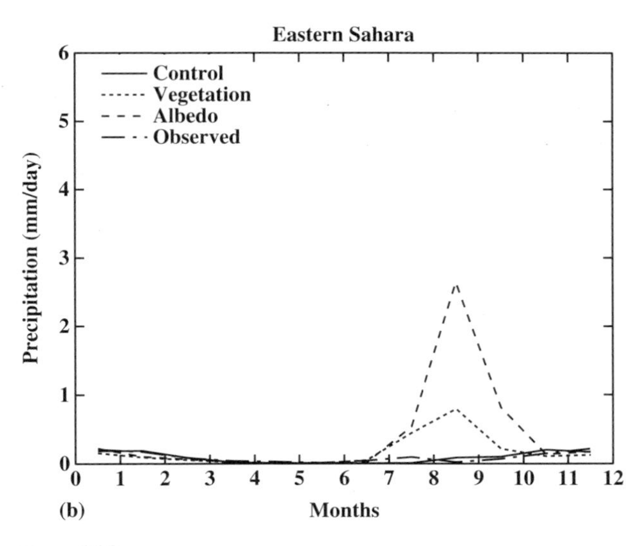

Figure 26.2
Annual cycle of precipitation for the control, vegetation experiment, albedo experiment, and observations for (a) western Sahara and (b) eastern Sahara. Observations are based on Legates and Willmott (1990), and model is based on second decade of simulations.

mer maximum in precipitation is usually larger in the albedo experiment than it is in the vegetation experiment.

In western Sahara (figure 26.4a), there is considerable interannual variability in the maximum summer precipitation in the control. Figure 26.2 showed that the model's maximum summer precipitation there for the present climate is twice the observed. Much of that can be accounted for by the model's high interannual variability, and specifically the three highest summer precipitation months during the 22-year record. In almost all years, the albedo experiment has greater precipitation than does the vegetation experiment, which in turn has much greater precipitation than the control. For the vegetation and albedo experiments, the relative magnitude of the interannual variability is significantly lower than that of the control.

In eastern Sahara (figure 26.4b), the changes are more irregular. One of the interesting differences between eastern and western Sahara is that the interannual variability of the control is large in the west and small in the east. However, for the vegetation and albedo experiments, this is reversed (i.e., higher interannual variability in the east and smaller in the west). The increase in the maximum precipitation in the albedo experiment also is much higher than in the vegetation experiment, although in three of the ten years shown, the vegetation experiment's maximum is higher.

Among the most interesting changes are those that occur in the soil moisture budget. Figure 26.5 shows that for the control in western Sahara, there is a small annual cycle in soil moisture with the maximum occurring in summer, after the maximum precipitation. When compared with figure 26.4, one can see that the larger peaks in soil moisture are associated with peaks in precipitation, as expected. For western Sahara, the soil moisture tends to go back down after the large peaks, whereas in eastern Sahara, it tends to remain high after the maximum precipitation. The soil moisture then decreases until the following rainy season. When vegetation is inserted into the region, the magnitude of the annual cycle of soil moisture for western Sahara increases, as does its mean, which increases from 45 cm to 73 cm in August. In the albedo experiment, there is an even larger increase in soil moisture, to 92 cm in August. Why are the changes in soil moisture so different between the vegetation and albedo cases? Consider first the albedo experiment; the lower surface albedo leads to the absorption of more solar radiation during clear days, which leads to increases in upward sensible heat, which warms the surface air, which rises, drawing moist air into the region at low altitudes. There will be increased evaporation only if there is sufficient water to evaporate—which there is, due to the significantly enhanced precipitation. Although the winter drawdown of water is much higher than in the control, it starts at a much higher value, and therefore the minimum monthly soil moisture is still higher than in the control.

Why does the above picture change when the full effects of vegetation are included, and not just the

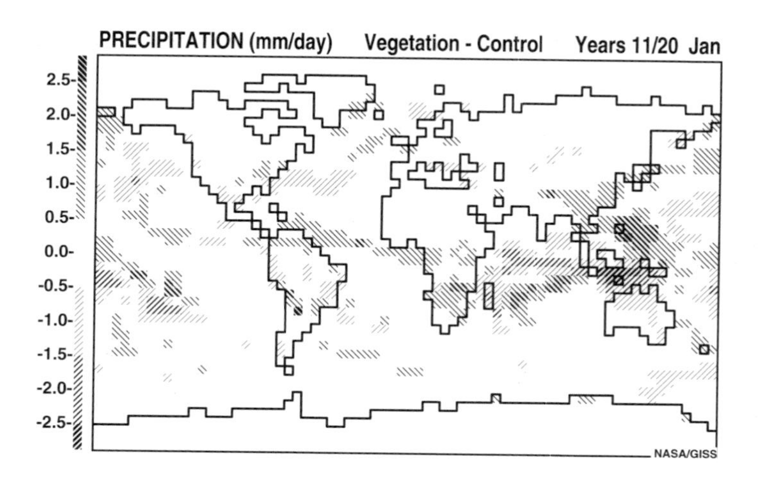

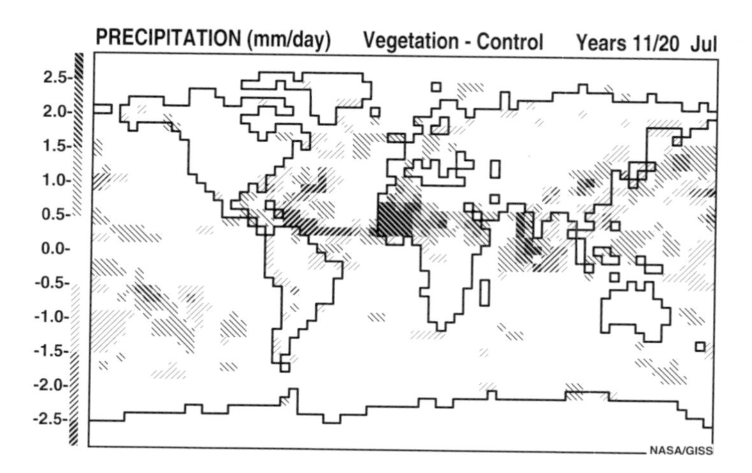

Figure 26.3
Changes in precipitation in the vegetation experiment (vegetation experiment minus control), based on second decade of model simulations.

albedo? The difference is that vegetation has roots. The analysis given above for the albedo case still holds, but now the drawdown of water in the dry season is larger because there are roots that can extract water from all six soil layers instead of only the surface layer in the albedo case. This drawdown causes the minimum soil moisture to approach that of the control. The conclusion is that the effect of the vegetation's albedo on soil moisture is much larger alone than when combined with the other effects of vegetation. This is because roots can remove water from deeper layers than is the case when only albedo is modified. The result for eastern Sahara is of considerable interest because the soil moisture in the vegetation experiment is actually lower than in the control for some months. As before, the albedo experiment leads to higher soil moisture in all months.

Discussion and Conclusions

In this study we used a global climate model to examine the effect of vegetation on the hydrologic cycle in the Sahara Desert. For the present climate and the present-day land cover, the model's representation of monthly precipitation in western and eastern Sahara is in reasonably good agreement with observations; the precipitation is low in winter and higher in summer. When the land surface is changed from desert to vegetation, there is little change in precipitation during the winter dry season, but there is a significant increase in precipitation and soil moisture during summer, when the precipitation is maximum. These changes in hydrologic variables have implications for other variables within the climate system, such as the surface heat budget and surface air temperature. Figure 26.6 and table 26.1 show that when the land

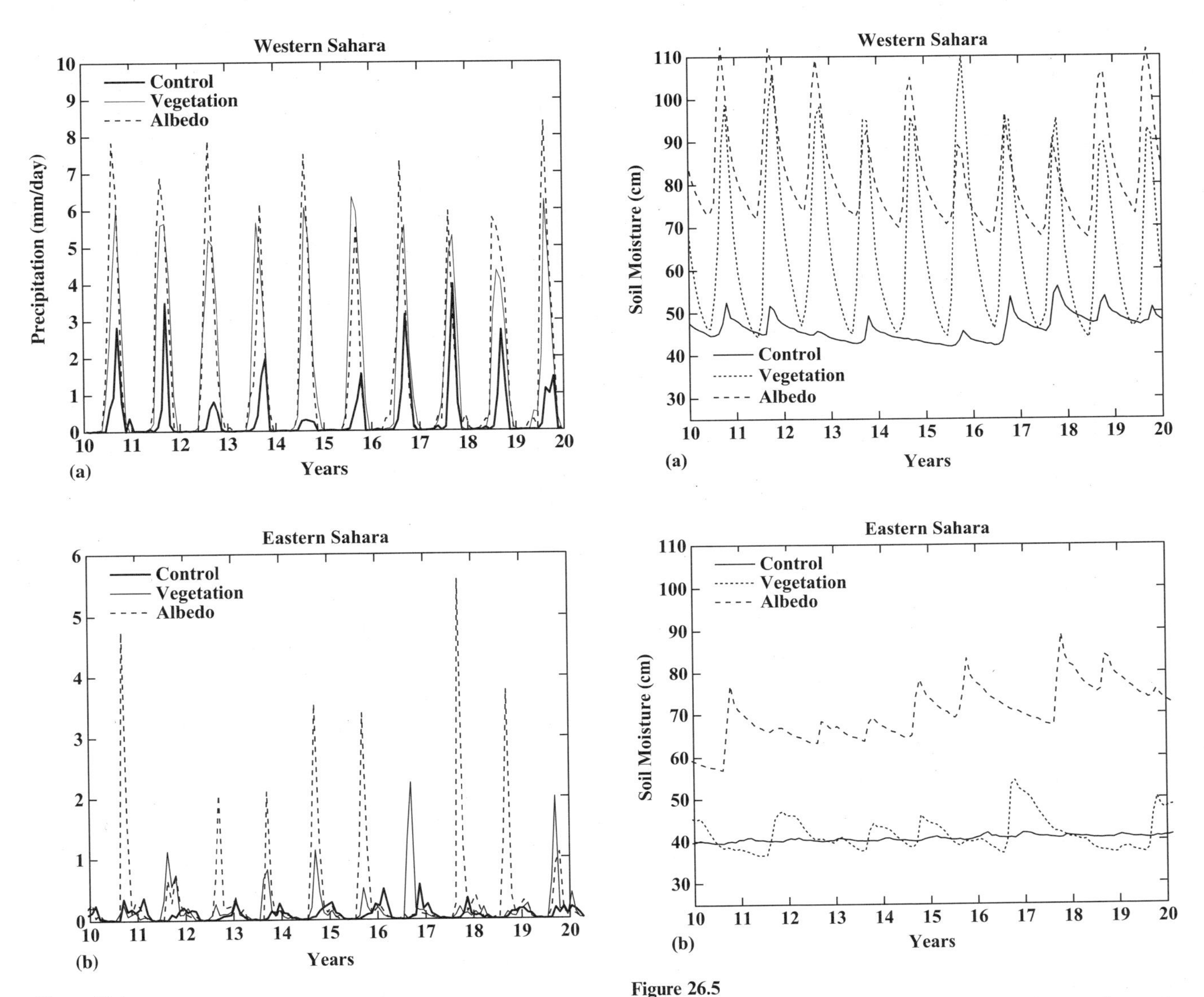

Figure 26.4
Monthly time series of precipitation for the second decade of the control, vegetation, and albedo experiments for (a) western Sahara and (b) eastern Sahara.

Figure 26.5
Monthly time series of soil moisture for the control, vegetation, and albedo experiments for (a) western Sahara and (b) eastern Sahara.

surface is vegetated, the surface air temperature in summer decreases in western Sahara but increases in eastern Sahara. The cooling in the west is due to increased cloud cover. In winter, the temperature is higher throughout the region.

The results here are consistent with other studies which demonstrate that the replacement of higher albedo desert with lower albedo vegetation in North Africa enhances the regional rainfall by increasing the intensity of the summer monsoon (Sud and Molod, 1988; Kutzbach et al., 1996; Knorr et al., 2001). In this chapter, we have sorted out the separate effects of albedo and roots in the vegetation experiments. The results indicate that one should be careful when comparing model simulations that change only the albedo in vegetation experiments with those that change both albedo and rooting depth. Although both mean annual precipitation and the amplitude of its seasonal cycle increase in the albedo-only and albedo plus roots experiments, the enhancement of the amplitude is much larger when only the albedo effect is included. The reduction of the amplitude when roots are added occurs because the roots extract water from deeper layers in summer.

Kirchner (2002) posits that a central property of Gaia is that biologically mediated feedbacks will

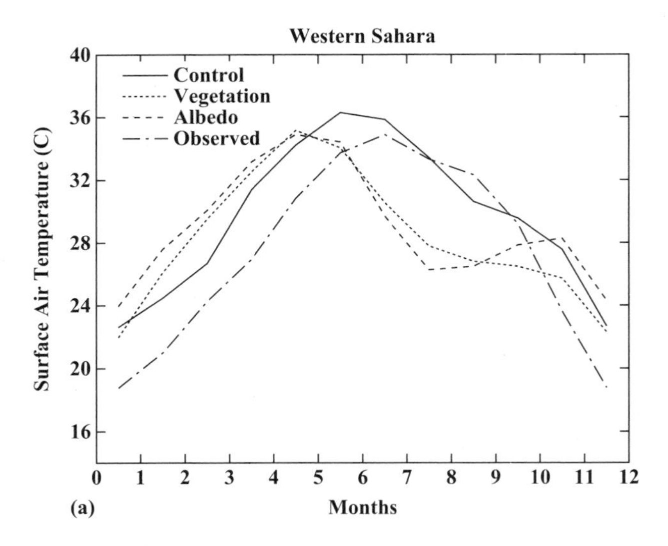

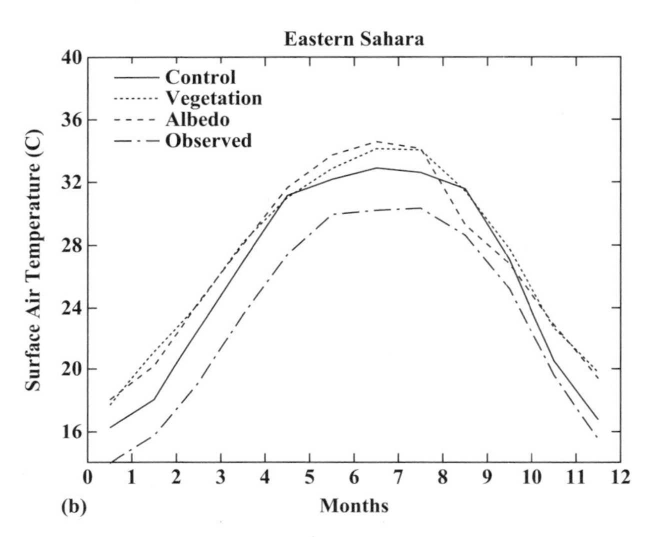

Figure 26.6
Annual cycle of surface air temperature for the control, vegetation experiment, albedo experiment, and observations for (a) western Sahara and (b) eastern Sahara. Observations are based on Shea (1986), and the model is based on the second decade of simulations.

make the environment more suitable for life. Our results would support this property in North Africa. Although we have examined a regional area only, we find that increased precipitation and soil moisture tend to enhance local productivity. These results support Kleidon's (2002) third null hypothesis, that the biota will enhance the environmental conditions for life. However, it is important to note that vegetation does not exist in most of North Africa today, even though results of many studies, including this one, indicate that the presence of vegetation would be self-enhancing. Perhaps Gaia needs some help to reach its optimal state (a vegetated Sahara). There is evidence that Gaia may have had this help in the past.

The pollen and fossil records in North Africa indicate that the region was wetter and more vegetated about 6000 years ago (Kutzbach et al., 1996). This is thought to be due to changes in the Earth's orbital parameters that led to increased solar radiation in summer. This enhanced the land-ocean temperature difference and increased the intensity of the summer monsoon, thus bringing in more water from the ocean. Kutzbach et al. used a climate model to show that the monsoon enhancement caused by changes in orbital parameters would be strengthened by an increase in vegetation. Bonfils et al. (2001) used a global model and showed that 6000 years ago the summer monsoon change in North Africa is significantly larger when the background desert albedo is the lowest.

Whether the global feedbacks between life and the hydrologic cycle tend to stabilize the climate system about some equilibrium level or tend to optimize the climate for the biota is still a difficult question to answer. At the global scale, Betts (1999) used a climate model to examine the effects of terrestrial vegetation by considering two experiments, one with present-day vegetation and one without vegetation, and then using the resulting climates in a vegetation model. He found that with contemporary vegetation, continental precipitation increases by 15 percent, net primary productivity over land increases by 28 percent, and 69 percent more vegetation biomass can be sustained. From a Gaian perspective, this chapter and other modeling studies on the effects of vegetation in the Sahara region lead to a number of conclusions related to stability and optimization. First, Gaia may need help from other feedbacks in the climate system to enhance the environment for life. A related point is that Gaian feedbacks are temporally dependent (i.e., they may have been operative in North Africa 6000 years ago, but not strong enough to have a significant impact today).

In addition to the temporal dependence, Gaian feedbacks are spatially dependent. Douville et al. (2001) show that the effects of vegetation are more important in enhancing the summer monsoon in North Africa than they are for the Asian monsoon. This appears to be, in part, because the Asian monsoon is more dynamic and chaotic than the North African monsoon. Claussen's (1998) modeling study investigating the stability of the atmosphere-vegetation system under present-day conditions of solar irradiation and sea surface temperature found that North Africa was the region most sensitive to the global-scale land surface changes. Hence, the questions about Gaia's ability to stabilize or optimize the environment for the biota may never be cleanly separated from Gaia's interactions with more physically based perturbations to the climate system. Catastrophic impacts are perhaps the most extreme forms of such perturbations.

In the context of the Gaia hypothesis, the results here support the idea that the addition of vegetation to the Sahara Desert would modify the climate in such a way as to encourage the existence of vegetation. The results support the hypothesis that there is a positive feedback between vegetation and soil moisture, at least locally. As noted, however, there are significant differences in how these feedbacks operate at different times and in different places, and in how the feedbacks mediated by the biota interact with other feedbacks within the climate system and with external forcing events.

References

Abramopoulos, F., C. Rosenzweig, and B. Choudhury. 1988. Improved ground hydrology calculations for global climate models (GCMs): Soil water movements and evapotranspiration. *Journal of Climate*, 1, 921–941.

Betts, R. A. 1999. Self-beneficial effects of vegetation on climate in an ocean–atmosphere general circulation model. *Geophysical Research Letters*, 26, 1457–1460.

Bonan, G. B. 2001. Observational evidence for reduction of daily maximum temperature by croplands in the midwest United States. *Journal of Climate*, 14(11), 2430–2442.

Bonan, G. B., D. Pollard, and S. L. Thompson. 1992. Effects of boreal forest vegetation on global climate. *Nature*, 359, 716–718.

Bonfils, C., N. de Noblet-Ducoudre, P. Braconnot, and S. Joussaume. 2001. Hot desert albedo and climate change: Mid-Holocene monsoon in North Africa. *Journal of Climate*, 14, 3724–3737.

Charney, J. 1975. Dynamics of deserts and drought in the Sahel. *Quarterly Journal of the Royal Meteorological Society*, 101, 193–202.

Claussen, M. 1998. On multiple solutions of the atmosphere–vegetation system in present-day climate. *Global Change Biology*, 4, 549–559.

Douville, H., F. Chauvin, and H. Broqua. 2001. Influence of soil moisture on the Asian and African monsoons. Part I: Mean monsoon and daily precipitation. *Journal of Climate*, 14, 2381–2403.

Hansen, J., G. Russell, D. Rind, P. Stone, A. Lacis, S. Lebedeff, R. Ruedy, and L. Travis. 1983. Efficient three-dimensional global models for climatic studies: Models I and II. *Monthly Weather Review*, 111, 609–662.

Kirchner, J. L. 2002. The Gaia hypothesis: Fact, theory, and wishful thinking. *Climatic Change*, 52, 391–408.

Kleidon, A. 2002. Testing the effect of life on earth's functioning: How Gaian is the earth system? *Climatic Change*, 52, 383–389.

Kleidon, A., K. Fraedrich, and M. Heimann. 2000. A green planet versus a desert world: Estimating the maximum effect of vegetation on the land surface climate. *Climatic Change*, 44, 471–493.

Kleidon, A., and M. Heimann. 1998. Optimized rooting depth and its impacts on the simulated climate of an atmospheric general circulation model. *Geophysical Research Letters*, 25, 345–348.

Knorr, W., K. Schnitzler, and Y. Govaerts. 2001. The role of bright desert regions in shaping North African climate. *Geophysical Research Letters*, 28, 3489–3492.

Kutzbach, J., G. Bonan, J. Foley, and S. P. Harrison. 1996. Vegetation and soil feedbacks on the response of the African monsoon to orbital forcing in the early to middle Holocene. *Nature*, 384, 623–626.

Legates, D., and C. Willmott. 1990. Mean seasonal and spatial variability in gauge-corrected global precipitation. *International Journal of Climatology*, 10, 111–127.

Lovelock, J. E., and L. M. Margulis. 1974. Atmospheric homeostasis by and for the biosphere: The Gaia hypothesis. *Tellus*, 26, 2–10.

Russell, G. L., J. R. Miller, and D. Rind. 1995. A coupled atmosphere–ocean model for transient climate change studies. *Atmosphere–Ocean*, 33, 683–730.

Shea, D. 1986. Climatological atlas: 1950–1979, surface air temperature, precipitation, sea-level pressure and sea-surface temperature (45°S–90°N). Technical Note/TN-269+STR. Boulder, CO: National Center for Atmospheric Research.

Shukla, J., and Y. Mintz. 1982. Influence of land-surface evapotranspiration on the earth's climate. *Science*, 215, 1498–1501.

Sud, Y. C., D. M. Mocko, G. K. Walker, and R. D. Koster. 2001. Influence of land surface fluxes on precipitation: Inferences from simulations forced with four ARM-CART SCM datasets. *Journal of Climate*, 14, 3666–3691.

Sud, Y. C., and A. Molod. 1988. GCM simulation study of the influence of Saharan evapotranspiration and surface–albedo anomalies on July circulation and rainfall. *Monthly Weather Review*, 116, 2388–2400.

Zheng, X., and E. A. B. Eltahir. 1998. The role of vegetation in the dynamics of the West African monsoons. *Journal of Climate*, 11, 2078–2095.

V
LIFE FORMS AND GAIA: MICROBES TO EXTRATERRESTRIALS

27
Extraterrestrial Gaias

Siegfried Franck, Werner von Bloh, Christine Bounama, and Hans-Joachim Schellnhuber

Abstract

We present an integrated Earth system model that can be applied to calculate the habitable zone (HZ) around any main-sequence central star. The HZ is defined as the region within which an Earth-like planet might enjoy the moderate surface temperatures required for advanced life forms. Therefore, the HZ defines the conditions for the existence of Gaia. In our view, Gaia is a globally acting geosphere-biosphere system with homeorrhesis (i.e., the self-regulation of this system around an evolving point by and for the benefit of the biota). We implicitly assume that extraterrestrial Gaias have characteristics common to life found on Earth: being carbon-based by photosynthesis and assuming liquid water. With the help of a subset of the Drake equation we can estimate the number of contemporary sisters of Gaia in the Milky Way, N_{Gaia}. For this estimation we need to know the total number of stars in the Milky Way, the fraction of stars with Earth-like planets, the average number of planets per planetary system which are in the HZ, and the fraction of habitable planets where life emerges. We find that N_{Gaia} is on the order of about half a million.

Introduction

The beginning of the Gaia hypothesis is related to the work of James Lovelock in the 1960s NASA space program. First ideas of the Gaia hypothesis were not derived from thinking about mother Earth itself, but from thinking about methods for detecting the presence of life on other planets. Lovelock recognized that a global-acting biosphere shifts its physical environment away from thermodynamic equilibrium (Lovelock, 1965; Hitchcock and Lovelock, 1967). The best-known example of this interaction is the change of atmospheric chemistry by the presence of life. Therefore, a global biosphere on an extraterrestrial planet may be detected at first not by biochemical experiments but by atmospheric analysis.

The Gaia hypothesis of Lovelock and Margulis (see, e.g., Lovelock and Margulis, 1974) states that the ecosphere of a planet with life is homeostated by and for the benefit of the biota. In recent investigations a Gaian system is characterized by the process of homeorrhesis (Guerzoni et al., 2001).

The new century offers the exciting possibility of discovering other planets bearing life. After the detection of more than 100 Jupiter-mass planets from the ground (J. Schneider, *The Extrasolar Planets Encyclopaedia*, ⟨http://www.obspm.fr/planets⟩), NASA and ESA are both designing space missions for the second decade of the twenty-first century that will use interferometry and nulling to search for Earth-like planets. If such planets could be detected, spectroscopic investigations of their atmospheres, to search for such gases like oxygen, ozone, and methane, will follow. This would be an excellent revival of James Lovelock's ideas from the 1960s.

In this chapter we want to introduce an integrated system approach to investigate the coevolution of biosphere and geosphere on a global scale (i.e., to investigate the conditions for the existence of Gaia). The integrated system approach is a special kind of the even more general method of Earth system analysis, concerning not only the ecosphere but also the so-called human factor (Schellnhuber, 1999). Our definition of Gaia as a globally acting geosphere-biosphere system with homeorhesis will be applied to the investigation of extraterrestrial planetary systems. We implicitly assume that extraterrestrial Gaias have characteristics common to life found on Earth: being carbon based by photosynthesis and using liquid water. About 90 percent of the recent surface biosphere is photosynthetically driven. As a first approximation we neglect other life forms (e.g., chemoautotrophes) and also the deep biosphere (Gold, 1998). The definition of habitability is closely related to the definition of life. Following Lissauer (1999), we define a habitable planet as one capable of supporting life on a global scale.

The histories and fates of the three "terrestrial planets"—Venus, Earth, and Mars—suggest that a combination of factors such as distance from the sun, planetary size, and geological and perhaps biological evolution will control the existence of liquid water at a planetary surface. Earth-like planets cannot have liquid surface water if they are much closer to the sun than one astronomical unit (as defined by Earth's orbit), because unfavorably high temperatures and loss of water by photodissociation would be unavoidable in this case. On the other hand, an Earth-like planet which is quite distant from the sun would have permanent surface temperatures below the freezing point of water, and therefore would not be habitable.

In general, the habitable zone (HZ) around the sun can be defined as the region within which an Earth-like planet might enjoy moderate surface conditions needed for advanced life forms. The more specific definition related to the existence of liquid water at the planet's surface was introduced by Huang (1959, 1960) and extended by Dole (1964) and Shklovskii and Sagan (1966). Hart (1978, 1979) calculated the evolution of the terrestrial atmosphere over geologic time at varying distances. He found that the HZ between "runaway greenhouse" and "runaway glaciation" is surprisingly narrow for G2 stars like our sun: $R_{inner} = 0.958$ AU, $R_{outer} = 1.004$ AU, where AU is the astronomical unit. A main disadvantage of those calculations is the neglect of the negative feedback between atmospheric CO_2 content and mean global surface temperature, as discovered by Walker et al. (1981). The full consideration of this feedback by Kasting et al. (1988) provided the interesting result of an almost constant inner boundary, yet a remarkable extension of the outer boundary. Later on, the calculations of the HZ were improved and extended to other main sequence stars (Kasting et al., 1993; Kasting, 1997; Williams, 1998). A comprehensive overview can be found in the proceedings of the first international conference on circumstellar habitable zones (Doyle, 1996).

Recent studies conducted by the authors of this chapter (see, e.g., Franck et al., 2000a, 2000c) have generated a rather general characterization of habitability, based on the possibility of photosynthetic biomass production under large-scale geodynamic conditions. Thus not only the availability of liquid water on a planetary surface, but also the suitability of CO_2 partial pressure, is taken into account. Our definition of habitability is described in detail in the next sections.

As a matter of fact, the same type of stability calculations sketched above for the solar system, with the sun as the central star, can be performed for other stars. The basic results for the HZ around these other central bodies are relatively simple: In order to have a surface temperature in the terrestrial range, a planet orbiting a central star with lower mass would have to be closer to the star than 1 AU, whereas a planet orbiting a brighter star having more mass than our sun would have to be farther away than 1 AU from the star. But the problem is a bit more complicated. One also has to take into account the different times that stars spend on the so-called main sequence. Such stars receive their energy mainly from hydrogen burning, the fusion of hydrogen to helium (see, e.g., Kippenhahn and Weigert, 1990; Sackmann et al., 1993).

Beyond the discussion of extraterrestrial life, there is an ongoing debate about other civilizations outside the solar system. The Drake equation, which was first presented by Drake in 1961 (see, e.g., Dick, 1998) and identifies the relevant factors for a statistical estimation, can provide further information about the abundance of possible extraterrestrial civilizations. Although several factors are highly speculative, a subset of them can be investigated rather rigorously. Based on investigations by Franck et al. (2001) on the habitable zone for extrasolar planets, one can calculate the probability for the existence of an Earth-like planet in the habitable zone as one factor of this subset.

In this chapter, we introduce our integrated system approach and review our calculations of the conditions for the existence of Gaia within the solar system and in extrasolar planetary systems. Based on these results, we calculate the number of extraterrestrial Gaias within the Milky Way with the help of a subset of the Drake equation.

Integrated Systems Approach for the Gaian System

Our model (Franck et al., 1999, 2000a, 2000c) couples the increasing solar luminosity S_{sun}, the silicate–rock weathering rate F_{wr}, and the global energy balance to estimate the partial pressure of atmospheric carbon dioxide P_{atm}, the mean global surface temperature T_s, and the biological productivity Π as a function of time t in the geological past and future.

The global energy balance of the planet's climate is usually expressed with the help of the Arrhenius equation (Arrhenius, 1896).

$$(1 - a)S_{sun} = 4\sigma T_{bbr}^4, \tag{1}$$

where a is the planetary albedo, σ is the Stefan–Boltzmann constant, and T_{bbr} is the effective black-body radiation temperature. The surface temperature of the planet T_s is related to T_{bbr} by the greenhouse warming factor ΔT:

$$T_s = T_{bbr} + \Delta T. \tag{2}$$

Usually ΔT is parameterized as a function of T_s and P_{atm} (Caldeira and Kasting, 1992; Franck et al., 1999). The main drawback of this parameterization is the limited range of applicability to high atmospheric CO_2 partial pressures P_{atm} above 0.1 bar. In our model the upper limit of P_{atm} can be as high as 10 bar carbon dioxide (Kasting and Ackerman, 1986; Tajika and Matsui, 1992). Therefore, we apply the global energy balance given by Williams (1998) and valid also for P_{atm} higher than 0.1 bar and implicitly including the greenhouse effect:

$$S_{sun}(1 - a(T_s, P_{atm})) = 4I(T_s, P_{atm}), \tag{3}$$

where I is the outgoing infrared flux. For I and a polynomial approximations of a radiative-convective climate model were used.

The total process of weathering embraces first the reaction of silicate minerals with carbon dioxide, then the transport of weathering products, and finally the deposition of carbonate minerals in sediments. The basic assumptions and limitations of this approach are given in Franck et al. (2000a). Combining the direct temperature effect on the weathering reaction, the weak temperature influence on river runoff, and the dependence of weathering on soil CO_2 concentration, the global mean silicate-rock weathering rate can be formulated via the following implicit equation (Walker et al., 1981; Caldeira and Kasting, 1992):

$$\frac{F_{wr}}{F_{wr,0}} = \left(\frac{a_{H^+}}{a_{H^+,0}}\right)^{0.5} \exp\left(\frac{T_s - T_{s,0}}{13.7K}\right). \tag{4}$$

Here the prefactor outlines the role of the CO_2 concentration in the soil P_{soil}, a_{H^+} is the activity of H^+ in fresh soil water and depends on P_{soil} and the global mean surface temperature T_s. The quantities $F_{wr,0}$, $a_{H^+,0}$, and $T_{s,0}$ are the present-day values for the weathering rate, the H^+ activity, and the surface temperature, respectively. The equilibrium constants for the chemical activities of the carbon and sulfur systems involved were taken from Stumm and Morgan (1981). The sulfur content in the soil also contributes to the global weathering rate, but its influence does not depend on temperature. It can be regarded as an overall weathering bias which has to be taken into account for the estimation of the present-day value.

Equation 4 is the key relation for our models. For any given weathering rate the surface temperature and the CO_2 concentration in the soil can be calculated self-consistently, as will be shown below. P_{soil} can be assumed to be linearly related to the terrestrial biological productivity Π (see Volk, 1987) and the atmospheric CO_2 concentration P_{atm}. Thus we have

$$\frac{P_{soil}}{P_{soil,0}} = \frac{\Pi}{\Pi_0}\left(1 - \frac{P_{atm,0}}{P_{soil,0}}\right) + \frac{P_{atm}}{P_{soil,0}}, \tag{5}$$

where $P_{soil,0}$, Π_0, and $P_{atm,0}$ are present-day values.

The main role of the biosphere in the context of our model is to increase P_{soil} in relation to the atmospheric CO_2 partial pressure and in proportional to the biologic productivity Π. Π is considered to be a function of temperature and CO_2 partial pressure in the atmosphere only. The pressure dependence corresponds to the so-called asymptotic model (Franck et al., 2000a).

$$\frac{\Pi}{\Pi_{max}} = \left(1 - \left(\frac{T_s - T_{opt}}{T_{opt}}\right)^2\right)\left(\frac{P_{atm} - P_{min}}{P_{1/2} + (P_{atm} - P_{min})}\right) \tag{6}$$

Π_{max} is the maximum productivity and is assumed to be twice the present value, Π_0 (Volk, 1987). $P_{1/2} + P_{min}$ is the value at which the pressure-dependent factor is equal to $1/2$, and $P_{min} = 10^{-5}$ bar is the minimum value for photosynthesis. For fixed P_{atm}, equation 6 produces maximum productivity at the optimum temperature ($T_s = T_{opt}$) and zero productivity outside the temperature tolerance interval $[0°C \cdots 2T_{opt}]$. The present biosphere can be described with $T_{opt} = 25°C$. For the description of a thermophilic or hyperthermophilic biosphere, we investigated models with maximum biological productivity at $T_s = 50°C$ and zero productivity for $T_s \leq 0°C$, $T_s \geq 100°C$.

First we solved the system of equations 1–6 under the assumption that the weathering rate F_{wr} is always equal to the present value $F_{wr,0}$. This is clearly a rather rough approximation. We call this approach the geostatic model (GSM). Franck et al. (1999) introduced the geodynamic model (GDM). In this case a balance between the CO_2 sink in the atmosphere-ocean system and the metamorphic (plate tectonic) sources is expressed with the help of dimensionless quantities (Kasting, 1984):

$$f_{wr} \cdot f_A = f_{sr}, \tag{7}$$

where $f_{wr} \equiv F_{wr}/F_{wr,0}$ is the weathering rate normalized by the present value, $f_A \equiv A_c/A_{c,0}$ is the continental area normalized by the present value, and $f_{sr} \equiv S/S_0$ is the spreading rate normalized by the present value. The ratio f_{sr}/f_A is the geophysical forcing ratio (GFR). It describes the influence of volcanic activity at midocean ridges ($\sim S$) and the continental area on the global climate (Volk, 1987). Midocean ridge volcanism is the main volcanic activity on Earth, and therefore the most important source of CO_2 in the atmosphere. Every year along the 75,000 km of midocean ridge more than 20 km^3 of magma is erupted, about 90 percent of the total global magma production (McKenzie and Bickle, 1988). Weathering is the most important sink. The balance equation 7 describes a stable equilibrium of the global geosphere-biosphere system, which is influenced by an evolving GFR and a slowly changing insolation. The process of weathering is mediated by the biosphere. A stronger biotic enhancement of weathering may lead to bistability, a biotic and an abiotic solution that could be stable (Lenton and von Bloh, 2001). The stable biotic solution ($\Pi > 0$) describes a Gaian system with homeorrhesis.

With the help of equation 7 we can calculate the weathering rate from geodynamical theory. The spreading rate $S(t)$ is related to the mantle heat flow $q_m(t)$ in the framework of boundary layer theory of mantle convection (Turcotte and Schubert, 1982):

$$S(t) = \frac{q_m^2(t)\pi\kappa A_o(t)}{[2k(T_m(t) - T_{s,0})]^2}. \quad (8)$$

Here k is the thermal conductivity, T_m is the average mantle temperature, κ is the thermal diffusivity, and $A_o(t)$ is the area of ocean basins at time t. $T_{s,0}$ is taken as the constant outer temperature of the upper boundary layer in the parameterized convection approximation (Franck, 1998). The area of the planet's surface A_e is obviously the sum of $A_o(t)$ and the area of continents $A_c(t)$, that is,

$$A_e = A_o(t) + A_c(t). \quad (9)$$

Equations 8 and 9 can be used to introduce continental growth models into the equations for the volatile cycle. In the case of the solar system, we apply an episodic continental growth model that is based on geological investigations (Condie, 1990). Nevertheless, it turns out, as a result of sensitivity tests for various continental growth models (episodic growth, delayed growth, linear growth, fixed continental area) that the corresponding HZs do not differ qualitatively

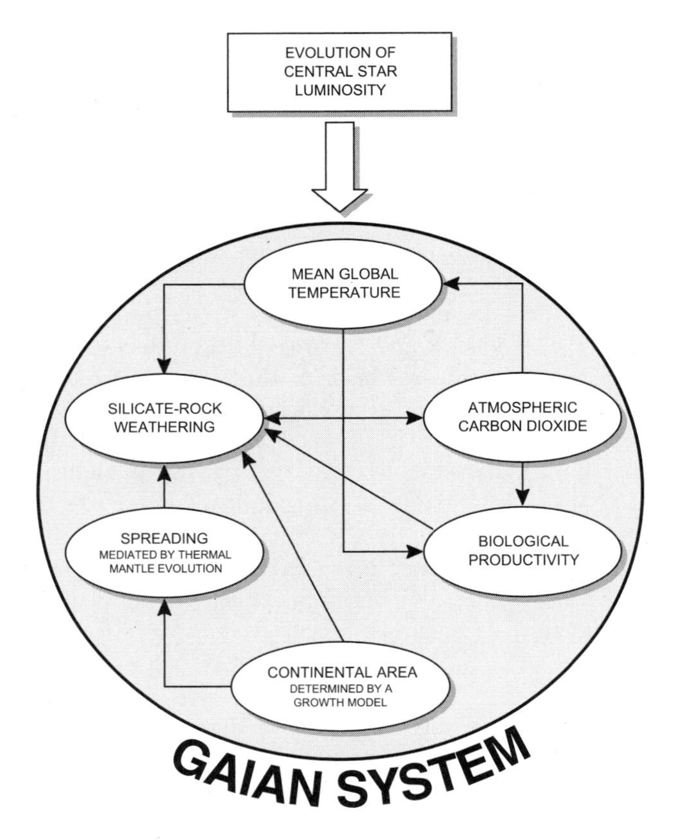

Figure 27.1
Box model for the Gaian system. The arrows indicate the different forcing and feedback mechanisms.

(Franck et al., 2000a). In our calculations for extraterrestrial Gaias we apply a less specific linear continental growth model. Now we have all the means to calculate the weathering rate for every time step of the evolution of an Earth-like planet with the help of equations 7–9, and to determine self-consistently the climate parameters and the biological productivity from the system of equations 1–6. Our integrated systems approach for the Gaian system is sketched in figure 27.1.

HZ for the Solar System: Was There an Ancient Gaia on Mars?

Since the early work of Hart (1978, 1979), there have been many improvements regarding the climatic constraints on the inner and outer boundaries of the HZ. One of the most comprehensive studies in this field is by Kasting et al. (1993). The authors define the boundaries of the HZ via so-called critical solar fluxes. For the inner radius of the HZ, they give three different estimations. The first assumes loss of planetary water by a moist greenhouse (Kasting, 1988); the second assumes loss of planetary water by a runaway greenhouse; and the third is based on the observation that there was no liquid water on Venus's surface at least for the last one Gyr. The outer radius of the HZ is also estimated in three different ways. The first one is based on arguments that early Mars had a warm and wet climate (see also Golombek, 1999; Malin and Edgett, 2000); the second one assumes a maximum possible CO_2 greenhouse heating; and the third one is related to first condensation limit of CO_2 clouds that increase the planetary albedo.

Based on the Earth system model described in the previous section, the HZ for the solar system can be determined in a different way. Here the HZ for an Earth-like planet is the region around the sun within which the surface temperature of the planet stays between 0°C and 100°C and the atmospheric CO_2 partial pressure is higher than 10^{-5} bar (i.e., suitable for photosynthesis-based life; biological productivity $\Pi > 0$):

$$HZ := \{R \,|\, \Pi(P_{atm}(R,t), T_s(R,t)) > 0\} = [R_{inner}, R_{outer}]. \quad (10)$$

The upper limit of the CO_2 partial pressure is 10 bar. This value is derived from the carbon cycle on an early ocean-covered Earth (Walker, 1985). In this case, much of Earth's carbon should have been in the atmosphere because silicate weathering would have been inhibited. The term *Earth-like* explicitly implies the occurrence of plate tectonics as a necessary condition for the operation of the carbonate-silicate cycle as the mechanism to compensate the gradual brightening of the sun during its "life" on the main sequence. The geodynamical evolution of the considered Earth-like planet provides an even stronger constraint. In the geological past, the volcanic input of CO_2 to the atmosphere was much higher and the continental area (available for weathering) was much smaller than today.

The HZ defined in equation 10 may be determined by calculating the behavior of our virtual Earth system at various distances from the sun, which gives different insolations. Our results (Franck et al., 2000b) for the estimation of the HZ are shown in figure 27.2. In this figure we show the HZ for the GSM and a GDM with an upper limit of atmospheric carbon content of 10 bar and $T_{opt} = 50°C$ (i.e., 100°C as the upper bound for the biological productivity). First we find that the width and the position of the HZ are completely different for the GSM and the GDM.

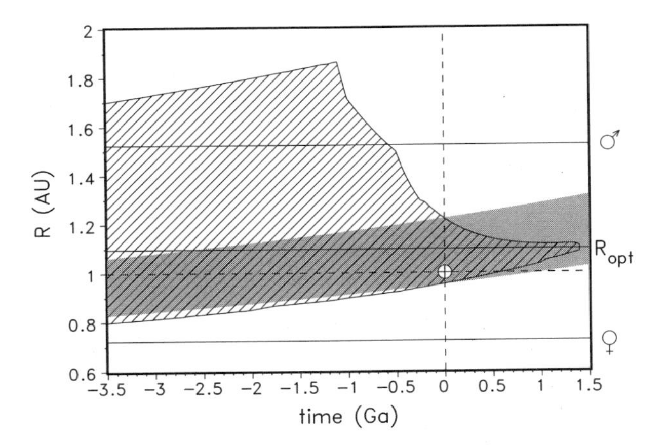

Figure 27.2
Evolution of the HZ for GSM (dark gray) and GDM (hatched). The optimum position of an Earth-like planet is at $R_{opt} = 1.08$ AU. In this case the life span of the biosphere is at maximum (Franck et al., 2000c).

For the geostatic case (GSM) the width of the HZ is nearly constant and shifts only slightly outward with time as the result of increasing insolation. In about 1,000 Ma the inner boundary of the HZ reaches the Earth distance ($R = 1$ AU) and the biosphere ceases to exist.

Our favorite geodynamic model (GDM) shows both a shift and a narrowing of the HZ. The inner boundary reaches the Earth distance in about 600 Ma in correspondence with the shortening of the life span of the biosphere by about 400 Ma compared to GSM. In the GDM the outer boundary shows the following behavior: the nearly linear increase of the outer edge of HZ from 1.7 AU at −3.5 Ga to the maximum greenhouse limit of 1.8 AU at −1.1 Ga results only from the increase in solar luminosity. After −1.1 Ga geodynamics comes directly into play via two processes: first, more effective weathering because of growing continental area, and second, decreasing CO_2 input because of decreasing volcanic activity caused by decreasing spreading rate. The effect of these two processes provokes a lowering of the outer edge of HZ down to 1.08 AU in about 1.4 Ga. At this point, the outer and inner boundaries of the HZ coincide and the HZ of our favored GDM vanishes.

Furthermore, we can state from figure 27.2 that in the framework of our favorite model GDM, the optimal distance of the Earth system would be about 1.08 AU. At such a distance the self-regulation mechanism would work optimally against increasing external forcing arising from increasing solar insolation, and the life span of the biosphere would be extended to 1.4 Ga. But after this time the biosphere would definitely cease to exist. For the model parameters of the GDM in figure 27.2 ($T_{opt} = 50°C$, $P_{CO_2}(max) = 10$ bar), an Earth-like planet at the position of Venus is always outside the HZ while such a planet at Martian distance is within the HZ from the Hadean up to about 500 Ma ago. Nevertheless, there is an ongoing discussion about the role of CO_2 clouds in early Martian climate (Pollack et al., 1987; Kasting, 1991, 1997; Squyres and Kasting, 1994; Forget and Pierrehumbert, 1997; Haberle, 1998) that will be not repeated here. If plate tectonics had acted on ancient Mars, all geodynamic processes caused by the internal cooling of the planet should have declined much faster than on the Earth because of its smaller size.

Nevertheless, we can speculate that our findings about the HZ are an upper bound for the time that a Gaia existed on Mars. This is in good agreement with investigations concerning an early warmer and wetter Martian environment (Golombek, 1999) and with recent observations that plate tectonics may have once operated on Mars (Connerney et al., 1999). Furthermore, there is evidence of subsurface hydrogen derived from experiments onboard *Mars Odyssey* (Feldman et al., 2002; Mitrofanov et al., 2002). This can be interpreted as H_2O ice buried beneath tens of centimeters of hydrogen-poor soil. Data from the Mars Global Surveyor Mission (Baker, 2001) suggest that brief episodes of water-related activity punctuated the geological history of Mars. The most recent of these episodes seems to have occurred within the past 10 million years. However, such episodes would not allow a long-term establishment of a self-regulating Gaian system. Concerning the search for signs of ancient life on Mars, there are ongoing discussions about chemical biomarkers in the meteorite called ALH84001 that are based on McKay et al. (1996).

HZ for Extrasolar Planetary Systems: The Search for Gaia's Contemporary Sisters

The same type of HZ calculations, on the basis of climatic constraints as well as on the basis of Earth system modeling, can be performed for stars with masses different from the solar mass. Kasting et al. (1993) restricted themselves to stellar lifetimes greater than 2 Gyr, which correspond to masses less than 1.5 M_s (1 M_s = 1 solar mass). At the low-mass end they restricted themselves to masses greater than 0.5 M_s because stars with masses ≤ 0.5 M_s show negligible evolutions. Stellar luminosities and temperatures were taken directly from Iben (1967a, 1967b). As expected, HZs for more massive stars are rather short because they have to be truncated at the end of the main sequence. HZs for low-mass stars remain essentially constant over time.

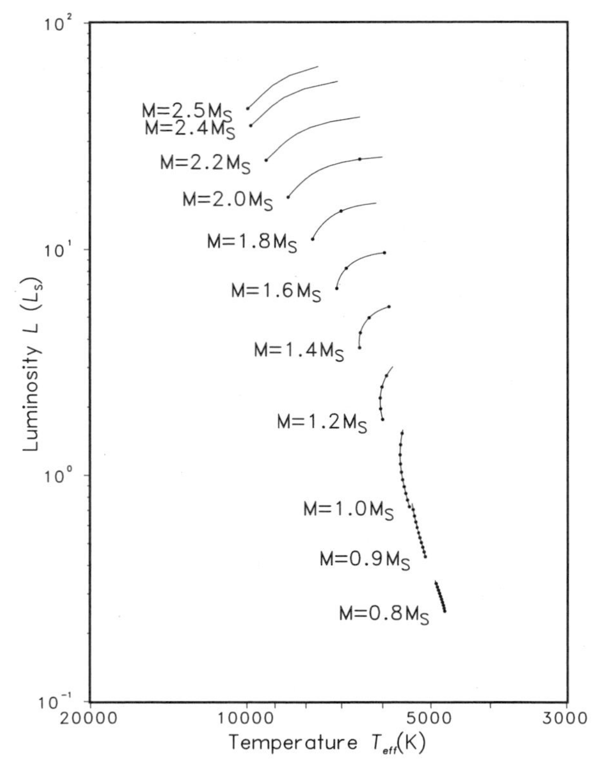

Figure 27.3
Hertzsprung–Russell diagram for central stars in the mass range between 0.8 and 2.5 M_S. Only the main sequence evolution is considered. Successive dots on the mass-specific branches are separated in time by 1 Gyr (Franck et al., 2000b).

In Franck et al. (2000b), the HZ in extrasolar planetary systems is calculated using the luminosity evolution of central stars on the main sequence in the mass range between 0.8 and 2.5 M_s. The results were obtained by polynomial fitting of detailed stellar evolution models by Schaller et al. (1992). The corresponding Hertzsprung–Russell diagram, a plot of luminosity versus effective radiating temperature, is shown in figure 27.3. The temperature tolerance window for the biological productivity was again in the range between 0°C and 100°C in order to incorporate thermophiles (Schwartzman et al., 1993), but for this study the linear continental growth model was employed.

To present the results of our modeling approach we have delineated the HZ for an Earth-like extrasolar

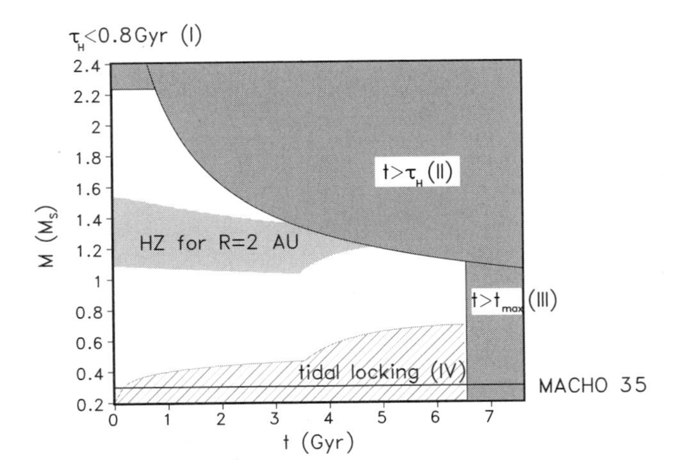

Figure 27.4
Shape of the GDM HZ (gray shading) in the mass–time plane for an Earth-like planet at distance $R = 2$ AU from the central star. The potential overall domain for accommodating the HZ for planets at some arbitrary distance is limited by a number of factors that are independent of R: (1) minimum time for biosphere development, (2) central star lifetime on the main sequence, (3) geodynamics of the Earth-like planet, and (4) tidal locking of the planet (nontrivial subdomain excluded). The excluded realms are marked by dark gray shading in the case of the first three factors and by gray hatching for the tidal-locking effect (Franck et al., 2000b).

planet at a given but arbitrary distance R in the stellar mass–time plane (figure 27.4). In this case, the HZ is limited by the following effects:

1. Stellar lifetime on the main sequence decreases strongly with mass. Using simple scaling laws (Kippenhahn and Weigert, 1990), we estimated the central hydrogen burning period and got $\tau_H < 0.8$ Gyr for $M > 2.2$ M_s. Thus there is no point in considering central stars with masses larger than 2.2 M_s because an Earth-like planet may need approximately 0.8 Gyr of habitable conditions for the development of life (Hart, 1978, 1979). Quite recently, smaller numbers for the time span required for the emergence of life have been discussed, for instance, 0.5 Gyr (Jakosky, 1998). If we perform calculations with $\tau_H < 0.5$ Gyr, we obtain qualitatively similar results, but the upper bound of central star masses is shifted to 2.6 M_s.

2. When a star leaves the main sequence to turn into a red giant, there clearly remains no HZ for an Earth-like planet. This limitation is relevant for stellar masses in the range between 1.1 and 2.2 M_s.

3. In the stellar mass range between 0.6 and 1.1 M_s the maximum life span of the biosphere is determined exclusively by planetary geodynamics, which is independent (in a first approximation, but see limiting effect 4) of R. Thus we obtain the limitation $t < t_{max}$.

4. There have been discussions about the habitability of tidally locked planets. We take this complication into account and indicate the domain where an Earth-like planet on a circular orbit experiences tidal locking. That domain consists of the set of (M, t) couples, which generate an outer HZ boundary below the tidal-locking radius. This limitation is relevant for $M < 0.6$ M_s.

As an illustration we depict the HZ for $R = 2$ AU in figure 27.4. Under these circumstances the HZ is restricted to a stellar mass range between 1.1 and 1.5 M_s and stellar ages below 4.7 Gyr.

The general question is whether an Earth-sized planet discovered outside the solar system can accommodate a self-regulating geosphere–biosphere system with homeorrhesis (i.e., a sister of Gaia). This question can be answered with the help of our results if the mass and age of the central star and the planet's orbit are known. Looking at figure 27.4, we find that the probability of finding an Earth-like planet within the HZ, which is of course related to the probability of finding one of Gaia's sisters, is significantly higher for younger and lighter central stars than for more massive and older stars. Let us emphasize again that we assume an Earth-like planet possesses plate tectonics; this is a crucial ingredient for our parameterization of the weathering rate. The present understanding of plate tectonics is not sufficient, however, to enable us to predict whether a given planet would exhibit such a phenomenon. First theoretical steps to tackle this problem were made by Solomatov and Moresi (1997).

Up to now more than 100 extrasolar planets have been discovered. Among them there is no candidate for a sister of Gaia. Future missions of NASA (Kepler) and ESA (Eddington) open the possibility to detect terrestrial planets. Our method provides a tool to determine whether such a detected terrestrial planet is in the habitable zone or not.

Estimating the Number of Gaias in the Milky Way

From the view of Earth system analysis, we will focus on an estimation of the contemporary sisters of Gaia in the Milky Way selected from the Drake equation. These are denoted by N_{Gaia}. According to the Drake equation, the number of extraterrestrial civilizations, N_{civ}, is given by

$$N_{civ} = N_{Gaia} \cdot f_{civ} \cdot \delta, \tag{11}$$

where N_{Gaia} is

$$N_{Gaia} := N_{MW} \cdot f_P \cdot n_{CHZ} \cdot f_L. \tag{12}$$

Let us discuss the specific factors in detail:

• N_{MW} denotes the total number of stars in the Milky Way.

• f_P is the fraction of stars with Earth-like planets.

• n_{CHZ} is the average number of planets per planetary system that are suitable for the development of life.

• f_L is the fraction of habitable planets where life emerges.

• f_{CIV} denotes the fraction of sisters of Gaia developing technical civilizations. Life on Earth began over 3.85 billion years ago (Jakosky, 1998). Intelligence took a long time to develop. On other life-bearing planets it may happen faster, it may take longer, or it may not develop at all.

• δ describes the average ratio of civilization lifetime to biosphere lifetime.

f_{CIV} and δ are highly speculative: there is no information about the *typical* evolutionary path of life or the characteristic "life span" of communicating civilizations. The typical lifetime of ancient advanced civilizations was limited by increasing environmental degradation or overexploitation of natural resources (Kasting, 1995). One can also speculate that the development and utilization of certain techniques which facilitate the arise of advanced civilizations may be accompanied with new vulnerabilities or hazard potentials that endanger the continuance of civilizations. As a consequence, the lifetime of any advanced (communicating) civilization may be limited to the range of a few hundred years, but this is really uncertain. f_L seems to be potentially assessable by geophysiological theory and observation, and the remaining factors are deducible from biogeophysical science.

Our contribution to the estimate of N_{Gaia} is the new calculation of n_{CHZ}. Based on equation 10, we introduce the continuously habitable zone (CHZ) (Kasting et al., 1993) as a band of orbital distances where a planet is within the HZ for a certain time interval, τ. The effect of the extension of the CHZ on the magnitude of galactic Gaia abundance can be estimated by considering the main sequence (hydrogen burning) stars. The integration over the stellar distributions for distances (R), masses (M), and ages (t), provides the geodynamic/geostatic abundance ratio as a function of the time of continuous residence in the HZ. This is done by defining the probability that the position of a planet is in the interval $[R, R + dR]$ according to $p(R)\,dR$, where R is the distance to the central star. The probable number of planets within the CHZ $[\tilde{R}_{inner}(\tau, t), \tilde{R}_{outer}(\tau, t)]$ of an extrasolar planetary system can be formulated as follows (Whitmire and Reynolds, 1996):

$$P_{hab}(M, t) = C \int_{\tilde{R}_{inner}(M,\tau,t)}^{\tilde{R}_{outer}(M,\tau,t)} p(R)\,dR. \tag{13}$$

In order to estimate n_{CHZ}, the following assumptions are made:

1. The distribution of planets can be parameterized by $p(R) \propto R^{-1}$, that is, their distribution is uniform on a logarithmic scale (Kasting, 1996), which does not contradict our knowledge of already discovered planetary systems.

2. The stellar masses, $M \in [0.4\ \mathrm{M_s} \cdots 2.2\ \mathrm{M_s}]$, are distributed according to a power law $M^{-2.5}$ (Scheffler and Elsässer, 1988).

3. The stellar ages, t, are equally distributed in $[0, \tau_H(M)]$.

4. The factor C is defined as $N_P / \int_{R_{min}}^{R_{max}} p(R)\,dR$, where $N_P = 10$ is the average number of planets per stellar system, and $R_{min} = 0.1$ AU and $R_{max} = 20$ AU define the boundaries of the planetary system.

Then, by integrating over masses, M, and their corresponding lifetime on the main sequence, $\tau_H(M)$, one gets

$$n_{CHZ} = C' \int_{0.4\ \mathrm{M_s}}^{2.2\ \mathrm{M_s}} \frac{M^{-2.5}}{\tau_H(M)} \int_0^{\tau_H(M)} P(M, t)\,dt\,dM, \tag{14}$$

where

$$C' = 1 \Big/ \left[\int_{0.4\ \mathrm{M_s}}^{2.2\ \mathrm{M_s}} M^{-2.5}\,dM \right]. \tag{15}$$

For the time interval $\tau = 500$ Myr, necessary for the development of life (Jakosky, 1998), and our favored geodynamic model, we get $n_{CHZ} = 0.012$. Figure 27.5 shows the geodynamic/geostatic abundance ratio of n_{CHZ}. It demonstrates a geodynamic correction of approximately 2 for a residence time up to 3 Gyr. Now we can start to calculate the number of Gaias with the help of equation 12.

For the total number of starts we assume $N_{MW} \approx 4 \times 10^{11}$ (Dick, 1998). According to Gonzalez et al. (2001), there also exists a galactic habitable zone,

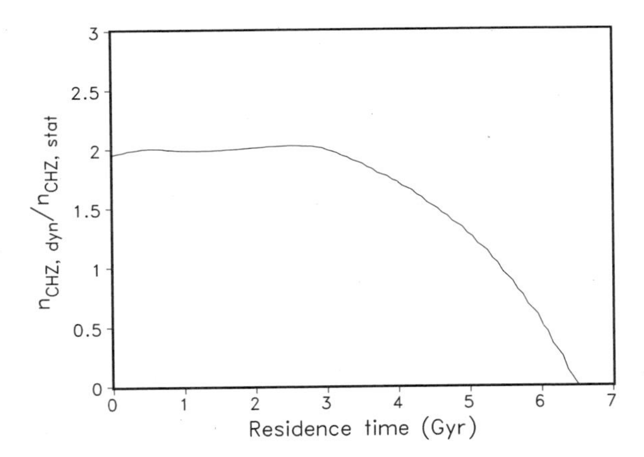

Figure 27.5
Geodynamic/geostatic abundance ratio as a function of time of continuous residence, τ, in the habitable zone.

which may reduce the number of stars where an Earth-like planet can be habitable at all. Until now, Gonzalez et al. have investigated only the outer limit quantitatively. Therefore, a quantitative estimate of the galactic habitable zone is not possible, and we are still using the total number of stars, N_{MW}.

Current extrasolar planet detection methods are sensitive only to giant planets. According to Marcy and Butler (2000) and Marcy et al. (2000), approximately 5 percent of sunlike stars surveyed possess giant planets. Up to now, the fraction of stars with Earth-like planets can be estimated only by theoretical considerations. In such an approach Lineweaver (2001) combines star and Earth formation rates based on the metallicity of the host star. Using his results, we can find a rough approximation for the fraction of stars with Earth-like planets from the ratio of Earth formation rates to star formation rates. Since the sun was formed, this ratio has always been between 0.01 and 0.014. As a conservative approximation, the value $f_P \approx 0.01$ has been used.

The fraction of habitable planets where life emerges is a topic of controversial discussions. The main question is whether biochemistry on a habitable planet would necessarily lead to replicating molecules ($f_L = 1$; see, e.g., Dick, 1998). On the other hand, there are also suggestions that f_L is an extremely low number (Hart, 1995). We use $f_L \approx 0.01$ as a midway point between the predominant optimistic view and pessimistic estimations sketched above. Combining these factors, we finally get

$$N_{Gaia} \approx 4.8 \times 10^5, \qquad (16)$$

which is a rather large number. The number of habitable planets in the Milky Way is even larger. We expect 50 million habitable extrasolar Earth-like planets.

Our value of N_{Gaia} indicates that Gaias should be abundant in our galaxy. It is evident, though, that each of the four factors in equation 12 is associated with large error bars. In this sense, our result must be perceived as a thoroughly educated guess.

The famous question "Is Gaia commonplace in the Milky Way?" can be answered with "Yes!" according to the considerations made above. This may be still premature, however, because our estimation does not include important factors (such as the presence of a large moon, the company of a giant planet, the abundance of long-lived radioisotopes, or, on the other hand, the occurrence of destructive cosmic events) that can significantly reduce the number of Gaia's sisters (Franck et al., 2001).

The ultimate question goes one step further and asks "Are we the only intelligent species in our galaxy?" This riddle is addressed by the systematic search for extraterrestrial brainpower (see, e.g., Drake and Sobel, 1992; and the SETI initiative) and formalized by the full Drake equation (equation 11). The last factor in this equation, the average ratio of civilization lifetime to biosphere lifetime, dominates everything else; its value may be as small as 10^{-7} (Franck et al., 2001). Thus the number of civilizations whose radio emissions might be detectable could be really minute, if not equal to zero. Therefore, our conclusions are in the line with the "Rare Earth Hypothesis" of Ward and Brownlee (2000): primitive life may be very common in planetary systems, but complex multicellular life (including intelligent life as the most advanced one) seems to be very rare in the universe.

Acknowledgments

This work was supported by the German Science Foundation (grant number IIC5-Fr910/7-4) and the Federal Government and the Länder agreement (HSPN, grant number 24-04/325; 2000).

References

Arrhenius, A. 1896. On the influence of carbonic acid in the air upon the temperature of the ground. *Philosophical Magazine*, 41(5), 237–276.

Baker, V. R. 2001. Water and the Martian landscape. *Nature*, 412, 228–236.

Caldeira, K., and J. F. Kasting. 1992. The life span of the biosphere revisited. *Nature*, 360, 721–723.

Condie, K. C. 1990. Growth and accretion of continental crust: Inferences based on Laurentia. *Chemical Geology*, 83, 183–194.

Connerney, J. E. P., M. H. Acuna, P. J. Wasilewski, N. F. Ness, H. Rème, C. Mazelle, D. Vignes, R. P. Lin, D. L. Mitchell, and P. A. Cloutier. 1999. Magnetic lineations in the ancient crust of Mars. *Science*, 284, 794–798.

Dick, S. J. 1998. *Life on Other Worlds*. Cambridge: Cambridge University Press.

Dole, S. H. 1964. *Habitable Planets for Man*. New York: Blaisdell.

Doyle, L. R., ed. 1996. *Circumstellar Habitable Zones. Proceedings of the First International Conference*. Menlo Park, CA: Travis House.

Drake, F., and D. Sobel. 1992. *Is Anyone Out There? The Scientific Search for Extraterrestrial Intelligence*. New York: Delta.

Feldman, W. C., W. V. Boynton, R. L. Tokar, T. H. Prettyman, O. Gasnault, S. W. Squyres, R. C. Elphic, D. J. Lawrence, S. L. Lawson, S. Maurice, G. W. McKinney, K. R. Moore, and R. C. Reedy. 2002. Global distribution of neutrons from Mars: Results from Mars Odyssey. *Science*, 297, 75–78.

Forget, F., and R. T. Pierrehumbert. 1997. Warming early Mars with carbon dioxide clouds that scatter infrared radiation. *Science*, 278, 1273–1276.

Franck, S. 1998. Evolution of the global heat flow over 4.6 Gyr. *Tectonophysics*, 291, 9–18.

Franck, S., A. Block, W. von Bloh, C. Bounama, H.-J. Schellnhuber, and Y. Svirezhev. 2000a. Reduction of biosphere life span as a consequence of geodynamics. *Tellus*, 52B(1), 94–107.

Franck, S., A. Block, W. von Bloh, C. Bounama, H.-J. Schellnhuber, and Y. Svirezhev. 2000b. Habitable zone for Earth-like planets in the solar system. *Planetary and Space Sciences*, 48, 1099–1105.

Franck, S., A. Block, W. von Bloh, C. Bounama, M. Steffen, D. Schönberner, and H.-J. Schellnhuber. 2000c. Determination of habitable zones in extrasolar planetary systems: Where are Gaia's sisters? *Journal of Geophysical Research—Planets*, 105(E1), 1651–1658.

Franck, S., A. Block, W. von Bloh, C. Bounama, and H.-J. Schellnhuber. 2001. Planetary habitability: Estimating the number of Gaias in the Milky Way. *ESA—Proceedings of the 1st European Workshop on Exo/Astrobiology, Frascati/Italy, 21–23 May 2001*. ESA SP-496, pp. 73–78. Noordwijk, The Netherlands: ESA Publications Division.

Franck, S., K. Kossacki, and C. Bounama. 1999. Modelling the global carbon cycle for the past and future evolution of the Earth system. *Chemical Geology*, 159, 305–317.

Gold, T. 1998. *The Deep Hot Biosphere*. New York: Copernicus.

Golombek, M. P. 1999. A message from warmer times. *Science*, 283, 1470–1471.

Gonzalez, G., D. Brownlee, and P. Ward. 2001. The galactic habitable zone: Galactic chemical evolution. *Icarus*, 152, 185–200.

Guerzoni, S., S. Harding, T. Lenton, and F. Ricci Lucci, eds. 2001. *Earth System Science, a New Subject for Study (Geophysiology) or a New Philosophy?* Proceedings of the International School of Earth and Planetary Sciences, Siena, 2001. Siena: ISEPS.

Haberle, R. M. 1998. Early Mars climate models. *Journal of Geophysical Research*, 103, 28467–28479.

Hart, M. H. 1978. The evolution of the atmosphere of the Earth. *Icarus*, 33, 23–39.

Hart, M. H. 1979. Habitable zones about main sequence stars. *Icarus*, 37, 351–357.

Hart, M. H. 1995. Atmospheric evolution, the Drake equation and DNA: Sparse life in an infinite universe. In B. Zuckerman and M. H. Hart, eds., *Extraterrestrials—Where Are They?*, pp. 215–225. Cambridge: Cambridge University Press.

Hitchcock, D. R., and J. E. Lovelock. 1967. Life detection by atmospheric analysis. *Icarus*, 7, 149–159.

Huang, S.-S. 1959. Occurrence of life in the universe. *American Scientist*, 47, 397–402.

Huang, S.-S. 1960. Life outside the solar system. *Scientific American*, 202(4), 55–63.

Iben, I. 1967a. Stellar evolution within and off the main sequence. *Annual Review of Astronomy and Astrophysics*, 5, 571–626.

Iben, I. 1967b. Stellar evolution. VI. Evolution from the main sequence to the red-giant branch for stars of mass 1 $M_\odot$, 1.25 $M_\odot$, and 1.5 $M_\odot$. *Astrophysical Journal*, 147, 624–649.

Jakosky, B. 1998. *The Search for Life on Other Planets*. Cambridge: Cambridge University Press.

Kasting, J. F. 1984. Comments on the BLAG model: The carbonate–silicate geochemical cycle and its effect on atmospheric carbon dioxide over the past 100 million years. *American Journal of Science*, 284, 1175–1182.

Kasting, J. F. 1988. Runaway and moist greenhouse atmospheres and the evolution of Earth and Venus. *Icarus*, 74, 472–494.

Kasting, J. F. 1991. CO_2 condensation and the climate of early Mars. *Icarus*, 94, 1–13.

Kasting, J. F. 1995. Fit for life. *Science Spectra*, 2, 32–36.

Kasting, J. F. 1996. Habitable zones around stars: An update. In L. R. Doyle, ed., *Circumstellar Habitable Zones*, pp. 17–18. Menlo Park, CA: Travis House.

Kasting, J. F. 1997. Habitable zones around low mass stars and the search for extraterrestrial life. *Origins of Life and Evolution of the Biosphere*, 27, 291–307.

Kasting, J. F., and T. P. Ackerman. 1986. Climatic consequences of very high carbon dioxide levels in the Earth's early atmosphere. *Science*, 234, 1383–1385.

Kasting, J. F., O. B. Toon, and J. B. Pollack. 1988. How climate evolved on the terrestrial planets. *Scientific American*, 256, 90–97.

Kasting, J. F., D. P. Whitmire, and R. T. Reynolds. 1993. Habitable zones around main sequence stars. *Icarus*, 101, 108–128.

Kippenhahn, R., and A. Weigert. 1990. *Stellar Structure and Evolution*. Berlin and Heidelberg: Springer-Verlag.

Lenton, T. M., and W. von Bloh. 2001. Biotic feedback extends the life span of the biosphere. *Geophysical Research Letters*, 28, 1715–1718.

Lineweaver, C. H. 2001. An estimate of the age distribution of terrestrial planets in the universe: Quantifying metallicity as a selection effect. *Icarus*, 151, 307–313.

Lissauer, J. J. 1999. How common are habitable planets? *Nature*, 402(Supp.), C11–C14.

Lovelock, J. E. 1965. A physical basis for life detection experiments. *Nature*, 207, 568–570.

Lovelock, J. E., and L. M. Margulis. 1974. Atmospheric homeostasis by and for the biosphere: The Gaia hypothesis. *Tellus*, 26, 2–10.

Malin, M. C., and K. S. Edgett. 2000. Evidence for recent groundwater seepage and surface runoff on Mars. *Science*, 288, 2330–2335.

Marcy, G. W., and R. P. Butler. 2000. Millennium essay: Planets orbiting other suns. *Publications of the Astronomical Society of the Pacific*, 112, 137–140.

Marcy, G. W., W. D. Cochran, and M. Mayor. 2000. Extrasolar planets around main-sequence stars. In V. Mannings, A. Boss, and S. Russel, eds., *Protostars and Planets IV*, pp. 1285–1311. Tucson: University of Arizona Press.

McKay, D. S., E. K. Gibson, K. L. Thomas-Keprta, H. Vali, C. S. Rommanek, S. J. Clemett, D. F. Chillier, C. R. Maechling, and R. H. Zare. 1996. Search for life on Mars: Possible relic biogenic activity in Martian meteorite ALH84001. *Science*, 273, 924–930.

McKenzie, D., and M. J. Bickle. 1988. The volume and composition of melt generated by extension of the lithosphere. *Journal of Petrology*, 29, 625–679.

Mitrofanov, I., D. Anfimov, A. Kozyrev, M. Litvak, A. Sanin, V. Tret'yakov, A. Krylov, V. Shvetsov, W. Boynton, C. Shinohara, D. Hamara, and R. S. Saunders. 2002. Maps of subsurface hydrogen from the high energy neutron detector, Mars Odyssey. *Science*, 297, 78–81.

Pollack, J. B., J. F. Kasting, S. M. Richardson, and K. Poliakoff. 1987. The case for a warm wet climate on early Mars. *Icarus*, 71, 203–224.

Sackmann, I.-J., A. I. Boothroyd, and K. E. Kraemer. 1993. Our sun. III. Present and future. *Astrophysical Journal*, 418, 457–468.

Schaller, G., D. Schaerer, G. Meynet, and A. Meader. 1992. New grids of stellar models from 0.8 to 120 solar masses at $Z = 0.020$ and $Z = 0.001$. *Astronomy and Astrophysics Supplement Series*, 96-2, 269–331.

Scheffler, H., and H. Elsässer. 1988. *Physics of the Galaxy and Interstellar Matter*. Berlin: Springer-Verlag.

Schellnhuber, H.-J. 1999. "Earth system" analysis and the second Copernican revolution. *Nature*, 402(Suppl.), C19–C23.

Schwartzman, D., M. McMenamin, and T. Volk. 1993. Did surface temperatures constrain microbial evolution? *Bioscience*, 43, 390–393.

Shklovskii, I. S., and C. Sagan. 1966. *Intelligent Life in the Universe*. San Francisco: Holden-Day.

Solomatov, V. S., and L.-N. Moresi. 1997. Three regimes of mantle convection with non-Newtonian viscosity and stagnant lid convection on the terrestrial planets. *Geophysical Research Letters*, 24, 1907–1910.

Squyres, S. W., and J. F. Kasting. 1994. Early Mars: How warm, and how wet? *Science*, 265, 744–748.

Stumm, W., and J. J. Morgan. 1981. *Aquatic Chemistry*. New York: Wiley.

Tajika, E., and T. Matsui. 1992. Evolution of terrestrial proto-CO_2 atmosphere coupled with thermal history of the Earth. *Earth and Planetary Science Letters*, 113, 251–266.

Turcotte, D. L., and G. Schubert. 1982. *Geodynamics*. New York: Wiley.

Volk, T. 1987. Feedbacks between weathering and atmospheric CO_2 over the last 100 million years. *American Journal of Science*, 287, 763–779.

Walker, J. C. G. 1985. Carbon dioxide on the early Earth. *Origins of Life and the Evolution of the Biosphere*, 16, 117–127.

Walker, J. C. G., P. B. Hays, and J. F. Kasting. 1981. A negative feedback mechanism for the long-term stabilization of Earth's surface temperature. *Journal of Geophysical Research*, 86, 9776–9782.

Ward, P. D., and D. Brownlee. 2000. *Rare Earth*. New York: Copernicus and Springer-Verlag.

Whitmire, D. P., and R. T. Reynolds. 1996. Circumstellar habitable zones: Astronomical considerations. In L. R. Doyle, ed., *Circumstellar Habitable Zones*, pp. 117–143. Menlo Park, CA: Travis House.

Williams, D. M. 1998. The stability of habitable planetary environments. Ph.D. thesis, Department of Astronomy and Astrophysics, Pennsylvania State University.

28
The Tinto River, an Extreme Gaian Environment

Felipe Gómez, David Fernández-Remolar, Elena González-Toril, and Ricardo Amils

Abstract

Extreme ecosystems have recently attracted considerable interest, not only because they prove that life is robust and adaptable, but also because their existence increases the probability of finding life elsewhere in the universe. Most of the best-characterized extreme habitats on Earth correspond to geophysical constraints (temperature, ionic strength, radiation, or pressure) to which opportunistic microorganisms have adapted. However, some extreme acidic environments are unique in that they are the product of biological activity (chemolithotrophy). The Tinto River (Iberian Pyritic Belt) is an unusual acidic ecosystem (100 km long, mean pH 2.3) containing a high concentration of heavy metals (Fe, Cu, Zn, As, and Cr) and an unexpected level of microbial diversity (bacteria, archaea, and eukaryotes). In the past, the extreme conditions of the river were considered the result of intense mining activity. Geomicrobiological analysis of the Tinto ecosystem strongly suggests that these conditions are the result of the metabolic activity of chemolithotrophic prokaryotes, mainly iron and sulfur oxidizers. The system seems to be controlled by iron, which is used not only as an electron donor but also as an electron acceptor, allowing a full iron cycle to operate. Furthermore, the ferric ion is responsible for the maintenance of the constant pH of the ecosystem and can protect the organisms thriving in its waters from radiation. Laminar, iron-rich stromatolitic formations are generated by the precipitation of different iron minerals on the surface of the biofilms that cover most of the rocks in the river. These structures are similar to ancient, massive, bio-induced laminated iron formations that appeared long before mining activities started in the area 5000 years ago. Given these characteristics, we postulate that the Tinto River operates as a Gaian system in which the (extreme) environmental conditions are generated and regulated by the biological components of the ecosystem.

Introduction

The Tinto River is a 100-km-long river with a characteristic red wine color due to the high concentration of ferric iron (up to 20 g/l) kept in solution by the acidity of its waters (pH between 0.8 and 3, mean 2.3 ± 0.2). The Tinto River headwaters rise at Peña de Hierro, in the core of the Iberian Pyritic Belt, and the river flows into the Atlantic Ocean at Huelva (figure 28.1). The river is subject to a Mediterranean-type climate regime and has an extremely variable flow (López-Archilla et al., 1993).

The river gives its name to an important mining district which has been in operation for more than 5000 years. Chalcolithic Iberians, Tartessians, Phoenicians, and Romans mined the area intensively for at least 3000 years. After the Romans, the mines were practically inactive for more than 1400 years, until a British company bought the mining rights in the late 1800s. Since then, British and, to a lesser extent, Spanish companies have been mining this rich metalliferous formation for copper, gold and silver, and sulfuric acid production (Avery, 1974; Leistel et al., 1998).

In spite of the harsh conditions of its waters, the Tinto River contains a significant level of microbial diversity, mainly eukaryotic (figure 28.2) (López-Archilla et al., 2001; Amaral Zettler et al., 2002). Moreover, the extreme conditions of the habitat (acidity and metal content) seem to be the product of the metabolic activity of chemolithotrophic microorganisms thriving in the Iberian Pyritic Belt, and not the consequence of the intensive mining activity carried out in the area, as has been suggested by several authors (Geen et al., 1997; Leblanc et al., 2000; Davis et al., 2000; Elbaz-Poulichet et al., 2001). The analysis of massive laminated iron formations that make up the old terraces of the river has shown that they predate the oldest mining activity reported in the area, and that they are similar to the laminar structures currently being formed in the river (Amils et al., 2001; González-Toril et al., 2003).

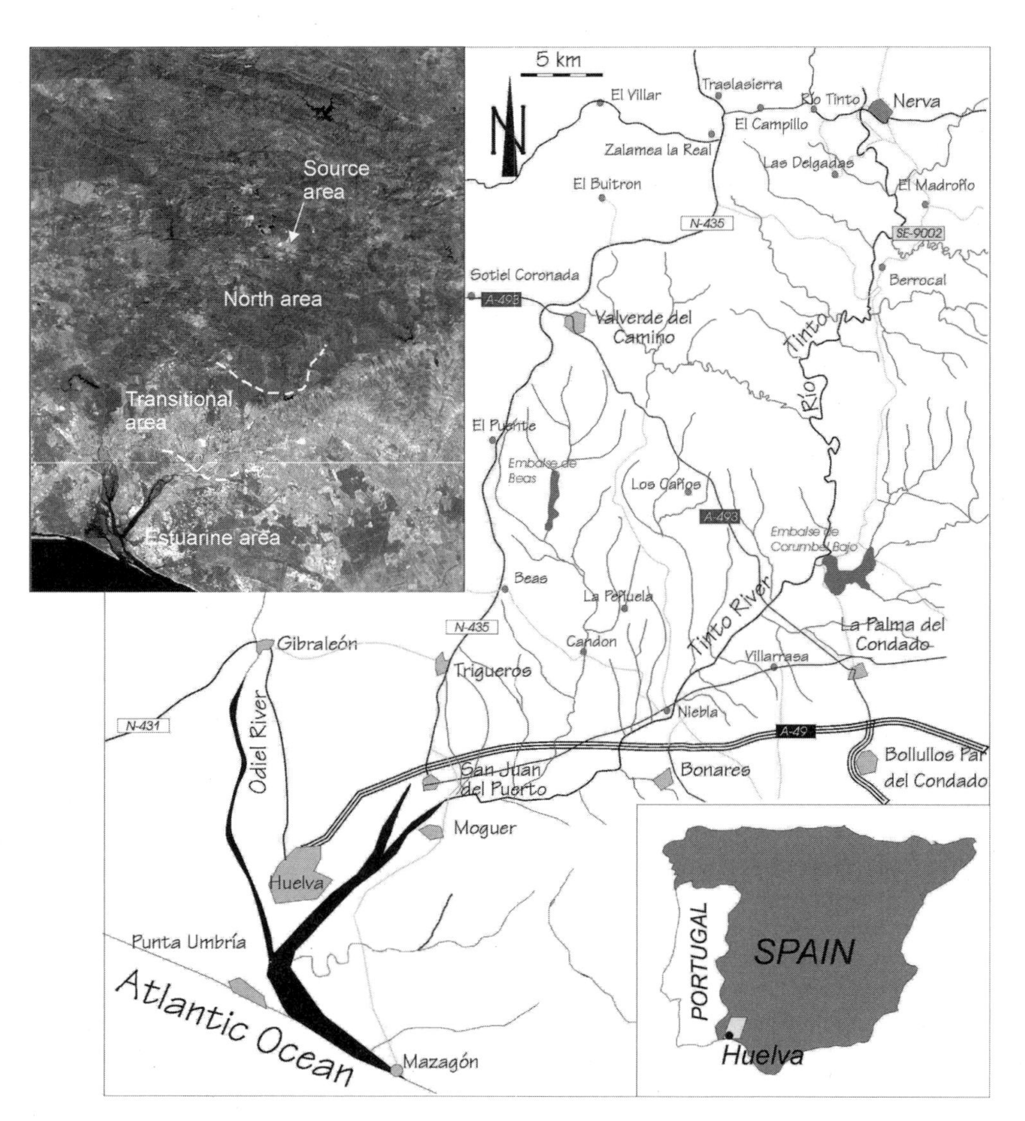

Figure 28.1
Geographical location of the Tinto River Basin.

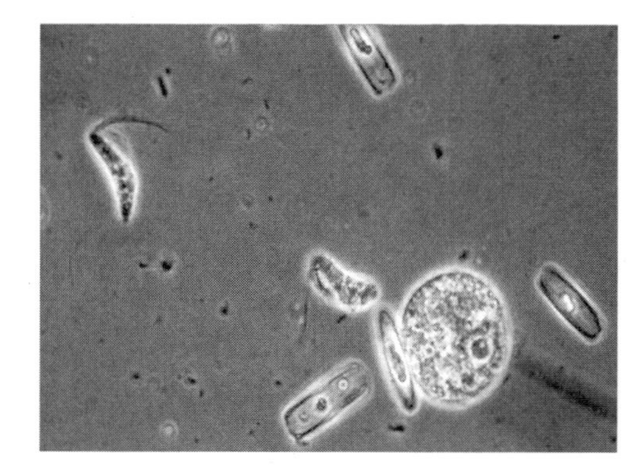

Figure 28.2
Eukaryotic diversity in the water column of the Tinto River.

In this chapter we present the current status of the geomicrobiological characterization of the Tinto ecosystem and an analysis of a Gaian system operating in the river with its possible implications in an Archean iron world.

The Iberian Pyritic Belt

The Iberian Pyritic Belt is a 250-km-long geological entity embedded in the South Portuguese geotectonical zone of the Iberian Peninsula. Metallic ores were formed during the Hercynian orogenesis by hydrothermalism (Boulter, 1996; Lescuyer et al., 1998). Massive bodies of iron and copper sulfides, as well as minor quantities of lead and zinc, constitute the main mineral ores (Leistel et al., 1998).

The basin of the river covers an area of around 1700 km^2 (figure 28.1) (López-Archilla et al., 1993). According to the topographical, geological, and chemical characteristics of the Tinto River Basin, three main zones can be defined: the northern (from Peña de Hierro to Niebla), the transitional (from Niebla to San Juan del Puerto), and the estuary (from San Juan del Puerto to the Atlantic Ocean).

Climate

A climatic gradient is superimposed on the main geological zones of the Tinto River Basin. In the northern area, thermal and rain parameters correspond to a subhumid lower meso-Mediterranean to upper thermo-Mediterranean climatic stage, whereas the transitional and the estuary areas can be included in a subhumid to dry thermo-Mediterranean climate with semiarid conditions (Rivas-Martínez, 1987; Asensi and Diaez, 1987). The climate favors the existence of a stable water table that maintains the river flow during the extremely hot summers, in the absence of rain and with a high rate of evaporation.

Climograms of the basin show a characteristic bimodality, with a humid temperate season alternating with a dry warm season. This bimodality influences microbial activity by affecting physicochemical parameters such as temperature, water flow, and the concentration of different elements. These parameters also have an affect on the sedimentation patterns of the inorganic metabolic products of chemolithotrophic microorganisms.

Redox potential and oxygen concentration are also important parameters that vary with the interactions among the microbial activity, the mineral substrates, and the climatic conditions. The measured redox potentials range from 280 mV to 610 mV, and the oxygen content varies from saturation to complete anoxic conditions. The Tinto Basin has several anoxic locations (e.g., the acidic lake of Peña de Hierro, Presa 3.2, and Berrocal), resulting from oxygen consumption by aerobic respirers, the decrease of oxygen diffusion along the water column, and the lack of mixing in deep waters.

Acidic Chemolithotrophy

The mechanisms of microbially mediated metal leaching have been controversial for many years (Ehrlich, 2001). Silverman and Ehrlich (1964) proposed two basic mechanisms for metal sulfide bioleaching: the direct and the indirect attack. In the direct mode, bacteria solubilize metal sulfides by attaching to their surface, facilitating an enzymatic oxidation by channeling electrons from the reduced moiety of the mineral to an appropriate electron acceptor. In the indirect mode, bacteria are mainly involved in the regeneration of ferric iron, a strong oxidant, in the solution (figure 28.3).

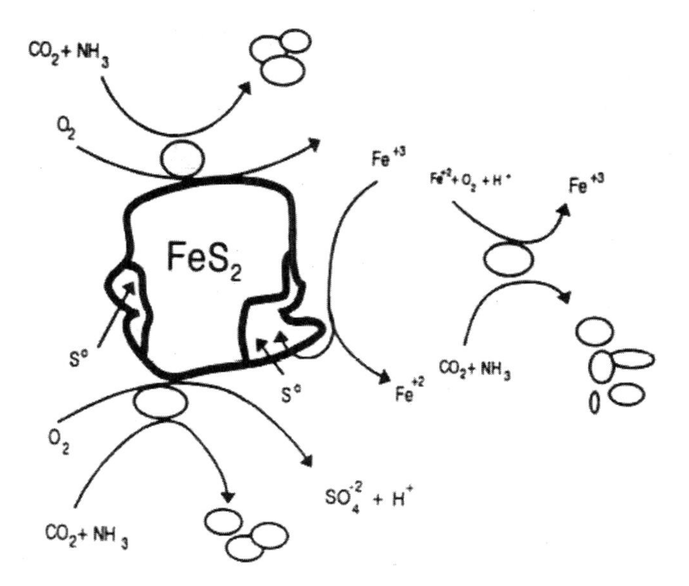

Figure 28.3
Scheme of the main mechanisms of microbial-mediated sulfide leaching: the direct and the indirect attack.

Many experiments have been performed since the 1960s to clarify this biotechnologically important issue. Some clarity emerged after it was found that ferric iron present in the cell envelope and/or in the exopolymers of leaching bacteria was involved in the electron transfer from the sulfidic mineral to the electron transport chain (Gehrke et al., 1995). Sand and collaborators (1995) proposed that because Fe^{3+} is responsible for the oxidation of metal sulfide in both the direct and the indirect attack, there is no basic difference between these mechanisms.

The difference seems to exist at the level of the chemical attack mechanisms, which depend on the nature of the sulfidic substrates. Three metal sulfides (pyrite, molybdenite, and tungstenite) undergo oxidation through the thiosulfate mechanism, while most other sulfides (e.g., sphalerite, chalcopyrite, galena, etc.) are oxidized through the complex polysulfide mechanism (Sand et al., 2001). Pyrite, molybdenite, and tungstenite are unique in their structure because they can be leached only by a strong oxidant attack (ferric ion in bioleaching). Most other sulfides are susceptible to a proton acid attack as well as to ferric ion oxidation.

The thiosulfate mechanism involves the following reactions:

$$FeS_2 + 6Fe^{3+} + 3H_2O \rightarrow S_2O_3^{2-} + 7Fe^{2+} + 6H^+ \quad \text{(reaction 1)}$$

$$S_2O_3^{2-} + 8Fe^{3+} + 5H_2O \rightarrow 2SO_4^{2-} + 8Fe^{2+} + 10H^+ \quad \text{(reaction 2)}$$

with sulfate as the main product of the chemical oxidation (Sand et al., 2001). The complex polysulfide mechanism can be summarized by the following equations:

$$8MS + 8Fe^{3+} + 8H^+ \rightarrow 8M^{2+} + 4H_2S_n + 8Fe^{2+} \quad (n \geqq 2) \quad \text{(reaction 3)}$$

$$4H_2S_n + 8Fe^{3+} \rightarrow S_8^0 + 8Fe^2 + 8H^+ \quad \text{(reaction 4)}$$

Elemental sulfur can then be microbially oxidized (*Acidithiobacillus thiooxidans*, *Acidithiobacillus ferrooxidans*, etc.) into sulfuric acid (Sand et al., 2001):

$$S_8^0 + 12O_2 + 8H_2O \rightarrow 8SO_4^{2-} + 16H^+ \quad \text{(reaction 5)}$$

Most authors currently agree with these basic mechanisms for metal sulfide bioleaching (Hansford and Vargas, 2001), in which iron plays a central and critical role.

pH Regulation

An important characteristic of the Tinto River is that the pH of the system remains constant, regardless of the temperature or the river flow. This is due to the acidity of ferric ion. When the river is diluted (by rain or tributaries), hydrolysis of ferric ion generates free protons:

$$Fe^{3+} + 3H_2O \Leftrightarrow Fe(OH)_3 + 3H^+ \quad \text{(reaction 6)}$$

When intense evaporation occurs during the summer, protons are consumed by the dissolution of ferric hydroxide precipitates. Due to this buffering capacity of ferric iron, the pH remains constant at around 2.3 all along the river, a property that may be key to the maintenance of the high level of biodiversity of the habitat. Although other metal cations may have similar buffering capacities, the high concentration of iron in the river points to this metal as the main pH regulator in the Tinto ecosystem (González-Toril et al., 2003).

Microbial Ecology of the Tinto River

Chemolithotrophs

The iron-oxidizing prokaryotes *Leptospirillum ferrooxidans*, *Acidithiobacillus ferrooxidans* (formerly *Thiobacillus ferrooxidans*) and *Ferroplasma* spp. have been isolated and identified in the Tinto ecosystem, using isolation from enrichment culture and molecular ecology techniques such as sequence comparison of amplified rDNAs resolved by denaturating gel electrophoresis and in situ hybridization. Quantification using specific fluorescent probes has shown that *L. ferrooxidans* and *At. ferrooxidans* are the main chemolithotrophs at most of the sampling stations along the river (Gonzalez-Toril, 2002). Iron-oxidizing *Archaea* were found in low numbers, and only in the upper part of the river near the origin. Of the sulfur-oxidizing prokaryotes only *At. ferroxidans* was found in high numbers. Other sulfur-oxidizing members of the *Acidithiobacillus* genus, such as *At. thiooxidans*, were detected only as minor components of the system (López-Archilla et al., 2001; González-Toril, 2002).

Most of these chemolithotrophic prokaryotes are autotrophic; thus, in addition to promoting the extreme conditions of the habitat, they are also primary producers. Different strains of *At. ferrooxidans* isolated from the Tinto are also able to reduce ferric iron through anaerobic respiration of reduced sulfur compounds, and they have been found in large numbers in some of the anaerobic zones of the river. Probably this is one of their main roles in the Tinto ecosystem. Anaerobic respiration of ferrous ion by *L. ferrooxidans* has also been observed (Mejias, personal communication), which strongly suggests fully operative aerobic and anaerobic iron cycles.

Molecular ecology techniques have also detected members of the *Ferrimicrobium* and *Acidimicrobium* genera. These gram-positive bacteria were originally described as iron oxidizers, and more recently as ferric iron reducers (anaerobic respirers) using reduced carbon compounds as electron donors (Bridge and Johnson, 2000). Their relative concentration in the water column is rather low. It remains to be seen whether they grow preferentially in unexplored anoxic sediments of the anaerobic parts of the river.

Photosynthetic Primary Producers

Algae accounted for the greatest proportion of biomass (over 65 percent) in the Tinto River. Due to their autotrophic metabolism they constitute, together with the chemolithoautotrophic prokaryotes, the primary producers of the system. Members of the Bacillariophyta (diatoms) Chlorophyta (*Chlamydomonas*, *Klebsormidium*, and *Zignema*), Euglenophyta (*Euglena*), and Rhodophyta (*Galdieria*) phyla have been identified, and some of them have been isolated (López-Archilla et al., 2001).

The relative abundance of diatoms in the river, together with their large size, makes them a major contributor to the algal biomass (41 percent). The second highest proportion of algal biomass (32 percent) is the Euglenophyta. Some of the euglenophytes have been identified as *Euglena mutabilis*. Some of the Chlorophyta (green algae) are filamentous algae of the order Ulotrichales, probably belonging to the genus *Klebsormidium*. Representatives of another filamentous genus, *Zignema* (class Conjugatophyceae) are sporadically observed. However, the most ubiquitous algae are the unicellular Chlorophyta. They are found at almost all the sampling stations throughout the year, although they contribute only 11 percent of the algal biomass. Some of them have been identified as *Chlamydomonas acidophila* and *Chlorella* spp. Unicellular and spherical algae corresponding to the phylum Rhodophyta (red algae) also have been isolated and identified as thermoacidophilic microalgae of the genus *Galdieria* (Moreira et al., 1994).

The use of molecular ecology techniques, mainly sequence analysis of amplified rRNA genes, identified photosynthetic protists closely related to those characterized phenotypically (*Chlamydomonas*, *Klebsormidium*, *Zignema*, *Euglena*, and *Chlorella*), emphasizing the high degree of eukaryotic diversity existing in the extreme conditions of the Tinto ecosystem (Amaral-Zettler et al., 2002).

Heterotrophs

A large number of heterotrophic bacteria were isolated from enrichment cultures. Some of the isolates were identified as members of the *Acidiphilium* genus. Members of this genus have frequently been found associated with chemolithotrophic bacteria, especially iron oxidizers. All known species of the genus *Acidiphilium* are also facultative anaerobic respirers, able to couple the oxidation of organic substrates with the reduction of ferric iron. In contrast to *At. ferrooxidans*, some *Acidophilium* spp. can reduce iron in the presence of concentrations up to 60 percent oxygen (Bridge and Johnson, 2000). Specific probes showed that this type of bacteria appears in significant numbers in the Tinto ecosystem, meaning that they may be important elements in the Tinto iron cycle, promoting the reductive dissolution of different ferric iron-containing minerals, such as ferrihydrite, jarosite, and goethite (Johnson, 1999). Other characterized heterotrophic bacterial isolates were gram-positive bacilli, aerobic spore formers of the genus *Bacillus*. These bacteria must exist in the river in the vegetative state and not as spores because they are sensitive to heat denaturation (López-Archilla et al., 2001).

Hybridization experiments strongly suggest the presence of sulfate-reducing bacteria in the Tinto ecosystem (González-Toril, 2002). To date, these microorganisms have not been isolated in the acidic part of the river, although they have been enriched and isolated in the estuary zone, the least acidic section (pH between 4 and 6, depending on the direction of the tide). Recent reports have described the isolation of acidophilic sulfate-reducing bacteria in acid mine drainage ecosystems, making it reasonable to assume that these important elements of the sulfur cycle may exist in the acidic zone of the Tinto River (Johnson, 1998, 1999).

Fungi, both yeasts and filamentous forms, are highly abundant and diverse in the Tinto River. Seven genera of yeasts (*Rhodotorula*, *Cryptococcus*, *Candida*, *Tremella*, *Holtermannia*, *Leucosporidium* and *Hansenula*) and seven genera of filamentous fungi have been isolated from the river and characterized. Most of the isolated yeast species also occur in other, less extreme aquatic environments. Two-thirds of the dematiaceous fungi, including *Scytalidium*, *Bahusakala*, *Phoma*, and *Heteroconium* species develop in the river's extreme conditions. Over 16 species of acidophilic *Penicillium* have been identified to date. In addition, strains of the genus *Lecythophora*, *Acremonium*, and several Zygomycetes belonging to the genus *Mortierella* have been isolated and characterized. Correlation analysis suggests that a close relationship exists between dematiaceous fungi and the more extreme conditions found in the habitat (López-Archilla et al., 1995, 2001). In this case, although some amplified rRNA gene sequences correspond to phenotypically identified fungi (*Scytalidium*, *Mortirella* and *Penicillium*), others have unique sequences that probably correspond to new, undescribed genera (Amaral-Zettler et al., 2002).

Of the extreme conditions found in the Tinto River, high heavy metal concentration is considered a significant challenge for both prokaryotes and eukaryotes. Massive screening for metal resistance, using approximately 200 acidophilic filamentous fungi, showed a high level of metal polyresistance, several orders of magnitude higher than the corresponding taxonomic counterparts from the culture collections used as reference. Many of the tested fungi were able to sequester metals from solution, some of them specifically, a property of interest for bioremediation (Durán et al., 1999, 2001). Also, some of the fungal isolates exhibited extreme sensitivity to heavy metals, suggesting that in the river they were not exposed to this challenge, probably because they were protected within the biofilms.

Among the eukaryotes, heterotrophic protists constitute the major consumer group in the Tinto ecosystem. Flagellates (phylum Zoomastigina), amoebas of the class Lobosea (phylum Rhizopoda), some representatives of the class Heliozoa (phylum Actinopoda) and ciliates (phylum Ciliophora) have been observed (figure 28.2), mainly associated with biofilms (López-Archilla et al., 2001). Molecular analysis detected nonphotosynthetic lineages, such as cercomonads, vahlkampfiid amoebas, and stramenopiles that eluded phenotypic detection. The phylogenetic placement of several sequences is consistent with the idea that they represent new eukaryotic lineages (Amaral-Zettler et al., 2002). New protocols for their enrichment are being designed to facilitate their isolation and physiological characterization.

Most of the biomass is located on the riverbed in dense biofilms, composed mainly of filamentous algae and fungi in which bacteria are trapped. Heterotrophic protists were also found associated with these biofilms. Significant mineral precipitation is observed on the surface of these biofilms, generating iron bioformations.

Iron Bioformations

The activity of chemolithotrophic microorganisms, mainly iron-oxidizing bacteria and archaea, can be appreciated not only by the high concentration of ferric iron in solution but also by the iron formations all along the basin of the river. These iron formations have been fundamental in demonstrating that the Tinto River corresponds to a natural, and not to an industrially contaminated, habitat.

Tree ironstone formations of hematitic and goethitic mineralogy have been detected in the Tinto River Basin (Amils et al., 2001; González-Toril et al., 2003). These units are located in the northern zone near the origin of the river. The most extensive formation appears at the top of the Alto de la Mesa, more than 35 m above today's river. Only the boundaries of this structure have survived after 5000 years of mining activities. It was first reported by Phillips (1881), who described the presence of a fossiliferous iron ore bog in this formation, underscoring its sedimentary origin. This lithostratigraphical unit seems to be made up of different subunits, with a diversity of conglomeratic, laminar, and massive facies showing a coarsening upward tendency. Plant fossils and fungal hyphae were found in some laminar structures. Facies analysis and the stratigraphical architecture suggest that an extensive fluvial environment produced this iron-rich formation.

The intermediate unit, 20–30 m above the river, crops out near Nerva. It consists of terrace sediments of goethitic composition that occupy the valley of the modern river. Their characteristic facies are thick, dark-greenish and reddish massive ironstones with laminar beds. The observed laminar and dome-shaped structures, several cm in diameter, and the fibrous frameworks may be of biogenic origin. These characteristics suggest that the ferruginous sediments of this unit were formed by an acidic meandering river.

The youngest unit, 1–2 m above the river, corresponds to abandoned river bars that are being entrenched by the modern fluvial system. Their facies are mainly represented by strata of conglomerates and paraconglomerates with a massive matrix of iron oxide and badly sorted boulders. Moreover, secondary massive laminar iron facies can be observed locally. These sediments seem to have been formed in an acidic high-energy flow that experienced a sequence of sedimentation events.

The age of the three ferruginous units is unknown because no reliable geochronological analysis or paleontological record is currently available. However, if the climatic parameters that drive the modern sediments are considered and the paleoclimatic record is taken into account, a preliminary estimate of its stratigraphical position can be established. Considering the chronostratigraphical position of a local alteration profile (Bonares laterite) (Rodríguez-Vidal et al., 1985), an upper Pliocene to lower Pleistocene age may be inferred for the first ironstone fluvial unit and the second fluvial terrace unit of the Tinto Basin.

From the stratigraphical situation of the third ironstone unit, which appears as laminated ferruginous sediments and iron-cemented bars eroded by the modern channel system, a late Pleistocene to early Holocene origin may be inferred. These sediments may have been formed by fluvial aggradation during the Flandrian sea level rise (Clemente et al., 1985).

Figure 28.4
Fossil iron bioformation from the Tinto River.

The study of the geomicrobial processes that form the modern river's iron laminar sediments allows us to understand the origin of these facies and their environmental significance. Apparently they are formed by the cyclic succession of seasonal periods of calm low water and turbulent high water with diluted solutions. The repetition of the cycle results in bio- and chemo-induced laminated structures (figure 28.4). The study of old and recent iron sediments, the water biogeochemistry, and the microbial communities present in the Tinto River can give us a vision of the temporal and spatial evolution of the geomicrobiological system.

As stated earlier, until recently it had been generally accepted that the extreme conditions found in the Tinto River were the direct consequence of the mining activities in the area during the last 5000 years. Different authors have reported the presence of high metal content in cores drilled in sediments of the Tinto estuary, which have been correlated with the Copper Age (4500 BP) and modern mining activities (120 BP) (Geen et al., 1997; Leblanc et al., 2000; Davis et al., 2000; Elbaz-Poulichet et al., 2001). Although, it is obvious that the mining activity must have had an important signature in the Tinto system, the existence of ancient iron-rich deposits formed prior to any known mining activity, under hydrochemical conditions similar to the modern deposits, is considered a strong argument in favor of a natural origin of the river.

Iron oxides and oxyhydroxides associated with sulfates are the characteristic minerals that are formed in the modern sediments. The Tinto River mineral parageneses results from the alternation of oxide and saline facies due to the annual climatic cycle. In addition, sulfates and oxides can coprecipitate, depending on the sulfate concentration of the water. The oxide facies are composed of combinations of ferrihydrite, goethite, hematite, and magnetite, their content depending on the degree of dehydration of the mineral. Hematite has been detected in both recent and old sediments at different dehydration degrees, while ferrihydrite exists only in hydrated and fresh sediments which have been recently precipitated.

Both microbial activity and high evaporation rates lead to the formation of acidic brines, which are composed mainly of sulfate complexes and metals. Two different parageneses, gypsum and jarosite, that are associated with different iron-bearing sulfates, are precipitated from these brines. A differential distribution

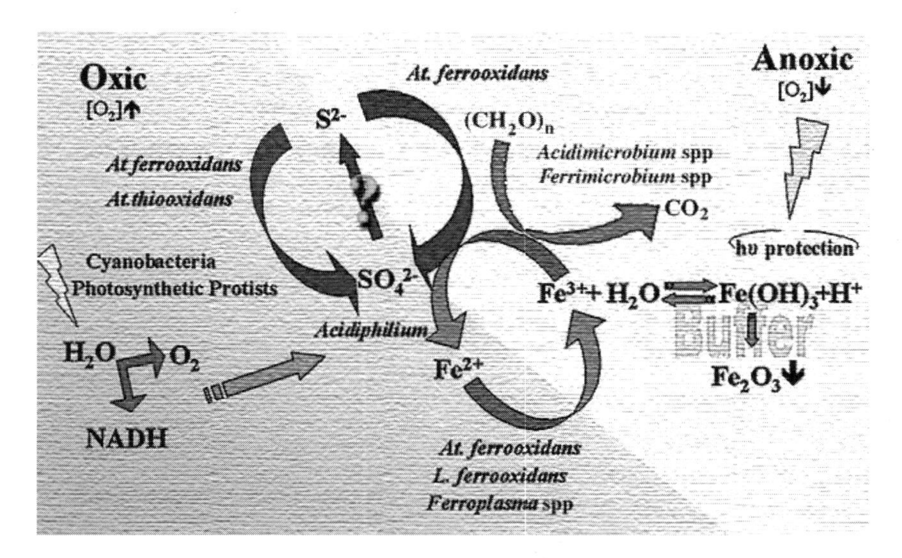

Figure 28.5
Geomicrobiological model of the Tinto River. The role of the different microorganisms identified and isolated from the river is shown associated to the iron and sulfur cycles operating in the Tinto Basin. The different properties of iron are displayed.

of these minerals has been observed along the river, with gypsum predominating in areas located near the source, whereas the jarosite content increases downstream, where the gypsum remains a secondary phase or even disappears. Iron sulfates, which are ubiquitous in modern sediments of the fluvial basin, decrease to negligible content in the ancient iron deposits. Moreover, ancient iron formations are devoid of any type of sulfur, probably as a consequence of the solubility of its most oxidized form.

The Tinto River Model System

As mentioned, both sulfur- and iron-metabolizing microorganisms have a fundamental role in this fluvial ecosystem. Until recently, much attention had been paid to the sulfur cycle microbiology. Fom a biochemical point of view, much more energy can be obtained from the sulfur- than from the iron-reduced compounds (Ingledew, 1982; Hazeu et al., 1986). Recently, however, the study of iron metabolism has taken on a more predominant role in microbial ecology. The demonstration that some anaerobic photosynthesizing bacteria can use ferrous ion reducing power to assimilate CO_2 (Widdel et al., 1993); the isolation of bacteria able to respire ferrous iron anaerobically, using nitrate as an electron acceptor (Benz et al., 1998); and the identification of strict iron oxidizers, such as *Leptospirillum* spp. and *Ferroplasma* spp., as critical organisms in biohydrometallurgy (Brierley, 1999; Sand et al., 2001) and in acid mine drainage generation (Johnson, 1999; Edwards et al., 2000) have completely changed this perspective.

The identification in the Tinto River of iron-oxidizing prokaryotes (*L. ferrooxidans*, *At. ferrooxidans*, and *Ferroplasma* spp.) and iron reducers (*Acidiphilium* spp., *At. ferrooxidans*, *Ferromicrobium* sp., and *Acidimicrobium* sp.) that can operate not only aerobically but anaerobically, highlight the existence of a fully operative iron cycle in the conditions found in the system (figure 28.5).

Obviously the sulfur cycle is also operative with the presence of *At. ferrooxidans*, *At. thioxidans*, and sulfate-reducing bacteria. Furthermore, the iron and the sulfur cycles operate in an interrelated manner (figure 28.5). Given the properties of the environment, we suggest that the Tinto River is an ecosystem controlled by iron (González-Toril et al., 2003). Iron has pleiotropic properties, which give it interesting ecological perspectives. Iron is not only a source of energy for iron-oxidizing prokaryotes, but also can be used as an electron acceptor for anaerobic respiration. In addition, ferric iron is responsible for the constant pH of the ecosystem, a property that may help maintain the high level of eukaryotic phylogenetic diversity found in the river (López-Archilla et al., 2001; Amaral Zettler et al., 2002). Furthermore, acidic ferric iron solutions readily absorb harmful UV radiation, thus protecting the organisms. All these properties are fully operative in the Tinto ecosystem (figure 28.5). They have been demonstrated in laboratory experiments with isolated native microorganisms, and have also been detected at different sampling stations along the river.

So far this model seems reasonable from a chemolithotrophic prokaryotic point of view. But, given the high level of eukaryotic phylogenetic diversity in the Tinto ecosystem, much higher than that of the prokaryotes (López-Archilla et al., 2001; Amaral Zettler et al., 2002), and the fact that most of the primary production (over 65 percent) seems to depend on photosynthetic protists, what is the advantage, if any, for eukaryotes to live in an extreme acidic environment, in the presence of such a high concentration of toxic metals? We think that a possible answer to this question is that freely available iron is highly concentrated in the Tinto system and rather limited in the neutral pH world.

We know that iron is an extremely important element for life. It has been established that iron is an important limiting factor for phytoplankton growth in the oceans (Martin, 1990; Boyd et al., 2000), for plants (Price, 1968), and for many other systems (Archibald, 1983). We know that most cell systems, both prokaryotes and, especially, eukaryotes, devote an important amount of genetic information and energy to iron scavenging. Organisms have developed very specific and sophisticated mechanisms to trap iron anywhere they find it (Reid et al., 1993; Braun and Killmann, 1999). Why is this so, if iron is one of the most abundant elements on Earth? A possible explanation is that in an oxygenic atmosphere at neutral pH, iron is rapidly oxidized into insoluble compounds so efficiently that even if mineral weathering releases important amounts of iron into the biosphere, most of it is not bioavailable because it is incorporated into anaerobic sediments where sulfate-reducing bacteria may further transform it into pyrite, an even more unreactive iron mineral. The geological recycling of these sediments and the microbiology associated with the iron cycle are different ways to reintroduce this critical element into the biosphere.

The availability of iron and other heavy metals is considered so important that a model has recently been proposed in which a Proterozoic anoxic ocean removed iron and other heavy metals, through the action of very active sulfate-reducing bacteria, so efficiently that a nitrogen cycle crisis occurred, possibly limiting photosynthetic protist distribution and the overall evolution of eukaryotes (Anbar and Knoll, 2002). The disturbance of this long-lasting steady state, probably due to massive oxygen production or some other catastrophic factor (Kirschvink et al., 2000), increased the bioavailability of iron in the Phanerozoic ocean, with all its evolutionary implications, although with the limitations imposed by the physicochemical properties of iron.

Has this iron-limiting scenario been a constant in the history of life on Earth, or was there a time when the soluble iron concentration was high enough to facilitate its widespread use in different biological activities? According to Holland (1973), the Archaean oceans held important concentrations of ferrous iron. And what about the atmospheric concentration of oxygen? Using Ohmoto's (1997) Archean oxygenic atmospheric model, most of the iron would have been precipitated due to the fast kinetics of iron oxidation at neutral pH. A different situation would be found at an acidic pH in which iron oxidation by oxygen is very slow. If we use Holland's conventional model (Holland, 1978, 1984; Kasting, 1993), in which most of the Archean was dominated by an atmosphere with an extremely low concentration of oxygen, the possibility of developing the concept of an iron world becomes feasible.

Iron World

In these conditions, and assuming that the absence or low concentration of oxygen precluded the appearence of sulfate-reducing activity, the iron concentrations in the Archean may have been as high as 50 μM (Holland, 1973; Anbar and Knoll, 2002), three orders of magnitude higher than exists in contemporary oxygenated seawater and much higher than that estimated for a Proterozoic sulfidic ocean (Anbar and Knoll, 2002). The presence of high concentrations of soluble iron would have facilitated the development of iron-oxidizing bacteria, which, as we know, can operate in the absence of oxygen. This activity, in turn, would have promoted the appearance of iron-reducing metabolism, thus maintaining an operative anaerobic Archean iron cycle independent of photosynthesis.

Is there any evidence for such a world? Geology suggests that the presence of high concentrations of iron in the Archean oceans was the result of intense submarine magmatic and hydrothermal activity (Holland, 1973; De Ronde et al., 1984; Barley et al., 1997; Vargas et al., 1998). A high concentration of ferrous ion in solution would have required a corresponding low concentration of hydrogen sulfide to avoid the formation of insoluble pyrite. The specific geological markers for the Archean, the controversial Banded Iron Formations (BIFs), are fully compatible with this iron world scenario (Cloud, 1973; Holland,

1973; de Duve, 1987; Trendall and Morris, 1993; Fenchel et al., 1998). Of the different interpretations offered for the BIFs' generation—(1) continental weathering of iron minerals, (2) photochemical oxidation of iron, (3) iron oxidation produced by oxygen, (4) iron as reducing power for CO_2 assimilation, and (5) anaerobic respiration of iron—those related to microbial iron metabolism are not only valid alternatives but also the direct consequence of an operative iron world. The massive BIFs deposited from about 3.8 to 2 Gyr are attributed to episodic oxidation of oceanic ferrous iron, resulting in precipitates of hematite and mixed oxide magnetite. "Red beds," extensive iron oxide deposits beginning at about 2 Gyr, provide evidence of continental weathering, and therefore of oxygen accumulation in the atmosphere to levels of perhaps 0.01 present atmospheric level. The roughly coincidental disappearance of BIFs and appearance of red beds marked a transition period toward the end of the Archaean and the beginning of the Proterozoic, and suggests the nearly complete disappearance of reduced metal species from seawater, resulting in a slightly oxidizing atmosphere. Interestingly, the iron world, based on an operative metabolic iron cycle, does not requires a high concentration of soluble sources of reduced iron, because chemolithotrophic microorganisms can efficiently oxidize insoluble iron.

The Tinto River as a Model for the Iron World

Concerning the existence of iron-oxidizing and iron-reducing microorganisms that could perform in the anoxic conditions prevailing at the bottom of the Archaean oceans, micropaleontology cannot offer much more information than the products of their metabolism. But we have an extant ecosystem, the Tinto River, in which most of the basic questions concerning the geomicrobiology of a putative iron world can be tested. Using specific hybridization probes, we can explain more than 80 percent of the prokaryotic activity of the system with well-characterized microorganisms which are basically involved with the iron cycle. In addition, these activities produce iron bioformations which, over time, become fossilized and show a mineral composition and a banding pattern reflecting both the geomicrobiology of the system and the climate change record, similar to that of the Archean BIFs.

Regarding oxygen, which was most probably absent or in very low concentration during the Archean period, we know that the Tinto's iron cycle can operate not only in the aerobic conditions that dominate the surface of the river and the well-mixed waters of the rocky sections of the river, but also in the anoxic conditions that prevail at many sites along the river in which the water column is deep enough (e.g., the acidic lake of Peña de Hierro, Presa 3.2, Berrocal). Furthermore, the Tinto River is an excellent model system to gain insight into how an iron world operated not only in the absence of oxygen (Archean), but also in its presence (Proterozoic and Phanerozoic). Probably the most interesting areas of the Tinto ecosystem are the ones at the fluctuating interface between oxic and anoxic conditions (figure 28.5).

As for the acidic pH, one of the most outstanding characteristics of the Tinto ecosystem is a direct consequence of the physicochemical properties of the metabolic product of pyrite oxidation: ferric ion. If ferrous iron, in soluble and/or insoluble mineral form, as well as iron-oxidizing bacteria are present in the same system, ferric ion will be produced in spite of the oxygen concentration. Ferric ion will precipitate as ferric hydroxide, releasing protons according to equation 6. This reaction maintains a constant acidic pH in the water column as long as enough ferric iron remains in solution. We have measured the threshold of the iron buffering capacity of a solution of 50 mM of ferric sulfate, which corresponds to the mean value of iron concentration in the river. The results showed that the solution maintains an acidic pH of around 3 after dilutions of up to three orders of magnitude. This property has also been observed with filtered Tinto River water from different locations. Ferric ion keeps the pH of the Tinto ecosystem constant until it dilutes in the waters of the Atlantic Ocean in the tidal section of the river. Tinto geomicrobiology is able to maintain sufficient iron in the system to keep its pH constant despite the dilutions introduced by neutral tributaries and seasonal floodings produced by its Mediterranean climate regime (González-Toril et al., 2003).

We suggest that the Tinto system, and other natural acidic environments, are relics of an iron world in which the basic metabolism of chemolithotrophic organisms controls the conditions of the system. Many chemolithotrophic microorganisms (*L. ferrooxidans*, *Ferroplasma* spp., *At. thiooxidans*) not only can tolerate, but also perform optimally, at a much lower pH than the one given by a solution of ferric ion. Sulfur-oxidizing organisms cannot control the pH of the system in which they grow, but iron-oxidizing bacteria can control this important environmental factor, using the final product of their metabolism. This pH control is probably necessary for the maintenance of the high level of eukaryotic diversity thriving in the system. Carbon fixed by the photosynthetic eukaryotes helps maintain the heterotrophic prokaryotes involved in the reduction of iron and/or the heterotrophic eukaryotes, such as filamentous fungi, which produce the basic network of the biofilms involved in mineral precipitation. We propose that this extant iron world is probably related to the ancient Archean system, behaving in a Gaian manner, because one of the most critical conditions in the system, the pH, is under biological regulation, thereby facilitating the development of a complex ecosystem with a remarkable level of diversity.

In conclusion, we have a rather peculiar acidic environment which is the consequence of chemolithotrophic metabolism and is controlled by iron, a product of this metabolism. It has produced iron bioformations, which prove that the system was in operation long before mining activities started in the area, thus underscoring the fact that the system is natural and not the product of industrial contamination. Probably the characteristics of the river allowed chalcolithic miners to find the Iberian Pyritic Belt in the first place. It was an easy track to follow. The chemolithotrophs involved in the production and maintenance of this environment are responsible for two critical cycles of nature: the iron and the sulfur cycles. In addition, some of these microorganisms are important in processes of environmental and biotechnological interest: acid mine drainage, biomining, and bioremediation. Understanding this model is vital not only to increase our knowledge of an unusual lifestyle with many astrobiological implications, but also to improve the performance of biotechnological processes and to search for new applications for the microorganisms that thrive in the red waters of the Tinto River.

Acknowledgments

This work was supported by grants BIO99-0184 (CICYT), BXX2000-1385 (DGICYT), and 07M/0038/2001 (CAM), and institutional grants to the Centro de Astrobiología and the Centro de Biología Molecular (F. Areces).

References

Amaral-Zettler, L. A., F. Gómez, E. Zettler, B. G. Keenan, R. Amils, and M. L. Sogin. 2002. Eukaryotic diversity in Spain's River of Fire. *Nature*, 417, 137.

Amils, R., E. González-Toril, F. Gómez, D. Fernández-Remolar, and N. Rodríguez. 2001. Geomicrobiology of an extreme acidic environment: The Tinto River case. *2000 Spring Meeting, American Geophysical Union*, abstract B22B-05. Washington, DC: American Geophysical Union.

Anbar, A. D., and A. H. Knoll. 2002. Proterozoic ocean chemistry and evolution: A bioinorganic bridge? *Science*, 297, 1137–1142.

Archibald, F. 1983. *Lactobacillus plantarum*, an organism not requiring iron. *FEMS Microbiology Letters*, 19, 29.

Asensi, A., and B. Diaez. 1987. Andalucía occidental. In *La Vegetación de España*, M. Peinado and S. Rivas-Martínez, eds., Colección Aula Abierta, 3, pp. 199–230. Alcalá de Henares: Universidad de Alcalá de Henares.

Avery, D. 1974. *Not on Queen Victoria's Birthday*. London: Collins.

Barley, M. E., A. L. Pickard, and P. J. Sylvester. 1997. Emplacement of a large igneous province as a possible cause of banded iron formation 2.45 billion years ago. *Nature*, 385, 55–58.

Benz, M., A. Brune, and B. Schink. 1998. Anaerobic and aerobic oxidation of ferrous iron at neutral pH by chemoheterotrophic nitrate-reducing bacteria. *Archives of Microbiology*, 169, 159–165.

Boulter, C. A. 1996. Did both extensional tectonics and magmas act as major drivers of convection cells during the formation of the Iberian Pyrite Belt massive sulfide deposits? *Journal of the Geological Society of London*, 153, 181–184.

Boyd, W. P., et al. 2000. A mesoscale phytoplankton bloom in the polar Southern Ocean stimulated by iron fertilization. *Nature*, 407, 695–702.

Braun, V., and H. Killmann. 1999. Bacterial solutions to the iron-supply problems. *Trends in Biochemical Sciences*, 24, 104–109.

Bridge, T. A. M., and D. B. Johnson. 2000. Reductive dissolution of ferric iron minerals by *Acidiphilium* SHJ. *Geomicrobiology Journal*, 17, 193–206.

Brierley, C. L. 1999. Bacterial succesion in bioheap leaching. In *Biohydrometallurgy and the Environment: Toward the Mining of the 21st Century*, vol. A, R. Amils and A. Ballester, eds., pp. 91–97. Amsterdam: Elsevier.

Clemente, L., L. Menanteau, and J. Rodríguez-Vidal. 1985. Los depósitos holocenos en el estuario de los ríos Tinto y Odiel (Huelva, España). In *Actas I Reunión del Cuaternario Ibérico*, vol. 1, pp. 339–353. Lisbon: Geological and Mining Institute.

Cloud, P. 1973. Paleoecological significance of the Banded Iron Formations. *Economic Geology*, 68, 1135–1143.

Davis, R. A., Jr., A. T. Welty, J. Borrego, J. A. Morales, J. G. Pendon, and J. G. Ryan. 2000. Rio Tinto estuary (Spain): 5000 years of pollution. *Environmental Geology*, 39, 1107–1116.

De Duve, C. 1987. Selection by differential molecular survival: A possible mechanism of early chemical evolution. *Proceedings of the National Academy of Science USA*, 84, 8253–8256.

De Ronde, C. E., M. J. De Wit, and E. T. C. Spooner. 1984. Early Archaean (< 3.2 Ga) Fe-oxide-rich, hydrothermal discharge vents in the Barbeton greenstone belt, South Africa. *Geological Society of America Bulletin*, 106, 86–104.

Durán, C., I. Marín, and R. Amils. 1999. Specific metal sequestering acidophilic fungi. In *Biohydrometallurgy and the Environment: Toward the Mining of the 21st century*, vol. B, R. Amils and A. Ballester, eds., pp. 521–530. Amsterdam: Elsevier.

Durán, C., C. Sargeant, N. Rodríguez, and R. Amils. 2001. Specific Cr(III) sequestering using an acidophilic fungal isolate. In *Biohydrometallurgy: Fundamentals, Technology and Sustainable Development*, Part B, V. S. T. Ciminelli and O. García, Jr., eds., pp. 237–246. Amsterdam: Elsevier.

Edwards, K. J., P. I. Bond, T. M. Gihrin, and J. F. Banfield. 2000. An Archaeal iron oxidizing extreme acidophile important in acidic mine drainage. *Science*, 287, 1796–1798.

Ehrlich, H. L. 2001. Past, present and future of biohydrometallurgy. *Hydrometallurgy*, 59, 127–134.

Elbaz-Poulichet, F., C. Braungardt, E. Achterberg, N. Morley, D. Cossa, J. M. Beckers, P. Nomérange, A. Cruzado, and M. Leblanc. 2001. Metal biogeochemistry in the Tinto–Odiel rivers (southern Spain) and in the Gulf of Cadiz: A synthesis of the results of TOROS project. *Continental Shelf Research*, 21, 1961–1973.

Fenchel, T., G. M. King, and T. H. Blackburn. 1998. *Bacterial Biogeochemistry: The Ecophysiology of Mineral Cycling*, 2nd ed. San Diego: Academic Press.

Geen, A. van, J. F. Adkins, E. A. Boyle, C. H. Nelson, and A. Palanques. 1997. A 120-year record of widespread contamination from mining of the Iberian pyrite belt. *Geology*, 25, 291–294.

Gehrke, T., R. Hallmann, and W. Sand. 1995. Importance of exopolymers from *Thiobacillus ferrooxidans* and *Leptospirillum ferrooxidans* for bioleaching. In *Biohydrometallurgical Processing*, vol. 1, T. Vargas, C. A. Jerez, K. V. Wiertz, and H. Toledo, eds., pp. 1–11. Santiago: Universidad de Chile.

González-Toril, E. 2002. Ecología molecular de la comunidad microbiana de un ambiente extremo: El Río Tinto. Ph.D. thesis, Universidad Autónoma de Madrid.

González-Toril, E., F. Gómez, N. Rodríguez, D. Fernández-Remolar, J. Zuluaga, I. Marín, and R. Amils. 2003. Geomicrobiology of the Tinto River, a model of interest for biohydrometallurgy. *Hydrometallurgy*, 71, 301–309.

Hansford, G. S., and T. Vargas. 2001. Chemical and electrochemical basis of bioleaching processes. *Hydrometallurgy*, 59, 135–145.

Hazeu, W., W. Bijleveld, J. T. C. Grotenhuis, E. Kakes, and J. G. Kuenen. 1986. Kinetics and energetics of reduced sulphur oxidation by chemostat cultures of *Thiobacillus ferrooxidans*. *Antonie van Leeuwenhoek Journal of Microbiology*, 52, 507–518.

Holland, H. D. 1973. The oceans: A possible source of iron in iron-formations. *Economic Geology*, 68, 1169–1172.

Holland, H. D. 1978. *The Chemistry of the Atmosphere and Oceans*. New York: Wiley.

Holland, H. D. 1984. *The Chemical Evolution of the Atmosphere and Oceans*. Princeton, NJ: Princeton University Press.

Ingledew, W. J. 1982. *Thiobacillus ferrooxidans*: The bioenergetics of an acidophilic chemolithotroph. *Biochemica et Biophysica Acta*, 683, 89–117.

Johnson, D. B. 1998. Biodiversity and ecology of acidophilic microorganisms. *FEMS Microbiological Ecology*, 27, 307–317.

Johnson, D. B. 1999. Importance of microbial ecology in the development of new mineral technologies. In *Biohydrometallurgy and the Environment: Toward the Mining of the 21st Century*, vol. A, R. Amils and A. Ballester, eds., pp. 645–656. Amsterdam: Elsevier.

Kasting, J. F. 1993. Earth's early atmosphere. *Science*, 259, 920–926.

Kirschvink, J. L., E. J. Gaidos, E. Bertani, N. J. Beukes, J. Gutzmer, L. N. Maepa, and R. E. Steinberg. 2000. Paleoproterozoic snowball Earth: Extreme climatic and geochemical global change and its biological consequences. *Proceedings of the National Academy of Sciences USA*, 97, 1400–1405.

Leblanc, M., J. A. Morales, J. Borrego, and F. Elbaz-Poulichet. 2000. A 4,500-year-old mining pollution in southwestern Spain: Long-term implications for modern mining pollution. *Economy Geology*, 95, 655–662.

Leistel, J. M., E. Marcoux, D. Thiéblemont, C. Quesada, A. Sánchez, G. R. Almodóvar, E. Pascual, and R. Sáez. 1998. The volcanic-hosted massive sulphide deposits of the Iberian Pyrite Belt. *Mineralium Deposita*, 33, 2–30.

Lescuyer, J. L., J. M. Leistel, E. Marcoux, J. P. Milési, and D. Thiéblemont. 1998. Late Devonian–early Carboniferous peak sulphide mineralization in the western Hercynides. *Mineralium Deposita*, 33, 208–220.

López-Archilla, A. I., I. Marín, and R. Amils. 1993. Bioleaching and interrelated acidophilic microorganisms from Río Tinto, Spain. *Geomicrobiology Journal*, 11, 223–233.

López-Archilla, A. I., I. Marín, and R. Amils. 2001. Microbial community composition and ecology of an acidic aquatic environment: The Tinto River, Spain. *Microbial Ecology*, 41, 20–35.

López-Archilla, A. I., I. Marín, A. González, and R. Amils. 1995. Identification of fungi from an acidic river. In *Fungal Identification Techniques*, L. Rossen, V. Rubio, M. T. Dawson, and J. Frisvard, eds., pp. 202–211. Luxembourg City: European Commission.

Martin, J. H. 1990. Glacial–interglacial CO_2 change: The iron hypothesis. *Paleoceanography*, 5, 1–13.

Moreira, D., A. I. López-Archilla, R. Amils, and I. Marín. 1994. Characterization of two new thermoacidophilic microalgae: Genome organization and comparison with *Galdieria sulphurica*. *FEMS Microbiology Letters*, 122, 109–114.

Ohmoto, H. 1997. When did the earth's atmosphere become oxic? *Geochemical News*, 93, 13.

Phillips, J. A. 1881. Note on the occurrence of remains of recent plants in brown iron-ore. *Journal of the Geological Society of London*, 27, 1–5.

Price, C. A. 1968. Iron compounds and plant nutrition. *Annual Review of Plant Physiology*, 19, 239–248.

Reid, R. T., D. H. Live, D. J. Faulkner, and A. Buttler. 1993. A siderophore from a marine bacterium with an exceptional ferric ion affinity constant. *Nature*, 366, 455–458.

Rivas-Martínez, S. 1987. Nociones sobre fitosociología, biogeografía y bioclimatología. In *La Vegetación de España*, M. Peinado and S. Rivas-Martínez, eds., Colección Aula Abierta, 3, pp. 19–45. Alcalá de Henares: Universidad de Alcalá de Henares.

Rodríguez-Vidal, J., E. Mayoral, and J. G. Pedón. 1985. Aportaciones paleoambientales al tránsito Plio-Pleistoceno en el litoral de Huelva. In *Actas I Reunión del Cuaternario Ibérico*, vol. 1, pp. 447–459. Lisbon: Geological and Mining Institute.

Sand, W., T. Gehrke, R. Hallmann, and A. Schippers. 1995. Sulfur chemistry, biofilm and the (in)direct attack mechanism—a critical evaluation of bacterial leaching. *Applied Microbiology and Biotechnology*, 43, 961–966.

Sand, W., T. Gehrke, P. G. Jozsa, and A. Schippers. 2001. (Bio)chemistry of bacterial leaching—direct vs. indirect bioleaching. *Hydrometallurgy*, 59, 159–175.

Silverman, M. P., and H. L. Ehrlich. 1964. Microbial formation and degradation of minerals. *Advances in Applied Microbiology*, 6, 153–206.

Trendall, A. F., and R. C. Morris. 1993. *Iron-Formation: Facts and Problems*. Amsterdam: Elsevier.

Vargas, M., K. Kashefi, E. L. Blunt-Harris, and D. R. Lovly. 1998. Microbiological evidence for Fe(III) reduction on early Earth. *Nature*, 395, 65–67.

Widdel, F., S. Schnell, S. Heising, A. Ehrenreich, B. Assmus, and B. Schink. 1993. Ferrous iron oxidation by anoxygenic phototrophic bacteria. *Nature*, 362, 834–836.

29
Climate and the Amazon—a Gaian System?

Peter Bunyard

Abstract

Models such as Daisyworld that support life's geophysiological role tend to be simplistic and can be dismissed on the grounds that they come nowhere near the complexity of interactions between living processes and the environment that occur in the real world. It has been suggested that the general circulation models (GCMs) used in predicting climate change over the next hundred years could provide the basis of models to test the validity of a "hard" Gaia hypothesis. However, in their current form GCMs are incapable of capturing regional atmospheric convection processes that lead to cloud formation and energy transfers, so critical to understanding climate. Such models therefore remain inadequate to the task.

On the other hand, mesoscale climate studies, with their higher resolution, can reveal interactions involving energy transfers between vegetation and the atmosphere as mediated through the hydrological cycle. Recent mesoscale studies on the powerful convection processes that take place over the Amazon Basin are providing evidence of the critical role played by vegetation in generating regional and even global climate. Such models therefore may provide us with the "correct" scale at which to test the potential of sub-Gaian systems to generate self-sustaining conditions that bear the hallmarks of a geophysiology.

Introduction

Owing to the limitations of our Earth science models, proof for Lovelock's "hard" Gaia hypothesis remains elusive. But what if we look at a sub-Gaian system that we know has a major impact on climate and that simultaneously demonstrates many of the regulatory phenomena, such as local temperature control and nutrient recycling which must feature in Gaia as a whole? And what better place to get the feel of life interacting with the atmosphere than a tropical rain forest, such as you'll find in the Colombian Amazon, and see for yourself the clouds of vapor being pumped up from the trees? The forest, held together by its intrinsic cloud-generating dynamics, as well as by the diversity of energy-partitioning organisms—bacteria, fungi, flowering plants, insects, vertebrates—is now known to contribute significantly to climate across the globe, just as it is affected by climate change and variability originating, for instance, in the Pacific Ocean or in the North Atlantic (Bunyard, 2001).

In the context of Lovelock's Gaia, the question is whether it matters one jot to the rest of living organisms on the planet if the forest survives in its current extent and form. Is the Amazon rain forest contributing in any measurable sense to the regulation of surface temperature? Is it benefiting life as a whole through enhancing biotic interactions with surface planetary phenomena by, for instance, affecting rainfall patterns in other regions? And can we adequately incorporate the mosaic of ecosystems that comprise the Amazon Basin into a climate model? If we were wise, we would not test the contribution of ecosystems to planetary phenomena such as climate by destroying them wholesale. Unfortunately, our current demands for natural resources are such that we do destroy, and are generally heedless of the long-term damage we may be causing. As a vast, largely unexploited region, the Amazon Basin is undoubtedly threatened with development activities, including turning millions of hectares of rain forest into soy plantations. As we shall see, such grandiose changes will play havoc with current hydrological processes and, consequently, have impacts much further afield.

Were we to perpetrate such a dangerous experiment with nature, we might conceivably find empirical evidence that the ecosystems of the Amazon Basin do play an essential role in stabilizing contemporary planetary conditions. Through such antievolutionary activities we might then, belatedly, prove the corollary to Jim Lovelock's thesis, in which he stated: "The evolution of organisms and their material environment proceeds as a single tight-coupled process from which self-regulation of the environment at a habitable state appears as an emergent phenomenon" (Lovelock, 1996, 2002).

Climate and the Amazon

The crucial work of the Brazilian physicist Eneas Salati in the 1980s reveals that recycling of the rainfall through evapotranspiration from the Amazon is essential for maintaining vegetation farther to the west of the basin, given that the prevailing winds of the Walker Circulation flow from east to west, from the Atlantic Ocean to the Andes. Carlos Nobre and others have since confirmed Salati's finding that approximately 50 percent of all the rainfall over the basin feeds into the Amazon River and out into the Atlantic. But whereas Salati believed that most of the 50 remaining percent stayed within the basin, as part of the dynamic of forest-induced cloud formation followed by precipitation, recent studies suggest that at most 20 to 30 percent of total rainfall stays. The rest gets carried out of the region, south toward São Paulo and Rio Grande do Sul and even across the South Atlantic toward South Africa, while to the north, another stream is deflected by the Andes toward the Caribbean. A small proportion rises over the Colombian Andes and goes on to the Pacific Ocean (Salati, 1987).

In fact, the Amazon Basin provides São Paulo with the greater part of its rainfall, some 70 percent, and southeastern Brazil and Argentina with as much as 50 percent. If Brazil allowed deforestation of the Amazon to occur to the point of collapse of the critical self-watering regime, it would be doing itself the biggest disservice imaginable—nor would its neighbors take kindly to seeing their source of rain dwindle away. As it happens, both São Paulo and Rio de Janeiro were subject to repeated blackouts in 2001, and are still suffering electricity rationing as a result of falling water levels in hydroelectric reservoirs in the non-Amazonian parts of Brazil. Those large cities had better ensure that their source of water remains unimpaired (Fearnside, 2000).

Equally dramatic, climatologists have become aware that the energy released in vast quantities when water vapor over the Amazon precipitates as rain does not remain within the region. Shafts of energy, conducted in wave trains of air moving through preferred routes across the globe, affect the climates of South Africa, North America, Southeast Asia, and even Europe. The Amazon Basin is truly connected—climatologists call it teleconnection—to global climate (Pielke, 2002).

Approximately 6.5 trillion metric tons of water (10^{12}) are evapotranspired over the Legal Amazon of Brazil each year, of which half is exported out of the region and the other half is recycled within the basin itself. In total that evapotranspiration takes up 1.63×10^{22} joules per year of the sun's energy, which is equivalent to nearly 520 terawatts (40 times the total energy used by humanity). A sizable proportion of that massive amount of energy is teleconnected across the globe, thereby affecting climate everywhere. Even if there were no forest, the water and energy transport would not be zero, because they are largely driven by the difference in the planetary energy balance between the equator and the poles. Nevertheless, it has become clear that the functioning of the Amazon Basin as a hydrological power engine is a critical component of contemporary climate (Bunyard, 2001).

Thunderstorms are the key to teleconnection. Most thunderstorms occur in a narrow band around the equator, some 1500 to 5000 a day, rising to considerable heights as precipitating water fuels them through the release of latent energy, and as localized differences in warming and heat transfer bring about powerful convective updrafts. Perhaps as much as two-thirds of precipitation around the planet is affected by the formation of cumulonimbus and stratiform cloud systems generated within the tropics. The heat, moisture, and kinetic energy, which are carried from the tropics to the middle and higher latitudes, have a profound impact on the ridge and trough pattern associated with the polar jet stream.

Changes in land use and in land cover over the humid tropics are therefore effecting climate by altering and transforming the dynamics of cloud formation. As R. Pielke points out, "These alterations in cumulus convection teleconnect to middle and higher latitudes, which alters the weather in those regions. This effect appears to be most clearly defined in the Winter Hemisphere" (Pielke, 2002).

At the present, our concerns about climate change in the future are based almost exclusively on the predictions emerging from general circulation models such as those used by the UK Met Office's Hadley Centre in Bracknell. Because of the enormous volume of data and number crunching required, the resolution used is too low to generate phenomena such as cloud formation at the local level. Climatologists who work with the general models maintain, almost as an article of faith, that by smoothing out localized phenomena into a broad average, the picture emerging from their computer iterations will finish up reasonably close to reality. For them, if the point of a journey is to get from A to B, then the outcome of following the road and getting to the destination is generally the same whether the road surface is smooth or marred by irregular bumps and indentations (Betts et al., 1997).

Another breed of climatologists hotly disputes the rationale that by applying broad generalities, you will necessarily arrive at the right conclusion. For them, each new bump and hole that forms on the general trajectory of the road may bring about a wholly different outcome, one quite contrary to that of the general model. Roni Avissar, professor of civil and environmental engineering at Duke University, and Pedro Silva Dias, at the Institute of Astronomy, Geophysics and Atmospheric Sciences at the University of São Paulo, work at the mesoscale, using models that are capable of realizing local phenomena, such as cloud formation. They are skeptical that even the best of today's general circulation models (GCMs) can be relied upon to give an accurate prediction of future climate (Werth and Avissar, 2002; Silva Dias et al., 2002).

As Avissar and his colleagues have pointed out, you cannot simulate thunderstorm formation explicitly with the GCMs because their resolution is not good enough. Indeed, thunderstorms are represented in GCMs only through a parameterization that in its present form is incapable of taking into account what is actually happening on the ground in terms of medium-scale variations to landscape. Though "unseen" in GCMs, powerful phenomena such as thunderstorms are also generated as a consequence of contrasting heat flows, which result from the juxtaposition of at least two fundamental landscape types, such as humid tropical forest abutting land that has been cleared for cattle, or that has a wide road or a natural feature of landscape, such as a river, cutting through it.

"We now know," says Avissar, "that convection activity is one of the key factors that trigger teleconnection between tropical regions and the midlatitudes and this essential cloud-forming activity is not captured in the GCMs. Consequently, the major impact of man-made landscape modification is not properly represented in GCMs, and the modeling results obtained so far probably underestimate the real effects of landscape alterations locally, regionally, and by teleconnection to other regions" (Avissar, 2002 Meeting at STRI, Panama).

Avissar also points out that we know far too little about the relationship between the hydrological cycle and climate to be able to make valid predictions about the future. "Water is the most important by far of all the greenhouse gases," he remarks, "and yet we are still uncertain as to how water will behave in a world with elevated carbon dioxide levels and a changing landscape. We cannot claim to any real powers of prediction: we are more in the realm of philosophy than of science" (Avissar, 2002, personal communication).

As the two climatologists point out, the mesoscale offers the opportunity to study what is happening to local climate over distances from 10 to 1000 kilometers as a result of differences in land cover. At that scale the questions that vitally need answers concern what happens to airflows if the clearing is smaller or bigger; under what circumstances convection patterns develop; whether such patterns lead to cloud formation and does that mean more rain; whether mathematical models can capture the actual physics of local meteorology; and how such models stand up against what is actually encountered in the field.

Finally, Avissar and Silva Dias hope that verification of their respective mathematical models by matching the results with those observed in the field will help them home in on the critical point at which the rain forest collapses because of a self-reinforcing drying out. One effect of drying out is to make the forest increasingly vulnerable to fire, especially during dry years, such as are associated with strong El Niños (e.g., that of 1998), when vast areas of the state of Roraima were ablaze.

Then, as William Laurance of the Smithsonian Tropical Research Institute puts it, "fires lit by small-scale farmers swept through an estimated 3.4 million hectares of fragmented and natural forest, savanna, regrowth and farmlands in the northern Amazonian state of Roraima. Even in the absence of drought," he continues, "Amazon forest remnants experience sharply elevated rates of tree mortality and damage, apparently as a result of increased desiccation and wind turbulence near forest edges. These changes lead to a substantial loss of forest biomass, which has been estimated to produce from 3 to 16 million tonnes of carbon emissions per year in the Brazilian Amazon alone. In drought years, the negative effects of fragmentation may well increase" (Laurance and Williamson, 2001).

The chances of lightning strikes starting fires in the forest are minimal, according to Mark Cochrane of Michigan State University and Daniel Nepstad of the Woods Hole Oceanographic Institute. Fires in the Amazon are a consequence of deforestation and land use change. Nepstad and his colleagues find that forests that have been subjected at least once to fires are far more vulnerable to subsequent fires in terms of tree mortality. Initial fires may cause up to 45 percent mortality in trees over 20 dbh (diameter breast height), and subsequent fires up to 98 percent mortality. Meanwhile, during observations of fires in December 1997 in the eastern part of the Amazon, in Tailândia, they found that initial fires led to the immediate release of 15 metric tons of carbon per hectare, and recurrent burns, up to 140 metric tons of carbon per hectare (Cochrane and Laurance, 2002).

Charcoal studies indicate that in lowland tropical rain forests, natural fires are rare events, perhaps involving a rotation of hundreds if not thousands of years. According to recent research by Cochrane and Laurance, "Fire-return intervals of less than 90 years can eliminate rain forest tree species, whereas intervals of less than 20 years may eradicate trees entirely.... Fragmented forests in the eastern Amazon are currently experiencing fire rotations of between 7 and 14 years. Previously burned forests are even more prone to burning, with calculated fire rotations of less than 5 years."

Successive dry years, such as a succession of El Niño years, also will make the forest extremely vulnerable to drying out and fires. During the exceedingly strong El Niño of 1998, says Nepstad, one third of Brazil's Amazon rain forest experienced the soil drying out down to 5 meters, close to the limits of water uptake through the roots. Consequently 3.5 million square kilometers were at risk, with some trees having to pull water up from as deep as 8 meters. During that period of stress, Nepstad noted that tree growth went down practically to zero as evidenced by canopy thinning rather than leaf-shedding.

Precipitation and the Forest

Many studies have shown the sharp differences in daily temperature between a natural forest and cleared land. In Nigeria, for example, the daytime temperature just above the soil in a clearing was 5°C higher than in the nearby forest, and humidity was 49 percent compared to the forest's 87 percent. Clearings also are far more likely to flood and, consequently, erode. Carlos Molion, of Brazil's Institute for Space Research, points out that the forest canopy in the Amazon intercepts on average about 15 percent of the rainfall, a large proportion of which evaporates directly back into the atmosphere. The removal of the canopy leads to as much as 4000 metric tons of water per hectare hitting the ground, causing selective erosion of finer clay particles and leaving behind increasingly coarse sand. Soil under intact forest absorbs ten times more water compared with pasture, where erosion rates may be 1000 times greater (Bunyard, 2001).

Given that evapotranspiration within the Amazon Basin is responsible for up to 30 percent of recycled rainfall, it is obvious to most meteorologists and climatologists that deforestation leads ultimately to a decline in rainfall. Roni Avissar points out that at least three general models of the relationship between deforestation and rainfall can be conceived, the most obvious being a straight-line curve in which overall precipitation declines more or less proportionately as forest is cleared, at least to 70 percent of the initial value. Another model indicates that rainfall declines sharply during the first third or so of forest clearance; from then on, the decline is steady and slow again until 70 percent is attained.

But what of a third possibility—that forest clearing initially leads to an increase rather than a decline in rainfall, until a critical point is reached in forest clearance and rainfall plummets to the 70 percent value? In terms of policy making, the third model, with its initial peak followed by precipitous decline, is likely to give a false sense of all being well—indeed, better than well—since rainfall initially goes up rather than down. But then, once the critical point has been passed, the rapid decline in rainfall becomes the driver of forest dieback, rather than the other way round. On the basis of mesoscale modeling and its correspondence to observations of rainfall changes, both Avissar and Pedro Silva Dias have few doubts that we had better reckon on the third model being the most likely to represent reality. In that regard, the above-normal increase in rainfall over Rondônia in the border lands between Brazil and Bolivia does not bode well, inasmuch as it suggests that deforestation has already led to fundamental disturbances to the hydrological cycle.

The differences in daytime ground temperatures and humidities of the forest compared to the clearing are critical components of the convective forces that lead to the rapid formation of rain-bearing clouds. Essentially, the air masses over the Amazon Basin are driven from east to west, but within that system, strong local convection is likely to arise as a result of the relatively dry, warmer air of pasture drawing in cooler, damper air from the surrounding forest. That process has its parallels with the strong onshore winds that blow from the sea over the land during late summer afternoons. Consequently cumulonimbus rain-bearing clouds begin to form and the net result is rainfall when, without the dynamic interface between forest and pasture, there might otherwise have been none. The models need elaborating further, says Avissar, but the general indications are that such convection is a function of clearings up to 100 kilometers or so in diameter.

When the clearings increase significantly in size and the surrounding forest becomes fragmented, the system of cloud formation abruptly breaks down. The precise relationship in terms of size between forest and clearing for enhancing rainfall has yet to be determined, if indeed it can. Avissar's hunch is that forest cover needs to be 60 percent or more overall, and the clearings not much greater than 100 kilometers across (Silva Dias and Avissar, 2002).

Together, mesoscale modeling and observations, including micrometeorology and satellite scanning, reveal that as the air rises over the clearing, it is counteracted by air sinking over the neighboring forest. In general, air high up in the lower atmosphere bears less moisture than air closer to the surface. Consequently, the sinking motion of the air mass brings drier air down to the surface, with the net result that the forest close to the margins of the clearing becomes vulnerable to edge effects and to dieback.

Both Roni Avissar and Pedro Silva Dias point out that the current climate over the Amazon Basin is far more a product of climate variability resulting from periodic changes in the Pacific and Atlantic oceans than it is of enhanced carbon dioxide levels and temperature rise. Both the southern Pacific El Niño/La Niña oscillation (ENSO) and the North Atlantic interdecadal oscillation are now known to have powerful influences on Amazonian climate. In fact, annual levels of carbon dioxide in the atmosphere are very much governed by phenomena such as El Niño which, by causing a fall in primary photosynthesis of terrestrial vegetation, can lead to a significant rise in carbon dioxide levels, as occurred in 1998 and no doubt, was augmented by human-induced forest fires in Brazil, Indonesia, and the boreal forests of Siberia (Silva Dias and Avissar, 2002).

In essence, the link between ENSO and the Amazon is of critical importance, especially if deforestation in Southeast Asia could be a contributing factor to the increase in strength and frequency of El Niños. More frequent, stronger El Niños, combined with the drying effects of excessive deforestation in the Amazon Basin, will undoubtedly have a pernicious impact on the remaining forest. Then the rapid collapse of the forest ecosystem will feed back onto global climate, reinforcing the severity and longevity of El Niños (Bunyard, 2001).

For a considerable period of modern human history, the Amazon rain forest has remained intact and presumably has functioned as a self-sustaining system in the context of the recycling of water through cumulonimbus formation. From a Gaian point of view, the self-sustaining recycling of water is an emergent property of the system, as is the extraordinary biodiversity associated with the Amazon Basin, which undoubtedly plays its role in maintaining the integrity of the entire system. Part of all that is the role of the forest in releasing cloud condensation nuclei in the form of volatile organic compounds such as isoprenes. As Pedro Silva Dias points out, during certain seasons the winds over certain regions of the Amazon switch to southwesterlies that are marked by the absence of convective cloud formation and instead bear rain as the result of vegetation-generated Cloud Condensation Nuclei (CCNs) (Silva Dias and Avissar, 2002).

Now, as a result of Avissar's recent mesoscale, high resolution climatological studies, we are beginning to get hints of the specific teleconnection relationships between the hydrology of the Amazon Basin and other parts of the world. Given the current U.S. administration's attitude to international conventions on climate change, not least the Kyoto Protocol, it should be of interest that the future of rainfall patterns over the corn belt of the United States may well be dependent on the "health" of Amazon rain forests. Indeed, in simulating the impact of the loss of Amazon rain forest on the United States, Avissar's models show a significant decline in rainfall, by as much as 15 to 20 percent over agricultural regions in the United States, during the critical growing months.

Such rainfall changes will have momentous implications for U.S. agriculture, given the importance of its contribution to global exports, not least for the meat and dairy business of western Europe. In fact, the sheer vastness of the Amazon Basin makes its teleconnection role uniquely important, far exceeding that of the remaining forests of tropical Africa or of Southeast Asia on Europe or on other parts of Asia and the Pacific.

Amazon Palaeoclimate

In recent years it has almost become dogma that the Amazon Basin loses its closed canopy forests during glacial periods because of the much drier climatic regime of a colder Earth. The contention is that forest is largely replaced by a grassland-dominated savanna until a switch in climate allows the forest to be regenerated from patches where the original ecosystem had survived—the so-called biological refugia. The inference therefore is of a forest system that is sensitive and even vulnerable to climate change.

That theory of forest loss and regeneration from refugia has received something of a setback from the work of Mark Maslin, Sharon Cowling, Martin Sykes, and others. They have studied pollen records going back at least 10 million years from fan material deposited as a series of annual strata in the mouth of the Amazon River. Their findings of tree species pollen dominating the paleovegetation record indicate that the biodiverse richness of today's Amazon Basin is most likely the result of physiological adaptations of the forest to colder conditions during glacial periods rather than of relatively small patches of forest ecosystem surviving in a sea of savanna grasses.

Cowling, Maslin, and Sykes have modeled the impact of each of the three physiological criteria of atmospheric carbon dioxide concentration, rainfall level, and temperature on the mean leaf area index, which is basically a measure of leaf coverage, and hence of whether the vegetation is forest with a closed canopy or is more savanna-like. The modeling strongly reinforces the palaeontological data taken from the mouth of the Amazon. It shows that forest can withstand low carbon dioxide levels and lower rainfall *only* when temperatures are also lower than today (Cowling and Sykes, 1999).

The main effect of the cooler temperatures is to reduce the photosynthetic losses brought about by photorespiration, in which oxygen competes with carbon dioxide for rubisco, the carbon-fixing enzyme in C_3 plants. In addition, lower temperatures reduce evapotranspiration, with the result that vegetation can make better use of the water available for carbon uptake into the leaves. As Sharon Cowling and her colleagues point out, "Cooler LGM (Last Glacial Maximum) temperatures may have helped to improve carbon and water balance in glacial-age tropical forests, thereby allowing them to out-compete grasslands and maintain dominance within most of the Amazon Basin" (Cowling et al., 2001, 143).

But what if temperatures rise over the forest and rainfall decreases? The higher carbon dioxide levels of modern times will certainly offset some of the photorespiration losses that will arise from higher temperatures, but the evidence is that the forest will suffer irremediably from the hotter internal conditions brought about through diminished availability of water for transpiration. The canopy-thinning that Nepstad noted during severe El Niño episodes indicates that, physiologically, the humid rain forests of the Amazon are close to their tolerance levels. They are now living close to the edge; hence, warmer temperatures and less precipitation are likely to serve as their coup de grace.

Global Warming Puts Amazon at Risk

Evidence that the future survival of the Amazonian rain forest could be at stake, were greenhouse gas emissions to increase at their current rate over the next 50 years, comes from the work of Richard Betts, Peter Cox, Matthew Collins, and others at the UK Met Office Hadley Centre and Reading University. From an elaborate, coupled climate–vegetation model they find that as global warming takes hold, the global climate becomes more like that of an El Niño year, leading to a substantial change in the airflow. Indeed, the model shows a warming in the eastern Pacific, with up to a 3°C rise in temperature, and precipitation over Amazonia down by 2 or more millimeters per day, on average, that mimics what happens during an El Niño event.

Those precipitation changes become even more marked when the fertilization effect of increased atmospheric carbon dioxide is taken into account, the reason being that the trees transpire less and so put less water back into the atmosphere. The net consequence is that the forest begins to die back, not only fueling the atmosphere with more greenhouse gases but also further diminishing evapotranspiration. That "positive" feedback leads to further dieback and ultimately to the loss of much of the natural forest of Amazonia. Under such circumstances, the net loss of carbon from vegetation and from soil in Amazonia will amount to more than 70 GtC. Worldwide the loss of carbon from soils could be double that, adding one-fifth to the current carbon dioxide in the atmosphere. Such emissions of carbon, combined with those from fossil fuel combustion given current trends, would raise average global land surface temperatures by 8°C, more than 50 percent higher again than reported by the Intergovernmental Panel on Climate Change in its 2000 report on global warming. Few doubt that such severe temperature rises would be catastrophic, taking the temperature to levels not seen for 40 million years, when the Earth had no permanent ice sheets (Cox et al., 2000; Betts et al., 2002).

Carbon Uptake Jeopardized by Deforestation

Other subtleties enter the picture that make the situation even graver. In a personal communication Richard Betts pointed out that inasmuch as deforestation causes an additional climatic effect through reducing evapotranspiration, carbon accounting will therefore underrepresent the effects of deforestation on climate: "In essence, the climate perturbation you would expect by only considering carbon emissions could be smaller than that which you would expect if you considered both carbon emissions and reduced evapotranspiration. You might therefore take the view that a tonne of carbon emitted by Amazon deforestation is even worse than a tonne of carbon emitted from fossil fuel burning, because of the extra climatic effects of deforestation through evapotranspiration reduction. Kyoto does not consider this."

Rossby Wave Teleconnection at Risk from Deforestation

Nicola Gedney and Paul Valdes, from the Department of Meteorology at University of Reading, show from their models that, independent of global warming, deforestation of the Amazon would lead to considerable disturbances to climate over the northeastern Atlantic and western Europe as well as the eastern seaboard of the United States, especially during the northern hemisphere winter, which consequently would become considerably wetter.

Normally, during those winter months, convection is at its strongest over the Amazon Basin. Such convection, based on the lifting of considerable quantities of vapor, propagates strong Rossby waves, some of which head in a northwesterly direction across the Atlantic, toward western Europe. The Rossby waves emanating from the Amazon tend to be suppressed by strong easterlies aloft; nevertheless, under normal circumstances, with the forest intact, the latent heat source for the Rossby waves is strong enough to override the easterlies. That situation reverses when the forest is replaced by grassland, because of reduced precipitation over the basin, which leads to a generalized weakening of the tropical air mass circulation —the Walker and Hadley cells. Under those circumstances the easterlies aloft bring about a suppression of the now weakened Rossby waves.

According to Gedney and Valdes, "Our results strongly suggest that there is a relatively direct physical link between changes over the deforested region and the climate of the North Atlantic and western Europe. Changes in Amazonian land cover result in less heating of the atmosphere above. This then weakens the local Hadley Circulation resulting in reduced descent and increased rainfall over the south eastern US. The result of this is a modification to the Rossby wave source which causes subsequent changes in the circulation at mid and high latitudes in the northern hemisphere winter. This in turn causes changes in precipitation, namely an increase over the North Atlantic and a suggestion of some change over Western Europe" (Gedney and Valdes, 2002).

Conclusion

As an extension of Darwin's theory of natural selection, Lovelock posits not only that life's resilience has resulted in new strategies for dealing with a changed and ever-changing situation, but also that it has brought about a regulated environment or *milieu extérieur* that more closely favors mainstream life in terms of metabolism and physiology. If Lovelock is correct, then the recovery of life following massive extinctions, as in the Permian, is not simply the result of life waiting in the wings until conditions improve but, on the contrary, the result of life being actively involved in the transformation of the environment such that conditions emerge that are favorable for a wide range of lifestyles and are conducive to the flourishing of a multitude of niche-adapting and niche-adapted organisms. In fact, the evidence for power-law relationships between evolution and extinctions, as described by Michael Boulter, indicates full well the extraordinary ability of life to radiate into the environment around it, thereby forming an intimate, transforming relationship with the physicochemical attributes of the planet (Boulter, 2002).

Furthermore, if we pursue Lovelock's "hard" Gaia to its logical conclusion, it becomes clear that life is an essential component in a dynamic planetary surface system that gives rise to a life-sustaining climate as an emergent property. Greenhouse gases, as we well know, are under the regulation of the biota, and one of the remarkable consequences of photosynthesis is the production and then destruction of ozone, during which virtually all UV-C and much of UV-B from the sun are converted to harmless infrared radiation. Life, too, through bacterial denitrification and nitrogen-fixing, is responsible for the remarkable levels of nitrogen in the atmosphere—courtesy of which life can now flourish on continents, thus truly making Gaia a planetary phenomenon (Volk, 1998; Kirchner, 1991).

It would seem to me that just as mesoscale climate studies bring out phenomena that are virtually missed in general circulation models, so a study of the climate dynamics of the Amazon Basin will reveal fundamentals of the biota's regulating hand on regional planetary processes that tend to be obscured when scaled up to cover the planet as a whole. We now have the tools to demonstrate Gaia in action (Bunyard, 1996).

References

Betts, R. A., P. M. Cox, S. E. Lee, and F. I. Woodward. 1997. Contrasting physiological structural vegetation feedbacks in climate change simulations. *Nature*, 387, 796–799.

Betts, Richard A., Peter M. Cox, Matthew Collins, John H. C. Gash, Philip P. Harris, Chris Huntingford, Chris D. Jones, and Keith D. Williams. 2002. *Amazonian Forest Die-back in the Hadley Centre Coupled Climate–Vegetation Model.* UK Met Office, Hadley Centre, Bracknell.

Boulter, Michael. 2002. *Extinction, Evolution and the End of Man.* London: Fourth Estate.

Bunyard, P. 1996. *Gaia in Action: Science of the Living Earth.* Edinburgh: Floris Books.

Bunyard, P. 2001. Climate Change. *The Ecologist, Report*, 31(9), 4–11.

Cochrane, M. A., and W. F. Laurance. 2002. Fire as a large-scale edge effect in Amazonian forests. *Journal of Tropical Ecology*, 18(3).

Cowling, Sharon A., and Martin T. Sykes. 1999. Physiological significance of low atmospheric CO_2 for plant–climate interactions. *Quaternary Research*, 52, 237–242.

Cowling, Sharon A., Mark A. Maslin, and Martin T. Sykes. 1999. Paleovegetation simulations of lowland Amazonia and implications for neotropical allopatry and speciation. *Quaternary Research*, 55, 140–149.

Cox, Peter M., Richard A. Betts, Chris Jones, Steven A. Spall, and Ian J. Totterdell. 2000. Acceleration of global-warming due to carbon-cycle feedbacks in a coupled climate model. *Nature Letters*, 408 (November 9).

Fearnside, P. M. 2000. Global warming and tropical land-use change: Greenhouse gas emissions from biomass burning, decomposition and soils in forest conversion, shifting cultivation and secondary vegetation. *Climatic Change*, 46, 115–158.

Gedney, Nicola, and Paul J. Valdes. 2000. The effect of Amazonian deforestation on the northern hemisphere circulation and climate. *Geophysical Research Letters*, 19, 3053–3056.

Kirchner, J. W. 1991. The Gaia hypotheses: Are they testable? Are they useful? In *Scientists on Gaia*, S. Schneider and P. J. Boston, eds. Cambridge, MA: MIT Press.

Laurance, W. F., and G. B. Williamson. 2001. Positive feedbacks among forest fragmentation, drought, and climate change in the Amazon. *Conservation Biology*, 15, 1529–1535.

Lovelock, J. 1996. What is Gaia? In *Gaia in Action: Science of the Living Earth*, P. Bunyard, ed. Edinburgh: Floris Books.

Lovelock, J. 2002. What is Gaia? *Resurgence*, 211 (March/April), 7–8.

Pielke, R. A. 2002. Mesoscale meteorological modeling, 2nd ed. San Diego: Academic Press.

Salati, E. 1987. The forest and the hydrological cycle. In *The Geophysiology of Amazonia*, R. E. Dickinson, ed. New York: Wiley Interscience.

Silva Dias, M. A. F., W. Petersen, P. Silva Dias, A. K. Betts, A. M. Gomes, G. F. Fisch, M. A. Lima, M. Longo, and M. A. Antonio. 2002. A case study of the process of organization of convection into precipitating convective lines in southwest Amazon. *Journal of Geophysical Research.*

Silva Dias, P., and R. Avissar. 2002. *The Future of the Amazon: Impacts of Deforestation and Climate.* Smithsonian Tropical Research Institute. Unpublished.

Volk, T. 1998. *Gaia's Body: Towards a Physiology of Earth.* New York: Copernicus and Springer-Verlag.

Werth, D., and R. Avissar. 2002. The local and global effects of Amazon deforestation. *Journal of Geophysical Research*, 107, 8037.

30

On the Co-evolution of Life and Its Environment

S. A. L. M. Kooijman

Abstract

Life interacted with climate and geochemical cycles during the evolution of system Earth. The only way to understand this interaction and appreciate its strength is by modeling it. Models, however, are useful only if their construction satisfies a number of basic requirements, which are briefly discussed in the context of the empirical cycle; many models do not satisfy these requirements. Modeling the interaction is only in its childhood; the biology in existing models for climate and geochemical cycles is generally weak. The main interaction mechanisms are briefly reviewed. The Dynamic Energy Budget (DEB) theory offers a useful framework for modeling the interactions; this theory specifies the uptake and use of substrates (including nutrients and light) by organisms and implies body-size scaling relationships for parameter values that quantify the processes of uptake and use of substrates. Current developments of the DEB theory are indicated, emphasizing the syntrophic interactions between species of organisms which are characteristic for life. The theory of adaptive dynamics is currently applied to DEB-structured theoretical ecosystems to understand the process of self-organization at an evolutionary timescale. The hope is that this will lead to insight into the structural and functional aspects of ecosystems which can be used to model the evolution of life and its environment.

Introduction

The subject of co-evolution of life on earth and its environment is fundamental to biology, and essential for understanding the global consequences of human actions. The idea that life has had, and still has, a substantial impact on geochemical cycles and climate is the essence of the Gaia hypothesis. Many varieties of this idea exist, but most of the ideas are rather qualitative.

An increasing number of global climate models where life plays some role are appearing in the literature. Most of these models are difficult to judge, because they do not result from a set of mechanistic assumptions, but are constructed to describe particular phenomena and observed patterns. Some experienced modelers seem to think that this is the main task of models (see, e.g., Harte, 1991); I disagree with this point of view.

Models have a lot in common with a story about quantities, phrased in the language of mathematics; they can have (and frequently do have) errors in grammar, they can tell nonsense and be uninteresting. However, they can also be exciting, depending on they way they are put together.

After identification of the scientific problem, the empirical cycle should start with a set of assumptions, a derivation of a mathematical model from these assumptions, a sequence of tests for consistency, coherence, parameter sensitivity, and relevance with respect to the problem (figure 30.1).

The second part of the empirical cycle consists of auxiliary theory for how variables in the model relate to things that can be measured, the setup of adequate experiments and/or sampling and measurement protocols to test model predictions, the collection of the measurements, and statistical tests of model predictions against measurements. These tests could reveal that the protocols have been less than adequate, and should be redesigned and executed; possible inadequacies should be detected in the auxiliary theory. Thus inconsistencies between data and model predictions do not necessarily point to inadequacies in the model.

If anywhere in this two-segment cycle the model needs to be improved, it should not be changed directly; the list of assumptions should be adapted, and the whole process should be repeated. It is a long and painstaking process, but sloppy procedures easily lead to useless results. Advocates of beginning the empirical cycle with observations, rather than with assumptions, are frequently unaware of the implicit assumptions that need to be made to give observations a meaning. The most important aspect of modeling is to make all assumptions explicit. If modeling

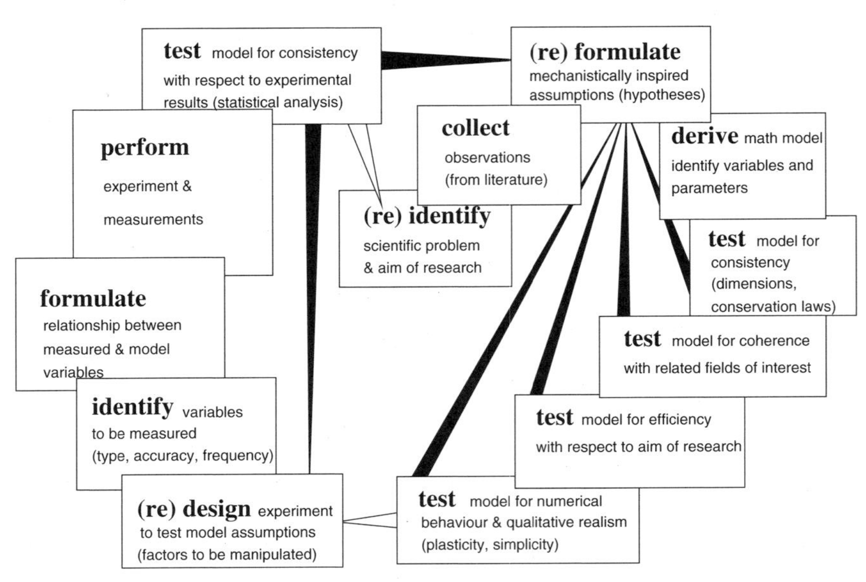

Figure 30.1
The empirical cycle as conceived by a theoretician; follow white arrows for "yes," black ones for "no." In the knowledge that nonsense models can easily fit any given set of data well, given enough flexibility in the parameters, realism is neither the first nor the most important criterion for useful models. Lack of fit (thus lack of realism) indicates that the modeling job is not completed. This discrepancy between prediction and observation can be used to guide further research, which is perhaps the most useful application of models. This application to improve understanding works only if the model meets the criteria indicated in the figure; few models meet these criteria, however.

procedures are followed in a sloppy way, by adapting models to fit data directly, it is likely that the result will be sloppy, too; one easily falls in the trap of curve-fitting. If it comes to fitting curves to data, the use of a pencil, rather than a model, is so much easier.

A lot more can be said about constraints on useful models, but a fair observation is that most existing climate and geochemical models are still weak in the role of biota and that geochemical cycling models are not well integrated with climate models. Reasons lie in the lack of training in biology among modelers, who frequently have a background in physics and/or chemistry; the lack of training in modeling and mathematics among biologists and geologists; and the complexity of life. The subject obviously calls for a multidisciplinary approach, for which we still need to find effective organizational structures.

The next section briefly reviews the main interactions between life and its environment that need consideration to reach a basic understanding of coevolution.

Effects of Life on Climate

Climate modeling mainly deals with energy (temperature) and water balances. Heat and water transport and redistribution, including radiation and convection in the atmospheres and oceans, depend on many chemical aspects, which means that climate modeling cannot be uncoupled from modeling biogeochemical cycling. I here focus on radiation, as affected via albedo and absorption by greenhouse gases.

Water

Because of its abundance, water is by far the most important greenhouse gas. Its origin is still unclear; some think it originates from degassing of the hot young planet (Krauskopf and Bird, 1995); others think from meteoric contributions in the form of carbonaceous chondrites (E. K. Berner and Berner, 1996), which possibly continues today.

Plants modify water transport in several ways. Although plants can extract water from the atmosphere, particularly in arid environments (by condensation at their surface as well as via the emission of condensation kernels), they generally pump water from the soil into the atmosphere, and increase the water capacity of terrestrial environments by promoting soil formation in bare environments (chemically, with help of bacteria; Berthelin, 1983), thereby reducing water runoff to the oceans. This became painfully clear during the flooding disasters in Bangladesh that followed the removal of Himalayan forests in India. On a short timescale, plants greatly reduce erosion; their roots prevent or reduce soil transport by common mild physical forces. In combination with rare strong and usually very short-term physical forces that remove vegetation (fires in combination with hurricanes or floods, for instance), however, plants increase erosion on a longer timescale, because they enhance soil formation in rocky environments. Because such "catastrophes" are rare, they have little impact on short timescales. The effects of plants on climate and geochemistry were perhaps most dramatic during their conquest of dry environments in the middle Devonian. It came with a massive discharge of nutrients and organic matter into the seas, which led to anoxia and massive extinctions in the oceans (Algeo et al., 2001).

Plants therefore affect the nutrient (nitrate, phosphate, silica, carbonates) supply to the oceans in complex ways, and thus the role of life in the oceans in the carbon cycle. Plants pump water from the soil into the atmosphere much faster in the tropics than in the temperate regions because of temperature (high temperature comes with large evaporation), seasonal torpor (seasons become more pronounced toward the poles, so plants are active during a shorter period of the year toward the poles), and nutrients in the soil (plants pump to get nutrients, which are rare in tropical soils).

Plants substantially influence their local environment, and facilitate colonization by other forms of life, which follows a sequence of ecosystem succession. As holds for most forms of life, plants, and especially the flowering plants, need other organisms (fungi, animals) for survival and propagation. Their massive appearance in the Carboniferous greatly affected global climate, via effects on the carbon cycle (Berner, 2001). Most climate models keep the mean global relative humidity constant at 50 percent (e.g., Crowley and North, 1991), but this assumption can be questioned.

Carbon Dioxide

Carbon dioxide is the second most important greenhouse gas. Its dynamics involves the global carbon cycle, which is still poorly known quantitatively. This is partly due to the coupling with other cycles.

Carbon dioxide is removed from the atmosphere by chemical weathering of silicate rocks, which couples the carbon and silica cycles. This weathering occurs via wet deposition, and gives a coupling between the carbon and the water cycle. When ocean downwashed calcium carbonate and silica oxide precipitate and become deeply buried by plate tectonics in earth's mantle, segregation occurs into calcium silicate and carbon dioxide; volcanic activity puts carbon dioxide back into the atmosphere. Geochemists generally hold this rock cycle to be the main long-term control of the climate system.

Westbroek (1991) argued that the role of life in the precipitation processes of carbonates and silica oxide became gradually more important during evolution. Mucus formers (by preventing spontaneous precipitation of supersaturated carbonates) and calcifiers have controlled carbonates since the Cretaceous. Diatoms (and radiolarians) have controlled silicates since the Jurassic (Krumbein and Werner, 1983). Corals and calcifying plankton (coccolithophores and foraminiferans) have an almost equal share in calcification. In freshwater, charophytes are in this guild. For every pair of bicarbonate ions, one is transformed into carbon dioxide for metabolism, and one into carbonate. Planktonic-derived carbonate partly dissolves, and contributes to the buildup of a concentration gradient of inorganic carbon in the ocean. This promotes the absorption of carbon dioxide from the atmosphere by seawater.

The dry deposition of carbon dioxide in the ocean is further enhanced by the organic carbon pump, where inorganic carbon is fixed into organic carbon, which travels down to deep layers by gravity. This process is accelerated by predation—unicellular algae are compacted into fecal pellets—and partial microbial decomposition recycles nutrients to the euphotic zone, boosting primary production. The secondary production also finds its way to the deep layers.

Most of the organic matter is decomposed in the deep ocean. The net effect is a depletion of inorganic carbon from the euphotic zone, which promotes the transport of carbon from the atmosphere into the oceans. This process is of importance on a timescale in the order of millennia (the cycle time for ocean's deep water), and thus is relevant for assessing effects of an increase of atmospheric carbon by humans. It is less important on much longer timescales.

In nutrient-rich shallow water, organic matter can accumulate fast enough to form anaerobic sediments, where decomposition is slow and incomplete, and fossilization into mineral oil occurs. Although textbooks on marine biogeochemistry do not always fully recognize the role of plants in the global carbon cycle (cf. Libes, 1992: 139), coal deposits in freshwater marshes are substantial enough to affect global climate. Oil formed by plankton and coal formed by plants occur mainly on continental edges, and affect climate on the multimillion timescale.

Methane
Methane is the third most important greenhouse gas; 85 percent of all emitted methane is (presently) produced by methanogens (in syntrophic relationships with other organisms, sometimes endosymbiotic) in anaerobic environments (sediments, guts) (Mah et al., 1977; Ferry, 1993). The flux is presently enhanced by large-scale deforestation by humans via termites. Apart from accumulation in the atmosphere, and in fossilized gas, big pools (2×10^3–5×10^6 Pg) of methane hydrates rest on near shore ocean sediments. Since, on a molar basis, methane can capture infrared radiation 25 times better than carbon dioxide, a release of the methane hydrates can potentially destabilize the climate system (Lashof, 1991). Oxidation of methane is a chief source of water in the stratosphere (Butcher et al., 1992), where it interferes with radiation.

Like carbon dioxide, the methane balance is part of the global carbon cycle. Since most of life's activity is limited by nutrients, the carbon cycle cannot be studied without involving other cycles. Nitrogen (nitrate, ammonia) is the primary limiting nutrient, but iron may be limiting as well in parts of the oceans (Baar and Boyd, 2000; Chai et al., 2000). After assuming that dinitrogen-fixing cyanobacteria could eventually relieve nitrogen limitation, Tyrell (1999) came to the conclusion that nitrogen was proximately limiting primary production, and phosphate was ultimately doing so. The question remains, however, whether cyanobacteria are active enough. Many important questions about the nitrogen cycle are still open, even if oceans represent a sink or a source of ammonia, nitrates, and nitrous oxide (Jaffe, 1992). The latter is, after methane, the next most important greenhouse gas, which can intercept infrared radiation 200 times better than carbon dioxide.

Most nutrients enter the oceans via rivers, which couples both systems and makes coastal zones very productive. The surface area of this habitat has obviously been under control of plate tectonics and seawater level changes, and therefore has been linked to ice formation and temperature. These remarks serve to show the link between climate and biochemical cycles.

Dioxygen
Complex relationships exist between the carbon and oxygen cycles. Dioxygen results from photosynthesis, so there is a direct relationship between dioxygen in the atmosphere and buried fossil carbon. The latter probably exceeds dioxygen on a molar basis, because of, for instance, the oxidation of iron and other reduced pools in the early history of the Earth. Photorespiration links dioxygen to carbon dioxide levels; both gases bind competitively to rubisco and drive carbohydrate synthesis in opposite directions. This effect of dioxygen is possibly an evolutionary accident that resulted from the anoxic origins of rubisco. Spontaneous fires require at least 75 percent of present-day dioxygen levels; oxygen probably now sets an upper boundary to the accumulation of organic matter in terrestrial environments, and thus partly controls the burial of fossilized carbon (Worsley et al., 1991). The extensive coal fires in China at 1 km depth, which have occurred since the beginning of human memory, illustrate the importance of this process. Model calculations by Berner (1991) suggest, however, that dioxygen was twice the present value during the Carboniferous. If true, this points to the control of fossil carbon accumulation by oxygen's being weak. The big question is, of course, to what extent humans are perturbing the climate system by enhancing the burning of biomass and fossil carbon. The massive burning of the worlds' rain forests after the latest El Niño event makes it clear that their rate of disappearance is accelerating, despite the worldwide concern.

Albedo
Apart from greenhouse gases, the radiation balance is affected by albedo. Ice and clouds are the main controlling components. Cloud formation is induced by microaerosols, which result from combustion processes, volcanoes and ocean spray-derived salt particles. Phytoplankton (diatoms, coccolithophorans) affects albedo via the production of dimethyl sulfide (DMS), which is transformed to sulfuric acid in the atmosphere, acting as condensation nuclei. The production is associated with cell death, because the precursor of DMS is mainly used in cell's osmoregulation. Plants, and especially conifers, which dominate in taiga and on mountain slopes, produce isoprenes and terpenes (Brasseur et al., 1999), which, after some oxidation transformations, also result in condensation nuclei. Since plants cover a main part of the continents, they change the color, and so the albedo, of the Earth in a direct way. Condensation nuclei derived from human-mediated sulfate emissions now seem to dominate natural sources, and possibly counterbalance the enhanced carbon dioxide emissions (Charlson, 1995).

Ice affects the climate system via the albedo and ocean level. If temperature drops, ice grows and increases the albedo, which makes it even colder. It also lowers the ocean level, however, which enhances weathering of fossil carbon and increases atmospheric carbon dioxide. This affects temperature in the opposite direction, and illustrates a coupling between albedo, and the carbon and water cycles.

Effects of Climate on Life

Climate affects life mainly through temperature and, in terrestrial environments, through precipitation and humidity. Nutrient supply and drain are usually directly coupled to water transport. The transport of organisms in water and in air also can be coupled to climate. The effects are in determining geographical distribution patterns, abundance, and activity rates.

The effects of body temperature on physiological rates are well described by the Arrhenius relationship, within a species-specific range of temperatures, which approximately results in a two-to-threefold increase in rate (respiration, feeding, reproduction, growth, etc.) for a 10-degrees increase in body temperature. At the lower temperature boundary, most organisms can switch to a torpor state, while instantaneous death results when temperature exceeds the upper boundary. Many species of organisms that do not switch to the torpor state escape bad seasons by migration, some of them traveling on a global scale. Endotherms (birds and mammals) are well-known examples of spectacular migrations; their energy budgets are tightly linked to the water balance. The capacity to survive periods of starvation has close links with body size; these periods tend to be proportional to volumetric body length.

Plant production increases in an approximately linear way with annual precipitation, which illustrates the importance of water availability in terrestrial environments. Plants use water for several purposes, one of them being the transport of nutrients from the soil to their roots. This is why the ratio of the surface areas of shoots and roots enters in the saturation constants for nutrient uptake by plants. Precipitation also affects nutrient availability via leakage.

Extensive pampa and savanna ecosystems require regular fires for existence. Many plant species require fire to trigger germination.

Local differences between seasons in temperate and polar areas are large with respect to global climate changes during the evolution of the Earth, which complicates the construction of simple models that aim to be realistic.

Synthesis: Metabolic Organization

The above-mentioned interactions between life and its environment serve as a kind of shopping list for models with a minimum complexity. When it comes to kinetics, we have to insert a minimum level of biological detail. The popular RKR model (Redfield et al., 1963) for plankton reads

$$106\ CO_2 + 16\ HNO_3 + H_3PO_4 + 122\ H_2O + \text{light}$$
$$\rightarrow ((CH_2O)_{106}(NH_3)_{16}(H_3PO_4)) + 138\ O_2$$

In the eyes of a biologist, however, a lot more needs to be done to understand life's activities at a global scale. More realistic modeling is still far away, however, but some progress has been made toward a basic understanding of the organization of metabolism at the ecosystem scale. This understanding is essential for making a healthy start incorporating the mentioned interactions.

I here focus on some recent developments in the dynamic energy budget (DEB) theory (Kooijman, 2000, 2001; Nisbet et al., 2000), which aims to identify the rules organisms follow for the uptake of nutrients, substrate, and food, and their use for maintenance, growth, development, and reproduction. It appears that all organisms (microorganisms, plants, animals) basically follow the same rules. The theory shows how organisms gradually reached an almost perfect state of homeostasis in evolutionary history, especially in animals, and how this poses stoichiometric constraints on life's activities, which organisms partially escape through the use of intracellular metabolic pools, combined with movement in spatially heterogeneous environments.

One of the difficulties in modeling life is that many factors determine its propagation, which easily leads to complex models that observe the conservation of mass and the second law of thermodynamics. Take, for instance, phytoplankton. The ocean's top layers are rich in light but poor in nutrients. The reverse applies at the bottom of the mixed layer. By making use of intracellular pools of carbohydrates and nutrients, cells manage to grow and divide if they are moved through the water column by wind action, even if the local nutrient/light environment does not support growth. Apart from nutrients and light, this involves wind as a growth-controlling factor. Model details have been worked out (Kooijman et al., 2002), and will soon be applied as a module in ocean circulation models, in an attempt to quantify ocean's primary production more realistically.

The metabolic organization of organisms has been captured within a single framework by the DEB theory, which can be used to understand evolutionary aspects of metabolism. Different species of organisms, using different substrates for growth, reproduction, and maintenance, can become engaged in a symbiontic relationship with each other when they start using each others' products to supplement their substrates. This relationship can become obligatory when they require those products. It turns out to be possible to merge these initially independent populations by incremental changes of some parameter values such that the new endosymbiotic single population again follows the DEB rules (Kooijman, Auger, et al., 2003), including the various forms of homeostasis. The parameter changes may be under evolutionary control (see below). This helps us to understand the quantitative aspects of the evolution of eukaryotes (Margulis, 1970, 1993), and the intracellular dynamics of plastids. The process of self-organization can be studied at the ecosystem level as follows.

Take, for instance, a homogeneous body of water that is closed for mass, but open for energy. It contains certain amounts of inorganic carbon and nutrients; we here follow ammonia and inorganic carbon only. When we inoculate this with a single mixotroph, it will start to grow autotrophically, and detritus will appear as a result of the aging process (as specified by the DEB theory). Let us assume, for simplicity's sake, that the mixotroph can fully mineralize this detritus. Its appetite for detritus relative to autotrophic activity is set by parameter values, the "binding probabilities," which are coupled to the specific maximum uptake capacities and the costs for growth (due to the costs for the assimilation machineries).

All we have to do to study self-organization is to allow the binding probabilities to follow a random walk across the generations, keeping the parameters constant for each individual. One daughter inherits the parameter value of the mother; the other can deviate by small amounts with certain (small) probabilities. Depending on the ratio of total carbon and total nitrogen, the mixotroph community can segregate into populations of auto- and heterotrophs, and the latter can specialize in different compounds, and/or the community continues to exist as mixotrophs (Kooijman et al., 2002). The segregation process can be stimulated by allowing the binding probabilities to drop independently to zero with a small probability. The DEB theory takes care of all aspects of mineral recycling, stoichiometric couplings, and energy dissipation; no additional rules are required for selection or definitions for fitness measures. The system is self-selecting and self-organizing, just as Darwin meant it to be.

Under suitable conditions, symbiotic relationships can develop between auto- and heterotrophs that exchange carbohydrates for nutrients directly; the DEB model for plants has basically the structure of a symbiosis between root and shoot, the model for mixotrophs can be viewed as a further integration of the metabolism of both partners. The formation of symbiotic relationships, and the dynamics of prebiotic evolution that leads to the DEB-kinetics, are subjects of present theoretical research.

We can extend this thought experiment in several ways. One is allowing carnivorousness to occur, by including another binding probability. The mixotroph community can then develop into a canonical community (i.e., a community that consists of producers, consumers, and decomposers; Kooijman and Nisbet, 2000). A next step is to allow for body size segregation. The basic processes are already covered by the DEB theory, which shows why the primary parameters can be grouped into intensive parameters, which are independent of ultimate body size, and extensive ones, which tend to covary across species; if expressed in the proper units, they are all proportional to ultimate volumetric length. This covariation couples many traits, such as respiration, feeding, growth, reproduction rates, and life span. By inserting this coupling into the random walk of parameter values across generations, food webs can develop. Other steps toward realism are to include spatial structure (a one-dimensional axis can allow for light and redox gradients and gravitation) and opening the environment for leakage and nutrient inputs. (Opening a closed environment is simple theoretically; most existing models do not allow application in a closed environment, because they do not fully respect conservation of mass and energy dissipation).

Most of the existing population dynamic models suffer from major deficiencies in the context of the DEB theory, because they do not relate the properties of individuals to those of populations and they are not based on energy and mass (nutrient) balances at the various levels of organization. Individual-based population dynamic theory started in the 1980s (Metz and Diekmann, 1986), but is developing fast (Cushing, 1998; Diekmann et al., 1998; Diekmann et al., 2001). The incorporation of balances is more problematic. Most population dynamic models describe grazing, for instance, on the basis of simple assumptions about the growth of vegetation (or phytoplankton), in order to analyze the population dynamics of herbivores. Although primary production is usually nutrient-limited, these models do not account for the release of nutrients during grazing and at death of the herbivores. Herbivores do stimulate primary production, as do carnivores. The nutrient release that is associated with predation, and its indirect stimulatory effects on primary and secondary production, are usually not taken into account. Moreover, carnivores have a preference for the weak, which compete with the productive individuals for food. Thus, predation comes with an increase of food availability and a reduction of the probability of contracting contagious diseases for the healthy ones (Kooijman, Andersen, et al., 2004). The mutualistic aspects of the herbivore-carnivore symbiosis have been underestimated so far; this also applies to other metabolic relationships between organisms. Insight into the dynamics of food webs and the stabilizing role of omnivorousness and cannibalism is still in its childhood.

The indicated lines of thought can be used to understand how an ecosystem develops and interacts in a given physicochemical environment. It will not be easy to extract mathematical properties from this type of simulation, but the task is not hopeless, given the rather simple and straightforward principles on which the DEB theory is based. I expect that surface area/volume relationships, in combination with mass and energy balances and stoichiometric constraints, will turn out to be the major organization principles that operate on all space-time scales. Life is mainly restricted to a thin skin around the globe, which receives light and nutrients across its surface, while maintenance requirements involve biomass, and thus a measure of volume. This holds not only for living systems as a whole but also for each individual; observations confirm that substrate (food, nutrients) uptake is proportional to individuals' surface area, and maintenance to individuals' structural mass; surface area/volume relationships explain the main differences in metabolic rates between bacteria and whales. Cellular metabolism can be understood in terms of interactions between surface-bound transport and transformation (membranes), and volume-based intracellular accumulation of compounds. The interplay between surface areas and volumes is felt at all levels of biological organization. One has only to think about the weathering of rocks (which involves surface areas), for instance, to realize that the surface area/volume interplay is a major organizing factor for many aspects of Earth systems science.

Discussion

Plate tectonics seems to dominate long-term transport of energy and water, and thus climate. This applies not only to ocean circulation patterns but also to rainfall patterns as modified by mountains. Rainfall relates to weathering and nutrient cycling, so it seems fair to state that plate tectonics also dominates the long-term development of life. Atmospheric dynamics operates at a spatial scale that is two to three orders of magnitude larger, and a temporal scale that is two to three orders of magnitude smaller than ocean dynamics, which complicates its incorporation into a single model. Life dynamics operates at many scales simultaneously, implying that realistic modeling of the activities of life will not be easy.

The understanding of the role of life requires a deeper insight into ecosystem functioning. The systematic study of the development of community structure, as simulated along the lines sketched above, will be a huge amount of work. The hope is that aggregation methods will allow us to simplify the main features of the process, and that invariants will show up, between ecosystem structure and function, that can be used to study the intriguing process of co-evolution of life and its environment. No guarantee can be given for solving all problems that need to be solved, but partial results of the exercise will be helpful for many less far-reaching applications of the theory.

Being free of empirical assumptions, and respecting chemical and physical constraints of metabolic organization, the DEB theory offers a rigorous framework for understanding context-specific selection mechanisms which seem to be linked to the development of this planet. The strength of the approach is that core concepts, such as selection, optimization, and self-regulation, appear as emergent properties from low-level mechanisms. Although the application in global modeling is neither easy nor straightforward, a systematic approach is required for a deeper understanding of the underlying principles. This point of view seems to be well accepted for purely physical aspects, but less so for biological ones. The sensible incorporation of these processes into biogeochemical modeling of Earth's development requires a well-coordinated interdisciplinary research program.

More information about the research program on the DEB theory and its results is given at http://www.bio.vu.nl/thb/deb. You can freely download software packages from the electronic DEB laboratory to understand relationships between many variables in the context of this theory.

Acknowledgments

The work presented in this chapter was supported by the Dutch government, National Research Programme on Global Air Pollution and Climate Change, contract no. 013/1204.10. I thank Peter Westbroek, Jan van Hinte, Mark Budding, and Henk Dijkstra for their helpful discussions.

References

Algeo, T. J., S. E. Scheckler, and J. B. Maynard. 2001. Effects of the middle to late Devonian spread of vascular land plants on weathering regimes, marine biotas, and global climate. In P. C. Gensel and D. Edwards, eds., *Plants Invade the Land: Evolutionary and Environmental Perspectives*. New York: Columbia University Press.

Baar, H. J. W. d., and P. W. Boyd. 2000. The role of iron in plankton ecology and carbon dioxide transfer of the global oceans. In R. B. Hanson, H. W. Ducklow, and J. G. Field, eds., *The Changing Ocean Carbon Cycle*, pp. 61–140. Cambridge: Cambridge University Press.

Berner, E. K., and R. A. Berner. 1996. *Global Environment: Water, Air, and Geochemical Cycles*. Englewood Cliffs, NJ: Prentice Hall.

Berner, R. A. 1991. Atmospheric oxygen, tectonics and life. In S. H. Schneider and P. J. Boston, eds., *Scientists on Gaia*, pp. 161–173. Cambridge, MA: MIT Press.

Berner, R. A. 2001. The effects of the rise of land plants on atmospheric CO_2 during the Paleozoic. In P. G. Gensel and D. Edwards, eds., *Plants Invade the Land; Evolutionary and Environmental Perspectives*, pp. 173–178. New York: Columbia University Press.

Berthelin, J. 1983. Microbial weathering. In W. E. Krumbein, ed., *Microbial Geochemistry*, pp. 233–262. Oxford: Blackwell Scientific.

Brasseur, G. B., J. J. Orlando, and G. S. Tyndall. 1999. *Atmospheric Chemistry and Global Change*. Oxford: Oxford University Press.

Butcher, S. S., R. J. Charlson, G. H. Orians, and G. V. Wlofe. 1992. *Global Biochemical Cycles*. London: Academic Press.

Chai, F., S. T. Lindley, J. R. Toggweiler, and R. R. Barber. 2000. Testing the importance of iron and grazing in the maintenance of the high nitrate condition in the equatorial Pacific Ocean: A physical-biological model study. In *The Changing Ocean Carbon Cycle*, pp. 155–186. Cambridge: Cambridge University Press.

Charlson, R. 1995. The vanishing climatic role of dimethyl sulfide. In G. M. Woodwell and F. T. Mackenzie, eds., *Biotic Feedbacks in the Global Climatic System*, pp. 251–277. Oxford: Oxford University Press.

Crowley, T. J., and G. R. North. 1991. *Paleoclimatology*. Oxford: Oxford University Press.

Cushing, J. M. 1998. *An Introduction to Structured Population Dynamics*. Philadelphia: Society for Industrial and Applied Mathematics.

Diekmann, O., M. Gyllenberg, H. Huang, M. Kirkilionis, J. A. J. Metz, and H. R. Thieme. 2001. On the formulation and analysis of general deterministic structured population models: II. Nonlinear theory. *Journal of Mathematical Biology*, 43, 157–189.

Diekmann, O., M. Gyllenberg, J. A. J. Metz, and H. R. Thieme. 1998. On the formulation and analysis of general deterministic structured population models: I. Linear theory. *Journal of Mathematical Biology*, 36, 349–388.

Ferry, J. G. 1993. *Methanogenesis: Ecology, Physiology, Biochemistry and Genetics*. New York: Chapman and Hall.

Harte, J. 1991. Ecosystem stability and diversity. In S. H. Schneider and P. J. Boston, eds., *Scientists on Gaia*, pp. 77–79. Cambridge, MA: MIT Press.

Jaffe, D. A. 1992. The nitrogen cycle. In S. S. Butcher, R. J. Charlson, G. H. Orians, and G. V. Wolfe, eds., *Global Biogeochemical Cycles*, pp. 263–284. London: Academic Press.

Kooijman, S. A. L. M. 2000. *Dynamic Energy and Mass Budgets in Biological Systems*. Cambridge: Cambridge University Press.

Kooijman, S. A. L. M. 2001. Quantitative aspects of metabolic organization: A discussion of concepts. *Philosophical Transactions of the Royal Society of London*, B356, 331–349.

Kooijman, S. A. L. M., T. R. Andersen, and B. W. Kooi. 2004. Dynamic energy budget representations of stoichiometric constraints to population models. *Ecology*. To appear.

Kooijman, S. A. L. M., P. Auger, J. C. Poggiale, and B. W. Kooi. 2003. Quantitative steps in symbiogenesis and the evolution of homeostasis. *Biological Reviews*, 78, 435–463.

Kooijman, S. A. L. M., H. A. Dijkstra, and B. W. Kooi. 2002. Light-induced mass turnover in a mono-species community of mixotrophs. *Journal of Theoretical Biology*, 214, 233–254.

Kooijman, S. A. L. M., and R. M. Nisbet. 2000. How light and nutrients affect life in a closed bottle. In S. E. Jørgensen, ed., *Thermodynamics and Ecological Modelling*, pp. 19–60. Boca Raton, FL: CRC.

Krauskopf, K. B., and D. K. Bird. 1995. *Introduction to Geochemistry*. New York: McGraw-Hill.

Krumbein, W. E., and D. Werner. 1983. The microbial silica cycle. In W. E. Krumbein, ed., *Microbial Geochemistry*, pp. 125–157. Oxford: Blackwell Scientific.

Lashof, D. A. 1991. Gaia on the brink: Biogeochemical feedback processes in global warming. In S. H. Schneider and P. J. Boston, eds., *Scientists on Gaia*, pp. 393–404. Cambridge., MA: MIT Press.

Libes, S. M. 1992. *An Introduction to Marine Biogeochemistry*. New York: Wiley.

Mah, R. A., D. M. Ward, L. Baresi, and T. L. Glass. 1977. Biogenesis of methane. *Annual Reviews in Microbiology*, 31, 309–341.

Margulis, L. 1970. *Origins of Eukaryotic cells*. San Francisco: W. H. Freeman.

Margulis, L. 1993. *Symbiosis in Cell Evolution*. New York: Freeman.

Metz, J. A. J., and O. Diekmann. 1986. *The Dynamics of Physiologically Structured Populations*. Berlin: Springer-Verlag.

Nisbet, R. M., E. B. Muller, K. Lika, and S. A. L. M. Kooijman. 2000. From molecules to ecosystems through dynamic energy budget models. *Journal of Animal Ecology*, 69, 913–926.

Redfield, A. C., B. H. Ketchum, and F. A. Richards. 1963. The influence of organisms on the composition of seawater. In M. N. Hill, ed., *The Sea*, vol. 2, pp. 26–77. New York: Wiley-Interscience.

Tyrell, T. 1999. The relative influences of nitrogen and phosphorus on oceanic primary production. *Nature*, 400, 525–531.

Westbroek, P. 1991. *Life as a Geological Force: Dynamics of the Earth*. New York: W. W. Norton.

Worsley, T. R., R. D. Nance, and J. B. Moody. 1991. Tectonics, carbon, life, and climate for the last three billion years: A unified system? In S. H. Schneider and P. J. Boston, eds., *Scientists on Gaia*, pp. 200–210. Cambridge, MA: MIT Press.

31
Stability and Instability in Ecological Systems: Gaia Theory and Evolutionary Biology

Francesco Santini and Lodovico Galleni

Abstract

Gaia theory and its science, geophysiology, have always received strong criticism from neo-Darwinian evolutionary biologists. Given that until very recently evolutionary theory was dominated by neo-Darwinism, it is usually assumed that Gaia theory and evolutionary biology are not compatible. This is very much a false impression. In this chapter we want to show that evolutionary theory is much more heterogeneous than commonly assumed, and that various theories of biological evolution are not incompatible with Gaia theory. Many of these theories, which in no way deny Darwinian evolution even though they reject the extremist neo-Darwinian spin of it, seem to be perfectly complemented by and integrated with Gaia theory. We also argue that because one of the main reasons for the dominance of neo-Darwinian theory so far has lain in the ability of its proponents to frame their theories within a sound mathematical framework, any alternative theory of evolution that wants to obtain widespread consideration has to be developed within a sound mathematical framework as well. Using deep-sea hydrothermal ecosystems as a case study, we provide an example of how simple mathematical tools can be used to bridge the gap between Darwinian evolutionary theories and geophysiology.

Gaia and Evolutionary Biology

The Gaia theory, first proposed as the Gaia hypothesis by James Lovelock in the 1970s (Lovelock and Margulis, 1974; Lovelock, 1979), has now been accepted by most researchers in such disciplines as oceanography, biogeochemistry, and microbial ecology as a valid scientific theory, able to formulate scientific predictions and to give birth to a new branch of science, geophysiology. Evolutionary biologists, on the other side, are perhaps the category of scientists who so far have offered the greatest resistance to Gaia theory (see Barlow, 1991, for a critical review). This is due to a multiplicity of reasons, most significant, perhaps, the fact that evolutionary biology has been dominated until very recently by neo-Darwinian population geneticists who were studying very few selected groups of macroorganisms (mainly vertebrates and insects). As a result the neo-Darwinian evolutionary theory has been built on the assumption that the processes of genetic variation and natural selection seen in modern populations of birds and fruit flies explain everything that we need to know about the evolution of life. Within neo-Darwinism organisms have always been seen as distinct from their environment, the former being conceived basically as mechanical objects subject to the selective agents of the latter. The variation within organisms is also seen as independent from the selective pressure of the environment (which can weed out only the less "adapted" variants). All variations are accidental, and organisms are conceived of as "islands" that try to resist the influence of the environment (almost invariably conceived as a negative one).

The main implication of the Gaia theory, on the contrary, is the concept that living and nonliving components of the planet Earth constitute a single entity, and their relationships have to be taken into consideration in order to understand the processes that drive the evolution of life.

> Through Gaia theory I now see the system of the material Earth and the living organisms on it, evolving so that self-regulation is an emergent property. In such a system active feedback processes operate automatically and solar energy sustains comfortable conditions for life. The conditions are only constant in the short term and evolve in synchrony with the changing needs of the biota as it evolves. Life and its environment are so closely coupled that evolution concerns Gaia, not the organisms or the environment taken separately. (Lovelock, 1995, p. 19)

While it has generally been acknowledged even by strict neo-Darwinists that living organisms evolve within ecological associations (ecosystems, landscapes, biomes, the biosphere), and that organisms do have an effect on their surrounding environment, most evolutionary biologists immediately questioned the ability of life to modify environmental parameters for its own good (a main point of the Gaia hypothesis). Claims such as those regarding the ability of life to regulate the atmospheric composition and the salinity of the oceans to make them fit life's needs, which could not be explained in any way using a neo-Darwinian logic, and the recurrent use of metaphors such as that of "earth as a superorganism" made the Gaia hypothesis an easy target for accusations of mysticism and vitalism (Doolittle, 1981; Dawkins, 1982). Because the "birth" and growth of Gaia theory (originally as Gaia hypothesis) during the 1970s and early 1980s coincided with the peak of the popularity of the most extremist views of neo-Darwinism, most biologists rapidly dismissed the Gaia hypothesis, notwithstanding the many predictions that could later be verified experimentally and that led to the transformation of the Gaia hypothesis into the Gaia theory.

However, several authors have managed to underline various of the inherent problems in the neo-Darwinian view of evolution, such as a lack of attention to biological hierarchies, to the importance of adaptive mutational systems and the epigenetic inheritance systems, to the developmental constraints caused by morphogenetic fields, and to phenomena of self-organization and the emergent properties of complex systems (Eldredge, 1985, 1995; Goodwin, 1994; Ho and Saunders, 1984; Jablonka and Lamb, 1995; Kauffman, 1993, 1995). In addition, as pointed out by some of the champions of Gaia theory (Margulis and Sagan, 1991), the bias of the neo-Darwinian investigators toward multicellular organisms, which led to ignoring the microbial biota and the processes that drive its evolution, is finally being recognized (Markos, 1996). Nowadays most workers have been forced to recognize that neo-Darwinian orthodoxy is a very partial explanation of the evolutionary processes, and that a more pluralistic and holistic theoretical framework is needed (Carroll, 2000).

We share the belief that in the search for a more comprehensive theory of evolution, a more holistic approach is needed, and that much more attention should be paid to the way in which living organisms interact with the environment. However, we want to bring to the attention of evolutionary biologists the fact that many alternatives to the strict neo-Darwinian view of Richard Dawkins and its followers already exist, and several of these views are not at all at odds with Gaia theory. On the contrary, various theories that are not neo-Darwinian (in some cases not well known because they were developed outside the English-speaking scientific community and the proper literature has never been made available in English) do provide support for geophysiology and are perfectly integrated by the idea of Gaia and its science.

Given that a great part of the success of neo-Darwinism has been due to its ability to use mathematical tools and modeling to make testable predictions, we believe that any alternative evolutionary theory must have a solid mathematical component. Thus, while in the first part of this chapter we intend to present a short summary of the main theories of evolution, that are not neo-Darwinian, in the second part we will present mathematical tools that can be used to better integrate geophysiology and evolutionary biology. A more extensive description of the theories and of the mathematical aspects, and an epistemological discussion are outside the scope of this chapter, but can be found in Benci and Galleni (1998) and Galleni (2001).

New Theoretical Frameworks for Evolutionary Biology

Galleni (2001) revised and integrated earlier discussions reported in Kauffman (1993, 1995) and Goodwin (1994). Of paramount importance is the idea that these alternative theories, which Galleni broadly divides into three categories, do not have to be used separately in order to explain evolutionary processes, but can (and often should) be used in combination.

The Genocentric Theory

This corresponds to the mainstream neo-Darwinian view. The fundamental unity of evolution is the gene, considered as a segment of DNA that is coding for a particular type of protein necessary in order to obtain a particular phenotype. Genes bring the information for a particular character, and the phenotype is the result of the sum of the different genes. Genes are freely evolving according to their own rules, and the only force that can organize ordered structures is natural selection. A causal connection between natural selection and the creation of variability is absent. This theory can probably best be visualized with Darwin's (1868) metaphor of the architect: the architect has the task of building a house, using stones made available by a landslide. He uses the stones to build an ordered structure (the house), but the form of the stones is determined by physical and chemical forces which acted on the rocks, which were not purposely made for the task of building the house.

Natural selection (the architect of the metaphor) will act as an ordering factor on the heritable variability (the stones) with the mechanisms summarized by A. R. Wallace (1895) and J. Huxley (1942). A connection between natural selection and the origin of variability is absent. For this reason the theory could also be called the theory of disconnections. This is the philosophical key to the genocentric theory. The work of Fisher (see Gayon, 1992) allowed biologists to make models of the evolution of sets of genes in order to predict and to test for experimental results: many examples were analyzed and a special branch of genetics, ecological genetics, was established (Ford, 1971). Evolution is then represented by the metaphor of a river: a river of DNA, whose course is directed by external forces such as selection, and branched by external forces such as changes in the geographical distribution of a species. This is the metaphor that the extreme representative of this theory, Richard Dawkins (1995), uses. It is clearly a reductionistic approach, but historically has been the main tool used to develop the study of the mechanisms of evolution. Within this theoretical framework there is clearly no room for an integration of evolutionary biology and geophysiological theory.

The Organocentric Theory

According to this perspective, the order of natural systems is mainly due to phenomena of self-organization. This theory is useful to explain the presence of regularities in the mathematical and geometrical construction of living beings. According to the authors of this theory, these regularities cannot be explained by the disconnected interactions between variability and natural selection. One of the main examples presented is the presence of the Fibonacci series in many living organisms, such as in the shell of the mollusk *Nautilus* or in plant spiral phyllotaxis (Goodwin, 1994). The self-organization concept of Kauffman (1993, 1995) could also be inserted in this theory. Natural selection is responsible only for local adaptations, and works more as a fine-tuning device than as an "architect" for evolution. This approach, although mainly conceived with organisms in mind, can be extended to ecological associations (Kauffman, 1995). Gaia becomes acceptable, but further developments are needed in order to obtain a synthesis of geophysiology and evolutionary theory.

The Biospherocentric Theory

This theory looks for connections between the different "parts" forming the biosphere, and it considers the whole biosphere as the entity that is evolving. This theory allows us to use the techniques and the concepts of complexity, and a global approach to evolution. It could also be considered as a way of applying system theory to evolution.

The beginnings of the biospherocentric theory are traced back to James Hutton, Vladimir Vernadsky, and Pierre Teilhard de Chardin (Galleni, 1995, 1996, 2001; Grinevald, 1996). Teilhard de Chardin in particular attempted to develop a first approach to the problem of the study of the evolution of the biosphere by using a new branch of science, geobiology, which he defined as the science of continental evolution, meaning that a large-scale approach, preferably a continental one, is the only way to study evolution at the global level without distortion. Geobiology is the science that studies the evolution of living beings, but is also concerned with the connections between living and nonliving beings, and it is assumed that evolution is a product of these relations (the term "continental evolution" simply refers to the spatial scale over which evolution occurs; it does not refer to theories of evolution of the continents, i.e., continental drift). In this way characteristics that are missed or neglected by using the reductionistic neo-Darwinian method are put into evidence. When the history of life, as analyzed through paleontology, is investigated with instruments that take into consideration the evolution of the whole biosphere, particulars such as canalization and parallelisms toward complexity are evidenced (Galleni and Groessens-Van Dyck, 2001). Teilhard de Chardin's experimental approach and his biospherocentric theory initiated the discussion about biology as the science of complexity (Galleni and Groessens-Van Dyck, 2001).

The next step is the passage from geobiology, intended as the science of global evolution, to geophysiology, intended as the science of connections of living and nonliving beings in order to maintain the stability of the biosphere. The main point of Lovelock's Gaia theory is that it is not possible to talk about an independent evolution of the biological and physicochemical components of the biosphere, because both parts are strongly linked, and influence and direct the evolution of each other. Hence we should think about a unique integrated system (i.e., Gaia) in which the self-regulation that leads to the homeostasis of certain parameters is a self-emergent property obtained by the interactions between the several parts of the system (Lovelock, 1995). Geophysiology then becomes the science that studies ecological systems as if their components were co-evolving and influencing each other, and not just the sum of independently functioning parts. While Lovelock cited Theilard de Chardin as one of the precursors of Gaia theory, with Lovelock we have a further step. Not only does Gaia theory imply the co-evolution of life and physicochemical environment, but it also implies the novelty of the stability of the biosphere: taken as a whole, the biosphere is evolving in order to maintain the stability of its main parameters through negative feedbacks between the various components of the system. The complex object "biosphere" can then be considered as a system where parts are interacting in order to maintain the stability of the object itself. This is a major meaning of evolution put into evidence with the global approach (Galleni, 1995, 1996). Again the global approach introduces a new concept, that of the task of the whole. Moreover, a new way to look for adaptation is necessary: at the biosphere level we have not only living organisms adapted to the environment, but also the main parameters of the biosphere adapted to life.

However, if according to Gaia theory the biosphere is evolving to maintain its stability, but the fossil record shows that the history of life is rich in catastrophic events, present at different scales, the problem then becomes: When does a Gaian model allow evolution, if it does?

Mathematical Tools

While ecologists and evolutionary biologists have always massively used mathematical models in an attempt to simulate the dynamics of biological populations and communities, and hence predict the behavior of ecological systems, until recently most attempts have been directed toward producing quantitative predictions of the interactions between the biological components of ecological associations. There have been very few attempts to try to make qualitative predictions about natural phenomena, and to include the physicochemical environment and the interactions between the living and nonliving parts of the ecosystems in the mathematical modeling of temporal evolution (Benci and Galleni, 1998). The modeling work of Daisyworld has received considerable attention in the literature (Watson and Lovelock, 1983; Maddock, 1991; Schneider and Boston, 1991; Bunyard, 1996, and chapter 29, this volume), and we will not discuss it here. Of greater interest to us is the work presented by Benci and Galleni (1998). These authors discussed applying mathematical tools to make qualitative predictions regarding two of the fundamental problems of biological evolution: (1) stability or instability of natural systems (i.e., the tendency of natural systems such as ecosystems and the biosphere to persist for long spans of time with little or no significant changes in functioning, versus the tendency to undergo major modifications) and (2) gradualism and catastrophism (i.e., the pace at which the changes happen: many small and gradual changes versus a few major and sudden ones).

Using only the very basic assumptions of Darwinian theory and some basic concepts of geophysiology, a simple model was proposed. In this model the parameters of the biosphere are x factors, the living organisms and the interactions between them, and λ factors, the environmental parameters that are beyond the control of the biota, even though they are responsible, together with the interactions between the xs, for natural selection.

With this first general modeling that referred to the whole biosphere, it was possible to show some qualitative aspects of evolution. This model allowed us to identify three kinds of intervals during the evolution of biological systems. (It is important to note that we are concerned only with ecological systems, which are defined as systems composed of entities above the organismal level. We are not concerned with phenomena such as physiological processes within cells and tissues of multicellular organisms, even though striking similarities can sometimes be identified.)

1. Stability periods—the evolution is gradual and smoothly follows the environmental changes. They were named Lyell periods, in honor of Charles Lyell and his theory of slow, gradual modification of the Earth's crust. The length of these periods can be on the order of millions of years.

2. Catastrophe periods—the number of individuals per species and of species per clade changes drastically and dramatically. A mass extinction occurs in these conditions. They were named Cuvier periods, after Baron Cuvier and his theory of catastrophes, and have lengths of thousands of years or less.

3. Postcatastrophe periods—rapid evolution, adaptive radiations, rapid colonization of empty ecological niches. They were named as Darwin periods, in honor of Charles Darwin, and they have a length of the order of tens of thousands of years or less.

This kind of model, integrating modern evolutionary theory and Gaia theory, demonstrates the presence of evolution in the presence of strong links connecting living and nonliving beings. Evolution is not gradual, but presents different phases, each identifiable with one of the three periods described above. In this way the succession of long periods of stasis with short catastrophic and postcatastrophic periods finds an explication. While according to neo-Darwinian theory connections between living and nonliving objects are weak, the relationships between living organisms are highly unstable, and the evolution of life is characterized by a succession of very unstable states and of nonpredictable transitions, the model with feedbacks presents periods of stability that require further attention. This model makes it possible to test whether evolution is a step-by-step movement toward stability by diversification. The result is that stability is a characteristic derived from a global approach but is not total, as periods of catastrophe show (Benci and Galleni, 1998).

The Model Applied to Ecosystem Studies

The original model was formulated in a very general and highly theoretical way, and used to illustrate the shift of the atmosphere of an hypothetical Earth-like planet from a reductive (no free molecular oxygen present) to an oxidized (abundant free oxygen present) state. In order to obtain a more reliable test of the ability of this model to make useful predictions, we decided to apply this approach to a new kind of ecosystem, discovered in the 1970s: the hydrothermal vents. These vents are produced by the geological processes that drive plate tectonics and are responsible for the genesis of oceanic crust at spreading centers. The biological communities of these particular ecosystems depend not on photosynthesis but on bacterial chemosynthesis, which obtains carbohydrates and other biological molecules from inorganic compounds, such as CO_2, exploiting as electron donors the reduced inorganic compounds present in the heated fluid. Thus the source of energy is not solar light but geothermal energy produced by the Earth's interior. The main reduced compounds used in chemosynthetic reactions are dissolved sulfides (Jannasch and Mottl, 1985; Tunnicliffe, 1991).

Chemoautolithotrophic bacteria are the only primary producers in deep-sea hydrothermal ecosystems. They form the base of the trophic structure of the communities. Vent communities have a much bigger biomass than the classic abyssal communities even if they show a more limited diversity. Chemoautotrophy and symbiosis are the reason for the presence of a great amount of biomass. Heterotrophic bacteria, protists, and several species of invertebrates, many of which are in symbiosis with chemotrophic bacteria, make up the remainder of these communities. The fauna of the inhabitants of the hydrothermal vents is endemic and taxonomically distinct from that of the normal deep-sea environment (Tunnicliffe, 1991).

For the purposes of this study such an ecosystem presents many advantages, including a short lifetime that makes it easy to test theoretical predictions, a relative isolation that decreases the number of factors that need to be taken into consideration in the model, and the hypotheses that life based on such an ecosystem could be present elsewhere in the solar system (Santini and Galleni, 2000, 2001).

Our Geophysiological Model

Given that it is possible to identify, among the several components of ecosystems, both biotic and abiotic phenomena of "self-limitation," "competition," and "predation," the mathematical model that we propose is a Lotka-Volterra variation (Murray, 1989). We present a very simple model that allows the use of all the variables introduced in the text. We emphasize that this is still a qualitative model. Due to needs of model manageability we have decided to limit it to a set of four differential equations. This work has already been discussed elsewhere (Santini et al., 1999; Cerrai et al., 2002; Santini et al., in press). Here we give only a brief description of the model. Readers interested in a more detailed analysis of the modeling are referred to the literature cited.

In order to simplify the ecological interactions, we have divided the biota into producers, x_1, and consumers, x_2. Producers include not only the free-living chemoautotrophic bacteria, but also the chemosynthetic bacteria that live in symbiosis with macroinvertebrates and their metazoan hosts in cases where these have no other means of finding food if deprived of their associates (as in the case of the Vestimentifera). Consumers include both primary (grazers and filter feeders) and secondary consumers. We have the following notations:

x_1 = the density of autotrophic organisms

x_2 = the density of heterotrophic organisms

In our model we have also decided to take into account the main chemical compounds that perform the roles of electron donors and acceptors. These two roles are performed mainly by hydrogen sulfide (H_2S) and molecular oxygen (O_2). Molecular oxygen is produced by the activity of living organisms outside the thermal ecosystem, while the hydrogen sulfide is mainly produced by venting, making the production of these compounds beyond the control of living beings inhabiting the hydrothermal vents.

The environmental parameters are represented by λ_1, molecular oxygen, and λ_2, hydrogen sulfide.

λ_1 = the density of O_2

λ_2 = the density of H_2S

The set of equations is the following:

$$dx_1/dt = [-a_1 - b_{11}x_1 - b_{12}x_2 + b_{14}\lambda_2]x_1$$

$$dx_2/dt = [-a_2 + b_{21}x_1 - b_{22}x_2]x_2$$

$$d\lambda_1/dt = [a_3 - b_{33}\lambda_1 - b_{34}\lambda_2]\lambda_1$$

$$d\lambda_2/dt = [a_4 - b_{41}x_1 - b_{43}\lambda_1 - b_{44}\lambda_2]\lambda_2$$

where the growth coefficients $\forall i, j = 1, 2, 3, 4$ and a_i and b_{ij} are positive constants which vary between 0 and 1. The values of these in the model are determined by our knowledge of the events that occur during the life stages of the ecosystem (i.e., the growth coefficient of chemoautotrophic bacteria will be high when there are abundant emissions of hydrogen sulfide from the vents, and zero if no hydrogen sulfide is present. The time of the model is taken to be in arbitrary units, which in the real ecosystems could span periods varying from days to years.

The meaning of the first equation is as follows: the growing coefficient of the organisms x_1 is promoted by the H_2S concentration. The interactions between the organisms x_1 and x_2 are likewise prey (x_1)-predator (x_2). Thus the second equation is the classical predator equation of the Lotka-Volterra model with self-limiting factor. The third equation describes the "competition" between O_2 and H_2S. Without H_2S the growth of O_2 obeys the logistic equation. H_2S is antagonist to O_2. Finally, the fourth equation for H_2S is similar to the third, but there is another limiting factor: the organisms x_1, which are consumers of H_2S.

Results

By selecting the growth coefficients in accordance to the experimental evidence, it is now possible to simulate the evolution of these ecosystems and their associated biological communities.

Figures 31.1a and 31.1b show the dynamics of the four components that we have considered during the early stages in the life of a vent, the period that follows the beginning of emissions of thermal fluids. We can assume that before the beginning of thermal activity, molecular oxygen is present in the pre-vent environment in a concentration similar to that of seawater, so its growth coefficient is originally equal to 1 in our model; hydrogen sulfide is absent, so its growth coefficient is originally 0. Producers also are present in the environment, even if scarce—and thus are assigned a low growth coefficient—because some chemotrophic prokaryotes live in anoxic areas of the sediment, while heterotrophic organisms are extremely rare in the pre-vent deep-sea environment, and their growth coefficient is even lower than that of producers. At this stage the system can be described as being in a Lyell period characterized by low diversity and high instability, because every external influence can have a large impact on its dynamics. Suddenly, as soon as the vent starts emitting thermal fluid, H_2S appears in the environment, and starts to react with molecular oxygen. The two substances tend to react very rapidly each having a negative effect on the other's concentration. After a short period of time, however, the system reaches a stable state, during which the amounts of sulfide and oxygen present in the hydrothermal ecosystem remains constant (figure 31.1a).

The appearance of H_2S drives the system into a Darwin period, and figure 31.1b shows the dynamics of living organisms. As soon as the vent starts emitting fluids rich in potential sources of energy, producers, originally very scarce, start to increase in number quite rapidly, due to the rapid life cycle of bacteria that can reproduce every few hours; hence their growth coefficient tends to approach the maximum value of 1. Consumers are slower to follow, given that animals require much longer periods for reproduction and growth; thus their growth coefficient is increasing but still much lower that that of producers. After some time, however, x_2 start increasing in abundance, consuming significant amounts of x_1. Thus there is an initial drop in the abundance of x_1 and a subsequent lower growth rate of the producers. The population of producers, under the increasing predatory pressure of the rapidly growing consumers, starts to decline; when the population of x_2 reaches a certain density, it will be limited by the competition for resources, and its growth rate will be reduced. Thus, as time passes, the population of consumers starts declining until it reaches a stable climax state. The density of the population of producers also will stabilize at this point (figure 31.1b).

While the new stable state for the concentrations of the chemical compounds reached shortly after the beginning of the life of the vent is maintained for the rest of the vent's life, the biological components of

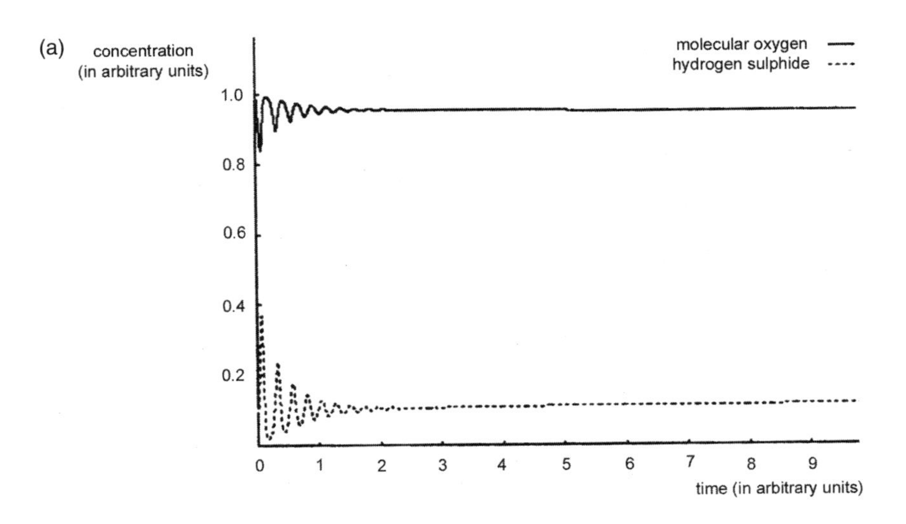

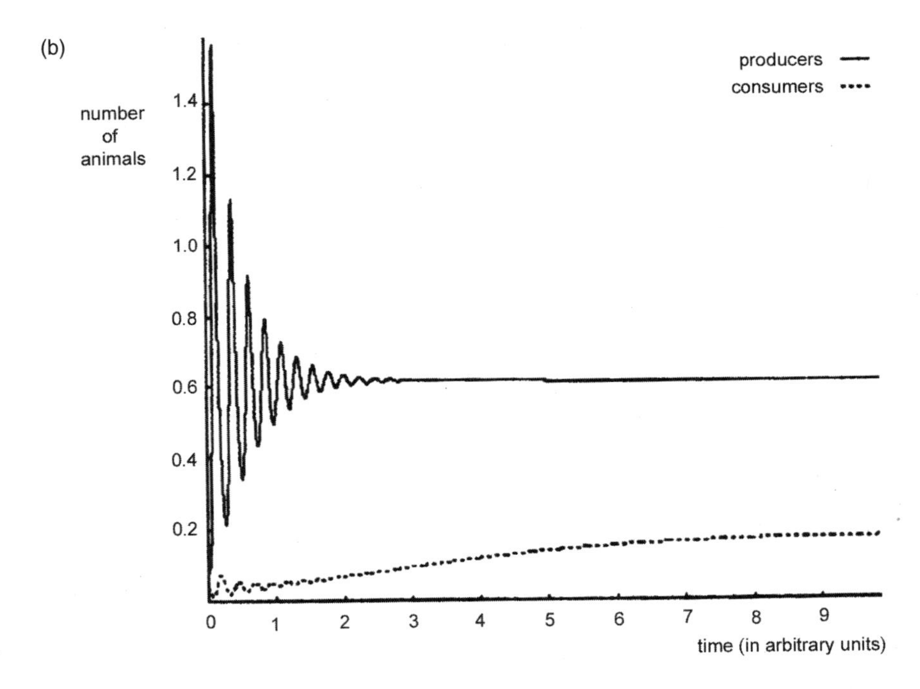

Figure 31.1
Dynamics of molecular oxygen (λ_1) and hydrogen sulfide (λ_2) (*a*) and producers (x_1) and consumers (x_2) (*b*) during the transition from early stages to mature stages of the life of the hydrothermal system.

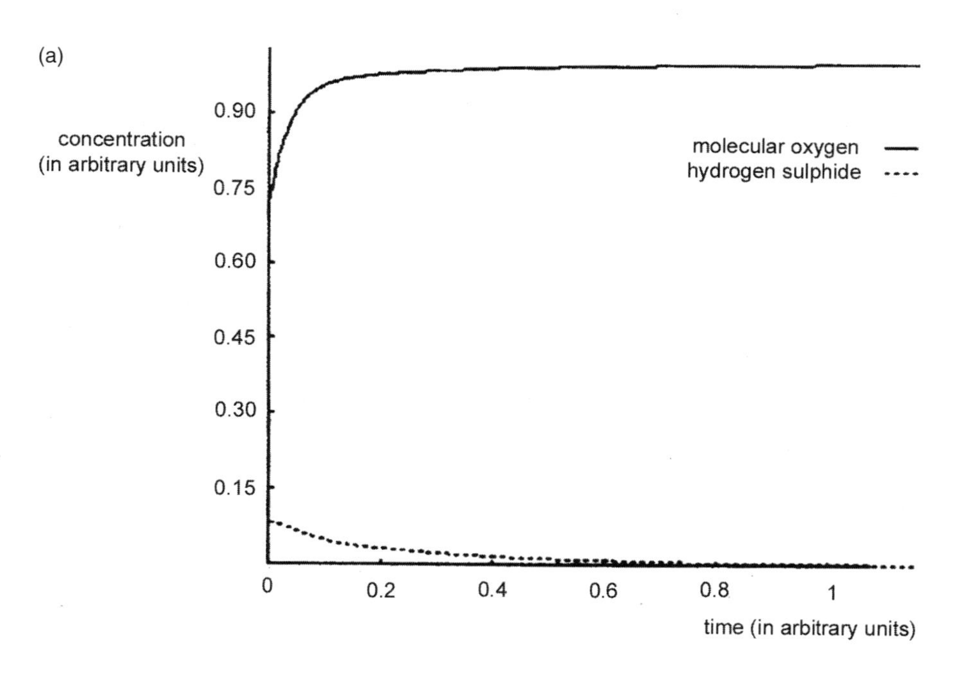

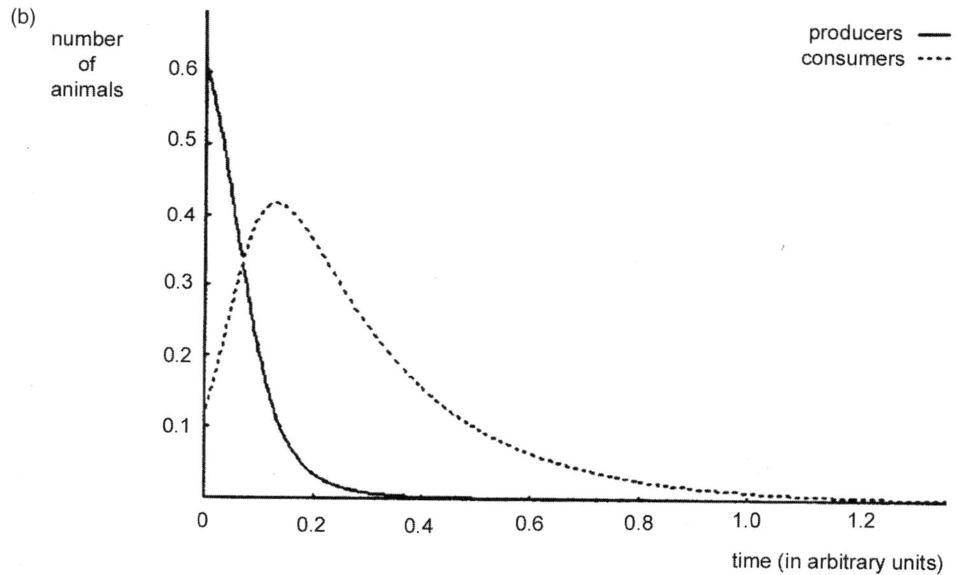

Figure 31.2
Dynamics of molecular oxygen (λ_1) and hydrogen sulfide (λ_2) (*a*) and producers (x_1) and consumers (x_2) (*b*) at the end of the life of the hydrothermal system, when the emission of thermal fluids stops.

the system need more time to reach a climax stage, the state of maturity and maximum complexity of the ecosystem. Eventually, however, the density of the population of producers also stabilizes, and a new, more stable (because more resistant to external perturbations) Lyell period begins. This period can persist indefinitely, until a major perturbation is caused by a λ factor.

Geological phenomena will eventually cause the end of the emission of thermal fluids from the vent (Tunnicliffe, 1991). At this point the growth coefficient of sulfides in the hydrothermal ecosystem drops to 0, given that no new sulfides are produced by the vent; and since all the sulfides already emitted are either consumed by the living organisms or oxidized by the molecular oxygen present in the seawater, sulfides basically disappear from the ecosystem (figure 31.2a). While the concentration of H_2S goes to 0, seawater represents a continuous source of oxygen; hence the concentration of O_2 in the hydrothermal ecosystem approaches the level present in the seawater, which in our model we have arbitrarily considered to be equal to 1 unit on our scale. When the hydrothermal emissions stop, a Cuvier period begins for the biota. Primary producers tend to collapse very rapidly, while the population of consumers converges towards zero more slowly (because they find prey to consume even within their own category; figure 31.2b).

Conclusions

Our model, which integrates Darwinian theory and Gaia theory, shows a dynamic very similar to that obtained by Benci and Galleni (1998). Our model, in which strong links are present between the various components and negative feedbacks exist among some of the key elements, shows long periods of stability which are at odds with neo-Darwinian thinking, where the connections between the various living and nonliving components are weak, and evolution is characterized by a continuous succession of highly unstable states and unpredictable transitions. While our model in no way denies that natural selection does play a major role in biological evolution (during Lyell periods natural selection probably drives evolutionary dynamics), there are periods of catastrophe and others of rapid evolution where other phenomena, such as contingency, may play a more dominant role. This kind of modeling can also help us understand how biological phenomena of diversification and complexification originate, and what drives them. Many of these phenomena seem to be triggered by what we label λ factors, elements that are outside of the control of the biota. These elements can drastically alter the dynamics of ecosystems, often in a way that may be subtle in the beginning but that produces cascading effects through its repercussions on the interactions between the x components of ecological systems.

Finally, our model illustrates the need to consider the interactions between all components of ecological systems and the biota when modeling ecosystem functioning. The fate of the living organisms of communities is determined not only by the classic biological phenomena, such as competition and predation, but also by the interactions between biological and chemical components of ecosystems. When in hydrothermal ecosystems the flow of thermal fluids is reduced, available sulfides are rapidly oxidized, and this event leads to the death of chemoautotrophs and the collapse of the biological community.

All this serves as a reminder that in order to be able to understand biological evolution, it is not possible to avoid considering the core idea of the Gaia theory. Evolution is not just a process that concerns living organisms. It is a phenomenon that should be investigated with a "biospherocentric" perspective, keeping in mind that living organisms cannot be taken into consideration without also considering the system that they form with their environment, an environment that they have a great influence in forming (Lovelock, 1995).

Acknowledgments

Dr. P. Cerrai, Department of Mathematics, University of Pisa, helped in the development of the model presented here. F. Santini was funded by the International Council for Canadian Studies through a Government of Canada Award, and by the Department of Zoology, University of Toronto. F. Santini thanks Dr. Tyler Volk for the encouragement and the suggestion to present the results of this work at the 2000 Chapman Conference on the Gaia Hypothesis. The suggestions of S. Scharf, University of Toronto, and two anonymous reviewers helped improve the manuscript.

References

Barlow, C., ed. 1991. *From Gaia to Selfish Genes: Selected Writings in the Life Sciences*. Cambridge, MA: MIT Press.

Benci, V., and L. Galleni. 1998. Stability and instability in evolution. *Journal of Theoretical Biology*, 194, 541–547.

Bunyard, P. 1996. *Gaia in Action*. Edinburgh: Floris Books.

Carroll, R. L. 2000. Towards a new evolutionary synthesis. *Trends in Ecology and Evolution*, 15, 27–32.

Cerrai, P., A. Cordelli, L. Galleni, and F. Santini. 2002. Geophysiological modeling: New ideas on modeling the evolution of ecosystems. *Biology Forum*, 95, 63–74.

Darwin, C. 1868. *The Variation of Animals and Plants Under Domestication*. London: John Murray.

Dawkins, R. 1982. *The Extended Phenotype*. London: Freeman.

Dawkins, R. 1995. *River Out of Eden*. London: Weidenfeld and Nicolson.

Doolittle, W. F. 1981. Is nature really motherly? *CoEvolution Quarterly*, 29, 58–63.

Eldredge, N. 1985. *Unfinished Synthesis*. New York: Columbia University Press.

Eldredge, N. 1995. *Reinventing Darwin: The Great Evolutionary Debate*. New York: John Wiley & Sons.

Ford, E. B. 1971. *Ecological Genetics*. London: Methuen.

Galleni, L. 1995. How does the Teilhardian vision of evolution compare with contemporary theories? *Zygon*, 30, 25–45.

Galleni, L. 1996. Levels of organization and evolution, from Teilhard de Chardin to Lovelock. *Studies in Science and Theology*, 4, 109–115.

Galleni, L. 2001. Is biosphere doing theology? *Zygon*, 36, 33–48.

Galleni, L., and M. Groessens-Van Dyck. 2001. Lettres d'un paléontologue. Neuf lettres inédites de Pierre Teilhard de Chardin a Marcellin Boule. *Revue des Questions Scientifiques*, 172, 5–104.

Gayon, J. 1992. *Darwin et l'Après-Darwin*. Paris: Editions Kimé.

Goodwin, B. 1994. *How the Leopard Changed Its Spots*. New York: Scribner's.

Grinevald, J. 1996. Sketch for a history of the idea of the biosphere. In P. Bunyard, ed., *Gaia in Action*. Edinburgh: Floris Books.

Ho, M. W., and P. Saunders. 1984. *Beyond Neo-Darwinism*. London: Academic Press.

Huxley, J. 1942. *Evolution, the Modern Synthesis*. London: George Allen and Unwin.

Kauffman, S. A. 1993. *The Origin of Order*. New York: Oxford University Press.

Kauffman, S. A. 1995. *At Home in the Universe*. Oxford: Oxford University Press.

Jablonka, E., and M. J. Lamb. 1995. *Epigenetic Inheritance and Evolution: The Lamarkian Dimension*. Oxford: Oxford University Press.

Jannasch, H. W., and M. J. Mottl. 1985. Geomicrobiology of deep-sea hydrothermal vents. *Science*, 229, 717–725.

Lovelock, J. 1979. *Gaia. A New Look at Life on Earth*. Oxford: Oxford University Press.

Lovelock, J. E. 1995. *The Ages of Gaia: A Biography of Our Living Earth Updated and Revised*. New York: Norton.

Lovelock, J. E., and L. Margulis. 1974. Atmospheric homeostasis by and for the biosphere: The Gaia hypothesis. *Tellus*, 26, 2–10.

Maddock, L. 1991. Effects of simple environmental feedback on some population models. *Tellus*, 43B, 331–337.

Margulis, L., and D. Sagan. 1991. *Microcosmos: Four Billion Years of Microbial Evolution*. New York: Touchstone Books.

Markos, A. 1996. The Gaia theory: The role of microorganisms in planetary information network. *Journal of Theoretical Biology*, 176, 175–180.

Murray, J. D. 1989. *Mathematical Biology*. Berlin: Springer-Verlag.

Santini, F., and L. Galleni. 2000. The significance of non-light based ecosystems in the search for life outside of the Earth. In *Bioastronomy '99—A New Era in Bioastronomy*, G. Lemarchand and K. J. Meech, eds., pp. 397–401. Chelsea, MI: Sheridan Books.

Santini, F., and L. Galleni. 2001. Non-light based ecosystems and bioastronomy. *Biology Forum*, 94, 427–442.

Santini, F., L. Galleni, and P. Cerrai. 1999. Geophysiological modeling: A new approach at modeling the evolution of ecosystems. *Interjournal*, manuscript 253, www.interjournal.org.

Santini, F., L. Galleni, and P. Cerrai. In press. Geophysiological modeling. In *Unifying Themes in Complex Systems II*, A. Minai and Y. Bar-Yam, eds., Perseus Press.

Schneider, S. H., and P. J. Boston. 1991. *Scientists on Gaia*. Cambridge, MA: MIT Press.

Tunnicliffe, V. 1991. The biology of hydrothermal vents: Ecology and evolution. *Oceanography and Marine Biology Annual Review*, 29, 319–407.

Wallace, Alfred Russel. 1895. *Natural Selection and Tropical Nature: Essays on Descriptive and Theoretical Biology*. London: Macmillan.

Watson, A. J., and J. E. Lovelock. 1983. Biological homeostasis of the global environment: The parable of Daisyworld. *Tellus*, 35B, 284–289.

Appendix
Studying Gaia: The NASA Planetary Biology Internship (PBI) Program

Andrew M. Wier and Jennifer M. Benson

Abstract

Gaian phenomena do not fall neatly within the established boundaries of traditional academic disciplines. The study of Gaia needs interdisciplinary research. The NASA Planetary Biology Internship (PBI) program has pioneered efforts to introduce graduate students, of any scientific background, to exobiology research at NASA centers and universities by granting the opportunity to travel outside their home institutions to participate in research related to NASA's planetary biology objectives. PBI places interns with NASA-funded researchers conducting investigations related to the vast interdisciplinary field of planetary biology. Typical internship topics include global ecology and remote sensing, microbial ecology and biomineralization, advanced life support, and origins and early evolution of life. PBI also provides support to students participating in the advanced microbiology course Microbial Diversity offered each summer by the Marine Biological Laboratory in Woods Hole, MA. PBI-funded research has been fruitful, as demonstrated by the large proportion of PBI projects resulting in publications. Students find that their internship experience adds to their general knowledge, complements their thesis research, and opens new opportunities about which they were previously unaware. The PBI program facilitates Gaian-related research and attracts new researchers to Earth systems sciences.

Planetary Biology Internships: The Support of Gaian Research

The worldview brought about by Lovelock's Gaia hypothesis has made apparent that the natural world cannot be categorized and broken up to fall neatly within the imaginary boundaries of traditional academic disciplines (Lovelock, 2003). These traditional boundaries have been hard to break down, and interdisciplinary research suffers from a lack of funding as a result (Margulis and Hinkle, 1991). Scientists and students interested in studying Gaian processes find themselves discouraged from traversing into other fields, caught in the trappings of their traditional disciplinary departmental program. In order for Gaian phenomena to be studied, an integrated, transdisciplinary approach is needed.

New, transdisciplinary fields, such as Earth systems science, are addressing this need. Bringing together and analyzing the interconnectedness of fields of research that previously were never studied together—such as ecology and evolution, atmospheric chemistry, biogeochemistry, mineralogy, geology, and paleontology—these transdisciplinary fields are opening up new avenues of inquiry and providing the resources needed to study the Earth as an integrated whole. The merging of these disciplines has allowed Gaian research to be conducted that would have been limited by the scope of the traditional fields. More programs, internships, scholarships, and grants that encourage interdisciplinary science studies need to be developed in order to encourage students to do research that bridges the traditional disciplinary boundaries instead of straining to fall neatly within them. One program that is addressing this need is the NASA Planetary Biology Internship (PBI) program.

The goal of the NASA Planetary Biology Internship (PBI) program is to provide graduate students with summer research experiences, for a period of eight weeks, on projects related to NASA's planetary biology mission. PBI places qualified graduate students with NASA-funded researchers at NASA research centers such as NASA Ames, the NASA Goddard Space Flight Center, the NASA Jet Propulsion Laboratory, and at various universities across the United States (Benson, 2001).

Tony Swain and Lynn Margulis, inspired by the Planetary Geology Internship (PGI) program, originally developed PBI. The program, currently directed by Lynn Margulis and Michael Dolan at the University of Massachusetts–Amherst and administered by the Marine Biological Laboratory in Woods Hole, Massachusetts, awards approximately ten internships each year to graduate students from around the world, in any field of science, including senior year undergraduates accepted to graduate school for the following fall. A stipend and a travel reimbursement are provided to each intern, affording them the opportunity to spend the summer focused entirely on their internship project. Most students conduct their internship research during summer months, although any eight-week period during the year is negotiable between the intern and the sponsor.

For more information on the NASA Planetary Biology Internship program, contact Michael Dolan, Department of Geosciences, University of Massachusetts–Amherst, Amherst, MA 01003 USA. Phone: 1-413-545-3244; fax: 1-413-545-1200; e-mail: pbi@geo.umass.edu. Application materials are available at http://www.geo.umass.edu/projects/pbi.

References

Benson, J. 2001. The Internship Experience 1996–2000: The NASA Planetary Biology Internship Program. NASA Life Sciences, Washington, D.C. Available from pbi@geo.umass.edu.

Lovelock, J. E. 2003. The Living Earth. *Nature*, 426, 769–770.

Margulis, L. and G. Hinkle. 1991. The Biota and Gaia: 150 Years of Support for Environmental Sciences. In *Scientists on Gaia*, S. H. Schneider and P. J. Boston, eds. Cambridge, MA: MIT Press.

Contributors

Ricardo Amils
Center for Molecular Biology
Autonomous University of Madrid
Madrid, Spain

Kaspar Arn
Institute of Geology
University of Neuchatel
Neuchâtel, Switzerland

Jennifer M. Benson
Department of Geosciences
University of Massachusetts–Amherst
Amherst, Massachusetts

Noam M. Bergman
School of Environmental Sciences
University of East Anglia
Norfolk, United Kingdom

Christine Bounama
Potsdam Institute for Climate Impact Research
Potsdam, Germany

Peter Bunyard
Lawellen Farm
Cornwall, United Kingdom

Alexandre Casanovas
Department of Thermodynamics
University of Valencia
Valencia, Spain

Thomas N. Chase
Cooperative Institute for Research in the
Environmental Sciences
University of Colorado
Boulder, Colorado

Eileen Crist
Department of Science and Technology in Society
Virginia Tech
Blacksburg, Virginia

Keith L. Downing
The Norwegian University of Science and Technology
Trondheim, Norway

David Fernández-Remolar
Center of Astrobiology
Madrid, Spain

Karl B. Föllmi
Institute of Geology
University of Neuchatel
Neuchâtel, Switzerland

Siegfried Franck
Potsdam Institute for Climate Impact Research
Potsdam, Germany

Lodovico Galleni
Department of Agricultural Chemistry and
Biotechnology
University of Pisa
Pisa, Italy

Felipe Gómez
Center of Astrobiology
Madrid, Spain

Vicent Gómez
Department of Thermodynamics
University of Valencia
Valencia, Spain

Elena González-Toril
Center for Molecular Biology
Autonomous University of Madrid
Madrid, Spain

Vijay K. Gupta
Cooperative Institute for Research in the
Environmental Sciences
University of Colorado
Boulder, Colorado

Stephan P. Harding
Programme in Holistic Science
Schumacher College
Devon, United Kingdom

Rachel Hosein
Institute of Geology
University of Neuchatel
Neuchâtel, Switzerland

Axel Kleidon
Department of Geography
University of Maryland
College Park, Maryland

Lee F. Klinger
10 Elm Ct.
San Anselmo, California

S. A. L. M. (Bas) Kooijman
Department of Theoretical Biology
Vrije Universiteit
Amsterdam, The Netherlands

Lee R. Kump
Department of Geosciences
The Pennsylvania State University
University Park, Pennsylvania

Timothy M. Lenton
Centre for Ecology and Hydrology
Midlothian, United Kingdom

James Lovelock
Cornwall, United Kingdom

Lynn Margulis
Department of Geosciences
University of Massachusetts
Amherst, Massachusetts

Mark A. S. McMenamin
Department of Earth and Environment
Mount Holyoke College
South Hadley, Massachusetts

James R. Miller
Department of Marine and Coastal Sciences
Rutgers University
New Brunswick, New Jersey

Keith M. Nordstrom
Cooperative Institute for Research in the Environmental Sciences
University of Colorado
Boulder, Colorado

Arthur C. Petersen
Netherlands Environmental Assessment Agency
National Institute for Public Health and the Environment
Bilthoven, The Netherlands

Claire Rambeau
Institute of Geology
University of Neuchatel
Neuchâtel, Switzerland

Jennifer M. Robinson
School of Environmental Science
Murdoch University
Murdoch, Australia

Gary L. Russell
NASA/Goddard Institute for Space Studies
New York, New York

Dorion Sagan
Sciencewriters
Amherst, Massachusetts

Francesco Santini
Department of Systematics and Evolution
National Museum of Natural History
Paris, France

Hans-Joachim Schellnhuber
Potsdam Institute for Climate Impact Research
Potsdam, Germany

Eric D. Schneider
Hawkwood Institute
Livingston, Montana

David Schwartzman
Department of Biology
Howard University
Washington, D.C.

Bruce Scofield
Kepler College
Amherst, Massachusetts

Federica Tamburini
Institute of Geology
University of Neuchatel
Neuchâtel, Switzerland

Crispin Tickell
Ablington Old Barn
Cirencester, United Kingdom

Pedro Ruiz Torres
University of Valencia
Valencia, Spain

J. Scott Turner
Department of Environmental and Forest Biology
State University of New York
Syracuse, New York

Toby Tyrrell
School of Ocean and Earth Science
Southampton University
Southampton, United Kingdom

Bas van de Schootbrugge
Institute of Geology
University of Neuchatel
Neuchâtel, Switzerland

Tyler Volk
Department of Biology
New York University
New York, New York

Werner von Bloh
Potsdam Institute for Climate Impact Research
Potsdam, Germany

Andrew J. Watson
School of Environmental Sciences
University of East Anglia
Norfolk, United Kingdom

Susanne L. Weber
Royal Netherlands Meteorological Institute
De Bilt, The Netherlands

Peter Westbroek
Department of Chemistry
Leiden University
Leiden, The Netherlands

Jessica Hope Whiteside
Department of Earth and Environmental Sciences
Columbia University
Palisades, New York

Andrew M. Wier
Department of Biological Sciences
University of Wisconsin–Milwaukee
Milwaukee, Wisconsin

David Wilkinson
Biological and Earth Sciences
Liverpool John Moores University
Liverpool, United Kingdom

Index